Edited by
Audrius Alkauskas, Peter Deák,
Jörg Neugebauer,
Alfredo Pasquarello, and
Chris G. Van de Walle

Advanced Calculations for Defects in Materials

Related Titles

Brillson, L. J.

Surfaces and Interfaces of Electronic Materials

2010

ISBN: 978-3-527-40915-0

Magnasco, V.

Methods of Molecular Quantum Mechanics
An Introduction to Electronic Molecular Structure

2009

ISBN: 978-0-470-68442-9

Friedrichs, P., Kimoto, T., Ley, L., Pensl, G. (eds.)

Silicon Carbide
Volume 1: Growth, Defects, and Novel Applications

2010

ISBN: 978-3-527-40953-2

Friedrichs, P., Kimoto, T., Ley, L., Pensl, G. (eds.)

Silicon Carbide
Volume 2: Power Devices and Sensors

2010

ISBN: 978-3-527-40997-6

Sholl, D., Steckel, J. A

Density Functional Theory
A Practical Introduction

2009

ISBN: 978-0-470-37317-0

Tilley, R. J. D.

Defects in Solids

2008

ISBN: 978-0-470-07794-8

Morkoc, H.

Handbook of Nitride Semiconductors and Devices

2008

ISBN: 978-3-527-40797-2

Edited by
Audrius Alkauskas, Peter Deák, Jörg Neugebauer,
Alfredo Pasquarello, and Chris G. Van de Walle

Advanced Calculations for Defects in Materials

Electronic Structure Methods

WILEY-VCH Verlag GmbH & Co. KGaA

The Editors

Dr. Audrius Alkauskas
EPFL 58, IPMC LSME
MX 136
Batiment MXC 12
1015 Lausanne
Schweiz

Prof. Dr. Peter Deák
Uni Bremen - Computational
Materials Science
Otto-Hahn-Allee 1
28359 Bremen

Prof. Dr. Jörg Neugebauer
Fritz-Haber-Institut
Max-Planck-Inst. f. Eisenfor.
Max-Planck-Str. 1
40237 Düsseldorf

Prof. Dr. Alfredo Pasquarello
EPFL-SB-ITP-CSEA
Station 3/ PH H2 467
1015 Lausanne
Schweiz

Prof. Dr. C. G. Van de Walle
Materials Department
University of California
Santa Barbara, CA 93106-5050
USA

All books published by **Wiley-VCH** are carefully produced. Nevertheless, authors, editors, and publisher do not warrant the information contained in these books, including this book, to be free of errors. Readers are advised to keep in mind that statements, data, illustrations, procedural details or other items may inadvertently be inaccurate.

Library of Congress Card No.: applied for

British Library Cataloguing-in-Publication Data
A catalogue record for this book is available from the British Library.

Bibliographic information published by the Deutsche Nationalbibliothek
The Deutsche Nationalbibliothek lists this publication in the Deutsche Nationalbibliografie; detailed bibliographic data are available on the Internet at http://dnb.d-nb.de.

© 2011 Wiley-VCH Verlag & Co. KGaA, Boschstr. 12, 69469 Weinheim, Germany

All rights reserved (including those of translation into other languages). No part of this book may be reproduced in any form – by photoprinting, microfilm, or any other means – nor transmitted or translated into a machine language without written permission from the publishers. Registered names, trademarks, etc. used in this book, even when not specifically marked as such, are not to be considered unprotected by law.

Composition   Thomson Digital, Noida, India
Printing and Binding   Strauss GmbH, Mörlenbach
Cover Design   Adam-Design, Weinheim

Printed in the Federal Republic of Germany
Printed on acid-free paper

ISBN: 978-3-527-41024-8

Contents

List of Contributors *XIII*

1	**Advances in Electronic Structure Methods for Defects and Impurities in Solids** *1*	
	Chris G. Van de Walle and Anderson Janotti	
1.1	Introduction *1*	
1.2	Formalism and Computational Approach *3*	
1.2.1	Defect Formation Energies and Concentrations *3*	
1.2.2	Transition Levels or Ionization Energies *4*	
1.2.3	Practical Aspects *5*	
1.3	The DFT-LDA/GGA Band-Gap Problem and Possible Approaches to Overcome It *6*	
1.3.1	LDA+U for Materials with Semicore States *6*	
1.3.2	Hybrid Functionals *9*	
1.3.3	Many-Body Perturbation Theory in the GW Approximation *12*	
1.3.4	Modified Pseudopotentials *12*	
1.4	Summary *13*	
	References *14*	
2	**Accuracy of Quantum Monte Carlo Methods for Point Defects in Solids** *17*	
	William D. Parker, John W. Wilkins, and Richard G. Hennig	
2.1	Introduction *17*	
2.2	Quantum Monte Carlo Method *18*	
2.2.1	Controlled Approximations *20*	
2.2.1.1	Time Step *20*	
2.2.1.2	Configuration Population *20*	
2.2.1.3	Basis Set *20*	
2.2.1.4	Simulation Cell *21*	
2.2.2	Uncontrolled Approximations *22*	
2.2.2.1	Fixed-Node Approximation *22*

2.2.2.2	Pseudopotential	22
2.2.2.3	Pseudopotential Locality	23
2.3	Review of Previous DMC Defect Calculations	23
2.3.1	Diamond Vacancy	23
2.3.2	MgO Schottky Defect	25
2.3.3	Si Interstitial Defects	25
2.4	Results	25
2.4.1	Time Step	26
2.4.2	Pseudopotential	26
2.4.3	Fixed-Node Approximation	26
2.5	Conclusion	29
	References	29

3 **Electronic Properties of Interfaces and Defects from Many-body Perturbation Theory: Recent Developments and Applications** 33
Matteo Giantomassi, Martin Stankovski, Riad Shaltaf, Myrta Grüning, Fabien Bruneval, Patrick Rinke, and Gian-Marco Rignanese

3.1	Introduction	33
3.2	Many-Body Perturbation Theory	34
3.2.1	Hedin's Equations	34
3.2.2	GW Approximation	36
3.2.3	Beyond the GW Approximation	37
3.3	Practical Implementation of GW and Recent Developments Beyond	38
3.3.1	Perturbative Approach	38
3.3.2	QP Self-Consistent GW	40
3.3.3	Plasmon Pole Models *Versus* Direct Calculation of the Frequency Integral	41
3.3.4	The Extrapolar Method	44
3.3.4.1	Polarizability with a Limited Number of Empty States	45
3.3.4.2	Self-Energy with a Limited Number of Empty States	46
3.3.5	MBPT in the PAW Framework	46
3.4	QP Corrections to the BOs at Interfaces	48
3.5	QP Corrections for Defects	54
3.6	Conclusions and Prospects	57
	References	58

4 **Accelerating GW Calculations with Optimal Polarizability Basis** 61
Paolo Umari, Xiaofeng Qian, Nicola Marzari, Geoffrey Stenuit, Luigi Giacomazzi, and Stefano Baroni

4.1	Introduction	61
4.2	The GW Approximation	62
4.3	The Method: Optimal Polarizability Basis	64
4.4	Implementation and Validation	68
4.4.1	Benzene	69

4.4.2	Bulk Si 70	
4.4.3	Vitreous Silica 70	
4.5	Example: Point Defects in a-Si_3N_4 72	
4.5.1	Model Generation 72	
4.5.2	Model Structure 73	
4.5.3	Electronic Structure 74	
4.6	Conclusions 77	
	References 77	

5 Calculation of Semiconductor Band Structures and Defects by the Screened Exchange Density Functional 79
S. J. Clark and John Robertson

5.1	Introduction 79	
5.2	Screened Exchange Functional 80	
5.3	Bulk Band Structures and Defects 82	
5.3.1	Band Structure of ZnO 83	
5.3.2	Defects of ZnO 85	
5.3.3	Band Structure of MgO 89	
5.3.4	Band Structures of SnO_2 and CdO 90	
5.3.5	Band Structure and Defects of HfO_2 91	
5.3.6	$BiFeO_3$ 92	
5.4	Summary 93	
	References 94	

6 Accurate Treatment of Solids with the HSE Screened Hybrid 97
Thomas M. Henderson, Joachim Paier, and Gustavo E. Scuseria

6.1	Introduction and Basics of Density Functional Theory 97	
6.2	Band Gaps 100	
6.3	Screened Exchange 103	
6.4	Applications 104	
6.5	Conclusions 107	
	References 108	

7 Defect Levels Through Hybrid Density Functionals: Insights and Applications 111
Audrius Alkauskas, Peter Broqvist, and Alfredo Pasquarello

7.1	Introduction 111	
7.2	Computational Toolbox 112	
7.2.1	Defect Formation Energies and Charge Transition Levels 113	
7.2.2	Hybrid Density Functionals 114	
7.2.2.1	Integrable Divergence 115	
7.3	General Results from Hybrid Functional Calculations 117	
7.3.1	Alignment of Bulk Band Structures 118	
7.3.2	Alignment of Defect Levels 120	

7.3.3	Effect of Alignment on Defect Formation Energies	122
7.3.4	"The Band-Edge Problem"	124
7.4	Hybrid Functionals with Empirically Adjusted Parameters	125
7.5	Representative Case Studies	129
7.5.1	Si Dangling Bond	129
7.5.2	Charge State of O_2 During Silicon Oxidation	131
7.6	Conclusion	132
	References	134

8   Accurate Gap Levels and Their Role in the Reliability of Other Calculated Defect Properties *139*
Peter Deák, Adam Gali, Bálint Aradi, and Thomas Frauenheim

8.1	Introduction	139
8.2	Empirical Correction Schemes for the KS Levels	141
8.3	The Role of the Gap Level Positions in the Relative Energies of Various Defect Configurations	143
8.4	Correction of the Total Energy Based on the Corrected Gap Level Positions	146
8.5	Accurate Gap Levels and Total Energy Differences by Screened Hybrid Functionals	148
8.6	Summary	151
	References	152

9   LDA + U and Hybrid Functional Calculations for Defects in ZnO, SnO_2, and TiO_2 *155*
Anderson Janotti and Chris G. Van de Walle

9.1	Introduction	155
9.2	Methods	156
9.2.1	ZnO	158
9.2.2	SnO_2	160
9.2.3	TiO_2	161
9.3	Summary	163
	References	163

10   Critical Evaluation of the LDA + U Approach for Band Gap Corrections in Point Defect Calculations: The Oxygen Vacancy in ZnO Case Study *165*
Adisak Boonchun and Walter R. L. Lambrecht

10.1	Introduction	165
10.2	LDA + U Basics	166
10.3	LDA + U Band Structures Compared to GW	168
10.4	Improved LDA + U Model	170
10.5	Finite Size Corrections	172
10.6	The Alignment Issue	173

10.7	Results for New LDA + U	174
10.8	Comparison with Other Results	176
10.9	Discussion of Experimental Results	178
10.10	Conclusions	179
	References	180

11 Predicting Polaronic Defect States by Means of Generalized Koopmans Density Functional Calculations 183
Stephan Lany

11.1	Introduction	183
11.2	The Generalized Koopmans Condition	185
11.3	Adjusting the Koopmans Condition using Parameterized On-Site Functionals	187
11.4	Koopmans Behavior in Hybrid-functionals: The Nitrogen Acceptor in ZnO	189
11.5	The Balance Between Localization and Delocalization	193
11.6	Conclusions	196
	References	197

12 SiO_2 in Density Functional Theory and Beyond 201
L. Martin-Samos, G. Bussi, A. Ruini, E. Molinari, and M.J. Caldas

12.1	Introduction	201
12.2	The Band Gap Problem	202
12.3	Which Gap?	204
12.4	Deep Defect States	207
12.5	Conclusions	209
	References	210

13 Overcoming Bipolar Doping Difficulty in Wide Gap Semiconductors 213
Su-Huai Wei and Yanfa Yan

13.1	Introduction	213
13.2	Method of Calculation	214
13.3	Symmetry and Occupation of Defect Levels	217
13.4	Origins of Doping Difficulty and the Doping Limit Rule	218
13.5	Approaches to Overcome the Doping Limit	220
13.5.1	Optimization of Chemical Potentials	220
13.5.1.1	Chemical Potential of Host Elements	220
13.5.1.2	Chemical Potential of Dopant Sources	222
13.5.2	H-Assisted Doping	223
13.5.3	Surfactant Enhanced Doping	224
13.5.4	Appropriate Selection of Dopants	226
13.5.5	Reduction of Transition Energy Levels	229
13.5.6	Universal Approaches Through Impurity-Band Doping	232

13.6	Summary 237	
	References 238	
14	**Electrostatic Interactions between Charged Defects in Supercells** 241	
	Christoph Freysoldt, Jörg Neugebauer, and Chris G. Van de Walle	
14.1	Introduction 241	
14.2	Electrostatics in Real Materials 243	
14.2.1	Potential-based Formulation of Electrostatics 245	
14.2.2	Derivation of the Correction Scheme 246	
14.2.3	Dielectric Constants 249	
14.3	Practical Examples 250	
14.3.1	Ga Vacancy in GaAs 250	
14.3.2	Vacancy in Diamond 252	
14.4	Conclusions 254	
	References 257	
15	**Formation Energies of Point Defects at Finite Temperatures** 259	
	Blazej Grabowski, Tilmann Hickel, and Jörg Neugebauer	
15.1	Introduction 259	
15.2	Methodology 261	
15.2.1	Analysis of Approaches to Correct for the Spurious Elastic Interaction in a Supercell Approach 261	
15.2.1.1	The Volume Optimized Aapproach to Point Defect Properties 262	
15.2.1.2	Derivation of the Constant Pressure and Rescaled Volume Approach 264	
15.2.2	Electronic, Quasiharmonic, and Anharmonic Contributions to the Formation Free Energy 266	
15.2.2.1	Free Energy Born–Oppenheimer Approximation 266	
15.2.2.2	Electronic Excitations 269	
15.2.2.3	Quasiharmonic Atomic Excitations 271	
15.2.2.4	Anharmonic Atomic Excitations: Thermodynamic Integration 272	
15.2.2.5	Anharmonic Atomic Excitations: Beyond the Thermodynamic Integration 274	
15.3	Results: Electronic, Quasiharmonic, and Anharmonic Excitations in Vacancy Properties 278	
15.4	Conclusions 282	
	References 282	
16	**Accurate Kohn–Sham DFT With the Speed of Tight Binding: Current Techniques and Future Directions in Materials Modelling** 285	
	Patrick R. Briddon and Mark J. Rayson	
16.1	Introduction 285	

16.2	The AIMPRO Kohn–Sham Kernel: Methods and Implementation *286*	
16.2.1	Gaussian-Type Orbitals *286*	
16.2.2	The Matrix Build *288*	
16.2.3	The Energy Kernel: Parallel Diagonalisation and Iterative Methods *288*	
16.2.4	Forces and Structural Relaxation *289*	
16.2.5	Parallelism *289*	
16.3	Functionality *290*	
16.3.1	Energetics: Equilibrium and Kinetics *290*	
16.3.2	Hyperfine Couplings and Dynamic Reorientation *291*	
16.3.3	D-Tensors *291*	
16.3.4	Vibrational Modes and Infrared Absorption *291*	
16.3.5	Piezospectroscopic and Uniaxial Stress Experiments *291*	
16.3.6	Electron Energy Loss Spectroscopy (EELS) *292*	
16.4	Filter Diagonalisation with Localisation Constraints *292*	
16.4.1	Performance *294*	
16.4.2	Accuracy *296*	
16.5	Future Research Directions and Perspectives *298*	
16.5.1	Types of Calculations *299*	
16.5.1.1	Thousands of Atoms on a Desktop PC *299*	
16.5.1.2	One Atom Per Processor *299*	
16.5.2	Prevailing Application Trends *299*	
16.5.3	Methodological Developments *300*	
16.6	Conclusions *302*	
	References *302*	

17   *Ab Initio* **Green's Function Calculation of Hyperfine Interactions for Shallow Defects in Semiconductors** *305*
       Uwe Gerstmann

17.1	Introduction *305*	
17.2	From DFT to Hyperfine Interactions *306*	
17.2.1	DFT and Local Spin Density Approximation *306*	
17.2.2	Scalar Relativistic Hyperfine Interactions *308*	
17.3	Modeling Defect Structures *311*	
17.3.1	The Green's Function Method and Dyson's Equation *311*	
17.3.2	The Linear Muffin-Tin Orbital (LMTO) Method *313*	
17.3.3	The Size of The Perturbed Region *315*	
17.3.4	Lattice Relaxation: The As_{Ga}-Family *317*	
17.4	Shallow Defects: Effective Mass Approximation (EMA) and Beyond *319*	
17.4.1	The EMA Formalism *320*	
17.4.2	Conduction Bands with Several Equivalent Minima *322*	
17.4.3	Empirical Pseudopotential Extensions to the EMA *322*	
17.4.4	*Ab Initio* Green's Function Approach to Shallow Donors *324*	

17.5	Phosphorus Donors in Highly Strained Silicon 328
17.5.1	Predictions of EMA 329
17.5.2	*Ab Initio* Treatment via Green's Functions 330
17.6	n-Type Doping of SiC with Phosphorus 332
17.7	Conclusions 334
	References 336

18	**Time-Dependent Density Functional Study on the Excitation Spectrum of Point Defects in Semiconductors** 341
	Adam Gali
18.1	Introduction 341
18.1.1	Nitrogen-Vacancy Center in Diamond 342
18.1.2	Divacancy in Silicon Carbide 344
18.2	Method 345
18.2.1	Model, Geometry, and Electronic Structure 345
18.2.2	Time-Dependent Density Functional Theory with Practical Approximations 346
18.3	Results and Discussion 351
18.3.1	Nitrogen-Vacancy Center in Diamond 351
18.3.2	Divacancy in Silicon Carbide 353
18.4	Summary 356
	References 356

19	**Which Electronic Structure Method for The Study of Defects: A Commentary** 359
	Walter R. L. Lambrecht
19.1	Introduction: A Historic Perspective 359
19.2	Themes of the Workshop 362
19.2.1	Periodic Boundary Artifacts 362
19.2.2	Band Gap Corrections 367
19.2.3	Self-Interaction Errors 370
19.2.4	Beyond DFT 372
19.3	Conclusions 373
	References 375

Index 381

List of Contributors

Audrius Alkauskas
Ecole Polytechnique Fédérale de
Lausanne (EPFL)
Institute of Theoretical Physics
1015 Lausanne
Switzerland

and

Institut Romand de Recherche
Numérique en Physique des Matériaux
(IRRMA)
1015 Lausanne
Switzerland

Bálint Aradi
Universität Bremen
Bremen Center for Computational
Materials Science
Am Fallturm 1
28359 Bremen
Germany

Stefano Baroni
CNR-IOM DEMOCRITOS
Theory@Elettra Group
s.s. 14 km 163.5 in Area Science Park
34149 Basovizza (Trieste)
Italy

and

SISSA – Scuola Internazionale
Superiore di Studi Avanzati
via Bonomea 265
34126 Trieste
Italy

Adisak Boonchun
Case Western Reserve University
Department of Physics
10900 Euclid Avenue
Cleveland, OH 444106-7079
USA

Patrick R. Briddon
Newcastle University
School of Electrical, Electronic and
Computer Engineering
Newcastle NE1 7RU
UK

Peter Broqvist
Ecole Polytechnique Fédérale de
Lausanne (EPFL)
Institute of Theoretical Physics
1015 Lausanne
Switzerland

and

Institut Romand de Recherche
Numérique en Physique des Matériaux
(IRRMA)
1015 Lausanne
Switzerland

Fabien Bruneval
European Theoretical Spectroscopy
Facility (ETSF)

and

CEA, DEN, Service de Recherches de
Métallurgie Physique
91191 Gif-sur-Yvette
France

G. Bussi
Universita di Modena e Reggio Emilia
CNR-NANO, S3 and Dipartimento di
Fisica
via Campi 213/A
41100 Modena
Italy

and

CNR-IOM Democritos and SISSA
via Bonomea 265
34136 Trieste
Italy

Marilia J. Caldas
Universidade de São Paulo
Instituto de Física
05508-900 São Paulo, SP
Brazil

S. J. Clark
Durham University
Physics Department
Durham
UK

Peter Deák
Universität Bremen
Bremen Center for Computational
Materials Science
Am Fallturm 1
28359 Bremen
Germany

Thomas Frauenheim
Universität Bremen
Bremen Center for Computational
Materials Science
Aon Fallturon
28359 Bremen
Germany

Christoph Freysoldt
Max-Planck-Institut für Eisenforschung
GmbH
Max-Planck-Str. 1
40237 Düsseldorf
Germany

Adam Gali
Hungarian Academy of Sciences
Research Institute for Solid State
Physics and Optics
POB 49
1525 Budapest
Hungary

and

Budapest University of Technology and
Economics
Department of Atomic Physics
Budafoki út 8
1111 Budapest
Hungary

Uwe Gerstmann
Universität Paderborn
Lehrstuhl für Theoretische Physik
Warburger Str. 100
33098 Paderborn
Germany

and

Université Pierre et Marie Curie
Institut de Minéralogie et de Physique
des Milieux Condensés
Campus Boucicaut
140 rue de Lourmel
75015 Paris
France

Luigi Giacomazzi
CNR-IOM DEMOCRITOS
Theory@Elettra Group
s.s. 14 km 163.5 in Area Science Park
34149 Basovizza (Trieste)
Italy

and

SISSA – Scuola Internazionale
Superiore di Studi Avanzati
via Bonomea 265
34126 Trieste
Italy

Matteo Giantomassi
European Theoretical Spectroscopy
Facility (ETSF)

and

Université catholique de Louvain
Institute of Condensed Matter and
Nanosciences
1 Place Croix du Sud, 1 bte 3
1348 Louvain-la-Neuve
Belgium

Blazej Grabowski
Max-Planck-Institut für Eisenforschung
GmbH
Max-Planck-Str. 1
40237 Düsseldorf
Germany

Myrta Grüning
European Theoretical Spectroscopy
Facility (ETSF)

and

University of Coimbra
Centre for Computational Physics and
Physics Department
Rua Larga
3004-516 Coimbra
Portugal

Thomas M. Henderson
Rice University
Departments of Chemistry and
Department of Physics and Astronomy
Houston, TX 77005
USA

Richard G. Hennig
Cornell University
Department of Materials Science and
Engineering
126 Bard Hall
Ithaca, NY 14853-1501
USA

Tilmann Hickel
Max-Planck-Institut für Eisenforschung
GmbH
Max-Planck-Str. 1
40237 Düsseldorf
Germany

Anderson Janotti
University of California
Materials Department
Santa Barbara, CA 93106-5050
USA

Walter R. L. Lambrecht
Case Western Reserve University
Department of Physics
10900 Euclid Avenue
Cleveland, OH 444106-7079
USA

Stephan Lany
National Renewable Energy Laboratory
1617 Cole Blvd
Golden, CO 80401
USA

L. Martin-Samos
Universita di Modena e Reggio Emilia
CNR-NANO, S3 and Dipartimento di
Fisica
via Campi 213/A
41100 Modena
Italy

and

CNR-IOM Democritos and SISSA
via Bonomea 265
34136 Trieste
Italy

Nicola Marzari
Massachusetts Institute of Technology
Department of Materials Science and
Engineering
77 Massachusetts Avenue
Cambridge, MA 02139
USA

E. Molinari
Universita di Modena e Reggio Emilia
CNR-NANO, S3 and Dipartimento di
Fisica
via Campi 213/A
41100 Modena
Italy

Jörg Neugebauer
Max-Planck-Institut für Eisenforschung
GmbH
Max-Planck-Str. 1
40237 Düsseldorf
Germany

Joachim Paier
Rice University
Departments of Chemistry and
Department of Physics and Astronomy
Houston, TX 77005
USA

Present affiliation:
Humboldt-Universität
Zu Berlin
Institut für Chemie
Unter den Linden 6
10099 Berlin
Germany

William D. Parker
The Ohio State University
Department of Physics
191 W. Woodruff Ave.
Columbus, OH 43210
USA

Alfredo Pasquarello
Ecole Polytechnique Fédérale de
Lausanne (EPFL)
Institute of Theoretical Physics
1015 Lausanne
Switzerland

and

Institut Romand de Recherche
Numérique en Physique des Matériaux
(IRRMA)
1015 Lausanne
Switzerland

Xiaofeng Qian
Massachusetts Institute of Technology
Department of Materials Science and
Engineering
77 Massachusetts Avenue
Cambridge, MA 02139
USA

Mark J. Rayson
Max-Planck-Institut für Eisenforschung
GmbH
Max-Planck-Str. 1
40237 Düsseldorf
Germany

and

Luleå University of Technology
Department of Mathematics
97187 Luleå
Sweden

Gian-Marco Rignanese
European Theoretical Spectroscopy
Facility (ETSF)

and

Université catholique de Louvain
Institute of Condensed Matter and
Nanosciences
1 Place Croix du Sud, 1 bte 3
1348 Louvain-la-Neuve
Belgium

Patrick Rinke
European Theoretical Spectroscopy
Facility (ETSF)

and

University of California
Department of Materials
Santa Barbara, CA 93106-5050
USA

John Robertson
Cambridge University
Engineering Department
Cambridge CB2 1PZ
UK

A. Ruini
Universita di Modena e Reggio Emilia
CNR-NANO, S3 and Dipartimento di
Fisica
via Campi 213/A
41100 Modena
Italy

Gustavo E. Scuseria
Rice University
Departments of Chemistry and Physics
and Astronomy
Houston, TX 77005
USA

Riad Shaltaf
European Theoretical Spectroscopy
Facility (ETSF)

and

University of Jordan
Department of Physics
Amman 11942
Jordan

Martin Stankovski
European Theoretical Spectroscopy
Facility (ETSF)

and

Université catholique de Louvain
Institute of Condensed Matter and
Nanosciences
1 Place Croix du Sud, 1 bte 3
1348 Louvain-la-Neuve
Belgium

Geoffoey Stenuit
CNR-IOM DEMOCRITOS
Theory@Elettra Group
s.s. 14 km 163.5 in Area Science Park
34149 Basovizza (Trieste)
Italy

Paolo Umari
CNR-IOM DEMOCRITOS
Theory@Elettra Group
s.s. 14 km 163.5 in Area Science Park
34149 Basovizza (Trieste)
Italy

Chris G. Van de Walle
University of California
Materials Department
Santa Barbara, CA 93106-5050
USA

John W. Wilkins
The Ohio State University
Department of Physics
191 W. Woodruff Ave.
Columbus, OH 43210
USA

Su-Huai Wei
National Renewable Energy Laboratory
1617 Cole Blvd
Golden, CO 80401
USA

Yanfa Yan
National Renewable Energy Laboratory
1617 Cole Blvd
Golden, CO 80401
USA

1
Advances in Electronic Structure Methods for Defects and Impurities in Solids

Chris G. Van de Walle and Anderson Janotti

1.1
Introduction

First-principles studies of point defects and impurities in semiconductors, insulators, and metals have become an integral part of materials research over the last few decades [1–3]. Point defects and impurities often have decisive effects on materials properties. A prime example is doping of semiconductors: the addition of minute amounts (often at the ppm level) of donor or acceptor impurities renders the material n type or p type, enabling the functionality of electronic or optoelectronic devices [4, 5]. Control of doping is therefore essential, and all too often eludes experimental efforts. Sometimes high doping levels required for low-resistivity transport are limited by compensation effects; such compensation can be due to point defects that form spontaneously at high doping. In other cases, unintentional doping occurs. For instance, many oxides exhibit unintentional n-type doping, which due to its prevalence has often been attributed to intrinsic causes, i.e., to native point defects. Recent evidence indicates, however, that the concentration of native point defects may be lower than has conventionally been assumed, and that, instead, unintentional incorporation of impurities may cause the observed conductivity [6]. Last but not least, many materials resist attempts at ambipolar doping, i.e., they can be easily doped one type but not the other. Again, the oxides (or more generally, wide-band-gap semiconductors) that exhibit unintentional n-type doping often cannot be doped p-type. The question then is whether this is due to an intrinsic limitation that cannot be avoided, or whether specific doping techniques might be successful.

Aside from the issue of doping, the study of point defects is important because they are involved in the diffusion processes and act to mediate mass transport, hence contributing to equilibration during growth, and to diffusion of dopants or other impurities during growth or annealing [7–9]. In addition, an understanding of point defects is essential for characterizing or suppressing radiation damage, and for analyzing device degradation.

Experimental characterization techniques are available, but they are often limited in their application [10–12]. Impurity concentrations can be determined using secondary ion mass spectrometry (SIMS), but some impurities (such as hydrogen) are hard to detect in low concentrations. Point-defect concentrations are even harder to determine. Electron paramagnetic resonance is an excellent tool that can provide detailed information about concentrations, chemical identity, and lattice environment of a defect or impurity, but it is a technique that requires dedicated expertise and possibly for that reason has few practitioners [12]. Other tools, such as Hall measurements or photoluminescence, can provide information about the effect of point defects or impurities on electrical or optical properties, but cannot by themselves identify their nature or character. For all these reasons, the availability of first-principles calculations that can accurately address atomic and electronic structure of defects and impurities has had a great impact on the field.

Obviously, to make the information obtained from such calculations truly useful, the results should be as reliable and accurate as possible. Density functional theory (DFT) [13, 14] has proven its value as an immensely powerful technique for assessing the structural properties of defects [1]. (In the remainder of this article, we will use the term "defects" to generically cover both native point defects and impurities.) Minimization of the total energy as a function of atomic positions yields the stable structure, including all relaxations of the host atoms, and most functionals [including the still most widely used local density approximation (LDA)] all yield results within reasonable error bars [15]. Quite frequently, however, information about electronic structure is required, i.e., the position of defect levels that are introduced in the band gap of semiconductors or insulators. Since DFT in the LDA or generalized gradient approximation (GGA) severely underestimates the gap, the position of defect levels is subject to large error bars and cannot be directly compared with experiment [16–18]. In turn, this affects the calculated formation energy of the defect, which determines its concentration. This effect on the energy is still not generally appreciated, since it is often assumed that the formation energy is a ground-state property for which DFT should give reliable results. However, in the presence of gap levels that can be filled with varying numbers of electrons (corresponding to the charge state of the defect), the formation energy becomes subject to the same type of errors that would occur when trying to assess excitation energies based on total energy calculations with N or $N + 1$ electrons. Recently, major progress has been made in overcoming these inaccuracies, and the approaches for doing so will be discussed in Section 1.2.

Another type of error that may occur in defect calculations is related to the geometry in which the calculations are performed. Typically, one wishes to address the dilute limit in which the defect concentration is low and defect–defect interactions are negligible. Green's functions calculations would in principle be ideal, but in practice have proven quite cumbersome and difficult to implement. Another approach would be to use clusters, but surface effects are almost impossible to avoid, and quantum confinement effects may obscure electronic structure. Nowadays, point defect calculations are almost universally performed using the supercell geometry, in which the defect is embedded within a certain volume of material which is periodically repeated. This has the advantage of maintaining overall periodicity, which is particularly advantageous when using plane-wave basis sets which rely on

Fast Fourier Transforms to efficiently move between reciprocal- and real-space representations. The supercells should be large enough to minimize interactions between defects in neighboring supercells. This is relatively straightforward to accomplish for *neutral* defects, but due to the long-range nature of the Coulomb interaction, interactions between charged defects are almost impossible to eliminate. This problem was recognized some time ago, and a correction was suggested based on a Madelung-type interaction energy [19]. It had been observed, however, that in many cases the correction was unreliable or "overcorrected," making the result less accurate than the bare values [20]. Recently, an approach based on a rigorous treatment of the electrostatic problem has been developed that outlines the conditions of validity of certain approximations and provides explicit expression for the quantities to be evaluated [21]. Issues relating to supercell-size convergence are addressed in detail in the article by Freysoldt *et al.* [22] in this volume.

We note that it is not the intent of the present paper to provide a comprehensive review of the entirety of this large and growing field. Rather, we attempt to introduce the main concepts of present-day defect calculations illustrated with a few select examples, and do not aspire to cover the countless important contributions to the field by many different research groups.

1.2
Formalism and Computational Approach

The key quantities that characterize a defect in a semiconductor are its concentration and the position of the transition levels (or ionization energies) with respect to the band edges of the host material. Defects that occur in low concentrations will have a negligible impact on the properties of the material. Only those defects whose concentration exceeds a certain threshold will have observable effects. The position of the defect transition levels with respect to the host band edges determines the effects on the electrical and optical properties of the host. Defect formation energies and transition levels can be determined entirely from first principles [1], without resorting to any experimental data for the system under consideration.

1.2.1
Defect Formation Energies and Concentrations

In the dilute limit, the concentration of a defect is determined by the formation energy E^f through a Boltzmann expression:

$$c = N_{\text{sites}} \exp(-E^f / -E^f k_B T). \tag{1.1}$$

N_{sites} is the number of sites (including the symmetry-equivalent local configurations) on which the defect can be incorporated, k_B is the Boltzmann constant, and T the temperature. Note that this expression assumes thermodynamic equilibrium. While defects could also occur in nonequilibrium concentrations, in practice most of the existing bulk and epitaxial film growth techniques operate close to equilibrium conditions. Equilibration of defects is actually unavoidable if the diffusion barriers are low enough to allow easy diffusion at the temperatures of interest. In addition, even if kinetic barriers would be present, Eq. (1.1) is still relevant because obviously defects with a high formation energy are less likely to form.

Defect formation energies can be written as differences in total energies, and these can be obtained from first principles, i.e., without resorting to experimental parameters. The dependence on the chemical potentials (atomic reservoirs) and on the position of the Fermi level in the case of charged defects is explicitly taken into account [1, 5]. This is illustrated here with the specific example of an oxygen vacancy in a 2+ charge state in ZnO. The formation energy of V_O^{2+} is given by:

$$E^f(V_O^{2+}) = E_{tot}(V_O^{2+}) - E_{tot}(ZnO) + \mu_O + 2E_F, \qquad (1.2)$$

where $E_{tot}(V_O^q)$ is the total energy of the supercell containing the defect, and $E_{tot}(ZnO)$ is the total energy of the ZnO perfect crystal in the same supercell. E_F is the energy of the reservoir with which electrons are exchanged, i.e., the Fermi level. The O atom that is removed is placed in a reservoir, the energy of which is given by the oxygen chemical potential μ_O. Note that μ_O is a variable, corresponding to the notion that ZnO can in principle be grown or annealed under O-rich, O-poor, or any other condition in between. It is subject to an upper bound given by the energy of an O atom in an O_2 molecule. Similarly, the zinc chemical potential μ_{Zn} is subject to an upper bound given by the energy of a Zn atom in bulk Zn. The sum of μ_O and μ_{Zn} corresponds to the energy of ZnO, which is the stability condition of ZnO. An upper bound on μ_{Zn}, given by the energy of bulk Zn, therefore leads to a lower bound on μ_O, and *vice versa*. The chemical potentials thus vary over a range given by the formation enthalpy of the material being considered. Formation enthalpies are generally well described by first-principles calculations. For instance, the calculated formation enthalpy of $-3.50\,\text{eV}$ for ZnO [8] is in a good agreement with the experimental value of $-3.60\,\text{eV}$ [23].

Note that it is, in principle, the *free energy* that determines the defect concentration, and one should in principle take into account vibrational entropy contributions in Eq. (1.1). Such contributions are usually small, on the order of a few k_B, and there is often a significant cancellation between vibrational contributions in the solid and in the reservoir [1]. In rare instances, inclusion of vibration entropy has a distinct impact on which configuration is most stable for a given defect or impurity [24], but it hardly ever has a significant effect on the overall concentration. The reader is referred to Ref. [1] for a detailed discussion on the calculation of defect formation energies from first principles.

1.2.2
Transition Levels or Ionization Energies

Defects in semiconductors and insulators can occur in different charge states. For each position of the Fermi level, one particular charge state has the lowest energy for a given defect. The Fermi-level positions at which the lowest-energy charge state changes are called transition levels or ionization energies. The transition levels are thus determined by formation energy differences:

$$\varepsilon(q/q') = \frac{E^f(D^q; E_F = 0) - E^f(D^{q'}; E_F = 0)}{(q' - q)}, \quad (1.3)$$

where $E^f(D^q; E_F = 0)$ is the formation energy of the defect D in the charge state q for the Fermi level at the valence-band maximum ($E_F = 0$). These are thermodynamic transition levels, i.e., atomic relaxations around the defect are fully included; for Fermi-level positions below $\varepsilon(q/q')$ the defect is stable in charge state q, while for Fermi-level positions above $\varepsilon(q/q')$, the defect is stable charge state q'. The thermodynamic transition levels are not to be confused with the single-particle Kohn–Sham states that result from band-structure calculations for a single charge state. They are also not to be confused with optical transition levels derived, for example, from luminescence or absorption experiments. In this case, the final state may not be completely relaxed, and the optical transition levels may significantly differ from the thermodynamic transition levels, as discussed in Ref. [1].

For a defect to contribute to conductivity, it must be stable in a charge state that is consistent with the presence of free carriers. For instance, in order to contribute to n-type conductivity, the defect must be stable in a positive charge state and the transition level from the positive to the neutral charge state should occur close to or above the conduction-band minimum (CBM). A defect is a typical shallow donor when the transition level for a positive to the neutral charge state [e.g., the $\varepsilon(+/0)$ level], as defined based on formation energies, lies above the CBM. In this case, a neutral charge state in which the electron is localized in the immediate vicinity of the defect cannot be maintained if the corresponding electronic level is resonant with the conduction band; instead, the electron will be transferred to extended states, but may still be bound to the positive core of the defect in a hydrogenic effective-mass state. Similarly, shallow acceptors are defects in which the transition level from a negative to the neutral charge state [e.g., the $\varepsilon(-/0)$ level] is near or below the VBM. If the latter, the hole can be bound to the negative core of the defect in a hydrogenic effective-mass state [1, 25].

1.2.3
Practical Aspects

The total energies in Eq. (1.2) are often evaluated by performing DFT calculations within the LDA or its semi-local extension, the GGA [26, 27]. Defects are typically calculated by using a supercell geometry, in which the defect is placed in a cell that is a multiple of the primitive cell of the crystal. The supercell is then periodically repeated in three-dimensional space. The use of supercells also has the advantage that the underlying band structure of the host remains properly described, and integrations over the Brillouin zone are replaced by summations over a discrete and relatively small set of special k-points. Supercell-size corrections for charged defects are addressed in Refs. [21] and [22]. Convergence with respect to the supercell size, number of plane waves in the basis set, and the number of special k-points should always be checked, to make sure that the quantities that are derived are representative of the isolated defect.

The number of atoms or electrons in the calculations is limited by the available computer power. For typical defect calculations, supercells containing 32, 64, 128, 216, and 256 atoms are used for materials with the zinc-blende structure, whereas supercells containg 32, 48, 72, and 96 atom cells are used for materials in the wurtzite structure. These fairly large cell sizes call for efficient computational approaches. Ultrasoft pseudopotential [28–30] and projector-augmented-wave [31] methods to separate the chemically active valence electrons from the inert core electrons have proven ideal for tackling such large systems. First-principles methods based on plane-wave basis sets have been implemented in many codes such as the Vienna *Ab initio* Simulation Program (VASP) [32–34], ABINIT [35, 36], and Quantum Expresso [37].

1.3
The DFT-LDA/GGA Band-Gap Problem and Possible Approaches to Overcome It

The LDA and the GGA in the DFT are plagued by the problem of large band-gap errors in semiconductors and insulators, resulting in values that are typically less than 50% of the experimental values [38–42]. It has often been assumed that the band-gap problem is not an issue when studying defects in semiconductors, since each individual calculation for a specific charge state of the defect could be considered to be a ground-state calculation. However, this notion is not correct, in the same way that the assumption that LDA calculations could yield reliable total-energy differences between N-electron versus $(N + 1)$-electron systems is not correct [16]. Indeed, the change in the number of electrons elicits the issue of the lack of a discontinuity in the exchange-correlation potential, which is at the root of the band-gap problem [38–42]. Similarly, the formation energy expressed in Eq. (1.2) involves changes in the occupation of defect-induced states. In other words, if a specific charge state of a defect involves occupying a state in the band gap, and the band gap is incorrect in DFT–LDA/GGA, then the position of the defect state and hence the calculated total energy will suffer from the same problem [8, 16]. Careful practitioners have always been aware of this problem and refrained from drawing conclusions that might be affected by these uncertainties. The problem is exacerbated, of course, in the case of wide-band-gap semiconductors in which the band-gap errors can be particularly severe; for example, in ZnO the LDA band gap is only 0.8 eV, compared to an experimental value of 3.4 eV.

In the remainder of this section we address several approaches that have been, or are being, developed to overcome these problems.

1.3.1
LDA + U for Materials with Semicore States

Many of the wide-band-gap materials of interest have narrow bands, derived from semicore states, that play an important role in their electronic structure [43]. For

1.3 The DFT-LDA/GGA Band-Gap Problem and Possible Approaches to Overcome It

example, in ZnO narrow bands derived from the Zn 3d states occur at ~8 eV below the valence-band maximum (VBM) and strongly interact with the top of the valence band derived from O 2p states. Inclusion of the Zn d states as valence states (as opposed to treating them as core states) is therefore important for a proper description of the electronic structure of ZnO, as it affects structural parameters, band offsets, and deformation potentials [44, 45]. The DFT–LDA/GGA does not properly describe the energetic position of these narrow bands due to their higher degree of localization, as compared to the more delocalized s and p bands. One way to overcome this problem is to use an orbital-dependent potential that adds an extra Coulomb interaction U for these semicore states, as in the LDA + U (or GGA + U) approach [46, 47].

In the LDA + U the electrons are separated into localized electrons for which the Coulomb repulsion U is taken into account via a Hubbard-like term in a model Hamiltonian, and delocalized or itinerant electrons that are assumed to be well described by the usual orbital-independent one-electron potential in the LDA. Although this approach had been developed and applied for materials with partially filled d bands [46, 47], it has been recently demonstrated that it significantly improves the description of the electronic structure of materials with completely filled d bands such as GaN and InN, as well as ZnO and CdO [44, 45].

An important issue in the LDA + U approach is the choice of the parameter U. It has often been treated as a fitting parameter, with the goal of reproducing either the experimental band gap or the experimentally observed position of the d states in the band structure. Neither approach can be justified, because (a) LDA + U cannot be expected to correct for other shortcomings of DFT-LDA, specifically, the lack of a discontinuity in the exchange-correlation potential, and (b) experimental observations of semicore states may include additional ("final state") effects inherent in experiments such as photoemission spectroscopy. An approximate but consistent and unbiased approach has been proposed in which the calculated U for the isolated atom is divided (screened) by the optical dielectric constant of the solid under consideration [44]. Tests on a number of systems have shown that applying LDA + U effectively lowers the energy of the narrow d bands, thus reducing their coupling with the p states at the VBM; simultaneously, it increases the energy of the s states that compose the CBM, due to the improved screening by the more strongly bound d states, leading to further opening of the band gap. Such improvements have been described in detail in the case of ZnO, CdO, GaN, and InN [44, 45].

One can take advantage of the partial correction of the band gap by the LDA + U to study defects. Based on an extrapolation of LDA and LDA + U results, one can obtain transition levels and formation energies that can be directly compared with experiments. Such extrapolation schemes have been applied in other contexts as well; they are based on evaluation of defect properties for two different values of the band gap followed by a linear extrapolation to the experimental gap. A number of empirical extrapolation approaches were described by Zhang *et al.* [48], for instance based on use of different exchange and correlation potentials or different plane-wave cutoffs. Such extrapolation schemes are most likely to be successful if the calculations that produce different band gaps are physically motivated, ensuring that the shifts in

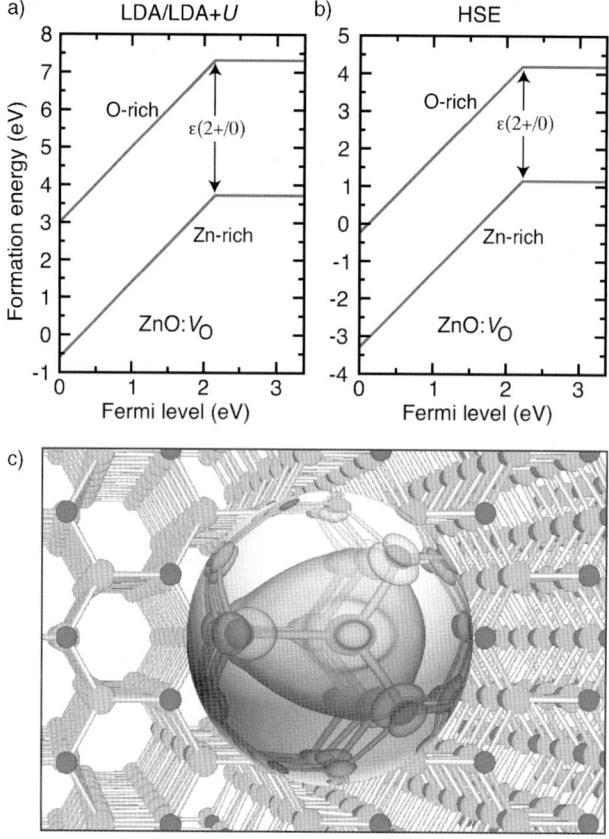

Figure 1.1 (online color at: www.pss-b.com) Formation energy as a function of Fermi level for an oxygen vacancy (V_O) in ZnO. (a) Energies according to the LDA/LDA + U scheme described in Section 1.3.1. (b) Energies according to the HSE approach [51]. The lower curve in each plot indicates Zn-rich conditions, and the upper curve O-rich conditions. The position of the transition level $\varepsilon(2+/0)$ is also indicated. (c) Charge density of the V_O^0 gap state, which is occupied with two electrons. The isosurface corresponds to 10% of the maximum.

defect states that give rise to changes in formation energies reflect the underlying physics of the system.

An extrapolation based on LDA and LDA + U calculations, as described in Refs. [8] and [17], has been shown to be particularly suitable for describing defect physics in materials with semicore d states. The LDA + U produces genuine improvements in the electronic structure related to the energetics of the semicore states; one of these effects is an increase in the band gap. The shifts in defect-induced states between LDA and LDA + U reflect their relative valence- and conduction-band character, and hence an extrapolation to the experimental gap is expected to produce reliable results. Such an approach has led to accurate predictions for point defects in ZnO, InN, and SnO_2 [8, 49, 50]. Figure 1.1(a) shows the result of this extrapolation scheme for the case of oxygen vacancy in ZnO. The success of this approach can be attributed to the fact that the defect states can in principle be described as a linear combination of host states, under the assumption that the latter form a complete basis. A defect state in the gap region will have contributions from both valence-band states and conduction-band states. The shift in transition levels with respect to the host band edges upon band-gap correction reflects the valence- versus conduction-band character of the defect-induced single-particle states. In the case of a shallow donor, the related transition level is expected to shift with the conduction band, i.e., the variation of the transition level is almost equal the band gap correction. For a shallow acceptor, the position of the transition level with respect to the valence band is expected to remain unchanged.

1.3.2
Hybrid Functionals

The use of hybrid functionals has been rapidly spreading in the study of defects in solids. In particular, hybrid functionals have proven reliable for describing the electronic and structural properties of defects in semiconductors. The method consists of mixing local (LDA) or semi-local (GGA) exchange potentials with the non-local Hartree–Fock exchange potential. The correlation potential is still described by the LDA or GGA. Hybrid functionals have been successful in describing structural properties and energetics of molecules in quantum chemistry, with Becke's three-parameter exchange functional (B3) with the Lee, Yang, and Parr (LYP) correlation (B3LYP) being the most popular choice [52]. However, the use of B3LYP for studying defects in solids has been limited due to its shortcomings in describing metals and narrow-gap semiconductors [53]. This issue is particularly important since formation enthalpies of metals usually enter the description of the chemical potential limits in the defect-formation-energy expressions (cf. Eq. (1.2)).

The introduction of a screening length in the exchange potential by Heyd, Scuseria, and Ernzerhof (HSE) [54, 55] and its implementation in a plane-wave code [56] have been instrumental in enabling the use of hybrid functionals in the study of defects in semiconductors. In the HSE the exchange potential is divided in short- and long-range parts. In the short-range part, the GGA exchange of Perdew, Burke, and Ernzerhof (PBE) [27] potential is mixed with non-local Hartree–Fock exchange potential in a ratio of 75/25. The long-range exchange potential as well as the correlation is described by the PBE functional. The range-separation is implemented through an Error function with a characteristic screening length set to ~ 10 Å [55], the variation of which can also affect band gaps [57]. The screening is essential for describing metals and insulators on the same footing. The HSE functional has been shown to accurately describe band gaps for many materials [56, 58]. We should note, however, that since the Hartree–Fock potential involves four-center integrals its implementation in plane-wave codes results in a high computational cost, and currently hybrid functional calculations take at least an order of magnitude more processing time than standard LDA calculations for systems with the same number of electrons.

As an example of hybrid functional calculations for defects in semiconductors, we show in Figure 1.1(b) the formation energy as a function of Fermi level for the oxygen vacancy (V_O) in ZnO using the HSE functional [51]. These calculations were performed by setting the mixing parameter to 37.5% so to reproduce the experimental value of the band gap of ZnO. We note that the position of the transition level $\varepsilon(2+/0)$ with respect to the band edges is in remarkably good agreement with the value obtained using the LDA/LDA + U approach in Figure 1.1(a). On the other hand, the absolute values of the formation energies are quite different, with the HSE results being more than 2 eV lower than the LDA/LDA + U results. This difference can be attributed to the effects of the HSE on the absolute position of the VBM in ZnO. In the LDA/LDA + U approach, U is applied only to the d states and the gap is corrected due to the effects of the coupling between the O 2p Zn d states, and the improved screening of the Zn 4s by the d states. Within this approach, it was assumed that the LDA + U would result in a correct position of the VBM. The HSE results show, however, that the position of the VBM on an absolute energy scale is affected by the inclusion of Hartree–Fock exchange [59]. That is HSE also corrects (at least in part) the self-interaction error in the LDA or GGA, which is still present in the LDA + U results, and this correction is significant for the O 2p bands that make up the VBM in ZnO. In Ref. [59] it was found that the VBM in ZnO is shifted down by 1.7 eV in HSE calculations, compared to PBE.

Other examples of the use of HSE include calculations for Si and Ge impurities in ZnO, which revealed that these impurities are shallow double donors when substituting on the Zn sites in ZnO, with relatively low formation energies [59]. Si can occur as a background impurity in ZnO, and these results indicate that it may give rise to unintentional n-type conductivity. Another example relates to p-type doping in ZnO. It has been long believed that incorporating N on the O site would lead to p-type ZnO. However, the effectiveness of N as a shallow acceptor dopant has never been firmly established. Despite many reports on p-type ZnO using N acceptors, the results have been difficult to reproduce, raising questions about the stability of the p-type doping and the position of the N ionization energy. Recent calculations for N in ZnO have shown that N is actually a very deep acceptor with a transition level at 1.3 eV above the VBM [60]. Therefore, it has been concluded that N cannot lead to p-type ZnO. For comparison and as a benchmark, HSE calculations correctly predicted that N in ZnSe is a shallow acceptor when substituting on Se sites, in agreement with experimental findings.

Hybrid functional calculations have also been performed for oxygen vacancies in TiO_2. Despite the fact that oxygen vacancies have frequently been invoked in the literature on TiO_2, their identification in bulk TiO_2 has remained elusive. First-principles calculations based on LDA or GGA suffer from band-gap problems and are unable to describe the neutral or the positively charged vacancy (V_O^+) in TiO_2 [61, 62]. In LDA or GGA, the Kohn–Sham single-particle states related to V_O are above the CBM, causing the electron(s) from V_O^0 or V_O^+ to occupy the CBM. Calculations based on the HSE, on the other hand, show that locally stable structures of V_O^0 and V_O^+ exist, in which the occupied single-article states lie within the band gap and the defect wave

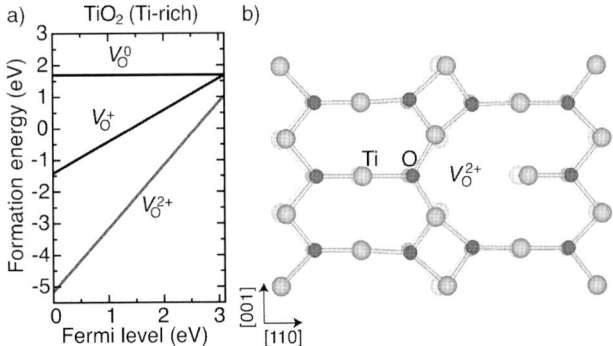

Figure 1.2 (online color at: www.pss-b.com) (a) Formation energy as a function of Fermi level for an oxygen vacancy (V_O) in TiO_2 in the Ti-rich limit, according to Ref. [62]. (b) Local lattice relaxations around V_O^{2+}. The positions of the atoms in the perfect crystal are also indicated (faded).

functions are localized within the vacancy. However, the formation energies of V_O^0 and V_O^+ are always higher in energy that of V_O^{2+} [62] as shown in Figure 1.2(a); The atoms around V_O^{2+} relax outward as indicated in Figure 1.2(b). Thus, oxygen vacancies are predicted to be shallow donors in TiO_2. This is in contrast to GGA + U calculations which indicate that V_O is a deep donor with transition levels in the gap [63]. The problem with GGA + U calculations for TiO_2 is that the conduction band in TiO_2 is derived from the Ti d states. The LDA/GGA + U approach was designed to be applied to narrow bands with localized electrons; hence its success when applied to semicore d states. The d states that constitute the conduction band of TiO_2, in contrast, are fairly delocalized, as evidenced by the high conductivity of this material. Applying LDA/GGA + U will always lead to an energy lowering of the occupied states, since that was what the approach was designed to do. Therefore, when the LDA/GGA + U approach is applied to a case in which electrons occupy the conduction band of TiO_2, localization will result. However, it is hard to distinguish whether this is a real physical effect or an artefact due to the nature of the LDA/GGA + U approach. We therefore feel that LDA/GGA + U should not be applied in cases where the states are intrinsically extended states, such as the d states that make up the conduction band of TiO_2.

An important issue regarding the use of hybrid functionals is the amount of Hartree–Fock exchange potential that is mixed with the GGA exchange [64]. Although a value of 25% was initially proposed, there is no *a priori* justification for this amount and this single value is not capable of correctly describing all semiconductors and insulators. For instance, in ZnO the experimental value of the band gap is obtained with HSE only when a mixing parameter of 37% is used. In GaN, a mixing parameter of 31% is necessary, and for MgO 32%. Since the position of transition levels in the band gap depends on the band-gap value, quantitative predictions require that the functional accurately describes band gaps, and an adjustment of the mixing parameter is the most straightforward way to achieve this.

1.3.3
Many-Body Perturbation Theory in the GW Approximation

Quasiparticle calculations in the GW approximation produce band structures that are in close agreement with experiments [65]. However, at present the calculation of total energies within the GW formalism [66] is still a subject of active research and currently not available for studying defects in solids. We note that the GW quasiparticle energies are defined as removal and addition energies. In the case of defects, the GW quasiparticle energies that appear in the band gap correspond to the transition levels, provided that the geometry of the defects remains unchanged. For instance, the highest occupied quasiparticle state in a calculation for a defect in charge state q represents the $\varepsilon(q+1/q)$ level, and the lowest unoccupied state represents the $\varepsilon(q/q-1)$ level for a fixed geometry of the defect. It is possible to combine these transition levels determined from GW calculations with relaxation energies from LDA or GGA calculations to extract thermodynamic transition levels for defects in semiconductors and insulators. Recent GW calculations for the self-interstitial in Si have demonstrated the effectiveness of this approach [67].

The LDA or GGA underestimates the formation energy of the self-interstitial in Si by more than 1 eV compared to values extracted from self-diffusion experiments. Calculations based on Quantum Monte Carlo can yield more accurate formation energies but are very expensive computationally. Calculating removal and addition energies for Si self-interstitials in GW and combining with relaxation energies from LDA calculations lead to formation energies that are in good agreement with Quantum Monte Carlo results [69]. The only assumption was that LDA gives correct formation energies for charge state configurations with no occupied states above the VBM, such as the 2+ charge state of the Si self-interstitial in the tetrahedral configuration. A similar approach has been used to study oxygen-related defects in SiO_2 [68].

As a drawback in the GW approach, it has been recently argued that for systems with semicore d states such as ZnO a very large number of unoccupied bands is necessary for a proper description of the band structure [70]. This result, if confirmed, indicates that GW calculations for defects in these systems may be prohibitivly expensive in practice. This unusually large number of unoccupied states required is likely related to the underbinding of the semicore d states which, as discussed in Section 1.3.1, can make a significant contribution to the band-gap error.

1.3.4
Modified Pseudopotentials

In the pseudopotential formalism, once a separation between valence electrons and the inert core electrons is adopted, there is still some flexibility in constructing the ionic cores. Indeed, within this approach, there is no unique scheme for generating pseudopotentials, and a number of different generation schemes have been proposed over the years, often aimed at creating computationally efficient, "softer" potentials which can be described with a smaller plane-wave basis set. This flexibility can in principle be exploited to generate potentials that produce a more accurate band structure. However, past attempts did not succeed in producing such improvements while still maintaining a proper description of atomic structure and energetics [71].

A new approach was recently proven to be remarkably successful in describing nitride semiconductors [72, 73]. It was based on a proposal by Christensen, first implemented within the linearized muffin-tin orbital method [74], to add a highly localized (delta-function-like) repulsive potential centered on the atomic nucleus of each atom. Such a potential only affects s states, and since the CBM in compound semiconductors has largely cation s character one expects an upward shift of the corresponding eigenstates. At the same time, the highly localized character of the added potential leads one to expect only minimal changes in other aspects of the pseudopotential. These expectations were indeed borne out in the case of GaN and InN, where the modified pseudopotentials produced atomic structures and energetics that are as reliable as those obtained with standard potentials, but simultaneously producing band structures in very good agreement with experiment [73]. Even though the fitting procedure only aimed to produce the experimental value of the direct gap, the modified potentials actually produced improvements for other aspects of the band structure as well, including the position of higher-lying indirect conduction-band minima as well as the position of semicore d states [73]. This leads us to believe that the seemingly *ad hoc* modifications introduced by the repulsive potential are capturing some essential physics, justifying the expectation that similarly good results can be obtained for other materials. An application of the modified pseudopotentials to the calculation of the electronic structure of nitride surfaces produced results in very good agreement with experiment [72, 75].

1.4
Summary

We have discussed recent progress in first-principles approaches to study defects in semiconductors and insulators. Emphasis was given to methods that overcome the band-gap problem in traditional DFT in the LDA; such approaches include $LDA + U$, hybrid functionals, GW, and modified pseudopotentials. While the $LDA + U$ approach is very efficient computationally, it should be limited to systems with semicore states for which LDA provides a poor description. Furthermore, the $LDA + U$ only partially corrects the band gap, and futher extrapolation is needed. The HSE hybrid functional on the other hand is general and has been demonstrated to be a reliable method that result in accurate band gaps and seems to be describing the properties of defects correctly. The HSE functional contains two parameters, the Hartree–Fock mixing ratio and the screening length, which offer some flexibility in obtaining correct band gaps; however, the consequences of changes in these parameters on the physics of the system has not been fully explored yet. The GW method offers a formal approach for describing excited-state properties and defect physics, but its applicability is limited by the lack of an efficient way to extract total energies. Combining GW excitation energies with LDA/GGA relaxation energies offers a promising way to address thermodynamic transition levels. Finally, modified pseud-potentials is an *ad hoc* but remarkably reliable approach, which has been demonstrated very effective at describing the properties of nitride semiconductors.

Acknowledgements

We acknowledge fruitful collaborations and discussions with C. Freysoldt, G. Kresse, J. Lyons, J. Neugebauer, P. Rinke, M. Scheffler, A. Singh, N. Umezawa, and J. Varley. This work was supported by the NSF MRSEC Program under Award No. DMR05-20415, by the UCSB Solid State Lighting and Energy Center, and by the MURI program of the Army Research Office under Grant No. W911-NF-09-1-0398. It made use of the CNSI Computing Facility under NSF grant No. CHE-0321368 and Teragrid.

References

1 Van de Walle, C.G. and Neugebauer, J. (2004) *J. Appl. Phys.*, **95**, 3851.
2 Drabold, D.A. and Estreicher, S.K. (eds) (2007) *Theory of Defects in Semiconductors*, Springer-Verlag, Berlin.
3 Asato, M., Mizuno, T., Hoshino, T., Masuda-Jindo, K., and Kawakami, K. (2001) *Mater. Sci. Eng. A*, **312**, 72.
4 Queisser, H.J. and Haller, E.E. (1998) *Science*, **281**, 945.
5 Van de Walle, C.G., Laks, D.B., Neumark, G.F., and Pantelides, S.T. (1993) *Phys. Rev. B*, **47**, 9425.
6 Janotti, A. and Van de Walle, C.G. (2009) *Rep. Prog. Phys.*, **72**, 126501.
7 Limpijumnong, S. and Van de Walle, C.G. (2004) *Phys. Rev. B*, **69**, 035207.
8 Janotti, A. and Van de Walle, C.G. (2007) *Phys. Rev. B*, **76**, 165202.
9 Janotti, A., Krčmar, M., Fu, C.L., and Reed, R.C. (2004) *Phys. Rev. Lett.*, **92**, 085901.
10 Lannoo, M. and Bourgoin, J. (1981) *Point Defects in Semiconductors I: Theoretical Aspects*, Springer-Verlag, Berlin; (1983) *Point Defects in Semiconductors II: Experimental Aspects*, Springer-Verlag, Berlin.
11 Pandelides, S.T. (ed.) (1992) *Deep Centers in Semiconductors: A State-of-the-Art Approach*, 2nd edn, Gordon and Breach Science, Yverdon.
12 Stavola, M. (ed.) (1999) *Identification of Defects in Semiconductors, Semiconductors and Semimetals*, vol. 51A, 51B Academic, San Diego.
13 Hohenberg, P. and Kohn, W. (1964) *Phys. Rev.*, **136**, B864; Kohn, W., Sham, L.J. (1965) *Phys. Rev.*, **140**, A1133.
14 Kohn, W. (1999) *Rev. Mod. Phys.*, **71**, 1253.
15 Payne, M., Teter, M.P., Allan, D.C., Arias, T.A., and Joannopoulos, J.D. (1992) *Rev. Mod. Phys.*, **64**, 1045.
16 Stampfl, C., Van de Walle, C.G., Vogel, D., Krüger, P., and Pollmann, J. (2000) *Phys. Rev. B*, **61**, R7846.
17 Janotti, A. and Van de Walle, C.G. (2005) *Appl. Phys. Lett.*, **87**, 122102.
18 Janotti, A. and Van de Walle, C.G. (2006) *J. Cryst. Growth*, **287**, 58.
19 Makov, G. and Payne, M.C. (1995) *Phys. Rev. B*, **51**, 4014.
20 Shim, J., Lee, E.-K., Lee, Y.J., and Nieminen, R.M. (2005) *Phys. Rev. B*, **71**, 035206.
21 Freysoldt, C., Neugebauer, J., and Van de Walle, C.G. (2009) *Phys. Rev. Lett.*, **102**, 016402.
22 Freysoldt, C., Neugebauer, J., and Van de Walle, C.G. (2010) *Phys. Status Solidi B*, doi: 10.1002/pssb.201046289.
23 Dean, J.A. (ed.) (1992) *Lange's Handbook of Chemistry*, 14th edn, (McGraw-Hill, Inc., New York.
24 Limpijumnong, S., Van de Walle, C.G., and Northrup, J.E. (2001) *Phys. Rev. Lett.*, **87**, 205505.

25 Neugebauer, J. and Van de Walle, C.G. (1999) *J. Appl. Phys.*, **85**, 3003.
26 Perdew, J.P. and Wang, Y. (1991) *Phys. Rev. Lett.*, **66**, 508.
27 Perdew, J.P., Burke, K., and Ernzerhof, M. (1996) *Phys. Rev. Lett.*, **77**, 3865.
28 Vanderbilt, D. (1990) *Phys. Rev. B*, **41**, 7892.
29 Laasonen, K., Pasquarello, A., Car, R., Lee, C., and Vanderbilt, D. (1993) *Phys. Rev. B*, **47**, 10142.
30 Kresse, G. and Hafner, J. (1994) *J. Phys.: Condens. Matter*, **6**, 8245.
31 Blöchl, P.E. (1994) *Phys. Rev. B*, **50**, 17953.
32 Kresse, G. and Hafner, J. (1993) *Phys. Rev. B*, **47**, 558.
33 Kresse, G. and Furthmüller, J. (1996) *Phys. Rev. B.*, **54**, 11169.
34 Kresse, G. and Furthmüller, J. (1996) *Comput. Mat. Sci.*, **6**, 15.
35 Gonze, X., Rignanese, G.-M., Verstraete, M., Beuken, J.-M., Pouillon, Y., Caracas, R., Jollet, F., Torrent, M., Zerah, G., Mikami, M., Ghosez, Ph., Veithen, M., Raty, J.-Y., Olevano, V., Bruneval, F., Reining, L., Godby, R., Onida, G., Hamann, D.R., and Allan, D.C. (2005) *Zeit. Kristallogr.*, **220**, 558.
36 Gonze, X., Amadon, B., Anglade, P.-M., Beuken, J.-M., Bottin, F., Boulanger, P., Bruneval, F., Caliste, D., Caracas, R., Cote, M., Deutsch, T., Genovese, L., Ghosez, Ph., Giantomassi, M., Goedecker, S., Hamann, D.R., Hermet, P., Jollet, F., Jomard, G., Leroux, S., Mancini, M., Mazevet, S., Oliveira, M.J.T., Onida, G., Pouillon, Y., Rangel, T., Rignanese, G.-M., Sangalli, D., Shaltaf, R., Torrent, M., Verstraete, M.J., Zerah, G., and Zwanziger, J.W. (2009) *Comput. Phys. Commun.*, **180**, 2582.
37 Giannozzi, P., Baroni, S., Bonini, N., Calandra, M., Car, R., Cavazzoni, C., Ceresoli, D., Chiarotti, G.L., Cococcioni, M., Dabo, I., Dal Corso, A., Gironcoli, S., Fabris, S., Fratesi, G., Gebauer, R., Gerstmann, U., Gougoussis, C., Kokalj, A., Lazzeri, M., Martin-Samos, L., Marzari, N., Mauri, F., Mazzarello, R., Paolini, S., Pasquarello, A., Paulatto, L., Sbraccia, C., Scandolo, S., Sclauzero, G., Seitsonen, A.P.,
Smogunov, A., Umari, P., and Wentzcovitch, R.M. (2009) *J. Phys.: Condens. Matter*, **21**, 395502.
38 Perdew, J.P. and Levy, M. (1983) *Phys. Rev. Lett.*, **51**, 1884.
39 Sham, L.J. and Schlüter, M. (1983) *Phys. Rev. Lett.*, **51**, 1888.
40 Perdew, J.P. (1985) *Int. J. Quant. Chem.*, **28**, 497.
41 Godby, R.W., Schlüter, M., and Sham, L.J. (1986) *Phys. Rev. Lett.*, **56**, 2415.
42 Mori-Sanchez, P., Cohen, A.J., and Yang, W. (2008) *Phys. Rev. Lett.*, **100**, 146401.
43 Wei, S.H. and Zunger, A. (1988) *Phys. Rev. B*, **37**, 8958.
44 Janotti, A., Segev, D., and Van de Walle, C.G. (2006) *Phys. Rev. B*, **74**, 045202.
45 Janotti, A. and Van de Walle, C.G. (2007) *Phys. Rev. B*, **75**, 121201.
46 Anisimov, V.I., Zaanen, J., and Andersen, O.K. (1991) *Phys. Rev. B*, **44**, 943.
47 Anisimov, V.I., Aryasetiawan, F., and Liechtenstein, A.I. (1997) *J. Phys.: Condens. Matter*, **9**, 767.
48 Zhang, S.B., Wei, S.H., and Zunger, A. (2001) *Phys. Rev. B*, **63**, 075205.
49 Janotti, A. and Van de Walle, C.G. (2008) *Appl. Phys. Lett.*, **92**, 032104.
50 Singh, A.K., Janotti, A., Scheffler, M., and Van de Walle, C.G. (2008) *Phys. Rev. Lett.*, **101**, 055502.
51 Oba, F., Togo, A., Tanaka, I., Paier, J., and Kresse, G. (2008) *Phys. Rev. B*, **77**, 245202.
52 Becke, A.D. (1993) *J. Chem. Phys.*, **98**, 1372.
53 Paier, J., Marsman, M., and Kresse, G. (2007) *J. Chem. Phys.*, **127**, 024103.
54 Heyd, J., Scuseria, G.E., and Ernzerhof, M. (2003) *J. Chem. Phys.*, **118**, 8207.
55 Heyd, J., Scuseria, G.E., and Ernzerhof, M. (2006) *J. Chem. Phys.*, **124**, 219906.
56 Paier, J., Marsman, M., Hummer, K., Kresse, G., Gerber, I.C., and Àngyàn, J.G. (2006) *J. Chem. Phys.*, **124**, 154709.
57 Komsa, H.-P., Broqvist, P., and Pasquarello, A. (2010) *Phys. Rev. B*, **81**, 205118.
58 Marsman, M., Paier, J., Stroppa, A., and Kresse, G. (2008) *J. Phys.: Condens. Matter*, **20**, 064201.

59 Lyons, J.L., Janotti, A., and Van de Walle, C.G. (2009) *Phys. Rev. B*, **80**, 205113.

60 Lyons, J.L., Janotti, A., and Van de Walle, C.G. (2009) *Appl. Phys. Lett.*, **95**, 252105.

61 Sullivan, J.M. and Erwin, E.C. (2003) *Phys. Rev. B*, **67**, 144415.

62 Janotti, A., Varley, J.B., Rinke, P., Umezawa, N., Kresse, G., and Van de Walle, C.G. (2010) *Phys. Rev. B*, **81**, 085212.

63 Osorio-Guillen, J., Lany, S., and Zunger, A. (2008) *Phys. Rev. Lett.*, **100**, 036601.

64 Alkauskas, A., Broqvist, P., and Pasquarello, A. (2010) *Phys. Status Solidi B*, doi: 10.1002/pssb.201046195.

65 Hybertsen, M.S. and Louie, S.G. (1986) *Phys. Rev. B*, **34**, 5390.

66 Sánchez-Friera, P. and Godby, R.W. (2000) *Phys. Rev. Lett.*, **85**, 5611.

67 Rinke, P., Janotti, A., Scheffler, M., and Van de Walle, C.G. (2009) *Phys. Rev. Lett.*, **102**, 026402.

68 Martin-Samos, L., Roma, G., Rinke, P., and Limoge, Y. (2010) *Phys. Rev. Lett.*, **104**, 075502.

69 Batista, E.R., Heyd, J., Hennig, R.G., Uberuaga, B.P., Martin, R.L., Scuseria, G.E., Umrigar, C.J., and Wilkins, J.W. (2006) *Phys. Rev. B*, **74**, 121102.

70 Zhang, P. and Shih, B., Abstract Q23, American Physical Society Meeting, March 2010, http://meetings.aps.org/Meeting/MAR10/Event/120825.

71 Wang, L.W. (2001) *Appl. Phys. Lett.*, **78**, 1565.

72 Segev, D. and Van de Walle, C.G. (2006) *Europhys. Lett.*, **76**, 305.

73 Segev, D., Janotti, A., and Van de Walle, C.G. (2007) *Phys. Rev. B*, **75**, 035201.

74 Christensen, N.E. (1984) *Phys. Rev. B*, **30**, 5753.

75 Van de Walle, C.G. and Segev, D. (2007) *J. Appl. Phys.*, **101**, 081704.

2
Accuracy of Quantum Monte Carlo Methods for Point Defects in Solids
William D. Parker, John W. Wilkins, and Richard G. Hennig

2.1
Introduction

Point defects, such as vacancies, interstitials and anti-site defects, are the only thermodynamically stable defects at finite temperatures [1]. The infinite slope of the entropy of mixing at infinitesimally small defect concentrations results in an infinite driving force for defect formation. As a result, at small defect concentrations, the entropy of mixing always overcomes the enthalpy of defect formations. In addition to being present in equilibrium, point defects often control the kinetics of materials, such as diffusion and phase transformations, and are important for materials processing. The presence of point defects in materials can fundamentally alter the electronic and mechanical properties of a material. This makes point defects technologically important for applications such as doping of semiconductors [2, 3], solid solution hardening of alloys [4, 5], controlling the transition temperature for shape-memory alloys [6], and the microstructural stabilization of two-phase superalloys.

However, the properties of defects, such as their structures and formation energies, are difficult to measure in some materials due to their small sizes, low concentrations, lack of suitable radioactive isotopes, *etc.* Quantum mechanical first-principles, or *ab initio*, theories make predictions to fill in the gaps left by experiment [7].

The most widely used method for the calculation of defect properties in solids is density functional theory (DFT). DFT replaces explicit many-body electron interactions with quasiparticles interacting via a mean-field potential, *i.e.*, the exchange-correlation potential, which is a functional of the electron density [8]. A universally true exchange-correlation functional is unknown, and DFT calculations employ various approximate functionals, either based on a model system or an empirical fit. The most commonly used functionals are based on diffusion Monte Carlo (DMC) simulations [9] for the uniform electron gas at different densities, *e.g.*, the local density approximation (LDA) [10, 11] and gradient expansions, *e.g.*, the generalized gradient approximation (GGA) [12–16]. These local and semi-local functionals suffer from a significant self-interaction error reflected in the variable accuracy of their predictions for defect formation energies, charge transition levels, and band gaps [17, 18]. Another class of functionals, called hybrid functionals, include a fraction of exact exchange to improve their accuracy [19, 20].

The seemingly simple system of Si self-interstitials exemplifies the varied accuracy of different density functionals and many-body methods. The diffusion and thermodynamics of silicon self-interstitial defects dominate the doping and subsequent annealing processes of crystalline silicon for electronics applications [3, 21, 22]. The mechanism of self-diffusion in silicon is still under debate. Open questions [23] include: (i) Are the interstitial atoms the prime mediators of self-diffusion? (ii) What is the specific mechanism by which the interstitials operate? (iii) What is the value of the interstitial formation energy? Quantum mechanical methods are well suited to determine defect formation energies. LDA, GGA, and hybrid functionals predict formation energies for these defects ranging from about 2 to 4.5 eV [24]. Quasiparticle methods such as the GW approximation reduce the self-interaction error in DFT and are expected to improve the accuracy of the interstitial formation energies. Recent G_0W_0 calculations [25] predict formation energies of about 4.5 eV in close agreement with HSE hybrid functional [24] and previous DMC calculations [24, 26]. Quantum Monte Carlo (QMC) methods provide an alternative to DFT and a benchmark for defect formation energies [27, 28].

In this paper, we review the approximations that are made in DMC calculations for solids and estimate how these approximations affect the accuracy of point defect calculations, using the Si self-interstitial defects as an example. Section 2.2 describes the QMC method and its approximations. Section 2.3 reviews previous QMC calculations for defects in solids, and Section 2.4 discusses the results of our calculations for interstitials in silicon and the accuracy of the various approximations.

2.2
Quantum Monte Carlo Method

QMC methods are among the most accurate electronic structure methods available and, in principle, have the potential to outperform current computational methods in both accuracy and cost for extended systems. QMC methods scale as $O(N^3)$ with system size and can handle large systems. At the present time, calculations for as many as 1000 electrons on 1000 processors make effective use of available computational resources [24]. Current work is under way to develop algorithms that extend the system size accessible by QMC methods to petascale computers [29].

Continuum electronic structure calculations primarily use two QMC methods [27]: the simpler variational Monte Carlo (VMC) and the more sophisticated DMC. In VMC, a Monte Carlo method evaluates the many-dimensional integral to calculate quantum mechanical expectation values. Accuracy of the results depends crucially on the quality of the trial wave function, which is controlled by the functional form of the wave function and the optimization of the wave functions parameters [30]. DMC removes most of the error in the trial wave function by stochastically projecting out the ground state using an integral form of the imaginary-time Schrödinger equation.

One of the most accurate forms of trial wave functions for QMC applications to problems in electronic structure is a sum of Slater determinants of single-particle orbitals multiplied by a Jastrow factor and modified by a backflow transformation:

$$\Psi(r_n) = e^{J(r_n, R_m)} \sum c_i \, \text{CSF}_i(x_n).$$

The Jastrow factor J typically consists of a low-order polynomial and a plane-wave expansion in electron coordinates r_n and nuclear coordinates R_m that efficiently describe the dynamic correlations between electrons and nuclei. Static (near-degeneracy) correlations are described by a sum of Slater determinants. Symmetry-adapted linear combinations of Slater determinants, so-called configuration state functions (CSF), reduce the number of determinant parameters c_i. For extended systems, the lack of size consistency for a finite sum of CSF's makes this form of trial wave functions impractical, and a single determinant is used instead. Finally, the backflow transformation $r_n \to x_n$ allows the nodes of the trial wave function to be moved, which can efficiently reduce the fixed-node error [31]. Since the backflow-transformed coordinate of an electron x_n depends on the coordinates of all other electrons, the Sherman–Morrison formula used to efficiently update the Slater determinant does not apply, increasing the scaling of QMC to $O(N^4)$. If a finite cutoff for the backflow transformation is used, the Sherman–Morrison-Woodbury formula [32] applies, and the scaling reduces to $O(N^3)$.

Optimization of the many-body trial wave function is crucial because accurate trial wave functions reduce statistical and systematic errors in both VMC and DMC. Much effort has been spent on developing improved methods for optimizing many-body wave functions, and this continues to be the subject of ongoing research. Energy and variance minimization methods can effectively optimize the wave function parameters in VMC calculations [30, 33]. Recently developed energy optimization methods enable the efficient optimization of CSF coefficients and orbital parameters in addition to the Jastrow parameters for small molecular systems, eliminating the dependence of the results on the input trial wave function [30].

VMC and DMC contain two categories of approximation to make the many-electron solution tractable: *controlled* approximations, whose errors can be made arbitrarily small through adjustable parameters, and *uncontrolled* approximations, whose errors are unknown exactly. The controlled approximations include the finite DMC time step, the finite number of many-electron configurations that represent the DMC wave function, the basis set approximation, *e.g.*, spline or plane-wave representation, for the single-particle orbitals of the trial wave function and the finite-sized simulation cell. The uncontrolled approximations include the fixed-node approximation, which constrains the nodes of the wave function in DMC to be the same as those of the trial wave function, the replacement of the core electrons around each atom with a pseudopotential to represent the core–valence electronic interaction and the locality approximation, which uses the trial wave function to project the nonlocal angular momentum components of the pseudopotential.

2.2.1
Controlled Approximations

2.2.1.1 Time Step
DMC is based on the transformation of the time-dependent Schrödinger equation into an imaginary-time diffusion equation with a source–sink term. The propagation of the 3N-dimensional electron configurations (walkers) that sample the wave function requires a finite imaginary time step, which introduces an error in the resulting energy [34, 35].

Controlling the time step error is simply a matter of performing calculations for a range of time steps, either to determine when the total energy or defect formation energy reaches the required accuracy or to perform an extrapolation to a zero time step using a low-order polynomial fit of the energy as a function of time step. Smaller time steps, however, require a larger total number of steps to sample sufficiently the probability space. Thus, the optimal time step should be small enough to add no significant error to the average while large enough to keep the total number of Monte Carlo steps manageable. In addition, the more accurate the trial wave function is the smaller the error due to the time step will be [35].

2.2.1.2 Configuration Population
In DMC, a finite number of electron configurations represent the many-body wave function. These configurations are the time-dependent Schrödinger equation's analogs to particles in the diffusion equation and have also been called psips [34] and walkers [27]. To improve the efficiency of sampling the many-body wave function, the number of configurations is allowed to fluctuate from time step to time step in DMC using a branching algorithm. However, the total number of configurations needs to be controlled to prevent the configuration population from diverging or vanishing [35]. This population control introduces a bias in the energy. In practice where tested [36], hundreds of configurations are sufficient to reduce the population control bias in the DMC total energy below the statistical uncertainty.

The VMC and DMC calculations parallelize easily over walkers. After an initial decorrelation run, the propagation of a larger number of walkers is computationally equivalent to performing more time steps. The variance of the total energy scales like

$$\sigma_E^2 \propto \frac{\tau_{corr}}{N_{conf} N_{step}},$$

where N_{conf} denotes the number of walkers, N_{step} the number of time steps, and τ_{corr} the auto-correlation time.

2.2.1.3 Basis Set
A sum of basis functions with coefficients represents the single-particle orbitals in the Slater determinant. A DFT calculation usually determines these coefficients. Plane waves provide a convenient basis for calculations of extended systems since they form an orthogonal basis that systematically improves with increase in number of plane waves that span the simulation cell. Increasing the number of plane waves

until the total energy converges within an acceptable threshold in DFT creates a basis set that has presumably the same accuracy in QMC.

Since the plane-wave basis functions extend throughout the simulation cell, the evaluation of an orbital at a given position requires a sum over all plane waves. Furthermore, the number of plane waves is proportional to the volume of the simulation cell. The computational cost of orbital evaluation can significantly be reduced by using a local basis, such as B-splines, which replaces the sum over plane waves with a sum over a small number of local basis functions. The resulting polynomial approximation reduces the computational cost of orbital evaluation at a single point from the number of plane waves (hundreds to thousands depending on the basis set) to the number of non-zero polynomials (64 for cubic splines) [37]. The wavelength of the highest frequency plane wave sets the resolution of the splines. Thus, the most important quantity to control in the basis set approximation is the size of the basis set.

2.2.1.4 Simulation Cell

Simulation cells with periodic boundary conditions are ideally suited to describe an infinite solid but result in undesirable finite-size errors that need correction. There are three types of finite-size errors. First, the single-particle finite-size error arises from the choice of a single k-point in the single-particle Bloch orbitals of the trial wave function. Second, the many-body finite-size error arises from the non-physical self-image interactions between electrons in neighboring cells. Third, the defect creates a strain field that results in an additional finite-size error for small simulation cells.

The single-particle finite-size error is greatly reduced by averaging DMC calculations for single-particle orbitals at different k-points that sample the first Brillouin zone of the simulation cell, so-called twist-averaging [38] Alternatively, the single-particle finite-size error can also be estimated from the DFT energy difference between a calculation with a dense k-point mesh and one with the same single k-point chosen for the orbitals of the QMC wave function.

For the many-body finite-size error, several methods aim to correct the fictitious periodic correlations between electrons in different simulation cells. The first approach, the model periodic Coulomb (MPC) interaction [39], revises the Ewald method [40] to account for the periodicity of the electrons by restoring the Coulomb interaction within the simulation cell and using the Ewald interaction to evaluate the Hartree energy. The second approach is based on the random phase approximation for long wavelengths. The resulting first-order, finite-size-correction term for both the kinetic and potential energies can be estimated from the electronic structure factor [41]. The third approach estimates the many-body finite-size error from the energy difference between DFT calculations using a finite-sized and an infinite-sized model exchange-correlation functional [42]. This approach relies on the exchange-correlation functional being a reasonable description of the system, whereas the other two approaches (MPC and structure factor) do not have this restriction. The MPC and structure factor corrections are fundamentally related and often result in similar energy corrections [43].

The defect strain finite-size error can be estimated at the DFT level using extrapolations of large simulation cells. Also, since QMC force calculations are expensive and still under development [44], QMC calculations for extended systems typically start with DFT-relaxed structures. Energy changes due to small errors in the ionic position as well as thermal disorder are expected to be quite small because of the quadratic nature of the minima and will largely cancel when taking energy differences for the defect energies.

2.2.2
Uncontrolled Approximations

2.2.2.1 Fixed-Node Approximation

The Monte Carlo algorithm requires a probability distribution, which is non-negative everywhere, but fermions, such as electrons, are antisymmetric under exchange. Therefore, any wave function of two or more fermions has regions of positive and negative value. For DMC to take the wave function as the probability distribution, Anderson [34] fixed the zeros or nodes of the wave function and took the absolute value of the wave function as the probability distribution. If the trial wave function has the nodes of the ground state, then DMC projects out the ground state. However, if the nodes differ from the ground state, then DMC finds the closest ground state of the system within the inexact nodal surface imposed by the fixed-node condition. This inexact solution has an energy higher than that of the ground state.

Three methods estimate the size of the fixed-node approximation: (i) In the Slater–Jastrow form of the wave function, the single-particle orbitals in the Slater determinant set the zeroes of the trial wave function. Since these orbitals come from DFT calculations, varying the exchange-correlation functional in DFT changes the trial wave function nodes and provides an estimate of the size of the fixed-node error. (ii) López Ríos et al. [31] applied backflow to the nodes by modifying the interparticle distances, enhancing electron–electron repulsion and electron–nucleus attraction. The expense of the method has thus far limited its application in the literature to studies of second- and third-row atoms, the water dimer and the 1D and 2D electron gases. (iii) Because the eigenfunction of the Hamiltonian has zero variance in DMC, a linear extrapolation from the variances of calculations with and without backflow to zero variance estimates the energy of the exact ground state of the Hamiltonian.

2.2.2.2 Pseudopotential

Valence electrons play the most significant roles in determining a composite system's properties. The core electrons remain close to the nucleus and are largely inert. The separation of valence and core electron energy scales allows the use of a pseudopotential to describe the core-valence interaction without explicitly simulating the core electrons. However, there is often no clear boundary between core and valence electrons, and the core–valence interaction is more complicated than a simple potential can describe. Nonetheless, the computational demands of explicitly simulating the core electrons and the practical success of calculations with pseudopotentials in reproducing experimental values promote their continued use in QMC.

Nearly all solid-state and many molecular QMC calculations to date rely on pseudopotentials to reduce the number of electrons and the time requirement of simulating the core–electron energy scales.

Comparing DMC energies using pseudopotentials constructed with different energy methods [DFT and Hartree-Fock (HF)] provides an estimate of the error incurred by the pseudopotential approximation. Additionally, the difference between density functional pseudopotential and all-electron energies estimates the size of the error introduced by the pseudopotential and is used as a correction term.

2.2.2.3 Pseudopotential Locality

DMC projects out the ground state of a trial wave function but does not produce a wave function, only a distribution of point-like configurations. However, the pseudopotential contains separate potentials (or channels) for different angular-momenta of electrons. One channel, identified as local, does not require the wave function to evaluate, but the nonlocal channels require an angular integration to evaluate, and such an integration requires a wave function. Mitáš et al. [45] introduced use of the trial wave function to evaluate the nonlocal components requiring integration. This locality approximation has an error that varies in sign. While there are no good estimates of the magnitude of this error, Casula [46] developed a lattice-based technique that makes the total energy using a nonlocal potential an upper bound on the ground-state energy. Pozzo and Alfè [47] found that, in magnesium and magnesium hydride, the errors of the locality approximation and the lattice-regularized method are comparably small, but the lattice method requires a much smaller time step (0.05 vs. 1.00 Ha^{-1} in Mg and 0.01 vs. 0.05 Ha^{-1} in MgH_2) to achieve the same energy. Thus, they chose the locality approximation.

While all-electron calculations would, in principle, make the pseudopotential and locality errors controllable, in practice, the increase in number of electrons, required variational parameters and variance of the local energy makes such calculations currently impractical for anything but small systems and light elements [48].

2.3
Review of Previous DMC Defect Calculations

To date, there have been DMC calculations for defects in three materials: the vacancy in diamond, the Schottky defect in MgO, and the self-interstitials in Si.

2.3.1
Diamond Vacancy

Diamond's high electron and hole mobility and its tolerance to high temperatures and radiation make it a technologically important semiconductor material. Diffusion in diamond is dominated by vacancy diffusion [53], and the vacancy is also associated with radiation damage [54]. Table 2.1 shows the range of vacancy formation and migration energies calculated by LDA [50] and DMC [49]. DMC used structures from

Table 2.1 DMC, GW, and DFT energies (eV) for neutral defects in three materials. DMC and experimental values have an estimated uncertainty indicated by numbers in parenthesis. For the diamond vacancy, DFT-LDA and DMC include a 0.36 eV Jahn–Teller relaxation energy. LDA relaxation produced the structures and transition path so the DMC value for migration energy is an upper bound on the true value. The Schottky energy in MgO is the energy to form a cation–anion vacancy pair. DFT-LDA produces a range from 6 to 7 eV depending on the representation of the orbitals and treatment of the core electrons. DMC using a plane-wave basis and pseudopotentials results in a value on the upper end of the experimental range. For Si interstitial defects, DFT values of the formation energy range from 2 eV below up to the DMC values, depending on the exchange-correlation functional (LDA, GGA [PBE], or hybrid [HSE]), and the GW values lie within the two-standard-deviation confidence level of DMC.

energy			DFT			GW	DMC	exp.	Refs.
		LDA	GGA	hybrid					
C	diamond vacancy	formation	6.98	7.51	-	-	5.96(34)	-	[49, 50]
		migration	2.83	-	-	-	4.40(36)	2.3(3)	
MgO	Schottky defect	formation	5.97, 6.99, 6.684	-	-	-	7.50(53)	5–7	[51, 52]
Si	self-interstitial defect	X	3.31	3.64	4.69	4.40	5.0(2), 4.94(5)	-	[24, 26]
		T	3.43	3.76	4.95	4.51	5.5(2), 5.13(5)	-	
		H	3.31	3.84	4.80	4.46	4.7(2), 5.05(5)	-	

LDA relaxation and single-particle orbitals employing a Gaussian basis. A LDA pseudopotential described the core electrons. The DMC calculations predict a lower formation energy than LDA. The DMC value for the migration energy is an upper bound on the actual number since the structures have not been relaxed in DMC. Furthermore, DMC estimates the experimentally observed dipole transition [49]. The GR1 optical transition is not a transition between one-electron states but between spin states ^1E and $^1T_2$. DMC calculates a transition energy of 1.5(3) eV from ^1E to $^1T_2$, close to the experimentally observed value of 1.673 eV. LDA cannot distinguish these states. For the cohesive energy, DMC predicts a value of 7.346(6) eV in excellent agreement with the experimental result of 7.371(5) eV while LDA overbinds and yields 8.61 eV.

2.3.2
MgO Schottky Defect

MgO is an important test material for understanding oxides. Its rock-salt crystal structure is simple, making it useful for computational study. Schottky defects are one of the main types of defects present after exposure to radiation, according to classical molecular dynamics simulations [55]. Table 2.1 shows that DMC predicts a Schottky defect formation energy in MgO at the upper end of the range of experimental values [51].

2.3.3
Si Interstitial Defects

Table 2.1 shows that DFT and DMC differ by up to 2 eV in their predictions of the formation energies of these defects [24, 26]. We compare the DMC values with our results including tests on the QMC approximations in Section 2.4.

2.4
Results

We specifically test the time step, pseudopotential and fixed-node approximations for the formation energies of three silicon self-interstitial defects, the split-$\langle 110 \rangle$ interstitial (X), the tetrahedral interstitial (T), and the hexagonal interstitial (H). The QMC calculations are performed using the CASINO [56] code. Density functional calculations in this work used the Quantum ESPRESSO [57] and WIEN2k [58] codes. The defect structures are identical to those of Batista et al. [24]. The orbitals of the trial wave function come from DFT calculations using the LDA exchange-correlation functional. The plane-wave basis set with a cutoff energy of 1088 eV (60 Ha) converges the DFT total energies to 1 meV. A $7 \times 7 \times 7$ Monkhorst-Pack k-point mesh centered at the L-point (0.5,0.5,0.5) converges the DFT total energy to 1 meV. A population of 1280 walkers ensured that the error introduced by the population control is negligibly small. Due to the computational cost of backflow, we perform the simulations for a supercell of 16(+1) atoms and estimate the finite-size corrections using the structure factor method [41]. The final corrected DMC energies for the X, T, and H defects are shown in the bottom line of Table 2.2.

2.4.1
Time Step

Figure 2.1 shows the total energies of bulk silicon and the X defect as a function of time step in DMC. A time step of 0.01 Ha^{-1} reduces the time step error to within the statistical uncertainty of the DMC total energy.

2.4.2
Pseudopotential

In our calculations, a Dirac-Fock (DF) pseudopotential represents the core electrons for each silicon atom [59–61]. To estimate the error introduced by the pseudopotential, we compare the defect formation energies in DFT using this pseudopotential with all-electron DFT calculations using the linearized augmented plane-wave method [58]. This comparison gives corrections of 0.083, −0.168, and 0.054 eV for the H, T, and X defects respectively.

2.4.3
Fixed-Node Approximation

The fixed-node approximation is the main source of error in DMC calculations. To estimate the size of the error introduced by the fixed-node approximation, we perform the calculations using the backflow transformation, which allows the nodes of the trial wave function to be moved and reduces the fixed-node error [31]. We estimate the error due to the fixed-node approximation by performing calculations with and without the backflow transformation and by extrapolating the resulting defect formation energies to zero variance.

Applying the backflow transformation to electron coordinates using polynomials of electron-electron, electron-nucleus and electron–electron–nucleus separations, we include polynomial terms to eighth-order for each spin type in electron–electron separation, to sixth-order in electron-nucleus separation and to third-order for each spin type in electron–electron–nucleus separation. Figure 2.2 shows the linear extrapolation of the DMC energies for the Slater–Jastrow and Slater-Jastrow-backflow trial wave function to zero variance. The total energy decrease for the bulk and interstitial cells due to the backflow transformation ranges from 0.20(5) to 0.62(5) eV. The backflow transformation results in a significantly improved nodal surface of the trial wave function, which is reflected in the reduced variance of the local energy.

Table 2.2 list the Si interstitials formation energies in DMC for the Slater-Jastrow, Slater-Jastrow-backflow wave function and the extrapolation. Applying the backflow transformation reduces the formation energies for the X, T, and H interstitials by 0.42(5), 0.05(5), and 0.15(5) eV, respectively. The linear extrapolation provides a simple

Table 2.2 DMC Si defect formation energies. Varying parameters and improved methods produce values for each defect that lie within two standard deviations of each other although the energetic ordering of the defects varies. All calculations use DFT-LDA to produce the orbitals in the Slater determinant.

defect formation energy (eV)			Jastrow factor				pseudo-potential method	plane-wave cutoff energy (eV)	finite-size correction	Ref.
			n-body terms							
X	T	H	e–e	e–n	e–e–n	plane-waves				
Slater-Jastrow										
5.0(2)	5.5(2)	4.7(2)	16	16	0	0	LDA	245	DFT k-pt.	[26]
4.94(5)	5.13(5)	5.05(5)	5	5	0	0	HF	435	DFT k-pt.	[24]
4.9(1)	5.2(1)	4.9(1)	8	8	3	19	DF	1088	DFT k-pt. + struc. fac.	(this work)
Slater-Jastrow backflow										
4.5(1)	5.1(1)	4.7(1)	8	8	3	19	DF	1088	DFT k-pt. + struc. fac.	(this work)
extrapolation										
4.4(1)	5.1(1)	4.7(1)	8	8	3	19	DF	1088	DFT k-pt. + struc. fac.	(this work)

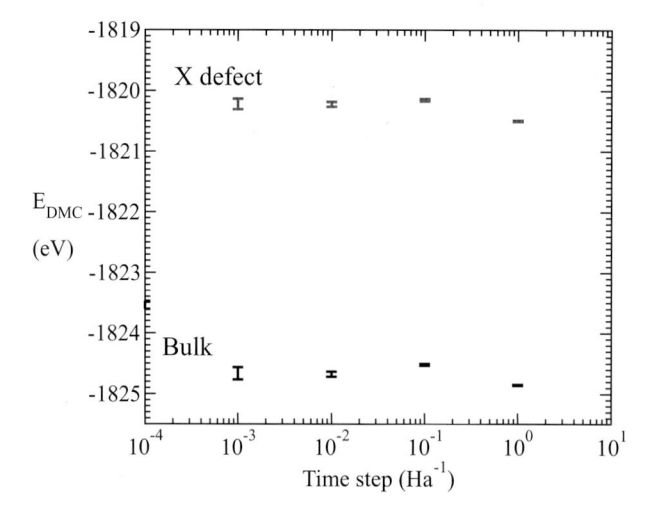

Figure 2.1 (online colour at: www.pss-b.com) DMC total energies with varying (imaginary) time steps for bulk silicon and the X defect. The error due to the finite-sized time step is smaller than the statistical uncertainty in the total energies for values from 10^{-3} to $0.1\,\text{Ha}^{-1}$. Note that these energies include no finite-size or pseudopotential corrections and thus differ in value from those in Fig. 2.2.

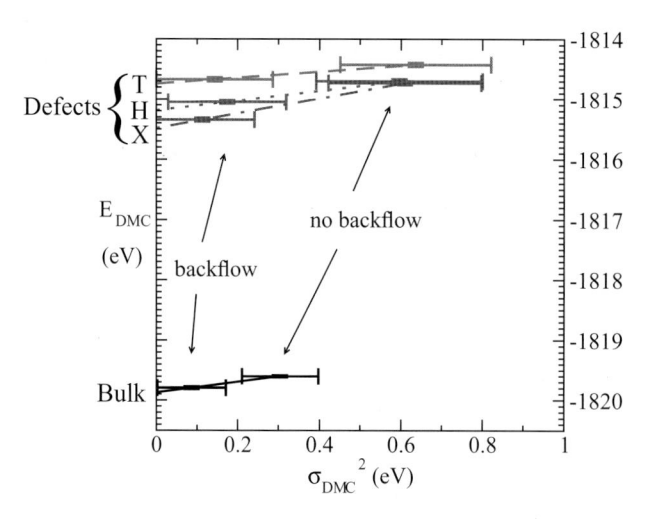

Figure 2.2 (online colour at: www.pss-b.com) DMC total energies calculated with and without backflow transformation and linearly extrapolated to zero variance. The backflow transformation reduces the total energies by 0.19(3), 0.62(5), 0.25(5), and 0.35(5) eV for bulk (solid line) and the X (dash-dot line), T (dashed line), and H (dotted line) defects respectively. The extrapolation to zero variance only slightly reduces the total energies by about 0.1 eV. The reduction of the total energy from the backflow transformation indicates the size of the errors due to the fixed-node and pseudopotential locality approximation.

estimate of the remaining fixed node error. The extrapolation lowers the interstitial formation energy by a negligible amount of 0.06(10), 0.00(10), 0.07(10) eV for the X, T, and H interstitials, respectively. The resulting Si interstitial formation energies are 4.4(1), 5.1(1), and 4.7(1) eV for the X, T, and H interstitial, respectively, in close agreement with recent $G_0 W_0$ [25] and previous HSE and DMC calculations [24, 26].

2.5
Conclusion

QMC methods present an accurate tool for the calculation of point defect formation energies, provided care is taken to control the accuracy of all the underlying approximations. Including corrections for the approximations yields DMC values for the Si interstitial defects on par with GW and hybrid-functional DFT calculations. While backflow transformation and zero-variance extrapolation to remove the fixed-node error modify the energies slightly, further work remains to carefully control for finite-size effects known to plague defect supercell calculations.

Acknowledgements

The work was supported by the U.S. Department of Energy under contract no. DE-FG02-99ER45795 and DE-FG05-08OR23339 and the National Science Foundation under contract no. EAR-0703226. This research used computational resources of the National Energy Research Scientific Computing Center, which is supported by the Office of Science of the U.S. Department of Energy under contract no. DE-AC02-05CH11231, the National Center for Supercomputing Applications under grant DMR050036, the Ohio Supercomputing Center and the Computation Center for Nanotechnology Innovation at Rensselaer Polytechnic Institute. We thank Cyrus Umrigar and Ann Mattsson for helpful discussion.

References

1. Tilley, R.J.D. (2008) *Defects in Solids*, Wiley, Hoboken, New Jersey.
2. Fahey, P.M., Griffin, P.B., and Plummer, J.D. (1989) *Rev. Mod. Phys.*, **61** (2), 289–384.
3. Eaglesham, D.J., Stolk, P.A., Gossmann, H.J., and Poate, J.M. (1994) *Appl. Phys. Lett.*, **65** (18), 2305–2307.
4. Pike, L.M., Chang, Y.A., and Liu, C.T. (1997) *Acta Mater.*, **45** (9), 3709–3719.
5. Zhu, J.H., Pike, L.M., Liu, C.T., and Liaw, P.K. (1999) *Acta Mater.*, **47** (7), 2003–2018.
6. Otsuka, K. and Ren, X. (1999) *Intermetallics*, **7** (5), 511–528.
7. Pelaz, L., Marqués, L.A., Aboy, M., López, P., and Santos, I. (2009) *Eur. Phys. J. B*, **72** (3), 323–359.
8. Jones, R.O. and Gunnarsson, O. (1989) *Rev. Mod. Phys.*, **61** (3), 689–746.
9. Ceperley, D.M. and Alder, B.J. (1980) *Phys. Rev. Lett.*, **45** (7), 566–569.
10. Perdew, J.P. and Zunger, A. (1981) *Phys. Rev. B*, **23** (10), 5048–5079.
11. Perdew, J.P. and Wang, Y. (1992) *Phys. Rev. B*, **45** (23), 13244–13249.

12 Perdew, J.P. (1991) Unified Theory of Exchange and Correlation Beyond the Local Density Approximation, in: *Electronic Structure of Solids '91* (eds P. Ziesche and H. Eschrig), Akademie Verlag, Berlin, pp. 11–20.

13 Perdew, J.P., Burke, K., and Ernzerhof, M. (1996) *Phys. Rev. Lett.*, **77** (18), 3865–3868.

14 Armiento, R. and Mattsson, A.E. (2005) *Phys. Rev. B*, **72** (8), 085108.

15 Wu, Z. and Cohen, R.E. (2006) *Phys. Rev. B*, **73** (23), 235116.

16 Perdew, J.P., Ruzsinszky, A., Csonka, G.I., Vydrov, O.A., Scuseria, G.E., Constantin, L.A., Zhou, X., and Burke, K. (2008) *Phys. Rev. Lett.*, **100** (13), 136406.

17 Heyd, J., Peralta, J.E., Scuseria, G.E., and Martin, R.L. (2005) *J. Chem. Phys.*, **123** (17), 174101.

18 Nieminen, R.M. (2009) *Model. Simul. Mater. Sci. Eng.*, **17** (8), 084001.

19 Becke, A.D. (1993) *J. Chem. Phys.*, **98** (7), 5648–5652.

20 Heyd, J., Scuseria, G.E., and Ernzerhof, M. (2003) *J. Chem. Phys.*, **118** (18), 8207–8215.

21 Richie, D.A., Kim, J., Barr, S.A., Hazzard, K.R.A., Hennig, R., and Wilkins, J.W. (2004) *Phys. Rev. Lett.*, **92** (4), 045501.

22 Du, Y.A., Hennig, R.G., Lenosky, T.J., and Wilkins, J.W. (2007) *Eur. Phys. J. B*, **57** (3), 229–234.

23 Vaidyanathan, R., Jung, M.Y.L., and Seebauer, E.G. (2007) *Phys. Rev. B*, **75**, 195209.

24 Batista, E.R., Heyd, J., Hennig, R.G., Uberuaga, B.P., Martin, R.L., Scuseria, G.E., Umrigar, C.J., and Wilkins, J.W. (2006) *Phys. Rev. B*, **74** (12), 121102.

25 Rinke, P., Janotti, A., Scheffler, M., and de Walle, C.G.V. (2009) *Phys. Rev. Lett.*, **102** (2), 026402.

26 Leung, W.K., Needs, R.J., Rajagopal, G., Itoh, S., and Ihara, S. (1999) *Phys. Rev. Lett.*, **83** (12), 2351–2354.

27 Foulkes, W.M.C., Mitas, L., Needs, R.J., and Rajagopal, G. (2001) *Rev. Mod. Phys.*, **73** (1), 33–83. Sections V and VI contrast QMC and DFT results. Section X.E discusses scaling with computer time. Section III introduces VMC and DMC.

28 Needs, R. (2006) Quantum Monte Carlo Techniques and Defects in Semiconductors, in: *Theory of Defects in Semiconductors* (eds D. Drabold and S. Estreicher, Topics Appl. Physics, Vol. 104, Springer Verlag, Berlin, Heidelberg, p. 141.

29 Esler, K.P., Kim, J., Ceperley, D.M., Purwanto, W., Walter, E.J., Krakauer, H., Zhang, S., Kent, P.R.C., Hennig, R.G., Umrigar, C., Bajdich, M., Kolorenč, J., Mitas, L., and Srinivasan, A. (2008) *J. Phys. Conf. Ser.*, **125**, 012057 (15pp).

30 Umrigar, C.J., Toulouse, J., Filippi, C., Sorella, S., and Hennig, R.G. (2007) *Phys. Rev. Lett*, **98** (11), 110201.

31 López Ríos, P., Ma, A., Drummond, N.D., Towler, M.D., and Needs, R.J. (2006) *Phys. Rev. E*, **74** (6), 066701.

32 Golub, G.H. and Loan, C.F.V. (1996) chap. 2, in: *Matrix Computations*, 3rd edn, The Johns Hopkins University Press, Baltimore, p. 51.

33 Umrigar, C.J., Wilson, K.G., and Wilkins, J.W. (1988) *Phys. Rev. Lett.*, **60** (17), 1719–1722.

34 Anderson, J.B. (1975) *J. Chem. Phys.*, **63** (4), 1499–1503.

35 Umrigar, C.J., Nightingale, M.P., and Runge, K.J. (1993) *J. Chem. Phys.*, **99** (4), 2865–2890.

36 Alfè, D., Gillan, M.J., Towler, M.D., and Needs, R.J. (2004) *Phys. Rev. B*, **70** (21), 214102.

37 Alfè, D. and Gillan, M.J. (2004) *Phys. Rev. B*, **70** (16), 161101.

38 Lin, C., Zong, F.H., and Ceperley, D.M. (2001) *Phys. Rev. E*, **64** (1), 016702.

39 Williamson, A.J., Rajagopal, G., Needs, R.J., Fraser, L.M., Foulkes, W.M.C., Wang, Y., and Chou, M.Y. (1997) *Phys. Rev. B*, **55** (8), R4851–R4854.

40 Ewald, P.P. (1921) *Annalen der Physik*, **369** (3), 253–287.

41 Chiesa, S., Ceperley, D.M., Martin, R.M., and Holzmann, M. (2006) *Phys. Rev. Lett.*, **97** (7), 076404.

42 Kwee, H., Zhang, S., and Krakauer, H. (2008) *Phys. Rev. Lett.*, **100** (12), 126404.

43 Drummond, N.D., Needs, R.J., Sorouri, A., and Foulkes, W.M.C. (2008) *Phys. Rev. B*, **78** (12), 125106.
44 Badinski, A., Haynes, P.D., Trail, J.R., and Needs, R.J. (2010) *J. Phys.: Condens. Matter*, **22** (7), 074202.
45 Mitáš, L. Shirley, E.L., and Ceperley, D.M. (1991) *J. Chem. Phys.*, **95** (5), 3467–3475.
46 Casula, M. (2006) *Phys. Rev. B*, **74** (16), 161102.
47 Pozzo, M. and Alfè, D. (2008) *Phys. Rev. B*, **77** (10), 104103.
48 Esler, K.P., Cohen, R.E., Militzer, B., Kim, J., Needs, R.J., and Towler, M.D. (2010) *Phys. Rev. Lett.*, **104** (18), 185702.
49 Hood, R.Q., Kent, P.R.C., Needs, R.J., and Briddon, P.R. (2003) *Phys. Rev. Lett.*, **91** (7), 076403.
50 Shim, J., Lee, E.K., Lee, Y.J., and Nieminen, R.M. (2005) *Phys. Rev. B*, **71** (3), 035206.
51 Alfè, D. and Gillan, M.J. (2005) *Phys. Rev. B*, **71** (22), 220101.
52 Gilbert, C.A., Kenny, S.D., Smith, R., and Sanville, E. (2007) *Phys. Rev. B*, **76** (18), 184103.
53 Bernholc, J., Antonelli, A., Del Sole, T.M., Bar-Yam, Y., and Pantelides, S.T. (1988) *Phys. Rev. Lett.*, **61** (23), 2689–2692.
54 Collins, A.T. and Kiflawi, I. (2009) *J. Phys.: Condens. Matter*, **21** (36), 364209.
55 Uberuaga, B.P., Smith, R., Cleave, A.R., Henkelman, G., Grimes, R.W., Voter, A.F., and Sickafus, K.E. (2005) *Phys. Rev. B*, **71** (10), 104102.
56 Needs, R.J., Towler, M.D., Drummond, N.D., and López Ríos, P. (2010) *J. Phys.: Condens. Matter*, **22** (2), 023201 (15pp).
57 Giannozzi, P., Baroni, S., Bonini, N., Calandra, M., Car, R., Cavazzoni, C., Ceresoli, D., Chiarotti, G.L., Cococcioni, M., Dabo, I., Corso, A.D., de Gironcoli, S., Fabris, S., Fratesi, G., Gebauer, R., Gerstmann, U., Gougoussis, C., Kokalj, A., Lazzeri, M., Martin-Samos, L., Marzari, N., Mauri, F., Mazzarello, R., Paolini, S., Pasquarello, A., Paulatto, L., Sbraccia, C., Scandolo, S., Sclauzero, G., Seitsonen, A.P., Smogunov, A., Umari, P., and Wentzcovitch, R.M. (2009) *J. Phys.: Condens. Matter*, **21** (39), 395502 (19pp).
58 Blaha, P., Schwarz, K., Madsen, G.K.H., Kvasnicka, D., and Luitz, J. (2001) *WIEN2K, An Augmented Plane Wave + Local Orbitals Program for Calculating Crystal Properties*, Karlheinz Schwarz, Techn. Universität Wien, Austria.
59 Trail, J.R. and Needs, R.J. (2005) *J. Chem. Phys.*, **122** (1), 014112.
60 Trail, J.R. and Needs, R.J. (2005) *J. Chem. Phys.*, **122** (17), 174109.
61 Trail, J.R. and Needs, R.J., CASINO pseudopotential library, http://www.tcm.phy.cam.ac.uk/~mdt26/casino2_pseudopotentials.html.

3
Electronic Properties of Interfaces and Defects from Many-body Perturbation Theory: Recent Developments and Applications

Matteo Giantomassi, Martin Stankovski, Riad Shaltaf, Myrta Grüning, Fabien Bruneval, Patrick Rinke, and Gian-Marco Rignanese

3.1
Introduction

Almost all electronic and optoelectronic devices (such as MOS transistors, photovoltaic cells, semiconductor lasers, etc.) contain metal–semiconductor, insulator–semiconductor, insulator–metal, and/or semiconductor–semiconductor interfaces. The electronic properties of such heterojunctions determine the device characteristics [1, 2]. The band gaps of the participating materials are usually different, hence, at least one of the band edges is different. The energy of charge carriers must then change when passing through the heterojunction. Most often, there will be discontinuities in both the conduction and valence bands. These so-called band offsets (BOs) are the origin of most of the useful properties of heterojunctions.

Defects also play a critical role for the functionality of devices [3–5]. They can have both positive as well as detrimental effects. As dopants they provide charge carriers in semiconductors, which can contribute to a current, but these carriers can also recombine at defect sites and are then lost. Problems like flat-band and threshold voltage shifts, carrier mobility degradation, charge trapping, gate dielectric wear-out, and breakdown, as well as temperature instabilities are believed to mainly originate from defects forming at (or close to) the heterojunction interface. A deep understanding of the defects concerned is thus highly desirable for the enhancement of device performance.

However, experimental characterization of defect energy levels at interfaces is often very difficult to achieve, so theoretical simulation can provide extremely useful information for further improvement of devices. In this framework, density functional theory (DFT) has been, and still is, widely used to investigate the electronic properties of various defective interfaces. Unfortunately, the semi-local approximations to DFT – such as the local density approximation (LDA) or the generalized-gradient approximation (GGA) – suffer from a well-known substantial underestimation of band gaps, which hinders a precise prediction of the energy-level alignment at interfaces. For this reason, hybrid density functionals have recently increased in popularity [6–11]. These functionals, which incorporate a fraction of Hartree-Fock (HF) exchange, lead to higher accuracies [12] and improved band gaps [13, 14] compared to corresponding results using semilocal functionals. The fraction of HF exchange to be included cannot be known in advance for all materials and its optimal value could even be property dependent [15, 16]. Therefore the reliability of hybrid density functionals cannot be assessed *a priori* [17].

In contrast, many-body perturbation theory (MBPT) [18–22] offers an approach for obtaining quasiparticle (QP) energies in solids which is controlled and amenable to systematic improvement. However, the cost of such calculations is generally higher than that of their DFT counterparts. Recently, considerable effort has been devoted to finding reliable techniques to speed up MBPT calculations and make them tractable for the larger systems needed to simulate defects and interfaces.

In this chapter, we will review the recent developments in MBPT calculations and the results obtained for interfaces and defects. Section 3.2 is devoted to the theoretical basis of MBPT. Hedin's equations are presented in Section 3.2.1. The GW approximation is introduced in Section 3.2.2, while approximations going beyond GW are discussed in Section 3.2.3. Section 3.3 focuses on the practical implementation and the recent developments of MBPT. In Section 3.3.1, we describe the perturbative approach, that is usually employed to obtain QP energies. The methods used to take into account the frequency dependence of the self-energy operators are presented in Section 3.3.2. In order to allow for a reduction of the number of unoccupied states that need to be included explicitly in the calculations, the extrapolar method is introduced in Section 3.3.3. We discuss the combination of MBPT with the projector-augmented wave (PAW) method in Section 3.3.4. Sections 3.4 and 3.5 are dedicated to MBPT results obtained for BOs at interfaces and for defects, respectively. Special emphasis is put on the caveats of the methods.

3.2
Many-Body Perturbation Theory

3.2.1
Hedin's Equations

A rigorous formulation for the properties of QPs is based on a Green's function approach [18]. The QP energies E_i^{QP} and wavefunctions ψ_i^{QP} are obtained by solving the QP equation:

$$\left\{-\frac{1}{2}\nabla^2 + V_{\text{ext}}(\boldsymbol{r}) + V_{\text{H}}(\boldsymbol{r})\right\}\psi_i^{QP}(\boldsymbol{r}) + \int \Sigma(\boldsymbol{r},\boldsymbol{r}';E_i^{QP})\psi_i^{QP}(\boldsymbol{r}')d\boldsymbol{r}' = E_i^{QP}\psi_i^{QP}(\boldsymbol{r}), \tag{3.1}$$

where V_ext and V_H are the external and Hartree potentials, respectively. In this equation, the exchange and correlation effects are described by the electron self-energy operator $\Sigma(r,r',E_i^{QP})$ which is non-local, energy dependent, and non-Hermitian. Hence, the eigenvalues E_i^{QP} are generally complex: their real part is the energy of the QP, while their imaginary part gives its lifetime.

The main difficulty is to find an adequate approximation for the self-energy operator Σ. Hedin [23] proposed a perturbation series expansion in the fully screened (as opposed to bare) Coulomb interaction. The Green's function, G_0, of a "zeroth-order" system of non-interacting electrons is first constructed from the one-particle wavefunctions ψ_i and energies E_i of the "zeroth-order" Hamiltonian, as:

$$G_0(r,r',E) = \sum_i \frac{\psi_i(r)\psi_i^*(r')}{E - E_i + i\eta\,\text{sgn}(E_i - \mu)}, \qquad (3.2)$$

where μ is the chemical potential and η is a positive infinitesimal. The exact one-body Green's function G is thus written using the Dyson equation:[1]

$$G(12) = G_0(12) + \int G_0(13)\Sigma(34)G(42)d(34). \qquad (3.3)$$

Here, the self-energy Σ is obtained by self-consistently solving Hedin's closed set of coupled integro-differential equations:

$$\Gamma(12;3) = \delta(12)\delta(13)$$
$$+ \int \frac{\delta\Sigma(12)}{\delta G(45)} G(46)G(75)\Gamma(67;3)d(4567), \qquad (3.4)$$

$$P(12) = -i \int G(23)G(42^+)\Gamma(34;1)d(34), \qquad (3.5)$$

$$W(12) = v(12) + \int W(13)P(34)v(42)d(34), \qquad (3.6)$$

$$\Sigma(12) = i \int G(14)W(1^+3)\Gamma(42;3)d(34), \qquad (3.7)$$

where P is the polarizability, W the screened and v the unscreened Coulomb interaction and Γ the *vertex* function, which describes higher-order corrections to

---

1) In Section 3.1, Hedin's simplified notation $1 \equiv (x_1,\sigma_1,t_1)$ is used to denote space, spin, and time variables and the integral sign stands for summation or integration of all of these where appropriate. 1^+ denotes $t_1 + \eta$ where η is a positive infinitesimal in the time argument. Atomic units are used in all equations throughout this paper.

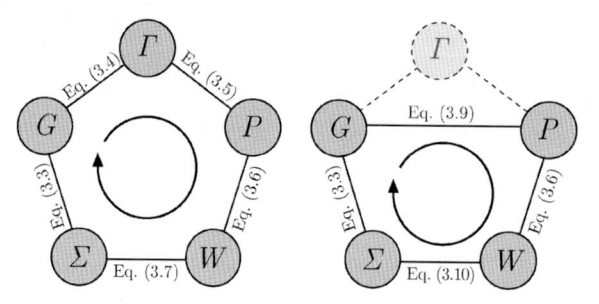

Figure 3.1 Graphical illustration of the self-consistent process required to solve the complete set of Hedin's equations (left panel) and the four coupled integro-differential equations resulting from the GW approximation (right panel). The so-called G_0W_0 approximation consists of performing the loop only once starting from $G = G_0$.

the interaction between quasiholes and quasielectrons. The self-consistent iterative process is illustrated in the left panel of Figure 3.1.

The most complicated term in these equations is Γ, which contains a functional derivative and hence cannot in general be evaluated numerically. The vertex is the usual target of simplification for an approximate scheme.

3.2.2
GW Approximation

Hedin's GW method [23] is the most widely used approximation for the self-energy, Σ. The approximation is defined by neglecting the variation of the self-energy with respect to the Green's function $\delta\Sigma(12)/\delta G(45) = 0$ in Eq. (3.4), leading to:

$$\Gamma(12;3) = \delta(12)\delta(13). \tag{3.8}$$

Thus, the polarizability in Eq. (3.5) is given by:

$$P(12) = -iG(12^+)G(21), \tag{3.9}$$

which corresponds to the random phase approximation (RPA) for the dielectric matrix. The self-energy in Eq. (3.7) becomes simply a product of the Green's function and the screened Coulomb interaction:

$$\Sigma(12) = iG(12)W(1^+2), \tag{3.10}$$

where the Green's function used is consistent with that returned by Dyson's equation.

Since the self-energy depends on G, this procedure should be carried out iteratively, beginning with $G = G_0$, until the input Green's function equals the output one. This yields the self-consistent GW approximation, in which the self-consistent cycle is restricted to Eqs. (3.3), (3.9), (3.6), and (3.10), as illustrated in the right panel of Figure 3.1.

In practice, it is customary to use the first iteration only, often called one-shot GW or G_0W_0, to approximate the self-energy operator. Here, W_0 is perhaps the simplest possible screened interaction, which in terms of Feynman diagrams involves an infinite geometric series over non-interacting electron–hole pair excitations as in the usual definition of the RPA.[2] This approximation for W, although tremendously successful for weakly correlated solids, is not free of self-screening errors [24, 25].

When using only a single iteration, it is important to make that one as accurate as possible, so an initial G_0 calculated using Kohn–Sham DFT is normally used. The logic is that the Kohn–Sham orbitals should produce an input G_0 much closer to the self-consistent solution, thus rendering a single iteration sufficient. This choice of G_0 has in the past produced accurate results for QP energies (i.e., the correct electron addition and removal energies, in contrast to the DFT eigenvalues [26]) for a wide range of s–p bonded systems [27]. However, because this choice of G_0 corresponds to a non-zero initial approximation for Σ_0, there is no longer a theoretical justification for the usual practice of setting the vertex to a product of delta functions before the decoupling. Also, different choices for the exchange-correlation functional may lead to different Green's functions [28, 29], making G_0W_0 results dependent on the starting point.

3.2.3
Beyond the GW Approximation

Since G_0 is often constructed from DFT orbitals, the self-energy and its derivative are not zero for the first iteration. Using the static exchange-correlation kernel, K_{xc}, (which is the functional derivative of the DFT exchange-correlation potential, V_{xc}, with respect to density, n) Del Sole et al. [30] demonstrated how G_0W_0 may be modified with a vertex function to make Σ consistent with the DFT starting point. They added the contribution of the vertex – decoupled after the first evaluation of $\delta\Sigma(12)/\delta G(45)$ in Eq. (3.4) – into both the self-energy, Σ (3.7), and the polarization, P (3.5). The result is a self-energy of the form $G_0W_0\Gamma$. Instead, the $G_0\widetilde{W}_0$ approximation is obtained when the vertex function is included in P only. As commented by Hybertsen and Louie [31] and Del Sole et al., both these results take the *form of* GW, but with W representing the Coulomb interaction screened by the test-charge-electron dielectric function and the test-charge-test-charge dielectric function, respectively, and with electronic exchange and correlation included through a time-dependent DFT (TDDFT) kernel.

Using the LDA for the exchange-correlation potential and kernel, Del Sole et al. found that $G_0W_0\Gamma$ yields final results almost equal to those of G_0W_0 for the band gap of crystalline silicon and that the equivalent results from $G_0\widetilde{W}_0$ were shown to close the gap slightly compared to standard G_0W_0. However, in this previous study the

---

2) In contrast to the common use of the RPA, there is no integration over the interaction strength, since the perturbation expansion itself takes care of the switching on of interactions.

3.3
Practical Implementation of GW and Recent Developments Beyond

3.3.1
Perturbative Approach

Often, it is more efficient to obtain the QP energies from Eq. (3.1) rather than solving the Dyson equation (Eq. 3.3) and searching for the poles of the Green's function. The approach consists of using perturbation theory with respect to the results of DFT. Despite some fundamental differences, the formal similarity is striking between the QP equation and the Kohn–Sham equation:

$$\left\{-\frac{1}{2}\nabla^2 + V_{ext}(r) + V_H(r)\right\}\psi_i^{DFT}(r) + V_{xc}(r)\psi_i^{DFT}(r) = E_i^{DFT}\psi_i^{DFT}(r), \quad (3.11)$$

where V_{xc} is the DFT exchange-correlation potential.[3] In many cases, the DFT energies E_i^{DFT} already provide a reasonable estimate of the band structure and are usually in qualitative agreement with experiment. Furthermore, in the simple systems for which the true QP amplitudes ψ_i^{QP} have been calculated, it was found that the DFT wave functions ψ_i^{DFT} are usually very close to the QP results [31, 32]. In silicon, for instance, the overlap between DFT-LDA and QP wave functions has been reported to be close to 99.9%, but for certain surface [33, 34] and cluster states [35, 36] the overlap is far less (see also Ref. [37] for comments and criticisms). This indicates that in the basis of Kohn–Sham wave functions, the self-energy can be considered a diagonally dominant matrix with negligible off-diagonal elements.

Hence, E_i^{DFT} and ψ_i^{DFT} for the ith state are used as a zeroth-order approximation for their QP counterparts. The QP energy E_i^{QP} is then calculated by adding to E_i^{DFT} the first-order perturbation correction which comes from replacing the DFT exchange-correlation potential V_{xc} with the self-energy operator Σ:

$$E_i^{QP} = E_i^{DFT} + \langle \psi_i^{DFT} | \Sigma(E_i^{QP}) - V_{xc} | \psi_i^{DFT} \rangle. \quad (3.12)$$

To solve Eq. (3.12), the energy dependence of Σ must be known analytically, which is usually not the case. Under the assumption that the difference between QP and DFT energies is relatively small, the matrix elements of the self-energy operator can

---

[3] Note that V_{xc} can be seen as a static, local, and hermitian approximation to $\Sigma(12)$.

be Taylor expanded to first-order around E_i^{DFT} in order to be evaluated at E_i^{QP}:

$$\Sigma(E_i^{QP}) \approx \Sigma(E_i^{DFT}) + (E_i^{QP} - E_i^{DFT}) \frac{\partial \Sigma(E)}{\partial E}\bigg|_{E=E_i^{DFT}}. \tag{3.13}$$

In this expression, the QP energy, E_i^{QP}, can be solved for:

$$E_i^{QP} = E_i^{DFT} + Z_i \langle \psi_i^{DFT} | \Sigma(E_i^{DFT}) - V_{xc} | \psi_i^{DFT} \rangle, \tag{3.14}$$

where Z_i is the *renormalization factor* defined by:

$$Z_i^{-1} = 1 - \langle \psi_i^{DFT} | \frac{\partial \Sigma(E)}{\partial E} \bigg|_{E=E_i^{DFT}} | \psi_i^{DFT} \rangle. \tag{3.15}$$

The principle is illustrated in Figure 3.2.

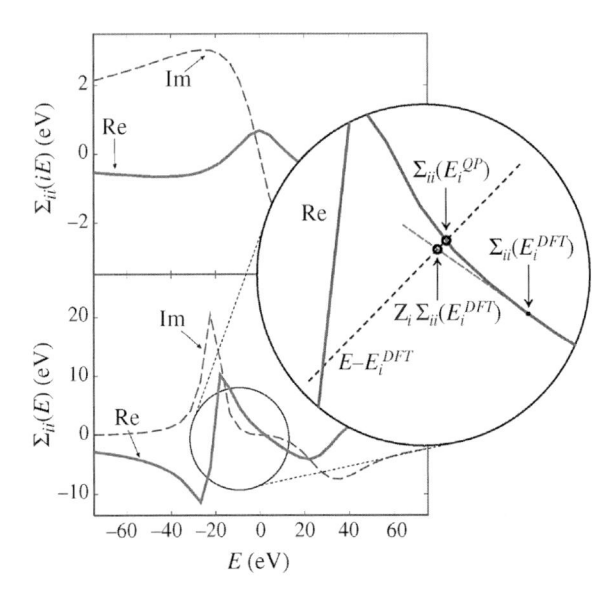

Figure 3.2 (online color at: www.pss-b.com) Schematic illustration (adapted from Ref. [38]) of the perturbative approach to finding the QP correction. In principle, the self-energy matrix element, $\Sigma_{ii}(E) = \langle \psi_i^{DFT} | \Sigma(E) - V_{xc} | \psi_i^{DFT} \rangle$, and the true QP correction, $\Sigma(E_i^{QP})$, is found from the solution of $E - E_i^{DFT} = \Sigma_{ii}(E)$, i.e., at the crossing of the dashed black line and $\Sigma_{ii}(E)$ in the circular zoom-in. In practice, the perturbative approach exploits the fact that it is more computationally feasible to use the Taylor expansion around $\Sigma(E_i^{DFT})$ [Eqs. (3.14) and (3.15)], and find an approximate value for the QP correction at the crossing of the red and black dashed lines.

3.3.2
QP Self-Consistent GW

The procedure described above has proven very efficient [27], but several questions inevitably arise: How much does the G_0W_0 result depend on the starting point? What happens if the starting DFT band structure is qualitatively wrong[4]? A self-consistent GW self-energy calculation should be free of such concerns.

However, performing self-consistency in GW is everything but straightforward, since Σ, being non-Hermitian and energy-dependent, should have non-orthogonal and energy-dependent left and right eigenvectors. In practice, for large systems, the solution of this Hamiltonian is not tractable without approximations. Furthermore, fully self-consistent GW calculations have been shown to worsen results compared to the standard one-shot G_0W_0 method [39–41].

A different solution to the self-consistency issue is the so-called QP self-consistent GW approximation (QSGW) developed by Faleev et al. [42] and co-workers [43, 44]. For a set of trial QP energies and amplitudes $\{E_i, \psi_i\}$ (for instance, the eigensolutions of the DFT or Hartree problem), the one-particle Green's function, G, and in turn the GW self-energy can be calculated. These authors proposed to constrain the dynamical GW self-energy to be static and Hermitian and as close as possible to the one-shot self-energy (G_0W_0) of a non-interacting reference system. Their model QSGW self-energy $\tilde{\Sigma}$ reads:

$$\langle\psi_i|\tilde{\Sigma}|\psi_j\rangle = \frac{1}{2}\mathcal{H}[\langle\psi_i|\Sigma(E_i)|\psi_j\rangle + \langle\psi_j|\Sigma(E_j)|\psi_i\rangle], \quad (3.16)$$

where $\mathcal{H}$ means that only the Hermitian part of the matrix is considered.

The approximated self-energy matrix, $\tilde{\Sigma}$, is diagonalized yielding a new set of orthogonal QP amplitudes and real-valued QP energies. From this new set of orbitals, a new density $n(r)$ and the corresponding Hartree potential is generated, a new $\tilde{\Sigma}$ is constructed and the procedure is iterated to self-consistency. Ideally, the final result should not depend on the initial Hamiltonian, though no firm mathematical proof for this has been reported so far. The QSGW approach improves the G_0W_0 results, giving band gaps very close to experiments with errors that are small and highly systematic [43].

Following the same spirit, Bruneval et al. [37] proposed using an alternative Hermitian and static approximation to the GW self-energy: the COHSEX approximation, derived by Hedin in 1965 [23]. COHSEX is a simple approximation which consists of two terms, the COulomb Hole part and the Screened EXchange part:

$$\Sigma_{COHSEX}(r,r') = \Sigma_{COH}(r,r') + \Sigma_{SEX}(r,r')$$
$$\Sigma_{COH}(r,r') = \delta(r,r')[W(r,r',\omega=0) - v(r-r')] \quad (3.17)$$
$$\Sigma_{SEX}(r,r') = -\sum_v \psi_v(r)\psi_v^*(r')W(r,r',\omega=0).$$

---

[4] For example, if DFT erroneously predicts a system to be metallic, when it is not.

These terms do not involve any summation over empty states (v runs only over occupied states). Performing self-consistency for the COHSEX approximation is hence more tractable than for the QSGW self-energy of Faleev and coworkers, although Σ_{COHSEX} may be a cruder approximation than $\tilde{\Sigma}$.

An alternative is to constrain the QP amplitudes in $\tilde{\Sigma}$ to their DFT counterparts and only update the QP energies until convergence. This method is referred to as the eigenvalue-only QSGW (e-QSGW).

3.3.3
Plasmon Pole Models *Versus* Direct Calculation of the Frequency Integral

In the frequency domain, the GW self-energy is given by the convolution

$$\Sigma(r,r',\omega) = \frac{i}{2\pi} \int e^{i\omega'\eta} G(r,r',\omega+\omega') W(r,r',\omega') d\omega', \qquad (3.18)$$

where η is a positive infinitesimal. Evaluating this expression requires, in principle, the knowledge of the full frequency dependence of $W(r,r',\omega')$. Moreover a fine frequency grid would be required, since $G(r,r',\omega)$ and $W(r,r',\omega)$ exhibit a fairly complex and rapidly changing frequency dependence on the real axis. There are, however, two different and more efficient techniques to evaluate Eq. (3.18): (i) integration with a PPM and (ii) integration through contour deformation (CD). In the former case, the frequency dependence of $\varepsilon^{-1}(\omega)$ is modeled with a simple analytic form, and the frequency convolution is carried out analytically.

In the latter approach, the integral is evaluated numerically by extending the functions into the complex plane, where the integrand is smoother. Since the fine details of $W(r,r',\omega)$ are integrated over in Eq. (3.18), it is reasonable to expect that approximated models, able to capture the main physical features of $W(r,r',\omega)$, should give sufficiently accurate results at considerably reduced computational effort. This is the basic idea behind the PPM, in which the frequency dependence of $W(r,r',\omega')$ is modeled in terms of analytic expressions. The coefficients of the model are derived from first principles, i.e., without any adjustable external parameters, either by enforcing exact relations or by anchoring the scheme on quantities that are calculated *ab initio*.

It is more convenient to Fourier transform all quantities to a frequency and wavevector basis using the following convention:

$$W(r,r',\omega) = \sum_{qGG'} e^{i(q+G)\cdot r} W_{GG'}(q,\omega) e^{-i(q+G')\cdot r'}, \qquad (3.19)$$

where G is a reciprocal lattice vector and q is a vector in the first Brillouin zone. The screened interaction is related to the dielectric matrix by:

$$W_{GG'}(q,\omega) = \varepsilon^{-1}_{GG'}(q,\omega) v(q+G'), \qquad (3.20)$$

where the Fourier transform of the bare Coulomb interaction takes the usual form $v(q) = 4\pi/(V|q|^2)$, V being the crystal volume. Adopting this formalism, the components with $G \neq G'$ generate the local fields.

Finally, when the vertex is neglected as in Eq. (3.8), the dielectric matrix is related to the polarizability, P, by:

$$\varepsilon_{GG'}(q, \omega) = \delta_{GG'} - v(q+G) P_{GG'}(q, \omega), \tag{3.21}$$

which is nothing but the usual RPA when Eq. (3.9) is used to compute P.

In the PPMs of Godby and Needs [45] (GN) and Hybertsen and Louie [31] (HL), the imaginary part of $\varepsilon_{GG'}^{-1}(q, \omega)$ is approximated in terms of a delta function centered at the plasmon frequency $\tilde{\omega}_{GG'}(q)$ with amplitude $A_{GG'}(q)$, i.e.:

$$\Im[\varepsilon_{GG'}^{-1}(q, \omega)] = A_{GG'}(q) \times [\delta(\omega - \tilde{\omega}_{GG'}(q)) - \delta(\omega + \tilde{\omega}_{GG'}(q))]. \tag{3.22}$$

The real part can then obtained by means of a Kramers–Kronig relation, and becomes:

$$\Re[\varepsilon_{GG'}^{-1}(q, \omega)] = \delta_{GG'} + \frac{\Omega_{GG'}^2(q)}{\omega^2 - \tilde{\omega}_{GG'}^2(q)}. \tag{3.23}$$

where $\Omega_{GG'}^2(q) = -A_{GG'}(q)\,\tilde{\omega}_{GG'}^2(q)$.

The approximation given by Eq. (3.22) is quite reasonable, since experiments and first-principles analysis reveals that $\Im[\Omega_{G,G'}(q, \omega)]$ is generally characterized by a sharp peak in correspondence to a plasmon excitation at the plasmon frequency, at least for low momentum transfers, q.

At this point, one defines a set of physical constraints to determine the parameters entering Eqs. (3.22) and (3.23). The GN and HL PPMs differ in the choice of the particular physical properties or exact relations they aim to reproduce.

In the GN approach, the parameters of the model are derived so that $\varepsilon_{GG'}(q, \omega)$ is correctly reproduced at two different frequencies: the static limit ($\omega = 0$) and an additional imaginary point located at the Sommerfeld plasma frequency $i\omega_p$, where $\omega_p = \sqrt{4\pi\varrho}$ with ϱ the number of electrons per volume [46]. After some algebra, the following set of equations defining the plasmon-pole coefficients can be derived:

$$\begin{cases} A_{GG'}(q) = \varepsilon_{GG'}^{-1}(q, \omega = 0) - \delta_{GG'} \\ \tilde{\omega}_{GG'}^2 = \omega_p^2 \left[\dfrac{A_{GG'}(q)}{\varepsilon_{GG'}^{-1}(q, \omega = 0) - \varepsilon_{GG'}^{-1}(q, i\omega_p)} - 1 \right]. \\ \Omega_{GG'}^2(q) = -A_{GG'}(q)\,\tilde{\omega}_{GG'}^2(q) \end{cases} \tag{3.24}$$

In the HL model, the PPM parameters are calculated so as to reproduce the static limit exactly and to fulfill a generalized f-sum rule relating the imaginary part of the exact $\varepsilon_{GG'}^{-1}(q,\omega)$ to the plasma frequency and the charge density [47, 48]. The final expression for the PPM parameters are:

$$\begin{cases} \Omega_{GG'}^2(q) = \omega_p^2 \dfrac{(q+G)\cdot(q+G')}{|q+G|^2}\dfrac{n(G-G')}{n(0)} \\ \tilde{\omega}_{GG'}^2(q) = \dfrac{\Omega_{GG'}^2(q)}{\delta_{GG'}-\varepsilon_{GG'}^{-1}(q,\omega=0)} \\ A_{GG'}(q) = -\dfrac{\pi}{2}\dfrac{\Omega_{GG'}(q)}{\tilde{\omega}_{GG'}(q)}. \end{cases} \quad (3.25)$$

Models based on Eqs. (3.22) and (3.23) have a number of undesirable features, despite their success. For instance, for some elements with $G \neq G'$, the plasmon poles $\tilde{\omega}_{GG'}(q)$ can become very small or even imaginary which is somewhat unphysical [31].

Two more recent PPM approaches due to Von der Linden and Horsch [49] (vdLH) and Engel and Farid [50] (EF) are expected to be more accurate. The vdLH PPM is derived starting from the spectral decomposition of the symmetrized inverse dielectric matrix:

$$\tilde{\varepsilon}_{GG'}^{-1}(q,\omega) = \frac{|q+G'|}{|q+G|}\varepsilon_{GG'}^{-1}(q,\omega), \quad (3.26)$$

by assuming that the frequency dependence is solely contained in the eigenvalues (see Ref. [49]). The disadvantage of the vdLH approach is that it satisfies the f-sum rule only for the diagonal elements. In the EF PPM, the eigenvalues and the eigenvectors are frequency dependent, and derived from an approximation to the reducible polarizability which is exact both in the static- and high-frequency limit. For further details on this plasmon-pole technique, see Ref. [50].

Since the frequency convolution in Eq. (3.18) can be carried out analytically once the plasmon-pole parameters are known, the PPM technique is an ideal tool for initial convergence studies. It usually proves to be accurate to within 0.1–0.2 eV for states close to the Fermi level, when compared to results obtained with a costly numerical integration of Σ [27]. On the other hand, the accuracy worsens for states far from the gap, especially for low-lying states. To analyze physical properties depending on these, it is necessary to avoid PPM methods, and calculate the frequency dependence of W explicitly.

A straightforward numerical evaluation of Eq. (3.18) is problematic due to the fact that G and W both have poles infinitesimally above and below the real axis. Therefore, a straightforward integration algorithm along the real axis would need evaluations of the integrand precisely in the region where it is ill-behaved. An alternative route to evaluating Eq. (3.18) traces back to the earliest GW calculations for the homogeneous electron gas [51]. The Green's function G and the screened Coulomb interaction W are analytic functions (except along the real axis) and can consequently be analytically continued to the full complex plane. The strategy is to use a deformation of the contour of integration in order to avoid

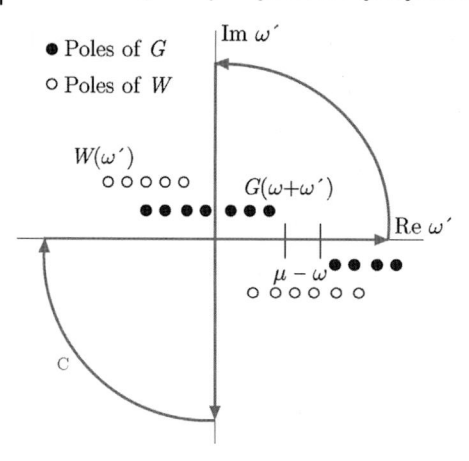

Figure 3.3 (online color at: www.pss-b.com) Schematic representation of the contour of integration in the complex ω' plane used to evaluate $\Sigma(\omega)$. The poles of the integrand are shown as circles. Only the poles due to Green's function that lie inside the path contribute to the final result.

having to deal with quantities close to the real axis as much as possible. Instead of evaluating the integral along the real axis, one evaluates the integral along the imaginary axis, and then adds the residues arising from the poles enclosed in the contour depicted in red in Figure 3.3.

3.3.4
The Extrapolar Method

GW calculations are computationally very demanding. Two major steps in these can be distinguished: the calculation of the polarizability and the evaluation of matrix elements of the self-energy. The quantities involved are not only non-local (two plane-wave indices), but also involve summations over all states (occupied and empty). Recently, Bruneval and Gonze [52] proposed an acceleration scheme to improve the convergence with respect to the number of states. The main idea is to replace the poles arising from the eigenvalues of empty high-energy states with a single (average) pole, which carries all the spectral weight above a certain cutoff for states. Note that the extrapolar technique was first introduced in the optimized effective potential framework [17] and in a preconditioning scheme [53].

Both in the polarizability and in the self-energy, the expressions to be evaluated contain a sum over wavefunctions in a numerator and energy differences in a denominator. If we were able to factor a simple common denominator out of the sum, it would be straightforward to eliminate the wavefunctions in the numerator, above some cutoff band index, N_b, by using the closure relation:

$$\sum_{b>N_b} |b\rangle\langle b| = 1 - \sum_{b\leq N_b} |b\rangle\langle b|. \tag{3.27}$$

Treating the denominator of the remainder (now dependent on all states $b \leq N_b$) is the delicate part, which requires careful consideration.

3.3.4.1 Polarizability with a Limited Number of Empty States

Using time-reversal symmetry, the truncated expression for the independent-particle polarizability in reciprocal and frequency space reads

$$P_{0GG'}(\boldsymbol{q},\omega) = \frac{2}{N_k \Omega} \sum_{k} \sum_{\substack{N_v < b \leq N_b \\ v \leq N_v}} M_k^{bv}(\boldsymbol{q}+\boldsymbol{G})[M_k^{bv}(\boldsymbol{q}+\boldsymbol{G'})]^* \qquad (3.28)$$

$$\times \left[\frac{1}{\omega-(\varepsilon_{vk}-\varepsilon_{bk-q})-i\eta} - \frac{1}{\omega-(\varepsilon_{bk-q}-\varepsilon_{vk})+i\eta} \right],$$

where Ω is the volume of the unit cell, η is a positive infinitesimal, N_v is the number of valence states, N_k is the number of *k*-points in the Brillouin zone, and the index *k* runs over the *k*-points of the Brillouin zone. The matrix elements:

$$M_k^{bb'}(\boldsymbol{q}+\boldsymbol{G}) = \langle \psi_{bk-q} | e^{-i(\boldsymbol{q}+\boldsymbol{G})\cdot \boldsymbol{r}} | \psi_{b'k} \rangle, \qquad (3.29)$$

are the so-called oscillator strengths.

The extrapolar method proposes that the empty states above the truncation index, N_b, all have the same energy. In this case, the dependence with respect to index *b* is removed in the denominator and one can apply the closure relation to the numerator in order to get rid of any dependence on this index. This procedure adds a term to the usual truncated expression for P_0. The correction consists of two terms:

$$\Delta_{GG'}(\boldsymbol{q},\omega) = \frac{2}{N_k \Omega} \sum_k \sum_{v \leq N_v} \langle \psi_{vk} | e^{i(\boldsymbol{G'}-\boldsymbol{G})\cdot \boldsymbol{r}} | \psi_{vk} \rangle$$

$$\times \left[\frac{1}{\omega-(\varepsilon_{vk}-\bar{\varepsilon}_{P_0})-i\eta} - \frac{1}{\omega-(\bar{\varepsilon}_{P_0}-\varepsilon_{vk})+i\eta} \right] \qquad (3.30)$$

$$-\frac{2}{N_k \Omega} \sum_k \sum_{\substack{b \leq N_b \\ v \leq N_v}} M_k^{bv}(\boldsymbol{q}+\boldsymbol{G})[M_k^{bv}(\boldsymbol{q}+\boldsymbol{G'})]^*$$

$$\times \frac{1}{\omega-(\varepsilon_{vk}-\bar{\varepsilon}_{P_0})-i\eta} - \frac{1}{\omega-(\bar{\varepsilon}_{P_0}-\varepsilon_{vk})+i\eta},$$

which are now free of any dependence on states above N_b. Instead they contain an "average" energy $\bar{\varepsilon}_{P_0}$ which represents the omitted part of the eigenvalue spectrum by a mean value. The best value for $\bar{\varepsilon}_{P_0}$ can be easily determined by a trial-and-error procedure or in a more elegant manner by considering the fulfillment of the *f*-sum rule for $P_0(\omega)$.

3.3.4.2 Self-Energy with a Limited Number of Empty States

An analogous procedure can be applied to the correlation part of the self-energy:

$$\langle \psi_{bk}|\Sigma_c(\varepsilon_{bk})|\psi_{bk}\rangle = \frac{i}{2\pi N_k \Omega} \int \sum_{b' \leq N_b} \sum_{qGG'} [W_{GG'}(q,\omega') - \delta_{GG'} v(q+G)]$$
$$\times \frac{M_k^{bb'}(q+G)[M_k^{bb'}(q+G')]^*}{\omega' - \varepsilon_{b'k-q} + \varepsilon_{bk} \pm i\eta} d\omega', \quad (3.31)$$

where η is a positive infinitesimal. The sign in front of η is plus when the state b' is empty, and minus otherwise.

Unlike for the polarizability, a PPM becomes necessary for the self-energy to make the extrapolar correction tractable. In this context, the PPM is a very good approximation. The final correction reads:

$$\Delta_{bk} = \frac{1}{N_k\Omega} \sum_{qGG'} \frac{\Omega^2_{GG'}(q) v(q+G)}{2\tilde{\omega}_{GG'}(q)[\tilde{\omega}_{GG'}(q) + \bar{\varepsilon}_\Sigma - \varepsilon_{bk} - i\eta]}$$
$$\times \left\{ \langle \psi_{bk}|e^{i(G'-G)\cdot r}|\psi_{bk}\rangle - \sum_{b' \leq N_b} M_k^{bb'}(q+G)\left[M_k^{bb'}(q+G')\right]^* \right\}. \quad (3.32)$$

Again, it consists of two terms that do not depend on any state above N_b. The introduced average energy $\bar{\varepsilon}_\Sigma$ in the denominators can safely be taken to be equal to the previously introduced $\bar{\varepsilon}_{P_0}$.

3.3.5
MBPT in the PAW Framework

Thanks to the excellent agreement obtained with respect to experiments, pseudopotential (PP)-based methods have for several decades represented a *de facto* standard for MBPT calculations. In recent years, however, results obtained with all-electron (AE) approaches [44, 54] have revealed that a fully consistent treatment of the electronic degrees of freedom produces GW band gaps that are systematically smaller than PP results, thus worsening the agreement between $G_0 W_0$ and experiments. These findings have led to quite an intense debate in the scientific literature concerning the reliability of the PP approach for MBPT calculations (see, for instance, Refs. [55–58]).

Systems with shallow cores or localized *d*- or *f*-electrons present severe challenges to PP GW calculations [28, 56, 58–60]. Core-valence exchange is large in these systems, due to the large overlap of the localized *d* or *f*-states (or semicore states) with lower-lying core states in the same atomic shell. To treat core-valence exchange consistently it is therefore either important to let the exchange part of the GW self-energy act on all electrons of one shell [56, 58, 59] – which can be very expensive computationally – or to build the exchange interaction into the PP [28, 29].

The PAW formalism introduced by Blöchl in 1994 [61] presents a flexible and efficient alternative to PPs in GW calculations. It combines the PP framework with an

AE description and allows for results on a par with AE accuracy at considerably reduced computational cost. The method takes advantage of several ideas and techniques developed in the past decades both in the PP and in the AE community. From the PP approach [62] it inherits the idea of substituting the true Kohn–Sham wave function $\psi(r)$ with a pseudized image $\tilde{\psi}(r)$ which can be efficiently expanded in an extended basis set (e.g., plane-waves). Similar to many AE approaches, PAW employs atomic orbitals to describe the AE wave function $\psi(r)$ inside non-overlapping atom-centered spheres, thus retaining information about the correct nodal structure of electronic orbitals.

The mapping between the true wave function, $|\psi\rangle$, with its complete and complex nodal structure around the nuclei, and the fictitious smooth pseudo wave functions, $|\tilde{\psi}\rangle$, is defined by the linear transformation: $|\psi\rangle = \hat{T}|\tilde{\psi}\rangle$. $\hat{T}$ is given by the identity operator plus a sum of localized terms, $\hat{T}_a$, only acting within the atomic spheres Ω_a centered on atomic sites a:

$$\hat{T} = \hat{1} + \sum_a \hat{T}_a. \tag{3.33}$$

A schematic representation of the division of the unit cell employed in the PAW method is shown in Figure 3.4.

The linear transformation within each augmentation region Ω_a is defined by specifying a set of functions, $\{\phi_i^a\}$, which form a complete basis set within Ω_a. This set of functions serves as a basis set for the expansion of the true electronic wave function in each augmentation region with coefficients c_i^a:

$$|\psi\rangle = \sum_i c_i^a |\phi_i^a\rangle \quad \text{in} \quad \Omega_a. \tag{3.34}$$

A possible and natural choice for the basis set $\{\phi_i^a\}$ are the solutions of the radial Schrödinger equation for the isolated atom. In this case the index i is a contracted notation for the atomic position R_a, the angular momentum quantum numbers (l, m), and an additional index n used to label solutions with different energy. The final expression for the linear transformation is given by [61, 63]

$$\hat{T} = \hat{1} + \sum_a \sum_i (|\phi_i^a\rangle - |\tilde{\phi}_i^a\rangle)\langle \tilde{p}_i^a|. \tag{3.35}$$

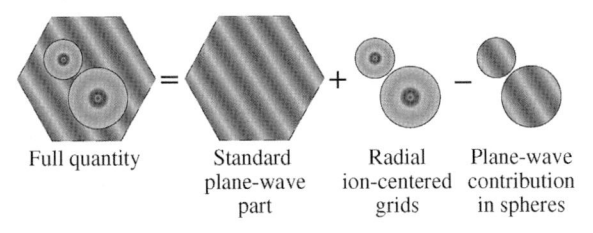

Full quantity = Standard plane-wave part + Radial ion-centered grids − Plane-wave contribution in spheres

Figure 3.4 (online color at: www.pss-b.com) Schematic representation of the division of the unit cell employed in the PAW method.

where the auxiliary pseudo partial waves $|\tilde{\phi}_i^a\rangle$ equal the AE counterparts $|\phi_i^a\rangle$ beyond the radius r_c^a of the PAW sphere, and are used to expand the pseudized function $|\tilde{\psi}\rangle$ inside the augmentation sphere. The atom-centered projector functions $|\tilde{p}_i^a\rangle$ are strictly localized inside the spheres and obey the orthogonality property:

$$\langle \tilde{p}_i^a | \tilde{\phi}_j^a \rangle = \delta_{ij}. \tag{3.36}$$

The matrix elements of a local or semilocal operator $\hat{A}$ between two AE wave functions can be efficiently and accurately evaluated by employing the linear transformation $\hat{T}$ given in Eq. (3.35). After some algebra one obtains:

$$\langle \psi | \hat{A} | \psi \rangle = \langle \tilde{\psi} | \hat{A} | \tilde{\psi} \rangle + \sum_{ij} \langle \tilde{\psi} | \tilde{p}_i \rangle [\langle \phi_i | \hat{A} | \phi_j \rangle - \langle \tilde{\phi}_i | \hat{A} | \tilde{\phi}_j \rangle] \langle \tilde{p}_j | \tilde{\psi} \rangle. \tag{3.37}$$

The first term in Eq. (3.37) has the same mathematical structure as the expression present in the PP formalism. As it involves only the "smooth" part of the wave function, it can be evaluated either in real or reciprocal space, depending on the nature of $\hat{A}$, by changing representation through fast Fourier transform techniques. The second term involves the onsite matrix elements of the $\hat{A}$ operator between AE and pseudo partial waves. It can be evaluated either by employing radial and angular meshes in real space or by expanding the operator $\hat{A}$ in terms of angular momenta.

Within the PAW formalism, the oscillator strengths, – i.e., the basic ingredients required to evaluate $P^0(\omega)$, and the matrix elements of $\Sigma(\omega)$ – can be obtained by means of the following equation [64]:

$$\langle \psi_{bk-q} | e^{-i(q+G)\cdot r} | \psi_{b'k} \rangle = \langle \tilde{\psi}_{bk-q} | e^{-i(q+G)\cdot r} | \tilde{\psi}_{b'k} \rangle + \sum_{ij} \langle \tilde{\psi}_{bk-q} | \tilde{p}_i \rangle \langle \tilde{p}_j | \tilde{\psi}_{b'k} \rangle e^{-i(q+G)\cdot R_i}$$

$$\times \left[\langle \phi_i | e^{-i(q+G)\cdot (r-r_i)} | \phi_j \rangle - \langle \tilde{\phi}_i | e^{-i(q+G)\cdot (r-r_i)} | \tilde{\phi}_j \rangle \right]$$

$$\times 4\pi \sum_{lm} (-i)^l \mathcal{Y}_m^l(q+G) \, G^{lm}_{l_i m_i l_j m_j} \int j_l(|q+G|r)(\phi_{n_i l_i} \phi_{n_j l_j} - \tilde{\phi}_{n_i l_i} \tilde{\phi}_{n_j l_j}) dr, \tag{3.38}$$

where the plane wave has been expressed in terms of Bessel functions $j_l(x)$ and real spherical harmonics $\mathcal{Y}_m^l(\hat{G})$ via the Rayleigh expansion. The symbol $G^{lm}_{l_i m_i l_j m_j}$ is used to denote the Gaunt coefficient [65], defined by:

$$G^{lm}_{l_i m_i l_j m_j} = \int \mathcal{Y}_{m_i}^{l_i} \mathcal{Y}_m^l \mathcal{Y}_{m_j}^{l_j} d\Omega. \tag{3.39}$$

3.4
QP Corrections to the BOs at Interfaces

In the DFT approach, the valence and conduction band offsets (VBO and CBO, respectively) are conveniently split into two terms:

$$\text{VBO} = \Delta E_v^{\text{DFT}} + \Delta V, \tag{3.40}$$

$$\text{CBO} = \Delta E_c^{\text{DFT}} + \Delta V. \tag{3.41}$$

The first term ΔE_v^{DFT} (resp. ΔE_c^{DFT}) on the right-hand side of Eq. (3.40) [resp. Eq. (3.41)] is referred to as the *band-structure contribution*. It is defined as the difference between the valence band maximum (VBM) (and the conduction band minimum (CBM), respectively) *relative to the average of the electrostatic potential in each material*. These are obtained from two independent standard bulk calculations on the two interface materials. Alternatively, these can be obtained from an analysis of the local density of states [66]. The second term ΔV, called the *lineup of the average of the electrostatic potential* across the interface, accounts for all the intrinsic interface effects. It is determined from a supercell calculation with a model interface.

Despite the limitations of DFT in finding accurate eigenenergies, the VBOs are often obtained with a very good precision, in particular for semiconductors [67]. This has opened an indirect route to computing the CBOs through the experimental band gaps using:

$$\text{CBO} = \Delta E_g^{\text{exp}} + \text{VBO}. \tag{3.42}$$

Note that this equation is equivalent to applying a scissor correction to the conduction bands on both sides of the interface, as can be seen by inserting Eqs. (3.40) and (3.42):

$$\text{CBO} = \Delta E_c^{\text{DFT}} + \Delta V + \left(\Delta E_g^{\text{exp}} - \Delta E_g^{\text{DFT}}\right), \tag{3.43}$$

and then comparing with Eq. (3.41).

The first QP calculation of the band-offsets (BOs) goes back to the work of Zhang et al. [68] who were investigating the VBO at the AlAs−GaAs(001) interface. They assumed that the lineup of the potential ΔV is already well described within DFT, arguing that QP corrections would not affect ΔV since it only depends on the long range electrostatic potentials. The latter are well-known functions of the electronic densities, which are given quite accurately by DFT.

Recently, the many-body effects on ΔV have been explicitly investigated [69]. This was done by comparing the electronic density and the resulting ΔV calculated within DFT and QSGW for a small model of the Si/SiO$_2$ interface illustrated in Figure 3.5(a). It was found that the QSGW results differ only slightly from DFT. The change in planar average of the electronic density, $\bar{\varrho}$, was at most 1 me/a.u. in the interface region, as illustrated in Figure 3.5(b). This lead to a variation in the macroscopic average of the local potential $\bar{V}$ [Figure 3.5(c)] smaller than 45 meV in that region. However, the net difference between the bulk materials, which is relevant for the lineup of the potential ΔV, was less than 12 meV. It was thus concluded that the interfacial charge density and, consequently, the associated dipole moments are well described within DFT, justifying the assumption that the lineup of the potential can be taken to be the same as in DFT. For metal-insulator or metal-semiconductor interfaces, this assumption still needs to be carefully checked.

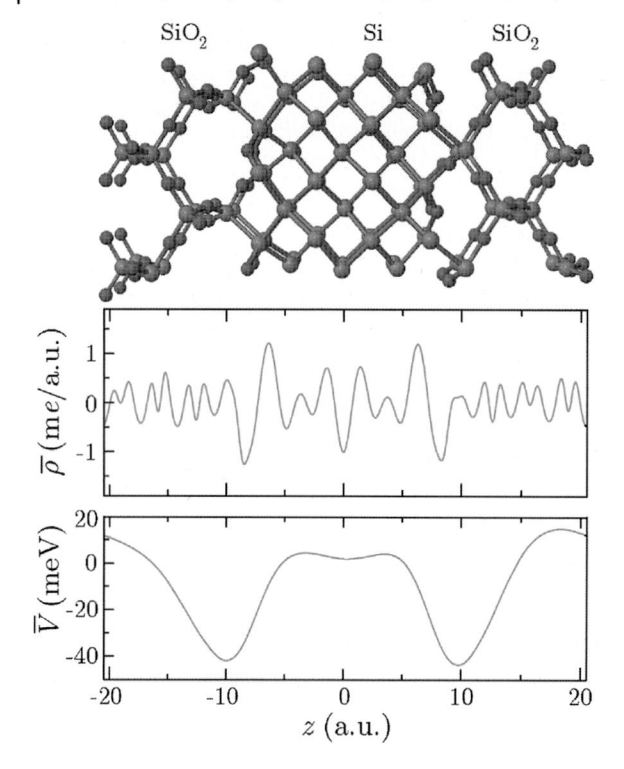

Figure 3.5 (online color at: www.pss-b.com) Small model of the Si/SiO$_2$ interface (upper panel) used in Ref. [69] to compute the difference between DFT and QSGW for the planar average of the electronic density $\bar{\varrho}$ (middle panel) and the macroscopic average of the local potential $\bar{V}$ (lower panel). The density and the potential are expressed in me/a.u. and in meV, respectively.

Assuming that ΔV can be taken from DFT, only the band-structure contribution is modified by QP corrections:

$$\text{VBO} = \Delta E_v^{QP} + \Delta V = \Delta E_v^{DFT} + \Delta(\delta E_v) + \Delta V, \qquad (3.44)$$

$$\text{CBO} = \Delta E_c^{QP} + \Delta V = \Delta E_c^{DFT} + \Delta(\delta E_c) + \Delta V, \qquad (3.45)$$

where $\delta E_v = E_v^{QP} - E_v^{DFT}$ (resp. $\delta E_c = E_c^{QP} - E_c^{DFT}$) is the QP correction at the VBM (resp. CBM) and $\Delta(\delta E_v)$ [resp. $\Delta(\delta E_c)$] is the corresponding difference between the two materials. It is important to stress that these corrections, which are obtained from bulk calculations, are the only additional ingredients that are required when DFT calculations of the VBO and CBO already exist.

Interestingly, for various semiconductor interfaces, the QP corrections of the band edges are found to be almost the same on both sides [68, 70] leading to $\Delta(\delta E_v) \leq 0.2$ eV in Eq. (3.44). As a result of this cancellation of errors, DFT is quite successful for these interfaces [67] with errors ranging from 0.1 to 0.5 eV, despite the limitations mentioned above. This relative success of DFT explains why it has been widely used to predict the VBO for a wide range of interfaces. And, when needed, the CBO was also predicted using a simple scissor operator to correct the band gap to the experimental value. This assumption was further motivated by the fact that MBPT calculations going beyond GW by including an approximate vertex correction ($GW\Gamma$) showed that the VBM remained at its DFT value for silicon, with the whole correction going to the conduction bands [30, 71].

However, when it comes to semiconductor–insulator or insulator–insulator interfaces, it appears that the errors to the VBO can be much more important in DFT. For instance, for the Si/SiO_2 interface, the VBOs are calculated to be 2.3–3.3 eV [66, 72–74] in noticeable disagreement with the experimental results of 4.3 eV [75, 76]. In contrast, for the Si/ZrO_2 and Si/HfO_2 interfaces, the calculated VBOs for the stable insulating O-terminated interfaces are around 2.5–3 eV [77–81], in reasonable agreement with experiment (2.7–3.4 eV) [82–90]. For these interfaces, scissor-corrected DFT has also been used to predict CBOs of about 1.7–2.2 eV, which compare quite well with the experimental values (1.5–2 eV) [87–90]. It seems that the cancellation of errors may vary strongly from one system to another, emphasizing the need to go beyond DFT by including QP corrections. Interestingly, hybrid functionals have been shown to give very good VBOs and CBOs compared to experiment for both the Si/SiO_2 and Si/HfO_2 interfaces by tuning the fraction of HF exchange for each bulk component to reproduce the experimental value of the band gap [9, 10].

For the Si/ZrO_2 interface, a QP correction of about 1.1 eV to the VBOs has been extracted from GW calculations for Si [70] and ZrO_2 [91] and used together with the experimental band gap to correct DFT BOs in several works [92, 93]. For the Si/HfO_2 interface, the same correction as for Si/ZrO_2 has been adopted [94] since there were no GW calculations available for HfO_2. Such an assumption seems quite reasonable given the analogous electronic structure of ZrO_2 and HfO_2. However, for both Si/ZrO_2 and Si/HfO_2 interfaces, the VBOs obtained by applying this correction are too large (and as a consequence the CBOs too small) with respect to the available experiments [92–94].

This discrepancy can be traced back to the fact that, while the QP corrections to the gap δE_g are *not* very sensitive to the choice of the PPM [64], the absolute values of δE_v and δE_c may vary from one PPM to another, as reported in Refs. [69, 95]. The results of Ref. [69] for Si and c-SiO_2 and those of Ref. [95] for c-ZrO_2 are summarized in Table 3.1. Since a precise knowledge of the QP corrections at the band edges is required for BO calculations, *it is necessary to go beyond PPMs*, by taking the frequency dependence of W into account explicitly. This can be done by using the CD method (see Section 2.3). The comparison between the CD and PPM results for a given system allows one to validate a PPM for further study of similar systems. Interestingly, the PPM proposed by GN [45] seems to lead to QP corrections in excellent agreement with those of the CD method (see Table 3.1), at variance with the other PPMs. Further investigation is still required to generalize this finding.

Table 3.1 QP corrections (in eV) at the VBM (δE_v), at the CBM (δE_c), and for the band gap (δE_g) for Si, c-SiO$_2$ (from Ref. [69]), and c-ZrO$_2$ (from Ref. [95]). The corrections are calculated within e-QSGW using the PPMs proposed by HL [31], vdLH [49], GN [45], EF [50], and without PPM using the CD method.

		HL	GN	vdLH	EF	CD
Si	δE_v	−0.6	−0.4	−0.6	−0.6	−0.4
	δE_c	+0.1	+0.2	+0.1	+0.1	+0.2
	δE_g	+0.7	+0.6	+0.7	+0.7	+0.6
c-SiO$_2$	δE_v	−2.6	−2.0	−2.5	−2.3	−1.9
	δE_c	+1.3	+1.5	+1.1	+1.2	+1.5
	δE_g	+3.9	+3.5	+3.6	+3.5	+3.4
c-ZrO$_2$	δE_v	−1.1	−0.5	–	–	−0.7
	δE_c	+1.3	+1.4	–	–	+1.4
	δE_g	+2.4	+1.9	–	–	+2.1

It is important to note that, rigorously, the QP corrections on the BOs should be calculated using the same PPs and the same exchange-correlation approximation as for the interface calculations. Indeed, the QP corrections are much more sensitive to these approximations than the band gap. Therefore, extreme caution should be applied when the QP corrections to DFT BOs are not calculated using the same approximations (e.g., in the PP, the exchange-correlation approximation, and the PPM).

Once this is carefully taken into account, the QP corrections can be calculated. It is also interesting to analyze the effect of including vertex corrections. The results reported in Ref. [69] for Si and c-SiO$_2$, and in Ref. [96] for c-HfO$_2$, are summarized in Table 3.2. While e-QSGW leads to a lowering of the VBM of Si compared to the DFT result ($\delta E_v \leq 0$), the inclusion of vertex corrections brings it back to roughly its original value with a small shift upwards of 0.1 eV, with all of the QP correction being on the conduction band. A similar result was also found previously [30, 71]. For HfO$_2$, the vertex correction acts in the same way though in this case the shift to the VBM is slightly larger (0.2 eV downwards). The results for Si and c-HfO$_2$ give some motivation to the use of a scissor operator to compute the CBO within DFT. However, for c-SiO$_2$, the results are very different. First, the VBM is also raised when including the vertex, but it definitely does not regress to the DFT level. This indicates that in Si and c-HfO$_2$, the recovery of the DFT VBM with the vertex is a coincidence. It also definitely rules out the use of a simple scissor operator for the computation of the BOs, unless further checks or refinements are made.

Finally, using Eqs. (3.44) and (3.45), the BOs can be computed within MBPT at the GW and $GW\Gamma$ levels. The results reported in Refs. [69, 95] compare very well with the experimental ones. Within e-QSGW the agreement is excellent for both the VBO and CBO (less than 0.3 eV difference). The effect of the vertex correction is less than 0.1 eV on the BOs. This results from a cancellation of the effects on each side of the interface. Indeed, in Eqs. (3.44) and (3.45), it is the difference between the QP corrections in both materials [$\Delta(\delta E_v)$ and $\Delta(\delta E_c)$] that matters. As

Table 3.2 QP corrections (in eV) at the VBM (δE_v), at the CBM (δE_v), and for the band gap (δE_g) for Si, c-SiO$_2$ (from Ref. 69), and c-HfO$_2$ (from Ref. [96]). The corrections are calculated using e-QSGW and e-QSGWΓ (the e-QS are omitted below).

	Si		c-SiO$_2$		c-HfO$_2$	
	GW	GWΓ	GW	GWΓ	GW	GWΓ
δE_v	−0.4	+0.1	−1.9	−1.3	−0.6	−0.2
δE_c	+0.2	+0.7	+1.5	+1.8	+1.1	+1.6
δE_g	+0.6	+0.6	+3.4	+3.1	+1.7	+1.8

can be seen in Table 3.2, this difference is typically less than 0.1 eV for the couples Si/SiO$_2$ and Si/HfO$_2$.

The effect of the vertex correction is very small compared to standard GW calculations. For the homogenous electron gas and atomic systems, it has been shown [97] that the local vertex correction of Del Sole et al. generally causes a large unphysical upward shift in the absolute values of band energies (and total energies). However, the relative changes in the QP energies obtained using $G_0 W_0 \Gamma$ are very small compared to $G_0 W_0$ results. The large shift can be attributed to an unphysical feature of the spectral function of the self-energy, which can come to have the wrong sign after a given energy. In the absence of non-trivial external electromagnetic fields, the spectral function of Σ should be strictly positive (negative) definite for frequencies below (above) the Fermi energy. A demonstration of this failure for the homogenous electron gas is given in Figure 3.6.

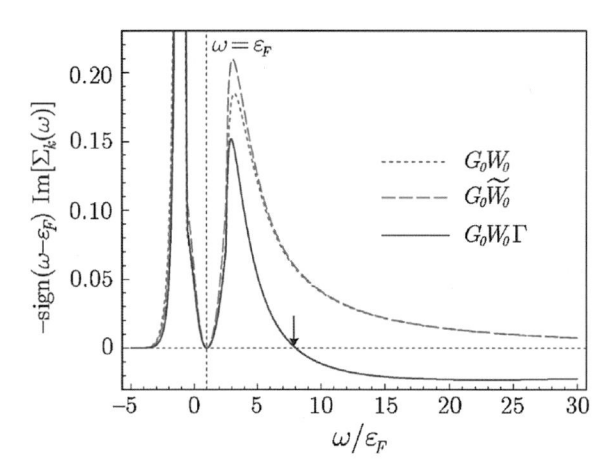

Figure 3.6 (online color at: www.pss-b.com) The imaginary part of the self-energy in jellium for $r_s = 2.0$ and $k = 0.5 k_F$. This is the spectral function of Σ. Plotted in this way, it should be positive definite everywhere. The curve for $G_0 W_0 \Gamma$ fails to be positive definite from the inset arrow onwards, after which it goes to a negative minimum and then slowly decays back to zero. The spectral function from $G_0 W_0$, in contrast, has no such behavior and has the proper limit.

In contrast, as can also be seen from Figure 3.6, there is no such failure for the implementation of the vertex only in the screened interaction, i.e., $G_0\tilde{W}_0$. Ironically, this is actually more time-consuming to implement in any existing code, as it requires an extra matrix multiplication before the calculation of the dielectric matrix. However, the bandwidth of metals compares better to experiments with this implementation, and it has also recently been used in Bethe-Salpeter calculations on molecules and metal clusters to good effect [98, 99]. These results indicate that the $G_0\tilde{W}_0$ vertex might yield better BOs, and the utility of this type of simple vertex correction certainly merits further study.

3.5
QP Corrections for Defects

Despite the methodological advancements discussed in previous sections it is still computationally challenging to compute total energies in GW and MBPT and no calculation for a defect has been reported so far.

The conventional way of obtaining defect formation energies, namely by calculating the total energy difference between the defective and a reference system [100], is therefore unavailable. Defect formation energies become accessible in GW again by realizing that QP energies correspond to electron addition and removal energies. Since the ionization potential and the electron affinity can be expressed in terms of total energy differences the formation energy of a defect can be formally rewritten as the successive charging of a lower (or if more convenient, higher) charge state [101, 102].

The formation of a neutral and a positive from a 2+ charge state is depicted schematically in Figure 3.7. For the example of the positive charge state this process reads mathematically

$$E_D^f(+,\varepsilon_F) = \Delta(+,\boldsymbol{R}_+^D,\boldsymbol{R}_{2+}^D) + A(2+,\boldsymbol{R}_{2+}^D) \\ + E_D^f(2+,\varepsilon_F=0) + \varepsilon_F, \quad (3.46)$$

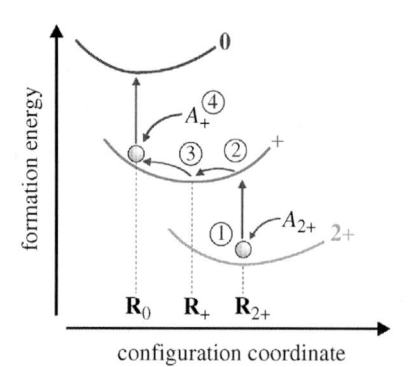

Figure 3.7 (online color at: www.pss-b.com) Formation of the neutral Si$_i$ from the 2+ charge state. A_+ and A_{2+} are short for the electron affinities $A(+,\boldsymbol{R}_0)$ and $A(2+,\boldsymbol{R}_{2+})$ (see text), respectively, and $\boldsymbol{R}_q$ denotes the atomic positions in charge state q.

where $E_D^f(+,\varepsilon_F)$ and $E_D^f(2+,\varepsilon_F=0)$ are the formation energies of the + and 2+ and ε_F the Fermi energy. R_q^D denotes the atomic coordinates of defect D in charge state q. $A(2+,R_{2+}^D)$ defines the vertical electron affinity of the 2+ state $E(+,R_{2+}^D)-E(2+,R_{2+}^D)$ (step 1 in Figure 3.7), referenced to the top of the valence band, whereas $\Delta(+,R_+^D,R_{2+}^D)$ gives the subsequent relaxation energy in the positive charge state $E(+,R_+^D)-E(+,R_{2+}^D)$ (step 2). The formation energy of higher charge states follows analogously.

Having split the formation energy into an electron addition and a lattice part, the most suitable computational technique can be employed for each part. For electron affinities (change in charge state at fixed geometry) we apply the GW approach. For relaxation energies (change of geometry in the same charge state) we retain DFT. Since the scheme has to be anchored on the formation energy of at least one charge state that cannot be corrected by GW (the 2+ in our example) the GW-correction approach depends on the quality of this formation energy and its associated valence band maximum (the reference for the Fermi energy). This is a weakness of the scheme and implies that relative formation energies (i.e., charge transition levels for which this dependence cancels exactly) are more accurate than absolute formation energies.

Applied to the self-interstitial in silicon the GW scheme corrects the DFT-LDA formation energy of different neutral configurations (see Figure 3.8) by ~1.1 eV [102] in good agreement with diffusion Monte Carlo calculations [103, 104]. For the $+ \to 0$ charge transition level of a phosphor vacancy at the InP(110) surface the GW-corrected value of 0.82 eV is in much better agreement with the experimental value of 0.75 ± 0.1 eV than the DFT-LDA charge transition level of

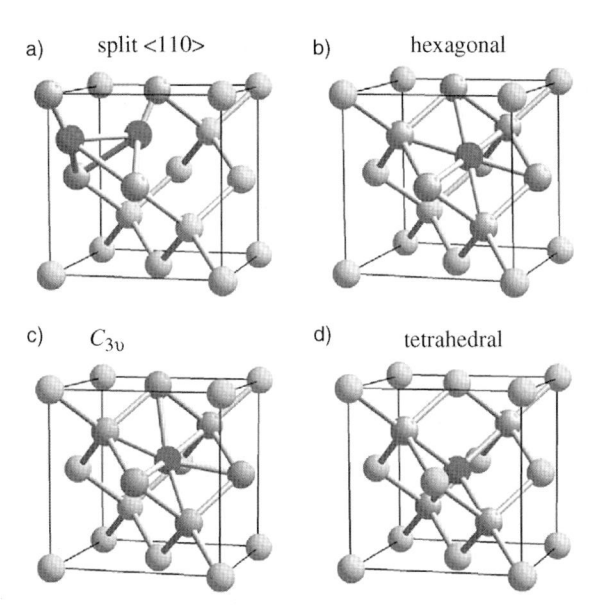

Figure 3.8 (online color at: www.pss-b.com) (a) Split <110>, (b) hexagonal, (c) C_{3v}, and (d) tetrahedral configuration of the Si_i. Defect atoms are shown in red and nearest neighbors in gray.

0.47 eV [101]. A long-standing problem was solved for silicon dioxide, where DFT-LDA favors the diffusion of charged oxygen interstitials, in clear disagreement with available experimental results. Agreement with experiment is recovered by applying the G_0W_0-correction approach, which substantially increases the formation energies of the negatively charged interstitials, leaving as dominant self-diffusion mechanism the neutral one [105].

The decomposition presented in the previous paragraphs supposes that the electronic states calculated in GW correspond to total energy differences. At present this assumption is not verifiable numerically, because GW total energies cannot be calculated for the defect systems at hand. The lowest excitation energies can, however, be expressed in two different ways. The electron affinities that enter in our discussion above can alternatively be seen as the ionization potential of a system with one extra electron:

$$I_0 = E_D(+) - E_D(0) = A_+. \tag{4.47}$$

While excitation energies calculated with the exact self-energy would satisfy Eq. (4.47) those from approximate self-energies – like GW – do not. This is shown in Figure 3.9 for the C split interstitial in silicon carbide (3C-SiC) in the neutral geometry. The first red dashed line is the electron affinity of the interstitial in the 2+ charge state (A_{2+}), while the second red dashed line corresponds to the ionization potential of the + charge state (I_+). If Eq. (4.47) was satisfied in these GW defect calculations the two lines would be equal. Instead they differ by 0.19 eV [106]. The excitations A_+ and I_0 (represented by the orange dashed lines) differ by 0.25 eV. This is not much, but noticeable.

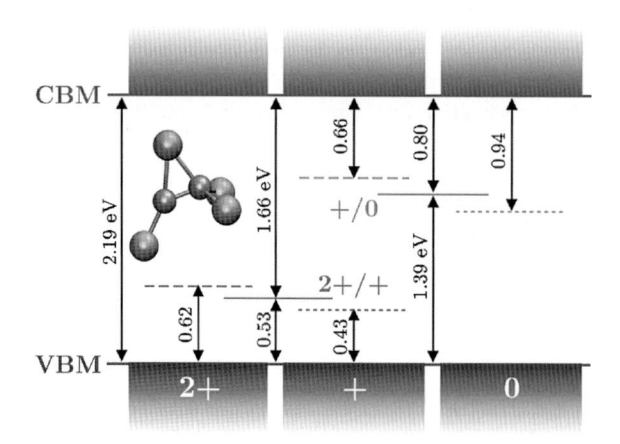

Figure 3.9 (online color at: www.pss-b.com) Defect level position for C split interstitial in 3C-SiC evaluated within GW for different charge states (2+, +, and 0). The atomic structure is fixed to the geometry of the neutral state (see inset). The occupied levels are depicted as a dotted line, the empty ones as a dashed line, and their mid values as a solid line. The transition energy is indicated in red for 2+/+ and in orange for +/0.

How can we reconcile this discrepancy? Slater [107] already identified this issue in the 1970s in DFT and HF calculations. He proposed to consider the total energy as a continuous function of electron number and to expand up to second order between integral numbers of electrons. The total energy difference in Eq. (4.47) can then be written as the energy of the highest occupied (or lowest unoccupied) state at half occupation. Equivalently, one could write this as the mean value between the energy of the highest occupied state of the neutral system and the energy of the lowest unoccupied state of a positively charged system. Since GW calculations at half occupation are not straightforward, the latter is more applicable. For the point defect from Figure 3.9 this then gives the following transition energies $E_{GW}(2+/+) = E_v + 0.53\,\text{eV}$ and $E_{GW}(+/0) = E_c - 0.80\,\text{eV} = E_v + 1.39$ eV. We see that the last equality only holds when performing the mean value technique, which reconciles the slight discrepancy between total energies and QP levels that exists in the GW formalism.

3.6
Conclusions and Prospects

The improvements discussed in this review are now available in several popular simulation packages, and have increased the speed of MBPT calculations. This has enabled the study of larger systems and more complex problems, such as interfaces and defects. We have illustrated this by presenting some recent MBPT results, while trying to highlight the main difficulties and caveats. It is to be expected that many more calculations on interfaces and defects relying on MBPT will follow.

As a final remark, it should be mentioned that DFT (or popular flavors of DFT) may fail to predict the correct geometry of certain interfaces or defects. In such cases, the energy levels (be it the VBM, the CBM, or defect levels) computed from MBPT could also be wrong. In order to avoid such problems, it is highly desirable to be able to compute the energy and the forces self-consistently from many-body theories for supercells ranging from 100 to 200 atoms. This is an important aim of future developments in MBPT implementations.

Acknowledgements

M. Giantomassi and G.-M. Rignanese acknowledge the support of the Fonds de la Recherche Scientifique-FNRS Belgium. M. Stankovski and G.-M. Rignanese are grateful to the Vlaanderen Agentschap voor Innovatie door Wetenschap en Technologie (IWT) for financial support. P. Rinke acknowledges the support of the Deutsche Forschungsgemeinschaft. M. Grüning thanks the Fundação para a Ciência e a Tecnologia (FCT) for its support through the Ciência 2008 program.

References

1 Capasso, F. and Magaritondo, G. (eds) (1987) *Heterojunction Band Discontinuities: Physics and Device Applications*, North-Holland, Amsterdam.
2 Sze, S.M. and Ng, K.K. (2007) *Physics of Semiconductor Devices*, 3rd edn, John Wiley & Sons, New York.
3 Fleetwood, D.M., Pantelides, S.T., and Schrimpf, R.D.C. (eds) (2008) *Defects in Microelectronic Materials and Devices*, CRC Press, Boca Raton, Florida.
4 Wada, K. and Pang, S.W. (eds) (2001) *Defects in Optoelectronic Materials*, CRC Press, Boca Raton, Florida.
5 Gusev, E. (ed.) (2006) *Defects in High-k Gate Dielectric Stacks*, Springer, Dordrecht, The Netherlands.
6 Knaup, J.M., Deák, P., Frauenheim, Th., Gali, A., Hajnal, Z., and Choyke, W.J. (2005) *Phys. Rev. B*, **72**, 115323.
7 Xiong, K., Robertson, J., Gibson, M.C., and Clark, S.J. (2005) *Appl. Phys. Lett.*, **87**, 183505.
8 Broqvist, P. and Pasquarello, A. (2007) *Microelectron. Eng.*, **84**, 2022.
9 Broqvist, P., Alkauskas, A., and Pasquarello, A. (2008) *Appl. Phys. Lett.*, **92**, 132911.
10 Akauskas, A., Broqvist, P., Devynck, F., and Pasquarello, A. (2008) *Phys. Rev. Lett.*, **101**, 106802.
11 Alkauskas, A., Broqvist, P., and Pasquarello, A. (2008) *Phys. Rev. Lett.*, **101**, 046405.
12 Curtiss, L.A., Redfern, P.C., Raghavachari, K., and Pople, J.A. (1998) *J. Chem. Phys.*, **109**, 42.
13 Muscat, J., Wander, A., and Harrison, N.M. (2001) *Chem. Phys. Lett.*, **342**, 397.
14 Paier, J., Marsman, M., Hummer, K., Kresse, G., Gerber, I.C., and Ángyán, Th. (2006) *J. Chem. Phys.*, **124**, 154709; (2006) *J. Chem. Phys.*, **125**, 249901.
15 Ernzerhof, M., Perdew, J.P., and Burke, K. (1997) *Int. J. Quantum Chem.*, **64**, 285.
16 Ernzerhof, M. and Scuseria, G. (1999) *J. Chem. Phys.*, **110**, 5029.
17 Kümmel, S. and Kronik, L. (2008) *Rev. Mod. Phys.*, **80**, 3.
18 Hedin, L. and Lundqvist, S. (1969) *Solid State Physics*, vol. 23 (eds H. Ehrenreich, F. Seitz, and D. Turnbull), Academic Press, New York, p. 1.
19 Fetter, A.L. and Walecka, J.D. (1971) *Quantum Theory of Many-Particle Systems*, McGraw-Hill, New York.
20 Abrikosov, A.A., Gorkov, L.P., and Dzyaloshinskii, E. (1975) *Methods of Quantum Field Theory in Statistical Physics*, Dover, New York.
21 Landau, L.D. and Lifschitz, E.M. (1980) *Statistical Physics Part II*, Pergamon, Oxford.
22 Onida, G., Reining, L., and Rubio, A. (2002) *Rev. Mod. Phys.*, **74**, 601.
23 Hedin, L. (1965) *Phys. Rev. A*, **139**, 796.
24 Nelson, W., Bokes, P., Rinke, P., and Godby, R.W. (2007) *Phys. Rev. A*, **75**, 032505.
25 Romaniello, P., Guyot, S., and Reining, L. (2009) *J. Chem. Phys.*, **131**, 154111.
26 Godby, R.W., Schlüter, M., and Sham, L.J. (1986) *Phys. Rev. Lett.*, **56**, 2415.
27 Aulbur, W.G., Jonsson, L., and Wilkins, J.W. (2000) *Solid State Phys.*, **54**, 1.
28 Rinke, P., Qteish, A., Neugebauer, J., Freysoldt, C., and Scheffler, M. (2005) *New J. Phys.*, **7**, 126.
29 Rinke, P., Qteish, A., Neugebauer, J., and Scheffler, M. (2008) *Phys. Status Solidi B*, **245**, 929.
30 Del Sole, R., Reining, L., and Godby, R.W. (1994) *Phys. Rev. B*, **49**, 8024.
31 Hybertsen, M.S. and Louie, S.G. (1986) *Phys. Rev. B*, **34**, 5390.
32 Li, J.-L., Rignanese, G.-M., and Louie, S.G. (2005) *Phys. Rev. B*, **71**, 193102.
33 White, I.D., Godby, R.W., Rieger, M.M., and Needs, R.J. (1997) *Phys. Rev. Lett.*, **80**, 4265.
34 Rohling, M., Wang, N.P., Krüger, P., and Pollmann, J. (2003) *Phys. Rev. Lett.*, **91**, 256802.
35 Pulci, O., Reining, L., Onida, G., Del Sole, R., and Bechstedt, F. (2001) *Comput. Mater. Sci.*, **20**, 300.

References

36 Rinke, P., Delaney, K., García-González, P., and Godby, R.W. (2004) *Phys. Rev. A*, **70**, 063201.

37 Bruneval, F., Vast, M., and Reining, L. (2006) *Phys. Rev. B*, **74**, 045102.

38 Rieger, M.M., Steinbeck, L., White, I.D., Rojas, H.N., and Godby, R.W. (1999) *Comput. Phys. Commun.*, **117**, 211.

39 Holm, B. and von Barth, U. (1998) *Phys. Rev. B*, **57**, 2108.

40 Schöne, W.-D. and Eguiluz, A.G. (1998) *Phys. Rev. Lett*, **81**, 1662.

41 Ku, W. and Eguiluz, A.G. (2002) *Phys. Rev. Lett.*, **89**, 126401.

42 Faleev, S.V., van Schilfgaarde, M., and Kotani, T. (2004) *Phys. Rev. Lett.*, **93**, 126406.

43 van Schilfgaarde, M., Kotani, T., and Faleev, S.V. (2006) *Phys. Rev. Lett.*, **96**, 226402.

44 Kotani, T., van Schilfgaarde, M., and Faleev, S.V. (2007) *Phys. Rev. B*, **76**, 165106.

45 Godby, R.W. and Needs, R.J. (1989) *Phys. Rev. Lett.*, **62**, 1169.

46 Ashcroft, N.W. and Mermin, N.D. (1976) *Solid State Physics*, Cornell University Press, Ithaca, New York.

47 Johnson, D.L. (1974) *Phys. Rev. B*, **9**, 4475.

48 Taut, M. (1985) *J. Phys. C*, **18**, 2677.

49 von der Linden, W. and Horsch, P. (1988) *Phys. Rev. B*, **37**, 8351.

50 Engel, G.E. and Farid, B. (1993) *Phys. Rev. B*, **47**, 15931.

51 Lundqvist, B.I. (1968) *Phys. Kondens. Mater.*, **7**, 117.

52 Bruneval, F. and Gonze, X. (2008) *Phys. Rev. B*, **78**, 085125.

53 Anglade, P.-M. and Gonze, X. (2008) *Phys. Rev. B*, **78**, 045126.

54 Kotani, T. and van Schilfgaarde, M. (2002) *Solid State Commun.*, **121**, 461.

55 Gómez-Abal, R., Li, X., Scheffler, M., and Ambrosch-Draxl, C. (2008) *Phys. Rev. Lett.*, **101**, 106404.

56 Marini, A., Onida, G., and Del Sole, R. (2001) *Phys. Rev. Lett.*, **88**, 016403.

57 Delaney, K., García-González, P., Rubio, A., Rinke, P., and Godby, R.W. (2004) *Phys. Rev. Lett.*, **93**, 249701.

58 Tiago, M.L., Ismail-Beigi, S., and Louie, S.G. (2004) *Phys. Rev. B*, **69**, 125212.

59 Rohlfing, M., Krüger, P., and Pollmann, J. (1995) *Phys. Rev. Lett.*, **75**, 3489.

60 Jiang, H., Gómez-Abal, R., Rinke, P., and Scheffler, M. (2009) *Phys. Rev. Lett.*, **102**, 126403.

61 Blöchl, P.E. (1994) *Phys. Rev. B*, **50**, 17953.

62 Hamann, D.R., Schlüter, M., and Chiang, C. (1979) *Phys. Rev. Lett.*, **43**, 1494.

63 Torrent, M., Jollet, F., Bottin, F., Zerah, G., and Gonze, X. (2008) *Comput. Mater. Sci.*, **42**, 337.

64 Arnaud, B. and Alouani, M. (2000) *Phys. Rev. B*, **62**, 4464.

65 Rotenber, M., Bivins, R., Metropolis, N., and Wooten, J.K. (1959) *The 3-j and 6-j Symbols* Technology Press, Massachusetts Institute of Technology, Cambridge.

66 Yamasaki, T., Kaneta, C., Uchiyama, T., Uda, T., and Terakura, K. (2001) *Phys. Rev. B*, **63**, 115314.

67 Van de Walle, C.G. and Martin, R.M. (1987) *Phys. Rev. B*, **35**, 8154.

68 Zhang, S.B., Tománek, D., Louie, S.G., Cohen, M.L., and Hybertsen, M.S. (1988) *Solid State Commun.*, **66**, 585.

69 Shaltaf, R., Rignanese, G.-M., Gonze, X., Giustino, F., and Pasquarello, A. (2008) *Phys. Rev. Lett.*, **100**, 6401.

70 Zhu, X. and Louie, S.G. (1991) *Phys. Rev. B*, **43**, 14142.

71 Fleszar, A. and Hanke, W. (1997) *Phys. Rev. B*, **56**, 10228.

72 Watarai, M., Nakamura, J., and Natori, A. (2004) *Phys. Rev. B*, **69**, 035312.

73 Tuttle, B.R. (2004) *Phys. Rev. B*, **70**, 125322.

74 Giustino, F. and Pasquarello, A. (2005) *Surf. Sci.*, **586**, 183.

75 Keister, J.W., Rowe, J.E., Kolodziej, J.J., Niimi, H., Madey, T.E., and Lucovsky, G. (1999) *J. Vac. Sci. Technol. B*, **17**, 1831.

76 Afanas'ev, V.V., Houssa, M., and Stesman, A. (2001) *Appl. Phys. Lett.*, **78**, 3073.

77 Peacock, P.W. and Robertson, J. (2004) *Phys. Rev. Lett.*, **92**, 057601.

78 Peacock, P.W., Xiong, K., Tse, K.Y., and Robertson, J. (2006) *Phys. Rev. B*, **73**, 075328.

79 Puthenkovilakam, R., Carter, E.A., and Chang, J.P. (2004) *Phys. Rev. B*, **69**, 155329.

80 Puthenkovilakam, R. and Chang, J.P. (2004) *J. Appl. Phys.*, **96**, 2701.
81 Chen, G.H., Hou, Z.F., and Gong, X.G. (2009) *Appl. Phys. Lett.*, **95**, 102905.
82 Miyazaki, S., Narasaki, M., Ogasawara, M., and Hirose, M. (2001) *Microelectron. Eng.*, **59**, 373.
83 Oshima, M., Toyoda, S., Okumura, T., Okabayashi, J. Kumigashira, H. Ono, K., Niwa, M., Usuda, K., and Hirashita, N. (2003) *Appl. Phys. Lett.*, **83**, 2172.
84 Wang, S.J., Huan, A.C.H., Foo, Y.L., Chai, J.W., Pan, J.S., Li, Q., Dong, Y.F., Feng, Y.P., and Ong, C.K. (2004) *Appl. Phys. Lett.*, **85**, 4418.
85 Rayner, G.B., Kang, D., Zhang, Y., and Lucovsky, G. (2002) *J. Vac. Sci. Technol. B*, **20**, 1748.
86 Sayan, S., Garfunkel, E., and Suzer, S. (2002) *Appl. Phys. Lett.*, **80**, 2135.
87 Afanas'ev, V.V., Stesmans, A., Chen, F., Shi, X., and Campbell, S.A. (2002) *Appl. Phys. Lett.*, **81**, 1053.
88 Sayan, S., Bartynski, R.A., Zhao, X., Gusev, E.P., Vanderbilt, D., Croft, M., Banaszak Holl, M., and Garfunkel, E. (2004) *Phys. Status Solidi B*, **241**, 2246.
89 Renault, O., Barrett, N.T., Samour, D., and Quiais-Marthon, S. (2004) *Surf. Sci.*, **566**, 526.
90 Bersch, E., Rangan, S., Bartynski, R.A., Garfunkel, E., and Vescovo, E. (2008) *Phys. Rev. B*, **78**, 085114.
91 Kralik, B., Chang, E.K., and Louie, S.G. (1998) *Phys. Rev. B*, **57**, 7027.
92 Fiorentini, V. and Gulleri, G. (2002) *Phys. Rev. Lett.*, **89**, 266101.
93 Dong, Y.F., Feng, Y.P., Wang, S.J., and Huan, A.C.H. (2005) *Phys. Rev. B*, **72**, 045327.
94 Tuttle, B.R., Tang, C., and Ramprasad, R. (2007) *Phys. Rev. B*, **75**, 235324.
95 Grüning, M., Shaltaf, R., and Rignanese, G.-M. (2010) *Phys. Rev. B*, **81**, 035330.
96 Shaltaf, R., Grüning, M., Stankovski, M., and Rignanese, G.-M., unpublished.
97 Morris, A.J., Stankovski, M., Delaney, K.T., Rinke, P., García-González, P., and Godby, R.W. (2007) *Phys. Rev. B*, **76**, 155106.
98 Tiago, M.L. and Chelikowsky, J.R. (2006) *Phys. Rev. B*, **73**, 205334.
99 Tiago, M.L., Idrobo, J.C., Öğüt, S., Jellinek, J., and Chelikowsky, J.R. (2009) *Phys. Rev. B*, **79**, 155419.
100 Van de Walle, C.G. and Neugebauer, J. (2004) *J. Appl. Phys.*, **95**, 3851.
101 Hedström, M., Schindlmayr, A., Schwarz, G., and Scheffler, M. (2006) *Phys. Rev. Lett.*, **97**, 226401.
102 Rinke, P., Janotti, A., Scheffler, M., and Van de Walle, C.G. (2009) *Phys. Rev. Lett.*, **102**, 026402.
103 Leung, W.-K., Needs, R.J., Rajagopal, G., Itoh, S., and Ihara, S. (1999) *Phys. Rev. Lett.*, **83**, 2351.
104 Batista, E.R., Heyd, J., Hennig, R.G., Uberuaga, B.P., Martin, R.L., Scuseria, G.E., Umrigar, J., and Wilkins, J.W. (2006) *Phys. Rev. B*, **74**, R121102.
105 Martin-Samos, L., Roma, G., Rinke, P., and Limoge, Y. (2010) *Phys. Rev. Lett.*, **104**, 075502.
106 Bruneval, F. (2009) *Phys. Rev. Lett.*, **103**, 176403.
107 Slater, J.C. (1974) *The Self-Consistent Field for Molecules and Solids*, vol. 4, McGraw-Hill, New York.

4
Accelerating GW Calculations with Optimal Polarizability Basis

Paolo Umari, Xiaofeng Qian, Nicola Marzari, Geoffrey Stenuit, Luigi Giacomazzi, and Stefano Baroni

4.1
Introduction

Density-functional theory (DFT) has grown into a powerful tool for the numerical simulation of matter at the nanoscale, allowing one to study the structure and dynamics of realistic models of materials consisting of up to a few thousands atoms, these days [1]. The scope of standard DFT, however, is limited to those dynamical processes that do not involve electronic excitations. Moreover, its time-dependent extension [2], which has been conceived to cope with such processes, still displays conceptual and practical difficulties.

The most elementary excitation is the removal or the addition of an electron from a system originally in its ground state. These processes are accessible to direct/inverse photo-emission spectroscopies and can be described in terms of *quasi-particle* (QP) spectra [3]. In insulators, the energy difference between the lowest-lying quasi-electron state and the highest-lying quasi-hole state is the QP band gap, a quantity that is severely (and to some extent erratically) underestimated by DFT [4].

Many-body perturbation theory (MBPT), in turn, provides a general, though unwieldy, framework for calculating QP properties and other excitation (such as optical) spectra [3]. A numerically viable approach to QP energy (QPE) levels (known as the GW approximation, GWA) was introduced in the 1960s [5], but it took two decades for a realistic application of it to appear [6, 7], and even today the numerical effort required by MBPT is such that its scope is usually limited to systems of a few handfuls of inequivalent atoms. The two main difficulties are the necessity to calculate and manipulate large matrices representing the charge response of the system (electron polarizabilities or polarization propagators) [8], on the one hand, and that of expressing such response functions in terms of slowly converging sums over empty one-electron states [8, 9–11], on the other hand. Recently, we addressed both problems. In a first work [8] we introduced a method to significantly reduce the computational and memory loads of GWA calculations through the introduction of optimal basis sets for representing polarizability operators built upon Wannier-like

Advanced Calculations for Defects in Materials: Electronic Structure Methods, First Edition.
Edited by Audrius Alkauskas, Peter Deák, Jörg Neugebauer, Alfredo Pasquarello, and Chris G. Van de Walle.
© 2011 Wiley-VCH Verlag GmbH & Co. KGaA. Published 2011 by Wiley-VCH Verlag GmbH & Co. KGaA.

orbitals [12–14]. Then, in a following communication [15], we proposed an approach to obtain fully converged GWA calculation avoiding at the same time any sum over empty states. In the same work we explained how also optimal polarizability basis sets can be constructed without explicitly evaluating empty states.

In this review, we present the strategy we have conceived for obtaining optimal polarizability basis sets and for calculating QPE levels, still considering sums over empty states. The paper is organized in this way: in Section 4.2 we briefly introduce the GW approximation, in Section 4.3 we describe our method for constructing optimal polarizability basis sets and for performing GWA calculations in isolated and extended systems, in Section 4.4 we validate our method by considering the benzene molecule, crystalline silicon, and a model of vitreous silica, in Section 4.5 we use our method for studying the electronic properties of a model of quasi-stoichiometric amorphous silicon nitride and of its point defects. Conclusion and perspectives are drawn in Section 4.6.

4.2
The GW Approximation

QP energies and QP amplitudes (QPA) are eigenvalues and eigenvectors of a Schrödinger-like equation (QPEq), which is similar to the DFT Kohn–Sham equation with the exchange-correlation potential, $V_{xc}(\mathbf{r})$, replaced by the non-local, energy-dependent, and non-Hermitian self-energy operator, $\tilde{\Sigma}(\mathbf{r}, \mathbf{r}', E)$ (a tilde indicates the Fourier transform of a time-dependent function):

$$\left(-\frac{1}{2}\Delta + V_{ext} + V_H + \tilde{\Sigma}(E_n)\right)\xi_n = E_n \xi_n(\mathbf{r}), \tag{4.1}$$

where we are using atomic units ($\hbar = 1$, $m = 1$, and $e = 1$) and V_{ext} is the external (ionic) potential, V_H is the Hartree potential, and E_n and ξ_n are the n-th QPE and QPA, respectively. It is worth noting that the Hartree–Fock equation can be obtained from Eq. (4.1) by setting:

$$\tilde{\Sigma}(\mathbf{r}, \mathbf{r}'; E) = -e^2 \frac{\varrho(\mathbf{r}, \mathbf{r}')}{|\mathbf{r} - \mathbf{r}'|}, \tag{4.2}$$

where ϱ is the one-particle density matrix and e is the elementary charge.

The next level of approximation is the GWA [5] where Σ is the product in time of the one-electron propagator, G, and of the dynamically screened interaction, W:

$$\Sigma_{GW}(\mathbf{r}, \mathbf{r}'; t) = iG(\mathbf{r}, \mathbf{r}'; t+\eta)W(\mathbf{r}, \mathbf{r}'; t), \tag{4.3}$$

where η is a positive infinitesimal and W is expressed in terms of the bare Coulomb interaction $v(\mathbf{r}, \mathbf{r}')$ and of the reducible polarizability operator $\Pi(\mathbf{r}, \mathbf{r}'; t)$:

$$W = v + v \cdot \Pi \cdot v, \tag{4.4}$$

where we indicate with a dot the product of two operators, such as in $v \cdot \chi(\mathbf{r}, \mathbf{r}', t) = \int d\mathbf{r}'' v(\mathbf{r}, \mathbf{r}'') \chi(\mathbf{r}'', \mathbf{r}'; t)$.

Then, the reducible polarizability operator is obtained from the irreducible polarizability operator P through the following Dyson's equation:

$$\Pi = (1 - P \cdot v)^{-1} \cdot P. \tag{4.5}$$

Finally, the irreducible polarizability is given from the product in time of one-electron propagators:

$$P(\mathbf{r}, \mathbf{r}'; t) = -i G(\mathbf{r}, \mathbf{r}'; t) G(\mathbf{r}', \mathbf{r}; -t). \tag{4.6}$$

The GWA alone does not permit to solve the QPEq, unless G and W are known, possibly depending on the solution of the QPEq itself.

One of the most popular further approximations is the so-called $G°W°$ approximation, where the one-electron propagator G is obtained from the eigenfunctions $\psi_n(\mathbf{r})$ and eigenenergies ε_n of a one-electron (usually a Kohn–Sham) Hamiltonian:

$$\begin{aligned}G°(\mathbf{r}, \mathbf{r}'; \tau) = &i \sum_v \psi_v(\mathbf{r}) \psi_v^*(\mathbf{r}') \, e^{-i\varepsilon_v \tau} \theta(-\tau) \\ &- i \sum_c \psi_c(\mathbf{r}) \psi_c^*(\mathbf{r}') \, e^{-i\varepsilon_c \tau} \theta(\tau),\end{aligned} \tag{4.7}$$

where, referred to the Fermi energy, v and c suffixes indicate valence states below and conduction states above the Fermi energy, respectively, and θ is the Heaviside step function. Now, using the definition of $G°$ in Eq. (4.6) is equivalent to calculating the irreducible polarizability within the random-phase approximation (RPA) which we indicate with $P°$. Then, from Eqs. (4.5) and (4.4), we obtain the approximate reducible polarizability operator $\Pi°$ and dynamically screened Coulomb operator $W°$. Finally, the approximate self-energy operator in the $G°W°$ scheme is calculated through:

$$\Sigma_{G°W°}(\mathbf{r}, \mathbf{r}'; t) = i G°(\mathbf{r}, \mathbf{r}'; t + \eta) W°(\mathbf{r}, \mathbf{r}'; t). \tag{4.8}$$

A further approximation, usually referred to as the diagonal approximation, is introduced for solving the QPEq: the QPAs are approximated directly with the non-interacting eigenfunctions:

$$\xi_n(\mathbf{r}) \approx \psi_n(\mathbf{r}). \tag{4.9}$$

This permits to find the QPEs by solving the following self-consistent one-variable equation:

$$E_n \approx \varepsilon_n + \langle \tilde{\Sigma}_{G°W°}(E_n) \rangle_n - \langle V_{XC} \rangle_n, \tag{4.10}$$

where $\langle A \rangle_n = \langle \psi_n | A | \psi_n \rangle$.

The apparently simple $G°W°$ approximation still involves severe difficulties, mainly related to the calculation and manipulation of the polarizability that enters the definition of $W°$. These difficulties are often addressed using the so-called plasmon-pole approximation [6], which however introduces noticeable ambiguities and inaccuracies when applied to inhomogeneous systems [16]. A well-established technique to address QP spectra in real materials without any crude approximations on response functions is the *space–time method* (STM) by Godby and coworkers [17]. In the STM the time/energy dependence of the $G°W°$ operators is represented on the imaginary axis, thus making them smooth (in the imaginary frequency domain) or exponentially decaying (in the imaginary time domain). The various operators are represented on a real-space grid, a choice which is straightforward, but impractical for systems larger than a few handfuls of inequivalent atoms. In the STM, the self-energy expectation value in Eq. (4.10) is obtained by analytically continuing to the real frequency axis the Fourier transform of the expression:

$$\langle \Sigma_{G^\circ W^\circ}(i\tau) \rangle_n = \mp \sum_l e^{\varepsilon_l \tau} \int \psi_n(\mathbf{r}) \psi_l(\mathbf{r}) \psi_l(\mathbf{r}') \psi_n(\mathbf{r}') W^\circ(\mathbf{r},\mathbf{r}';i\tau) \, d\mathbf{r}\, d\mathbf{r}', \quad (4.11)$$

where the upper (lower) sign holds for positive (negative) times, the sum extends below (above) the Fermi energy, and QPAs are assumed to be real. For simplicity, in the rest of the paper, one-particle wavefunctions will be always considered to be real, which is always possible for time-symmetric systems. By substituting v for W, Eq. (4.11) yields the exchange self-energy, whereas $v \cdot \Pi \cdot v$ yields the correlation contribution, Σ_C, whose evaluation is the main size-limiting step of GW calculations.

4.3
The Method: Optimal Polarizability Basis

Let us suppose that a small, time-independent, orthonormal basis set $\{\Phi_\mu(\mathbf{r})\}$ exists for representing polarizability operators:

$$\Pi(\mathbf{r},\mathbf{r}';i\tau) \approx \sum_{\mu\nu} \Pi_{\mu\nu}(i\tau) \Phi_\mu(\mathbf{r}) \Phi_\nu(\mathbf{r}'). \quad (4.12)$$

Then, the correlation contribution Σ_C to the self-energy is given by Eq. (4.11):

$$\langle \Sigma_C(i\tau) \rangle_n \approx \mp \sum_{l\mu\nu} e^{\varepsilon_l \tau} \Pi_{\mu\nu}(i\tau) S_{nl,\mu} S_{nl,\nu} \theta(E_C^1 - \varepsilon_l), \quad (4.13)$$

where E_C^1 is an energy cutoff that limits the number of conduction states to be used in the calculation of the self-energy and:

$$S_{nl,\nu} = \int \psi_n(\mathbf{r}) \psi_l(\mathbf{r}) \frac{e^2}{|\mathbf{r}-\mathbf{r}'|} \Phi_\nu(\mathbf{r}') \, d\mathbf{r}\, d\mathbf{r}'. \quad (4.14)$$

Then a convenient representation of the polarizability would thus allow QPEs to be calculated from Eq. (4.10), by analytically continuing to the real axis the Fourier transform of Eq. (4.13). Our goal is to shrink the dimension of the polarizability basis set $\{\Phi_\mu(\mathbf{r})\}$ without loss of accuracy. Therefore, an optimal polarizability basis would allow fast and accurate GW calculations.

We construct an optimal representation in three steps:

i) we first express the Kohn–Sham orbitals, whose products enter the definition of P°, in terms of localized, Wannier-like, orbitals,

ii) we then construct a basis set of localized functions for the manifold spanned by products of Wannier orbitals,

iii) finally, this basis is further restricted to a set of approximate eigenvectors of P°, corresponding to eigenvalues larger than a given threshold.

Let us start from the RPA irreducible polarizability:

$$\tilde{P}^\circ(\mathbf{r}, \mathbf{r}'; i\omega) = \sum_{cv} \Phi_{cv}(\mathbf{r}) \Phi_{cv}(\mathbf{r}') \tilde{\chi}^\circ_{cv}(i\omega), \quad (4.15)$$

where

$$\tilde{\chi}^\circ_{cv}(i\omega) = 2\text{Re}\left(\frac{1}{i\omega - \varepsilon_c + \varepsilon_v}\right), \quad (4.16)$$

and

$$\Phi_{cv}(\mathbf{r}) = \psi_c(\mathbf{r})\psi_v(\mathbf{r}). \quad (4.17)$$

We express valence and conduction QPAs in terms of localized, orthonormal maximally localized Wannier functions [12, 14]:

$$\begin{aligned} u_s(\mathbf{r}) &= \sum_v \mathcal{U}_{vs}\psi_v(\mathbf{r})\theta(-\varepsilon_v) \\ v_s(\mathbf{r}) &= \sum_c \mathcal{V}_{cs}\psi_c(\mathbf{r})\theta(\varepsilon_c)\theta(E_C^2 - \varepsilon_c), \end{aligned} \quad (4.18)$$

where $E_C^2 \leq E_C^1$ is a second energy cutoff that limits a *lower conduction manifold* (LCM) to be used *only* in the construction of the polarizability basis and the $\mathcal{U}$ and $\mathcal{V}$ matrices are unitary.

We then reduce the number of product functions from the product, which scales quadratically with the system size, between the number of valence and the number of conduction states, to a number that scales linearly. Indeed, we have transformed the problem of calculating products in real space of delocalized (usually Kohn–Sham) orbitals in that of calculating products in real space of localized Wannier functions. We express the $\bar{\Phi}$'s as approximate linear combinations of products of the u's v's:

$$\bar{\Phi}_{cv}(\mathbf{r}) \approx \sum_{rs} \mathcal{O}_{cv,rs} W_{rs}(\mathbf{r}) \theta(|W_{rs}|^2 - s_1), \quad (4.19)$$

where:

$$\mathcal{O}_{cv,c'v'} = \mathcal{U}_{vv'} \mathcal{V}_{cc'}, \quad (4.20)$$

and the products in real space are given by:

$$W_{rs}(\mathbf{r}) = u_r(\mathbf{r}) v_s(\mathbf{r}), \quad (4.21)$$

and $|W_{cv}|$ is the L^2 norm of $W_{rs}(\mathbf{r})$, which is arbitrarily small when the centers of the u_r and v_s functions are sufficiently distant, and s_1 is an appropriate threshold.

The number of basis functions can be further reduced on account of the non-orthogonality of the W's. Indeed it is possible to obtain an orthonormal basis for representing the W's whose dimension can be significantly smaller that the number of retained W's. This is done through a procedure analogous to a singular value decomposition. We first define the overlap matrix:

$$Q_{\varrho\sigma} = \int W_\varrho(\mathbf{r}) W_\sigma(\mathbf{r}) \, d\mathbf{r}, \qquad (4.22)$$

where the ϱ and σ indices stand for pairs of rs indices. Then, we calculate the eigenvalues $\{q_v\}$ and eigenvectors $\{\mathcal{U}_v\}$ of the matrix Q. It should be noted that the matrix Q is always positive definite. The magnitude of the eigenvalues is a measure of the relevance of their corresponding eigenvectors. Indeed an orthonormal basis set which spans the space of the $\{W_\varrho\}$ is given by the states Φ:

$$\Phi_v(\mathbf{r}) = \frac{1}{\sqrt{q_v}} \sum_\varrho \mathcal{U}_{v\varrho} W_\varrho(\mathbf{r}). \qquad (4.23)$$

An optimal polarizability basis can be obtained by retaining those Φ's for which q_v is larger than a given threshold, s_2. We can now write:

$$\bar{\Phi}_{cv}(\mathbf{r}) \approx \sum_{\varrho' v'} \mathcal{O}_{cv,\varrho'} \sqrt{q_{v'}} \, \mathcal{U}_{v'\varrho'} \, \Phi_{v'}(\mathbf{r}), \qquad (4.24)$$

where the indices ϱ' and v' run only over the elements which have been retained according to the thresholds s_1 and s_2, respectively.

It is worth noting that the optimal polarizability basis vectors $\{\Phi_v\}$ are the (approximate) eigenvectors of the polarizability operator P' at zero time constructed with empty states only from the LCM:

$$P'(\mathbf{r}, \mathbf{r}') = \sum_{vc'} \bar{\Phi}_{vc'}(\mathbf{r}) \bar{\Phi}_{vc'}(\mathbf{r}'), \qquad (4.25)$$

where c' indicates the empty states belonging to the LCM. As the $\mathcal{U}$ and $\mathcal{V}$ matrices are unitary, it holds:

$$P'(\mathbf{r}, \mathbf{r}') \approx \sum_{\varrho'} W_{\varrho'}(\mathbf{r}) W_{\varrho'}(\mathbf{r}'). \qquad (4.26)$$

From this equation and from Eq. (4.24), it is easy to show that:

$$\int d\mathbf{r}' P'(\mathbf{r}, \mathbf{r}') \Phi_v(\mathbf{r}') \approx q_v \Phi_v(\mathbf{r}). \qquad (4.27)$$

This means that the construction of the polarizability basis selects the most important eigenvectors of the polarizability at least at zero time. We have verified, however, that the manifold spanned by the most important eigenvectors of P° in the (imaginary) time domain depends very little on time, which permits the use of a same basis at different frequencies. We have also verified that although the polarizability basis has been constructed only with empty states from the LCM, it behaves very well also for representing polarizability operators constructed with much more complete sets of empty states.

It should be noted that equivalent optimal polarizability basis sets could be constructed by choosing $s_1 = 0$ and by considering directly products of Kohn–Sham orbitals without trasforming them into localized Wannier functions. Going through Wannier functions and discarding small overlaps permits only to speed up the construction of the polarizability basis set. Indeed, this results into a $O(N^3)$ process instead of a $O(N^4)$ process. This means that also for systems presenting delocalized orbitals it will always be possible to obtain optimal polarizabilty basis sets. However, in the limit case in which Kohn–Sham orbitals are simply plane waves, the optimal polarizability basis will be simply a basis of plane-waves. Hence, we expect to find larger benefits from the use of optimal polarizability basis sets in the case of isolated materials and in that of extended insulators, while in the limit of small gap extended systems we do not expect to find significant improvements with respect to the use of plane-waves basis sets.

Once an optimal basis set has been identified, an explicit representation for the irreducible polarizability,

$$\tilde{P}^\circ(\mathbf{r},\mathbf{r}';i\omega) = \sum_{\mu\nu} \tilde{P}^\circ_{\mu\nu}(i\omega)\Phi_\mu(\mathbf{r})\Phi_\nu(\mathbf{r}'), \tag{4.28}$$

is obtained. By equating Eq. (4.15) to Eq. (4.28) and taking into account the orthonormality of the Φ's, one obtains:

$$\tilde{P}^\circ_{\nu\mu}(i\omega) = \sum_{cv} T_{cv,\mu} T_{cv,\nu} \tilde{\chi}^\circ_{cv}(i\omega) \theta(E_C^1 - \varepsilon_c), \tag{4.29}$$

with

$$T_{cv,\mu} = \int \bar{\Phi}_{cv}(\mathbf{r}) \Phi_\mu(\mathbf{r}) \, d\mathbf{r}, \tag{4.30}$$

where the index c runs over all the empty states defined by the cutoff E_C^1. Finally, a representation for Π is obtained by simple matrix manipulations.

While isolated system can be easily treated by applying in Eq. (4.14) a truncated form of the Coulomb potential [18], extended ones require some additional steps which we briefly introduce here. Note that in the present work the Brillouin zone is generally sampled at the Γ-point only. First, it is convenient to introduce the frequency dependent symmetric dielectric matrix [19]:

$$\tilde{\varepsilon}^{sym}(i\omega) = 1 - v^{1/2} \cdot \tilde{P}^\circ(i\omega) \cdot v^{1/2}, \tag{4.31}$$

where v is the Coulomb interaction. From ε^{sym} the screened Coulomb interaction W is given by:

$$\tilde{W}^\circ(i\omega) = v^{1/2} \cdot \tilde{\varepsilon}^{sym,-1}(i\omega) \cdot v^{1/2}. \tag{4.32}$$

Because of the long-range character of the Coulomb interaction, the long-wavelength components, the "head" ($\mathbf{G} = \mathbf{G}' = 0$) and "wings" ($\mathbf{G} = 0$, $\mathbf{G}' \neq 0$), of $\tilde{\varepsilon}^{sym}(i\omega)$ cannot be neglected. As the optimal polarizability basis is orthogonal to the $\mathbf{G} = 0$ component, we calculate $\varepsilon^{sym}(i\omega)$ on the representation of the optimal polarizability basis plus the **G**=0 vector. This is done by calculating the head and wings terms at frequency $i\omega$ using a linear response approach [20], where optionally the Brillouin zone can be sampled with denser meshes of k-points [21], and by projecting the wings over the polarizability basis functions. Then, we extract from $\tilde{W}$ the long-range part behaving as v:

$$\tilde{W}^\circ(i\omega) = \tilde{\varepsilon}^{sym,-1}_{G=0\,G'=0}(i\omega)v$$
$$+ \sum_{\mu\nu} v^{1/2}|\Phi_\mu\rangle(\tilde{\varepsilon}^{sym,-1}_{\mu\nu}(i\omega) - \delta_{\mu,\nu}\tilde{\varepsilon}^{sym,-1}_{G=0G'=0}(i\omega))\langle\Phi_\nu|v^{1/2}. \quad (4.33)$$

The contribution to Σ_C due to the long-range part of W is then given by:

$$\langle\Sigma_C^{lr}(i\tau)\rangle_n \approx \mp \sum_l \int d\mathbf{r}\,d\mathbf{r}'\,e^2 \frac{\psi_n(\mathbf{r})\psi_l(\mathbf{r})\psi_l(\mathbf{r}')\psi_n(\mathbf{r}')}{|\mathbf{r}-\mathbf{r}'|} \times (\varepsilon^{sym,-1}_{G=0G'=0}(i\tau)-1)\,e^{\varepsilon_l\tau}\theta(E_C^1-\varepsilon_l). \quad (4.34)$$

As the calculation of such terms closely resembles the evaluation of exchange terms, we calculate them using the scheme introduced in Ref. [22], optionally using a denser sampling of the BZ. Finally, the contribution to Σ_C due to the short-range part of W is given by:

$$\langle\Sigma_C^{sr}(i\tau)\rangle_n \approx \mp \sum_{\mu\nu'\mu\nu'} e^{\varepsilon_l\tau} S_{nl,\mu} S_{nl,\nu}\theta(E_C^1-\varepsilon_l)$$
$$\times v^{-1/2}_{\mu\mu'}(\tilde{\varepsilon}^{sym,-1}_{\mu'\nu'}(i\tau)-\delta_{\mu',\nu'}\tilde{\varepsilon}^{sym,-1}_{G=0G'=0}(i\tau))v^{-1/2}_{\nu'\nu}, \quad (4.35)$$

where the operator v is calculated first on the polarizability basis:

$$v_{\mu\nu} = \langle\Phi_\mu|v|\Phi_\nu\rangle. \quad (4.36)$$

The evaluation of Eq. (4.36) does not present any difficulty as the polarizability basis functions Φ's are orthogonal to the **G**=0 vector.

4.4
Implementation and Validation

Our scheme has been implemented in the QUANTUM-ESPRESSO density functional package [23], for norm-conserving as well as ultra-soft [24] pseudopotentials, resulting in a new module called *gww.x* which uses a Gauss–Legendre discretization of the imaginary time/frequencies half-axes, and that is parallelized accordingly. In the following examples, DFT calculations were performed using the energy functional from Ref. [25] and pseudo-potentials have been taken from the Quantum-Espresso tables [23]. We used an imaginary time cutoff of 10 a. u., an imaginary frequency cutoff of 20 Ry, and grids of 80 steps in both cases. The self-energy was analytically continued using a two poles formula [17].

4.4.1
Benzene

We first illustrate our scheme by considering an isolated benzene molecule in a periodically repeated cubic cell with an edge of 20 a.u. using a first conduction energy cutoff $E_C^1 = 56.7$ eV, corresponding to 1000 conduction states, and a threshold on the norm of Wannier products $s_1 = 0.1$ a.u. We used the norm-conserving pseudopotentials: C.pz-vbc and H.pz-vbc. The wavefunctions and the charge density were expanded on plane waves, defined by kinetic energy cutoffs of 40 and 160 Ry, respectively. In Figure 4.1 we display the dependence of the calculated ionization potential (IP) on the second conduction energy cutoff used to define the polarization basis, E_C^2, and on the threshold on the eigenvalues of the overlap matrix between Wannier products, s_2. Convergence within 0.01 eV is achieved with a conduction energy cutoff E_C^2 smaller than 30 eV (less than 300 states) and a polarizability basis set of only ~400 elements. The convergence of other QPEs is similar.

In Figure 4.2, we display the convergence of the IP with respect to E_C^1, which turns out to be quite slow. These data can be accurately fitted by the simple formula:

$$\text{IP}(E_C^1) = \text{IP}(\infty) + \frac{A}{E_C^1}, \tag{4.37}$$

resulting in a predicted ionization potential $\text{IP}(\infty) = 9.1$ eV, in good agreement with the experimental value of 9.3 eV [26].

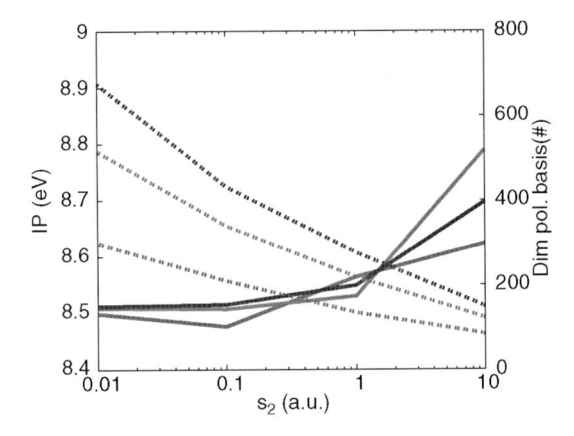

Figure 4.1 (online color at: www.pss-b.com) Calculated ionization potential of the benzene molecule (solid lines, left scale) and dimension of the polarization basis (dashed lines, right scale) *versus* the s_2 threshold. The polarization basis has been constructed with a conduction energy cutoff $E_C^2 = 16.7$ eV (red-grey, 100 states), $E_C^2 = 28.6$ eV (green-light grey, 300 states), and $E_C^2 = 38.3$ eV (blue-black, 500 states).

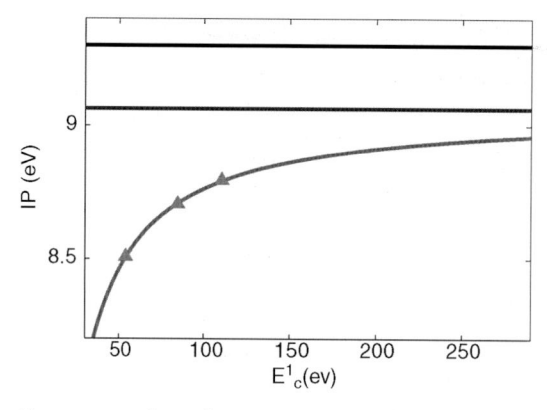

Figure 4.2 (online color at: www.pss-b.com) Calculated ionization potential as a function of the overall conduction energy cutoff, E_C^1. Black line: experimental value; red line: fit to the calculated values (green triangles); blue line extrapolated value. See text for more details.

4.4.2
Bulk Si

In order to demonstrate our scheme for extended systems, we consider crystalline silicon treated using a 64-atom simple cubic cell at the experimental lattice constant and sampling the corresponding Brillouin zone (BZ) using the Γ-point only. This gives the same sampling of the electronic states as would result from six points in the irreducible wedge of the BZ of the elementary 2-atom unit cell. We used an norm-conserving pseudopotential: Si.pz-rrkj. The wavefunctions and the charge density were expanded on plane waves, defined by kinetic energy cutoffs of 18 and 72 Ry, respectively. Then, the GW calculations were performed using $E_C^1 = 94.6$ eV (corresponding to 3200 conduction states) and $E_C^2 = 33.8$ eV (corresponding to 800 states in the LCM), $s_1 = 1.0$ a.u. and two distinct values for s_2 (0.01 and 0.001). For calculating the head and wing terms of the symmetric dielectric matrix we used a $4 \times 4 \times 4$ grid for sampling the BZ of the 64-atom cubic cell. Then, for calculating the long-range contribution to the self-energy given in Eq. (4.34), we used a $2 \times 2 \times 2$ grid. In Table 4.1 we summarize our results and compare them with previous theoretical results, as well as with experiments. An overall convergence within a few tens meV is achieved with a s_2 cutoff of 0.001 a.u., corresponding to a polarizability basis of ~6500 elements. The residual small discrepancy with respect to previous results [17] is likely due to our use of a supercell, rather than the more accurate k-point sampling used in previous works.

4.4.3
Vitreous Silica

Our ability to treat large supercells give us the possibility to deal with disordered systems that could hardly be addressed using conventional approaches. In Figure 4.3

Table 4.1 QPEs (eV) calculated in crystalline silicon and compared with experimental (as quoted in Ref. [17]) and previous theoretical results [17].

	Th$_1$	Th$_2$	prev th	expt
N_P	4847	6510		
Γ_{1v}	−11.45	−11.49	−11.57	−12.5 ± 0.6
X_{1v}	−7.56	−7.58	−7.67	
X_{4v}	−2.79	−2.80	−2.80	−2.9, −3.3 ± 0.2
Γ'_{25c}	0.	0.	0.	0.
X_{1c}	1.39	1.41	1.34	1.25
Γ_{15c}	3.22	3.24	3.24	3.40, 3.05
Γ'_{2c}	3.87	3.89	3.94	4.23, 4.1

"Th$_1$" and "Th$_2$" indicate calculations made with $s_2 = 0.01$ and $s_2 = 0.001$ a.u., respectively, while N_P is the dimension of the polarization basis.

we show the QPE density of states (DOS) as calculated for a 72-atom model of vitreous silica [27]. We used a norm-conserving pseudopotential for Si (Si.pz-vbc) and an ultrasoft [24] one (O.pz-rrkjus) for O. The wavefunctions and the charge density were expanded on plane waves, defined by kinetic energy cutoffs of 24 and 200 Ry, respectively. We used $E_C^1 = 48.8$ eV (corresponding to 1000 conduction states), $E_C^2 = 30.2$ eV (corresponding to 500 states in the LCM), $s_1 = 1$ a.u. and $s_2 = 0.1$ a.u. (giving rise to a polarization basis of 3152 elements). We checked the convergence with respect to the polarization basis by considering $s_2 = 0.01$ a.u. which leads to a basis of 3933 elements. Indeed, the calculated QPEs differ in average by only 0.01, eV with a maximum discrepancy of 0.07 eV. The QP band-gap resulting from our calculations is 8.5 eV, to be compared with an experimental value of ∼9 eV [28] and with a significantly lower value predicted by DFT in the local-density approximation (5.6 eV).

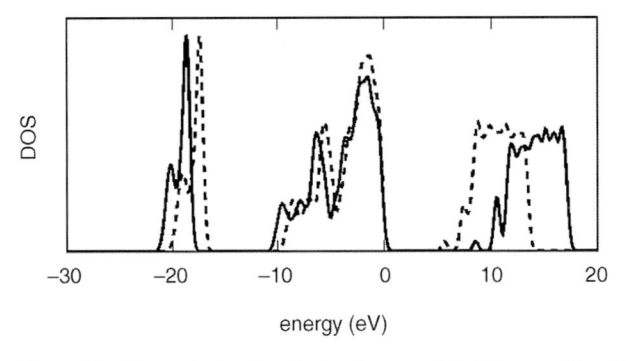

Figure 4.3 Electronic density of states for a model of vitreous silica: LDA (dashed line) and GW (solid line). A Gaussian broadening of 0.25 eV has been used. The top of the valence band has been aligned to 0 eV.

4.5
Example: Point Defects in a-Si$_3$N$_4$

Amorphous silicon nitride (a-Si$_3$N$_4$) is being widely studied as its mechanical and electronic properties lead to a wide range of applications [29] In microelectronics, amorphous silicon nitride (a-Si$_3$N$_4$) is used to fabricate insulating layers in triple oxide-nitride-oxide structures [30]. In particular, because of its high concentration of charge traps, a-Si$_3$N$_4$ is employed as charge storage layer in non-volatile memory devices [31]. Moreover, silicon nitride based materials are nowadays proposed for optoelectronic devices [32]. Due to the non-trivial nature of its structures, first-principles methods become very important for investigating its properties at the atomistic scale [33]. We review here how our gww method permitted to investigate the electronic structure of quasi-stoichiometric a-Si$_3$N$_4$ addressing a 152-atoms model structure [33].

4.5.1
Model Generation

In a-Si$_3$N$_4$ silicon atoms are fourfold coordinated forming almost regular SiN$_4$ tetrahedra. The latter are connected by corners in such a way that each N atom is shared by three tetrahedra. Nitrogen atoms are threefold coordinated, with the silicon neighbors arranged at the vertexes of a planar triangle. This results in a quite rigid network structure. Furthermore the a-Si$_3$N$_4$ network is supposed to contain not only corner-sharing but also edge-sharing SiN$_4$ tetrahedra [33, 34].

We generated a model of a-Si$_3$N$_4$ through first-principles molecular dynamics using the DFT approach and the exchange and correlation functional of Ref. [25]. Core-valence interactions were described through ultrasoft pseudopotentials [24] for N and H atoms and through a normconserving pseudopotential for Si atoms. The electronic wavefunctions and the charge density were expanded using plane waves basis sets defined by energy cutoffs of 25 and 200 Ry, respectively. The Brillouin's zone was sampled at the Γ-point. The model structure was generated through first-principles molecular dynamics starting from a diamond-cubic model of crystalline silicon which was changed into Si$_3$N$_4$ by addition of N atoms at intermediate distances between Si–Si neighbors. The initial model structure contained 64 Si and 86 N atoms in a periodically repeated cubic cell. A composition ratio $r=$ [N]/[Si] of 1.34 was chosen slightly differing from the ideal stoichiometry in order to trigger the formation of defects. We set up the density to the experimental value of 3.1 g/cm^3 [35]. Car and Parrinello [36] molecular dynamics runs were then performed for obtaining the model of a-Si$_3$N$_4$. First the system was thermalized at the temperature of 3500 K for 12 ps using a Nosé–Hoover thermostat [37]. Successively, the sample was quenched for 5 ps down to 2000 K below the theoretical melting point. Finally, the structural geometry was further optimized by a damped molecular dynamics run. As the model presented an empty state close to the top of the valence band, we passivated it by adding to the structure two H atoms in proximity of the two Si atoms which were threefold coordinated [38]. After structural relaxation, the H atoms moved close to two near N sites. We note that the structural and electronic properties of our model were only marginally affected by the addition of the two H atoms.

4.5.2
Model Structure

We report a picture of the final model structure in Figure 4.4. The main structural parameters are reported in Table 4.2. The average Si–N bond length equals to 1.730 Å with a standard deviation (std) of 0.060 Å. This value is found to be in excellent agreement with the experimental bond length of 1.729 Å [39]. The structure shows well-defined SiN_4 tetrahedral units. The average N–Si–N angle equals 109.1° with a standard deviation of 13°. This is very close to the ideal angle of 109.47° for regular tetrahedra. Moreover our structure shows also well-defined quasi-planar NSi_3 units. The average Si–N–Si angle equals 117.2° with a standard deviation of 15°. This is consistent with the value of 120° for regular planar NSi_3 units.

The amount of SiN_4 tetrahedra and NSi_3 triangular units is reported in Table 4.3 where we give the coordination numbers in the first-neighbor shells of Si and N atoms, together with the relative Si–N bond length averages. The majority of Si atoms is fourfold coordinated and shows an average Si–N bond length of 1.73 Å. Few Si

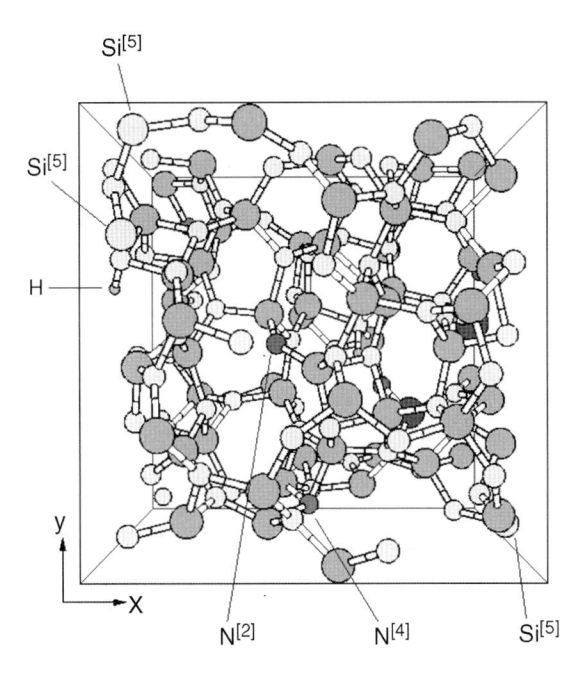

Figure 4.4 (online color at: www.pss-b.com) Balls-and-sticks picture of the $a\text{-}Si_3N_4$ model. Si atoms and N atoms are colored with dark and light gray, respectively. Threefold and fivefold coordinated Si atoms are colored in purple and yellow, respectively. Twofold and fourfold coordinated N atoms are colored in red and green. Hydrogen are colored in pink.

Table 4.2 Structural properties of our model of a-Si_3N_4 and reference values: average Si–N–Si and N–Si–N angles, and average bond length d_{SiN}.

	∠Si–N–Si	∠N–Si–N	d_{SiN} (Å)
model	117.2° (15.1°)	109.1° (13.0°)	1.73 (0.06)
ref.	120°·a)	109.47°·a)	1.729[b]

The respective standard deviations are given in parenthesis.
a) Ideal bonding geometry.
b) Expt. (Ref. [39]).

atoms are three- or fivefold coordinated. Correspondingly, almost all nitrogen atoms are bound to three silicon atoms and only a few show two- or fourfold coordination. Consequently, our model shows at short-range high topologic and chemical order. We want now to understand the role of the point defects on the electronic structure.

4.5.3
Electronic Structure

For obtaining the polarizability basis we used a cutoff $E_C^2 = 30$ eV, corresponding to 750 empty states in the LCM, and thresholds $s_1 = 2$. a.u. and $s_2 = 0.1$. This gave rise to a polarization basis of 5867 elements. Then for the obtaining the self-energy we chose a cutoff $E_C^1 = 45$ eV, corresponding to 1500 empty states.

We show in Figure 4.5a the electronic DOS for our model calculated with the GW approach together with the partial densities of s and p states for the Si and N atoms. The lowest part of the valence band mainly arises from N 2s states. While the low-energy side of the upper part of the valence band results from the Si–N bonds, formed by Si sp^3 and N 2p orbitals, the high-energy side, which defines the top of the valence band, consists of N 2p lone pairs. The low-energy side of the conduction band mainly

Table 4.3 Composition of first-neighbor shells in our model of a-Si_3N_4.

composition	n_{Si}	d_{SiN}
Si[3]	2	1.64 (0.05)
Si[4]	59	1.73 (0.05)
Si[5]	3	1.81 (0.09)

composition	n_N	d_{SiN}
N[2]	3	1.61 (0.03)
N[3]	79	1.73 (0.05)
N[4]	2	1.85 (0.09)
NSi_3H	2	1.78 (0.02)

Coordination numbers of Si and N atoms are indicated by the superscript number in square brackets. The number of Si and N atoms found in our model for each coordination are indicated by n_{Si} and n_N. Average Si–N bond length d_{SiN} (Å) together with its standard deviation (in parentheses) is given for each composition. We used cutoff radii of 2.2 Å.

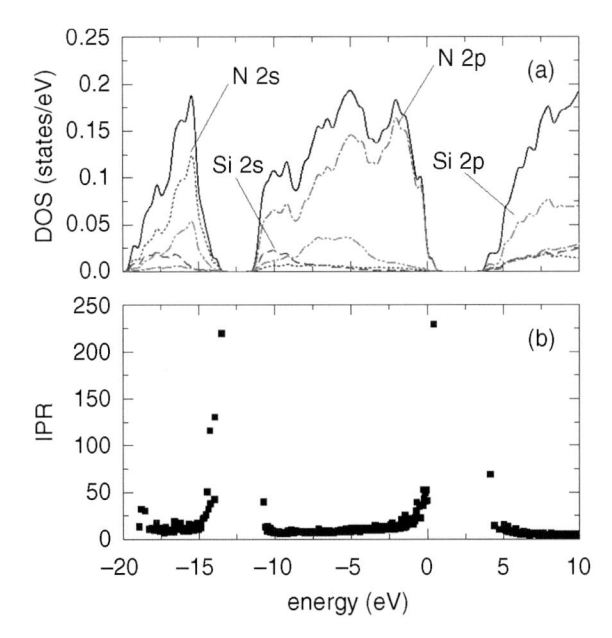

Figure 4.5 (online color at: www.pss-b.com) (a) Electronic density of states (black) and partial DOS obtained by projecting electronic states onto N 2s (blue/dotted), N 2p (red/dot-dashed), Si 2s (purple/dashed), and Si 2p (green/double dot-dashed). The highest occupied state is aligned at 0 eV. Gaussian broadening of 0.25 eV is used. GW energies are used. (b) Inverse participation ratio (IPR) of electronic states in silicon nitride.

consists of antibonding states associated to the Si–N bond. We note that the origin of the bands is analogous to the cases of SiO_2 (Ref. [40]) and GeO_2 (Ref. [41]), reflecting the common type of short-range arrangement of atoms based on the tetrahedral unit. Similar conclusions were obtained for the electronic DOS calculated through an approximate density-functional scheme [42] and through a tight binding approach [43]. Moreover, the calculated valence band is consistent with photoemission spectra [44].

We focus now on the role played by the defects in the DOS. In Figure 4.6a we give the partial DOS obtained by projecting the electronic states onto the 1s orbitals of the two H atoms of our model structure. The partial DOS of H atoms constitutes a very small contribution to the total DOS of Figure 4.5. In Figure 4.6b and c we show the partial DOS obtained by projecting the electronic states onto the 2p and 2s orbitals of the $N^{[2]}$ and $N^{[4]}$ atoms and of the $Si^{[3]}$ and $Si^{[5]}$ atoms. The partial DOS of $N^{[4]}$ atoms and $Si^{[5]}$ atoms do not show features localized near the band edges. At variance, the partial DOS of the $N^{[2]}$ atoms shows a sharp peak at the top of the valence band, while the partial DOS of $Si^{[3]}$ atoms exhibits sharp peaks close to the bottom of the conduction band [45]. As Figure 4.6 illustrates, these peaks are originated by N and Si 2p orbitals of $N^{[2]}$ and $Si^{[3]}$ atoms, respectively. Furthermore, the topmost occupied electronic state and the first empty electronic state are spatially localized around a $N^{[2]}$ atom and around a $Si^{[3]}$ atom, respectively. By excluding these two defect states, we

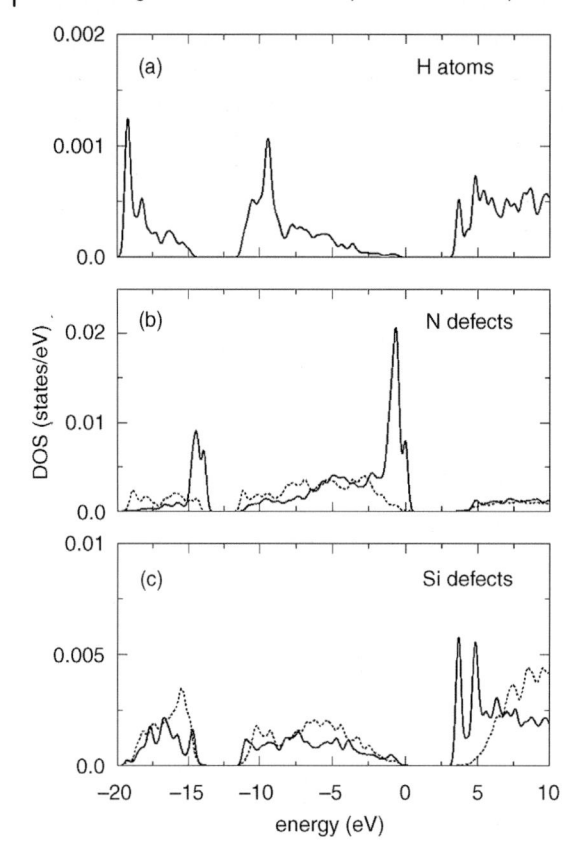

Figure 4.6 Partial DOS obtained by projecting the electronic states onto (a) 1s orbitals of H atoms, (b) 2p and 2s orbitals of twofold (solid) and fourfold (dotted) coordinated N atoms, (c) 2p and 2s orbitals of threefold (solid) and fivefold (dotted) coordinated Si atoms. The highest occupied state is aligned at 0 eV. Gaussian broadening of 0.25 eV is used. GW energies are used.

found a HOMO–LUMO band gap of 4.42 eV in excellent agreement with the experimental value of 4.55 eV of the optical band gap of sputtered a-Si_3N_4 given in Ref. [46]. However, we note that the band gap is quite sensitive to the adopted production method and for CVD samples is about 5.3 eV [46]. Yet, the GW method appears to correctly describe the electronic DOS where simpler LDA calculations fail giving for our model structure a HOMO–LUMO band-gap of only 2.9 eV, as typical for LDA calculations in silicon nitride [38, 42].

We now analyze the degree of localization of the electronic states. The localization of an electronic state ψ_n can be quantified by the inverse participation ratio (IPR) [38, 47]:

$$\text{IPR}_n = \Omega \frac{\int d\mathbf{r} |\psi_n(\mathbf{r})|^4}{|\int d\mathbf{r} |\psi_n(\mathbf{r})|^2|^2}, \tag{4.38}$$

where Ω is the volume of the simulation cell. The larger the IPR the more localized is the electronic state, so that highly localized/delocalized states show a large/small IPR. For completely delocalized states the IPR is equal to unity. In Figure 4.5b we show the IPR for the electronic states of our model of silicon nitride. We note that the states close to the band edges corresponding to $N^{[2]}$ and $Si^{[3]}$ defects result much more localized than the other electronic states. These results are consistent with the IPR data previously calculated for a-SiN_x in Ref. [38].

4.6
Conclusions

We have shown how the use of optimal basis sets for representing the polarizability operator permits to achieve a significant speed up of GW calculations, allowing the study of large model structures up to a few hundreds of atoms. Therefore it is appealing to use such scheme for investigating the electronic structure of defects as density-functional approaches result not to be adequate. The main limitation still present in our approach is the need of summing over a large number of empty states. For a discussion of this point and the presentation of a solution we indicate to the reader Ref. [15].

Acknowledgements

We thank C. Cavazzoni for his help in the parallelization of the code. Part of the calculations were performed using the computational facilities of CINECA.

References

1 See e.g., Martin, R.M. (2004) *Electronic Structure* Cambridge University Press, Cambridge, and references quoted therein.
2 Runge, E. and Gross, E.K.U. (1984) *Phys. Rev. Lett.*, **52**, 997. Marques, M.A.L., Ullrich, C.L., Nogueira, F., Rubio, A., Burke, K., and Gross, E.K.U. (2006) *Time-Dependent Density Functional Theory, Lecture Notes in Physics*, Vol. 706 Springer-Verlag, Berlin, Heidelberg.
3 Hedin, L. and Lundqvist, S. (1969) *Solid State Phys.*, **23**, 1.
4 Aryasetiawan, F. and Gunnarsson, O. (1998) *Rep. Prog. Phys.*, **61**, 237.
5 Hedin, L. (1965) *Phys. Rev.*, **139**, 796.
6 Hybertsen, M.S., and Louie, S.G. (1985) *Phys. Rev. Lett.*, **55**, 1418.
7 Onida, G., Reining, L., and Rubio, A. (2001) *Rev. Mod. Phys.*, **74**, 601.
8 Umari, P., Stenuit, G., and Baroni, S. (2009) *Phys. Rev. B*, **78**, 201104.
9 Reining, L., Onida, G., and Godby, R.W. (1997) *Phys. Rev. B*, **56**, R4301.
10 Steinbeck, L., Rubio, A., Reining, L., Torrent, M., Whited, I.D., and Godby, R.W. (2000) *Comput. Phys. Commun.*, **125**, 105.
11 Bruneval, F. and Gonze, X. (2008) *Phys. Rev. B*, **78**, 085125.
12 Marzari, N. and Vanderbilt, D. (1997) *Phys. Rev. B*, **56**, 12847.
13 Souza, I., Marzari, N., and Vanderbilt, D. (2001) *Phys. Rev. B*, **65**, 035109.
14 Gygi, F., Fattebert, J.-L., and Schwegler, E. (2003). *Comput. Phys. Commun.*, **155**, 1.
15 Umari, P., Stenuit, G., and Baroni, S. (2010) *Phys. Rev. B*, **81**, 115104.

16 Shaltaf, R., Rignanese, G.-M., Gonze, X., Giustino, F., and Pasquarello, A. (2008) *Phys. Rev. Lett.*, **100**, 186401.

17 Rojas, H.N., Godby, R.W., and Needs, R.J. (1995) *Phys. Rev. Lett.*, **74**, 1827. Rieger, M.M., Steinbeck, L., White, I.D., Rojas, H.N., and Godby, R.W. (1999) *Comput. Phys. Commun.*, **117**, 211.

18 Onida, G., Reining, L., Godby, R.W., Del Sole, R., and Andreoni, W. (1995) *Phys. Rev. Lett.*, **75**, 818.

19 Baroni, S. and Resta, R. (1986) *Phys. Rev. B*, **33**, 7017.

20 Baroni, S., Giannozzi, P., and Testa, A. (1987) *Phys. Rev. Lett.*, **58**, 1861.

21 Umari, P. and Pasquarello, A. (2003) *Phys. Rev. B*, **68**, 085114.

22 Spencer, J. and Alavi, A. (2008) *Phys. Rev. B*, **77**, 193110.

23 Giannozzi, P., Baroni, S., Bonini, N., Calandra, M., Car, R., Cavazzoni, C., Ceresoli, D., Chiarotti, G.L., Cococcioni, M., Dabo, I., Dal Corso, A., de Gironcoli, S., Fabris, S., Fratesi, G., Gebauer, R., Gerstmann, U., Gougoussis, C., Kokalj, A., Lazzeri, M., Martin-Samos, L., Marzari, N., Mauri, F., Mazzarello, R., Paolini, S., Pasquarello, A., Paulatto, L., Sbraccia, C., Scandolo, S., Sclauzero, G., Seitsonen, A.P., Smogunov, A., Umari, P., and Wentzcovitch, R.M. (2009) *J. Phys.: Condens. Matter*, **21**, 395502.

24 Vanderbilt, D. (1990) *Phys. Rev. B*, **41**, 7892.

25 Perdew, J.P. and Zunger, A. (1981) *Phys. Rev. B*, **23**, 5048.

26 Lipari, N.O., Duke, C.B., and Pietronero, L. (1976) *J. Chem. Phys.*, **65**, 1165.

27 Sarnthein, J., Pasquarello, A., and Car, R. (1995) *Phys. Rev. Lett.*, **74**, 4682; (1995) *Phys. Rev. B*, **52**, 12690.

28 Himpsel, F.J. (1986) *Surf. Sci.*, **168**, 764. Grunthaner, F.J., and Grunthaner, P.J. (1986) *Mater. Sci. Rep.*, **1**, 65.

29 Katz, R.N. (1980) *Science* **208**, 841. Liu, A.Y. and Cohen, M.L. (1990) *Phys. Rev. B*, **41**, 10727.

30 Ma, Y., Yasuda, T., and Lucovsky, G. (1994) *Appl. Phys. Lett.*, **64**, 2226.

31 Aozasa, H., Ujiwara, I.F., Nakamura, A., and Komatsu, Y. (1999) *Jpn. J. Appl. Phys.*, **38**, 1441. Bachhofer, H., Reisinger, H., Bertagnolli, E., and von Philipsborn, H. (2001) *J. Appl. Phys.*, **89**, 2791.

32 Dal Negro, L., Hamel, S., Zaitseva, N., Yi, J.H., Williamson, A., Stolfi, M., Michel, J., Galli, G., and Kimerling, L.C. (2006) *IEEE J. Sel. Top. Quantum Electron.*, **12**, 1151. Dal Negro, L., Yi, J.H., Michel, J., Kimerling, L.C., Hamel, S., Williamson, A., and Galli, G. (2006) *IEEE J. Sel. Top. Quantum Electon.* **12**, 1628. Dal Negro, L., Yi, J.H., Kimerling, L.C., Hamel, S., Williamson, A., and Galli, G. (2006) *Appl. Phys. Lett.*, **88**, 183103.

33 Giacomazzi, L. and Umari, P. (2009) *Phys. Rev. B*, **80**, 144201.

34 Kroll, P. (2001) *J. Non-Cryst. Solids* **293–295**, 238.

35 Sze, S.M. (1981) *Physics of Semiconductor Devices* John Wiley & Sons, Inc., New York.

36 Car, R. and Parrinello, M. (1985) *Phys. Rev. Lett.*, **55**, 2471.

37 Nosé, S. (1984) *Mol. Phys.*, **52**, 255. Hoover, W.G. (1985) *Phys. Rev. A*, **31**, 1965.

38 Justo, J.F., de BritoMota, F., and Fazzio, A. (2002) *Phys. Rev. B*, **65**, 073202. de Brito Mota, F., Justo, J.F., and Fazzio, A. (2002) *Braz. J. Phys.*, **32**, 436.

39 Misawa, M., Fukunaga, T., Niihara, K., Hirai, T., and Suzuki, K. (1979) *J. Non-Cryst. Solids* **34**, 313.

40 Binggeli, N., Troullier, N., Martins, J.L., and Chelikowsky, J.R. (1991) *Phys. Rev. B*, **44**, 4771. Giustino, F. and Pasquarello, A. (2006) *Phys. Rev. Lett.*, **96**, 216403.

41 Giacomazzi, L., Umari, P., and Pasquarello, A. (2006) *Phys. Rev. B*, **74**, 155208.

42 Alvarez, F. and Valladares, A.A. (2002) *Rev. Mex. Fís.*, **48**, 528.

43 Sorokin, A.N., Karpushin, A.A., Gritsenko, V.A., and Wong, H. (2008) *J. Non-Cryst. Solids,* **354**, 1531.

44 Kärcher, R., Ley, L., and Johnson, R.L. (1984) *Phys. Rev. B*, **30**, 1896.

45 Martn-Moreno, L., Martnez, E., Vergés, J.A., and Yndurain, F. (1987) *Phys. Rev. B*, **35**, 9683. Lin, S.-Y. (2003) *Opt. Mater.* **23**, 93.

46 Bauer, J. (1977) *Phys. Status Solidi*, **39**, 411.

47 Elliott, S.R. (1984) *Physics of Amorphous Materials* Longman, New York, p. 194.

5
Calculation of Semiconductor Band Structures and Defects by the Screened Exchange Density Functional

S. J. Clark and John Robertson

5.1
Introduction

The local density approximation (LDA) is an efficient method to solve the eigenvalue equation of the many-body electronic Hamiltonian. The simplicity of LDA is to represent the exchange–correlation energy E_{xc} as a functional of the electron density, not the wave function. However, DFT under estimates the band gap of semiconductors and insulators [1]. Typically, the error is 30%, but the error can be 70% in cases like ZnO or even negative gaps in cases like InAs. This causes severe problems in the treatment of point defects and for heterojunction interfaces.

Various methods can be used to correct the band gap problem of the LDA and generalized gradient approximations (GGA). The first method is the scissors operator [2], in which the conduction band is just rigidly shifted upwards to fit the experimental band gap. This is not suitable for defect calculations.

Another correction method is the self-interaction correction (SIC) method of Perdew and Zunger [3] but this is still difficult to implement [4]. The GW method is widely used to provide accurate band structures [5–10], but it is very expensive, is usually too costly for defect calculations using supercells, and it cannot be used variationally to find geometries.

The LDA + U method has been used to correct band structures in open shell systems, where an on-site repulsion energy U is used to open up a gap between spin-up and spin down electrons [11, 12]. However, this method is only valid for open-shell systems. It is not valid for the standard closed shell semiconductors and insulators. For semiconductors with shallow core states such as ZnO, the LDA + U method can be used to partly correct the band gap, by forcing the Zn 3d states down, and thus reducing their repulsion from below on the valence band maximum state [13]. Any use of LDA + U to fit the band gap of closed shell systems will require an unphysical value of U.

The local exchange and correlation functionals of LDA and GGA lead to a spurious electronic self-interaction. The HF method uses a non-local exchange, so that it can be

---

Advanced Calculations for Defects in Materials: Electronic Structure Methods, First Edition.
Edited by Audrius Alkauskas, Peter Deák, Jörg Neugebauer, Alfredo Pasquarello, and Chris G. Van de Walle.
© 2011 Wiley-VCH Verlag GmbH & Co. KGaA. Published 2011 by Wiley-VCH Verlag GmbH & Co. KGaA.

self-interaction free, but HF lacks electronic correlation, and its exchange is unrealistically long-ranged due to an absence of screening.

In the late 1990s, it was realized that mixing in a fraction of Hartree–Fock exchange into the LDA exchange–correlation energy could be used to correct the band gap error and make the eigenvalues of the Kohn–Sham equations equal to the quasi-particle energies. Becke [14] gave arguments based on an adiabatic linkage of the HF and LDA limits that 25% is an appropriate amount of HF exchange to mix into the local exchange–correlation functional. This gave rise to the so-called hybrid functionals, such as B3LYP [14] and the PBEh [15] hybrid, formerly known as PBE0, in which 25% of Fock exchange is substituted into the LDA of E_{xc}. Muscat et al. [16] found that B3LYP gave good values for the band gaps of various semiconductors. The PBEh also mixes in 25% of HF exchange, and gives reasonable band gaps [17].

The HF exchange is unrealistically long-range due to an absence of screening, and is divergent for a plane wave basis. This led to the development of the screened hybrid functionals of Heyd–Scuseria–Erzenhof (HSE) [18–21] by separating the non-local HF exchange into long and short-range parts, and replacing a fraction (again, $\alpha = 0.25$) of the short-range parts of the LDA exchange with HF exchange. This is based on the notion that the exchange and correlation terms cancel at long range. The retention of only short-range HF exchange allows faster calculations. HSE is a variational functional that can be used for energy minimization. HSE was implemented for a local orbital basis and it has been tested on various molecules and solids [20, 21], and later for a plane wave basis with projector augmented waves [22].

Earlier and in a similar way, Bylander and Kleinman [23] proposed a similar separation of long and short ranged parts of the screened exchange (SX) [5]. They represented the exchange interaction by a Thomas–Fermi screened Coulomb potential, and use the LDA correlation. It has similar attributes to HSE. Seidel et al. [24] realized that SX was a variational functional, and so it was suitable for geometry optimization. Previously, Freeman and coworkers [25, 26] have used SX extensively to calculate the band structures of semiconductors, some oxides and their optical properties. It was not used for energy minimization, which is done here.

5.2
Screened Exchange Functional

The non-local XC potential of SX is similar in form to the HF potential, but it also incorporates the effects of correlation by screening the long-range interactions of exchange [23]. The non-local contribution to the total energy of the system is

$$E_{nl}^{SX} = -\frac{1}{2} \sum_{ij,kq} \int\int \frac{\psi_{ik}^*(r)\psi_{ik}(r')\exp(-k_s|r-r'|)\psi_{jq}^*(r')\psi_{jq}(r)}{|r-r'|} \, dr \, dr', \quad (5.1)$$

5.2 Screened Exchange Functional

where i and j label electronic bands, k and q the k-points and k_s is a Thomas–Fermi screening length.

In order to maintain the exact expression for the homogeneous electron gas (HEG), a local (loc) contribution is also required, so that the total exchange–correlation energy in the screened-exchange method is

$$E^{SX} = E^{SX}_{nl} + E^{SX}_{loc},$$

where E^{SX}_{loc} is this additional contribution which is parameterized using Perdew's expression for the LDA [3]. Thus the local contribution to the exchange and correlation energy density is

$$\varepsilon^{SX}_{loc}(\varrho) = \varepsilon^{HEG}_{loc}(\varrho) - \varepsilon^{HEG}_{nl}(\varrho), \qquad (5.2)$$

where the local $\varepsilon^{HEG}_{loc}(\varrho)$ is the same as the LDA (HEG). The second term is obtained by applying the non-local functional to the HEG, which is given by

$$\varepsilon^{HEG}_{nl}(\varrho) = V^{HEG}_{X}(\varrho) F(\varrho), \qquad (5.3)$$

where V_x is the pure HF exchange of the HEG, $F(\varrho)$ is a screening function given in Eq. (5.3) in Bylander and Kleinman [23]. Thus the total exchange–correlation energy within the screened-exchange formalism is

$$E^{SX} = E^{SX}_{nl} + E^{HEG}_{loc} - \int V^{HEG}_{X}(\varrho) F(\varrho) \varrho(r)\, dr. \qquad (5.4)$$

This is analogous to the HSE term where the first term represents the '$\alpha = 0.25$' of HF, the second term is the long-ranged exchange and the final term is similar to the short-ranged-screened local PBE exchange of HSE. An important factor is that SX reproduces the correct asymptotic limits of XC in both the free electron gas and the HF limit.

The SX method has been implemented in the CASTEP code [27], a plane wave pseudopotential code. It uses norm-conserving pseudopotentials. In many cases, more transferable pseudopotential was generated using the Opium code [28]. The Thomas Fermi (TF) screening parameter is found from the valence electron density by $k_s = 2(k_F/\pi)^{1/2}$ where k_F is the Fermi wavenumber. There are two options, either k_s is set to the average valence electron density of the system, or it is given by a fixed value, such as the natural density of the HEG. In cases of shallow d core states, these can be included in the valence states, but those core electrons are *not* counted in the TF parameter. Non-local stresses are evaluated by the scheme of Gibson et al. [29], which now allows the efficient geometry optimization of the unit cell.

SX is efficient, so that it can be used to carry out the full geometry relaxations in realistic-sized defect supercells, and not just as a post-processing of geometries found by LDA or GGA. It is expected to have similar efficiency to HSE, but this depends on the implementation and the screening lengths used.

5.3
Bulk Band Structures and Defects

Table 5.1 compares the calculated SX band gaps with those calculated by GGA and the experimental values. It can be seen that there is a significant improvement. These are also shown in Figure 5.1. Notable cases are for Si the band gap improves from 0.69 to

Table 5.1 Comparison of the calculated GGA and SX bands gap to the experimental values are given. Also presented are the calculated GGA and SX lattice constants, which are compared to the experimental values.

compound	PBE band gap (eV)	SX band gap (eV)	exp band gap (eV)	GGA lattice constant (Å)	SX lattice constant (Å)	exp lattice constant (Å)
diamond	4.27	5.38	5.5	3.537	3.501	3.567
Si	0.69	1.07	1.12	5.401	5.397	5.431
Ge	0.59	0.69	0.7	5.478	5.414	5.657
c-SiC	1.47	2.25	2.42	4.320	4.262	4.348
AlP	1.66	2.21	2.45	5.438	5.386	5.451
AlAs	1.58	2.30	2.24	5.640	5.581	5.62
AlSb	1.57	1.83	1.70	6.063	6.021	6.13
GaP	1.70	1.85	1.9	5.502	5.374	5.45
GaAs	0.87	1.47	1.45	5.707	5.570	5.66
GaSb	1.00	1.13	0.82	6.066	5.905	6.09
ZnO(zb)	0.89	3.43	3.44	4.583	4.586	4.51
ZnO (wz)	0.8	3.41	3.44	3.268/5.299	3.27/5.25	3.25/5.21
ZnS	2.15	3.74	3.80	5.606	5.421	5.41
ZnSe	1.68	2.71	2.82	5.875	5.569	5.67
ZnTe	1.81	2.34	2.39	6.280	6.025	6.089
CdS	1.59	2.38	2.42	5.983	5.865	5.818
CdSe	1.33	1.88	1.84	6.245	6.113	6.05
CdTe	1.67	1.71	1.60	6.652	6.486	6.48
MgS (rs)	2.77	3.70	3.7	5.210	5.167	5.20
MgS (zb)	3.37	4.84	4.8	5.659	5.599	5.66
MgSe (zb)	2.95	3.91	4.0	5.949	5.893	5.91
CdO	−0.60	0.98	0.9	4.708	4.670	4.69
MgO	3.60	7.72	7.8	4.223	4.126	4.21
LiF	9.24	13.27	13.7	4.093	4.032	4.017
SiO_2	6.05	8.74	9	4.909/5.402	4.855/5.371	5.01/5.47
α-Al_2O_3	6.25	8.64	8.8	4.76/13.00	4.70/12.97	4.76/12.99
SnO_2	0.93	3.66	3.6	4.738/3.149	4.692/3.136	4.737/3.186
In_2O_3	0.90	3.03	2.9	10.118	10.016	10.12
Cu_2O	1.04	2.11	2.12	4.359	4.315	4.27
TiO_2	1.86	3.1	3.2	4.691/2.994	4.608/2.920	4.59/2.96
c-HfO_2	3.74	5.60	5.8	5.161	5.037	5.11
c-ZrO_2	3.43	5.76	5.7	5.131	5.022	5.07
$SrTiO3$	1.93	3.28	3.2	3.971	3.874	3.905
$PbTiO_3$	1.71	3.43	3.4	3.983	3.904	3.96

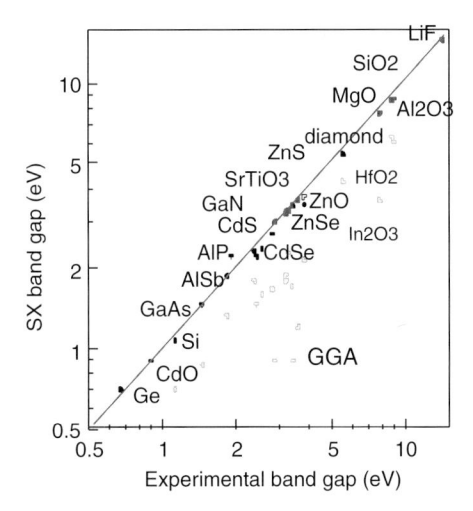

Figure 5.1 (online colour at: www.pss-b.com) GGA and SX band gaps, compared to experiment values. Closed Circles refer to the SX band gap, open circles refer to the GGA band gap.

1.07 eV compared to 1.19 eV experimentally. For GaAs, the band gap increases from 0.87 eV in GGA to 1.47 eV in SX. For insulators, SiO_2, the gap improves from 6.0 eV in GGA to 8.74 eV in SX, very close to the 9.0 eV experimental value. For wide gap insulators such as LiF, the gap is 13.27 eV in SX, 9.24eV in PBE, compared to 13.7 eV experimentally. It has been used before on HfO_2 and the multiferroic $BiFeO_3$ [30–32].

The improvement is most significant for the transparent conducting oxides such as ZnO, In_2O_3 and SnO_2. These band structures are characterized by a single, broad conduction band minimum at Γ. The minimum gap of SnO_2 is 0.9 eV in GGA, and this becomes 3.66 eV in SX, compared to 3.6 eV found experimentally. The reason is that direct gap at Γ is unrepresentative of the averaged gap. The average gap opens up by the typical 20%, but this translates into a very large fractional change at Γ.

5.3.1
Band Structure of ZnO

ZnO is an important semiconductor, which is widely used as a phosphor, for transparent electrodes in solar cells, and for ultraviolet light emission, spintronics, nanowires and for its high electron mobility [33–39]. It can be easily doped n-type but it is difficult to dope p-type [35]. This has been attributed to the nature of its intrinsic defects which cause a self-compensation of free carriers [40] and also to that common acceptors are deep [41]. It is therefore important to understand the energetics of its intrinsic defects.

There have been numerous first-principles studies of the bulk electronic structure of ZnO [13, 42–48] and its defect energies [49–63]. The LDA + U method has been used to improve the GGA band structure, by shifting the Zn 3d band downwards [13]. The SIC method has been used to find the band structure of ZnO [44]. Various types of GW methods have been used for ZnO [45–47]. The HSE hybrid has been used [47, 48, 52, 56, 61].

Defect calculations generally find that the O vacancy is the defect with lowest formation energy but it is deep, while the Zn interstitial is shallow but has higher formation energy. Nevertheless, there is a lack of consistency between the various results. This arise partly because of the band gap error of LDA and also sometimes because charge state corrections were not correctly included.

Patterson [54] used the B3LYP functional and localized orbitals to calculate the defect eigenvalues, but he did not calculate the defect formation energies from the total energies. Oba et al. [52] used the HSE functional to provide a complete set of defect formation energies, and tested the corrections for supercell size. Superficially, this is a well-defined calculation. However, it was necessary to increase the HF mixing parameter a from $a = 0.25$ to 0.375 in order to empirically fit the experimental gap. Agoston et al. [56] produced a valuable comparison of GGA and HSE results for O vacancies for all three conducting oxides.

Recently, we applied the SX method to the intrinsic defects of ZnO [61]. For our SX calculations, k_s is determined from the valence electron density, and for those elements like Zn with shallow filled d states, it is for s, p electrons only. The $k_{TF} = 2.27\,\text{Å}^{-1}$ for ZnO. A plane-wave cut-off energy of 800 eV is used, which converges total energy differences to better that 1 meV/atom. Integrations over the Brillouin zone are performed using the k-point sampling method of Monkhorst and Pack with a grid that converges the energies of the bulk unit cell to a similar accuracy. Geometry optimizations are performed self-consistently using a minimization scheme and the Hellmann–Feynman forces, and are converged when forces are below 0.04 eV/Å.

Table 5.2 shows the converged lattice parameters of ZnO, which are within 0.5% of experiment, whereas GGA (PBE) values are 1% too large. The free energy of ZnO per formula unit is found to be only 0.3 eV less than experiment, a 60% improvement over the PBE result.

Figure 5.2 shows the calculated band structure of bulk ZnO in the wurtzite and zincblende structure. The minimum gap is calculated to be 3.41 eV, and is very close to the 3.44 eV found experimentally [35]. Our value is much closer to experiment than HSE with the normal a parameter (2.87 eV) [48], or even with the expensive GW [45] which gets 2.7 eV.

Table 5.2 Bulk properties of wurzite ZnO, calculated compared to experiment.

	GGA	SX	exp
a (Å)	3.286	3.267	3.2495
c (Å)	5.299	5.245	5.2069
free energy (eV)	−2.82	−3.31	−3.63
direct gap (eV)	0.9	3.41	3.44
Zn 3d (eV)	−4.8	−7.0	−7.3

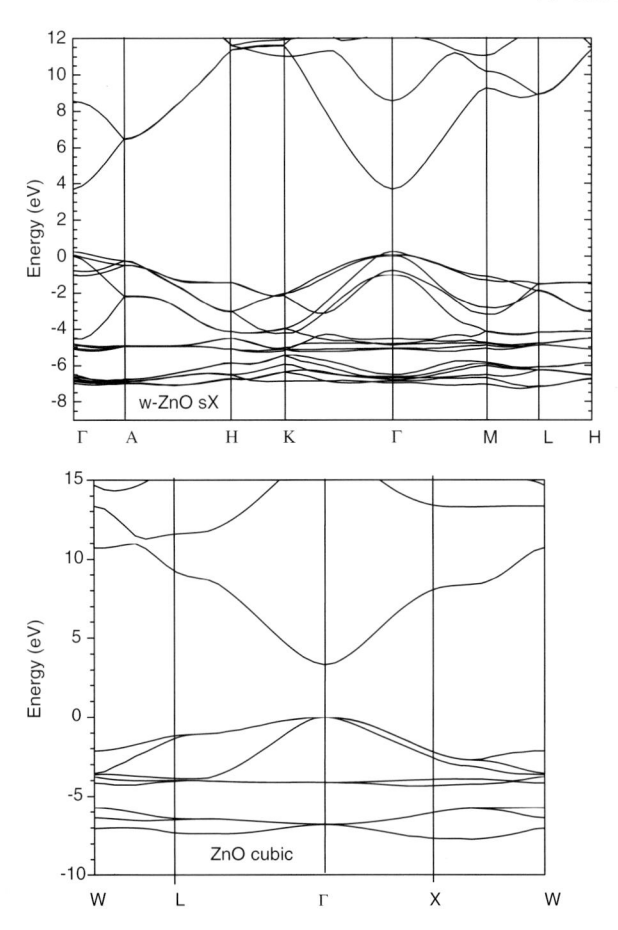

Figure 5.2 The band structure of ZnO in the wurtzite and zincblende structures evaluated using the SX functional.

Part of the LDA band gap error in ZnO arises from the Zn 3d (t_{2g}) levels lying too high, and their upwards repulsion of the Γ_{15} valence band maximum states. In SX, the Zn 3d states now lie at −7.0 below the VB maximum, very close to where they are found experimentally by angle-resolved photoemission [64, 65].

5.3.2
Defects of ZnO

The defect calculations are carried out using a 120 atom supercell, whose size is fixed at that of the defect-free cell, and the defect created. The internal geometry is relaxed within SX, using a single special k-point of (1/4, 1/4, 1/3) for Brillouin zone integrations, which converges the quantities faster than the Γ point with respect to supercell size [66].

The total energy (E_q) is calculated for the defect cell of charge q, for the perfect cell (E_H) of charge q, and for a perfect cell of charge 0. This allows us to calculate the defect formation energy, H_q, as a function of the relative Fermi energy (ΔE_F) from the valence band edge E_V and the relative chemical potential ($\Delta \mu$) of element α [60],

$$H_q(E_F, \mu) = [E_q - E_H] + q(E_V + \Delta E_F) + \sum_\alpha n_\alpha (\mu_\alpha^0 + \Delta \mu_\alpha),$$

where $q(E_V + \Delta E_F)$ is the change in energy when charge q is added to the system at the Fermi level and n_α is the number of atoms of species α. Essentially, this is the shift in the average electrostatic potential due to the charge of the system with respect to the uncharged system. The corrections for the background charge, band filling, etc., are included as described in Ref. [58]. The oxygen chemical potential (μ^0) is referred to that of the O_2 molecule, taken as zero, which is the O-rich limit. The O-poor limit corresponds to the Zn/ZnO equilibrium and is $\mu(O) = -3.31$ eV (the heat of formation of ZnO).

Figure 5.3a and b shows the calculated formation energies of the four intrinsic defects of ZnO in the O-rich and O-poor limits. We see that the O vacancy (V_O) has the lowest formation energy over a wide range of E_F, lower than the Zn interstitial. V_O is a deep defect and it has a transition state between its neutral and doubly positive states $E(0/2+)$ at 2.20 eV. The $+1$ state is never stable, so the O vacancy has a negative effective correlation energy (U), as found by others. We find $U = -2.0$ eV, the energy difference between the metastable $0/+$ and $+/2+$ transitions in Figure 5.4. The negative U arises because of the strong lattice relaxation with changing charge state, with the Zn-vacancy distance changing from 1.84 Å for V^0, to 2.16 Å for V^+ to 2.46 Å for V^{2+}, compared to a bulk Zn–O distance of 1.95 A.

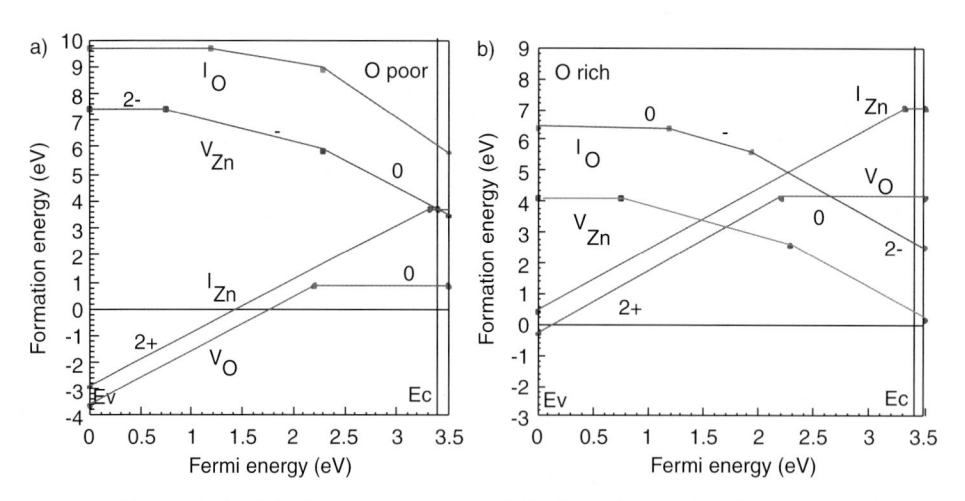

Figure 5.3 (online colour at: www.pss-b.com) The formation energies of native defects in ZnO evaluated using the SX functional under (a) oxygen poor and (b) oxygen rich conditions. The gradients of the lines give the charge state of the defect.

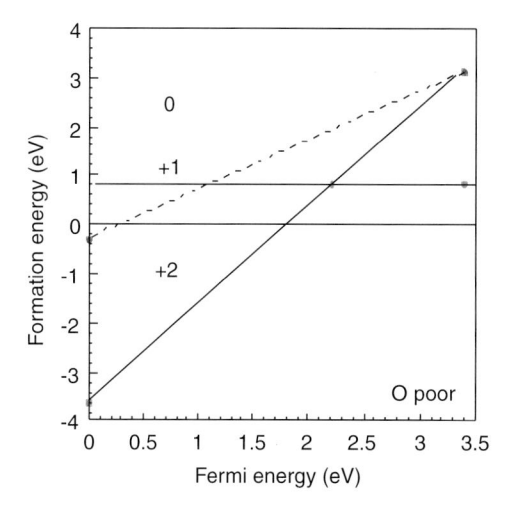

Figure 5.4 (online colour at: www.pss-b.com) The formation energy of the neutral, +1 and +2 charge states of the oxygen vacancy in ZnO under oxygen poor conditions.

These relaxations are seen in Figure 5.5a–c. Note that the relaxed structures found by SX are similar to those found by GGA, so in fact in this case, we could have used SX post-processing on GGA structures to find the defect formation energies.

We find that the Zn interstitial I_{Zn} is a shallow defect, with a slightly higher formation energy than V_O. Its transition state (0/2+) lies at 3.32 eV, essentially at the conduction band edge, consistent with experiment [37], and it has a $U=0$ eV. However, its neutral state has a large formation energy in O^- poor conditions.

The other two intrinsic defects I_O and V_{Zn} are more stable in O-rich conditions (compared to V_O and I_{Zn}), and have higher formation energies. They are both deep defects, and show two charge states in the gap, corresponding to positive U behaviour. The I_O forms the usual dumb-bell structure of O interstitials in its 0 and − charge states, while the I_{Zn} sits in the octahedral site as seen by others [50].

Figure 5.6 compares our formation energies of the O vacancy to those calculated by others. The calculated formation energy of V^0 of +0.85 eV in SX is similar to that found by Lany and Zunger [59], slightly less than the 1.0 eV found by Oba et al. [52], and similar to that found by Agoston et al. [56]. However, it is much less than the formation energy given by Janotti and van de Walle [50]. The +0.85 eV formation energy would correspond to a frozen-in vacancy concentration of 10^{19} cm^{-3} at 700 °C, which is consistent with the concentration found experimentally [67, 68]. However, being deep, it is not a source of free electrons. The large formation energy of neutral Zn interstitial means that this cannot be the source of free electrons in ZnO, as its concentration would be too low [69]. This means that in the absence of hydrogen, the source of free electrons must be a donor complex.

Our (0/2+) transition energy of 2.20 eV above the valence band top is the same as that found by Janotti and van de Walle [50, 51], by Oba et al. [52], and Agoston et al. [56], but higher than found by Lany in corrected GGA [59]. The metastable (0/+)

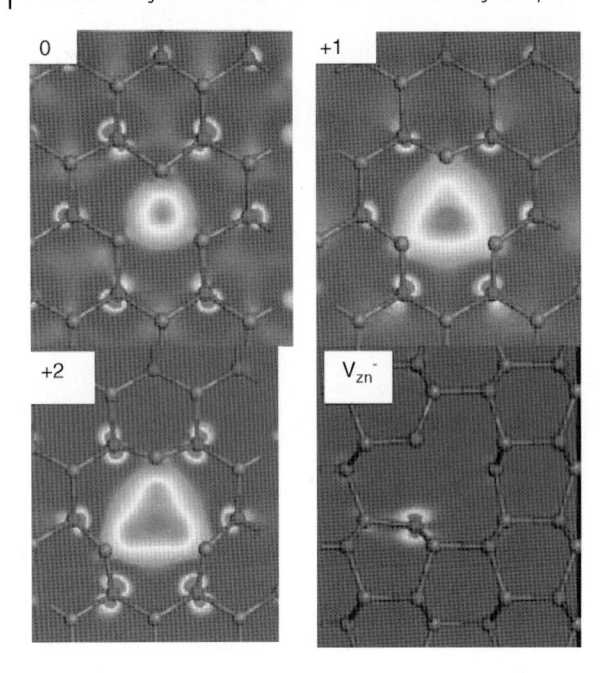

Figure 5.5 (online colour at: www.pss-b.com) In (a), (b) and (c), the electronic density of the defect state found in the band gap of ZnO is shown for the oxygen vacancy for the neutral, +1 and +2 charge states, respectively. In (d), the defect state of the Zn vacancy is shown and is found to be a p-like state located on only *one* of the adjacent oxygen atoms.

transition lies at 0.9 eV and this is consistent if the ODMR transition observed by Vlasenko and Watkins [70] is from valence band to the defect level. Thus, overall there is now a reasonable convergence in formation energies and transition energies between some of the calculations.

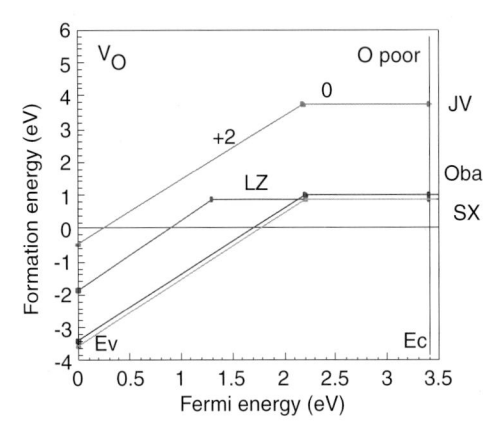

Figure 5.6 (online colour at: www.pss-b.com) The oxygen vacancy formation energy under oxygen poor conditions as a function of Fermi energy is shown for the present (SX) and other studies. Janotti and van de Walle (JV) [49], Lany and Zunger (LZ) [58], and Oba *et al.* [52].

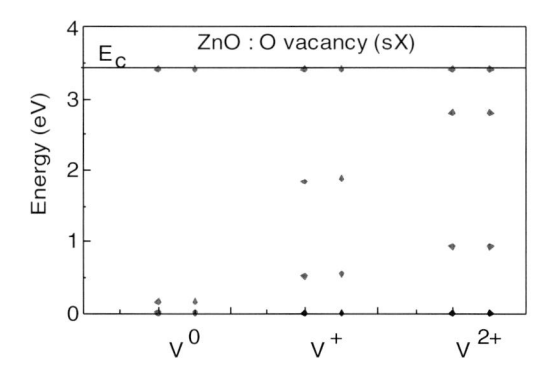

Figure 5.7 (online colour at: www.pss-b.com) The energy eigenvalues of the oxygen vacancy defect states lying in the band gap for SX.

It is interesting that for the oxygen vacancy, the SX and GGA relaxed atomic positions and wave functions are similar.

Figure 5.7 shows our calculated eigenvalues of the oxygen vacancy. We see that the eigenvalue of the neutral V_O lies at $+0.16$ eV above the VB top. In GGA, it lies in the lower gap. This is why V_O is able to display three charges even in GGA calculations despite the severe under-estimate of the gap. However, it does show that the calculated eigenvalues of V_O have little relationship to the transition energies in ZnO, due to the strong lattice relaxations. Hence the result of the B3LYP calculation of Patterson [54] is not particularly relevant, as it only gives transition energies.

Finally, the LDA is known to under-estimate the localization of hole states. For example, the trapped hole state of Al_{Si} in SiO_2 (smoky quartz) is well known to be trapped on a single oxygen, but LDA finds it localized on all four neighbouring oxygens [71]. There are other recent examples [71, 72]. Figure 5.5d shows the calculated SX charge density of the single trapped hole state of the Zn vacancy $V_{\overline{Zn}}$. We find it to be localized on one oxygen neighbour in SX, not four. In this case, the charge density in SX differs considerably from that of GGA. This is consistent with its spin resonance signature [73, 74]. Even LDA + U cannot localize it on one oxygen [51]. On the other hand, the wave functions of the three states of V_O are localized over all four Zn neighbours, consistent with a simple symmetric vacancy (Figure 5.5a–c). The localization is driven by distortion. Thus, in this case, the SX and GGA geometries are different, we could not have found defect formation energies by post-processing GGA geometries in SX.

5.3.3
Band Structure of MgO

Figure 5.8 shows the SX band structure of the clasical metal oxide MgO. Its calculated band gap is 7.7 eV, which is close to the experimental value of 7.8 eV. The valence band is formed of O 2p states and the conduction band is formed of Mg 3s states.

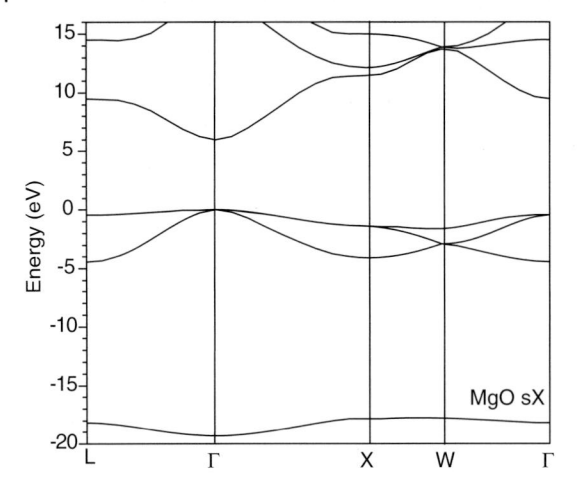

Figure 5.8 SX band structure of MgO in the rock salt structure.

5.3.4
Band Structures of SnO$_2$ and CdO

The three transparent conducting oxides ZnO, CdO and SnO$_2$ are good tests of band structure methods. ZnO has been treated already. CdO fails badly in GGA, where it is found to have a negative indirect band gap. In SX, the band gap is now positive, and 0.9 eV. This is close to the experimental value (Table 5.1). Note that CdO has an indirect gap from L to Γ, due to the effect of Cd d states on the upper valence band. Its conduction band is the standard free-electron like band, formed from Cd s states (Figure 5.9).

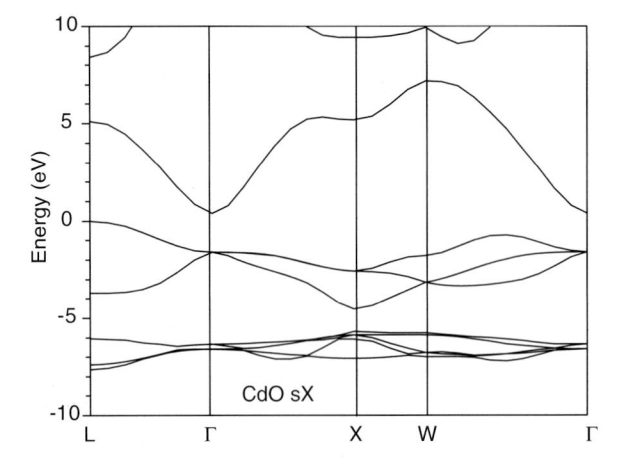

Figure 5.9 Calculated SX band structure of rock salt CdO. Note the real gap.

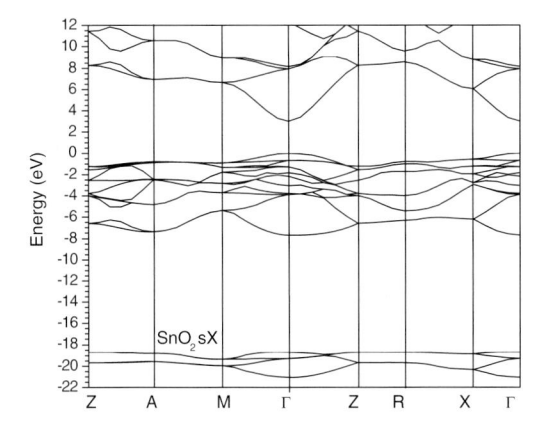

Figure 5.10 Calculated SX band structure of SnO$_2$ in the rutile structure.

SnO$_2$ has a simpler band structure, with a 3.6 eV direct-forbidden gap. However, in GGA the gap is typically only 0.9 eV. Figure 5.10 shows the calculated SX band structure, where the calculated gap is 3.6 eV, the experimental value. The band gaps between O 2p valence states and Sn s conduction band states.

5.3.5
Band Structure and Defects of HfO$_2$

We have carried out a SX calculations on various other oxides. HfO$_2$ is an important oxide in microelectronics as it is now used as the gate oxide in modern FETs. It is a closed shell transition metal oxide with a high dielectric constant. Figure 5.11 shows the band structure of cubic HfO$_2$ (fluorite structure) calculated by the SX functional [30]. The calculated band gap is slightly indirect and is 5.6 eV, which is

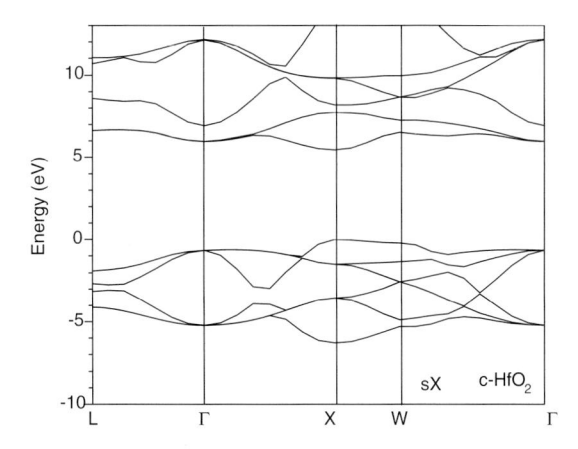

Figure 5.11 Band structure of cubic HfO$_2$ calculated by the SX method. The calculated band gap is 5.6 eV, compared to 5.8 eV experimentally.

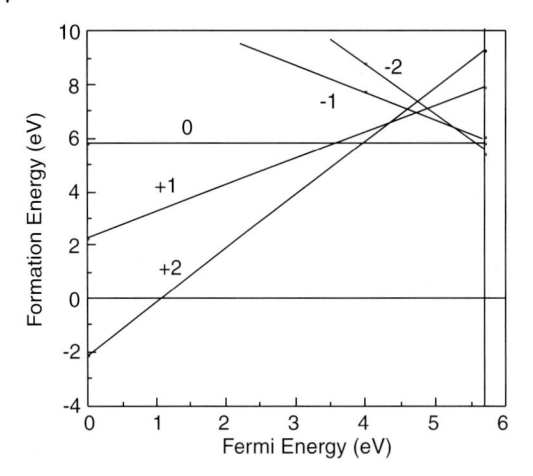

Figure 5.12 Defect formation energy versus Fermi energy for the oxygen vacancy in HfO_2, for an oxygen chemical potential of the O_2 molecule.

close to the experimental value of about 5.8 eV, whereas it is about 3.4–3.7 eV in LDA or GGA.

Defects are an important consideration in such oxides, as they lead to charge trapping, and an instability in the gate threshold voltage. The principle defect is now known to be the oxygen vacancy. From GGA calculations, it was unclear which defect was predominant, because GGA placed the vacancy levels either too low in the gap, or too high in the gap [75, 76], depending on which correction scheme was applied to the energy levels for the band gap error. SX played a critical role in resolving this question, as it was the first calculation of the oxygen vacancy levels which placed the energy levels correctly [30], and close to the experimental values observed by charge injection and optical absorption [77–81]. It was then realized that the oxygen vacancy was likely to be the main defect, as this is consistent with its behaviour in similar oxides such as ZrO_2.

Figure 5.12 shows the calculated SX formation energy of the oxygen vacancy versus the Fermi energy. The local structure was relaxed by SX. The lines represent the different charge states of the vacancy, and the transition states are where lines cross.

Figure 5.12 shows that the O vacancy is a negative U defect for both the -2 and $+1$ states. The calculated values are similar to those found by Gavartin et al. [82] and Broqvist and Pasquarello. [83] using the B3LYP and PBEh methods.

5.3.6
$BiFeO_3$

$BiFeO_3$ is the architypal multiferroic oxide, which shows both ferroelectric and antiferromagnetic properties. The interest arises from the possible electric field control of magnetic properties and vice versa. It is highly studied since it was made in thin film form on Si [84]. It has the R3c structure in its ferroelectric phase, which is

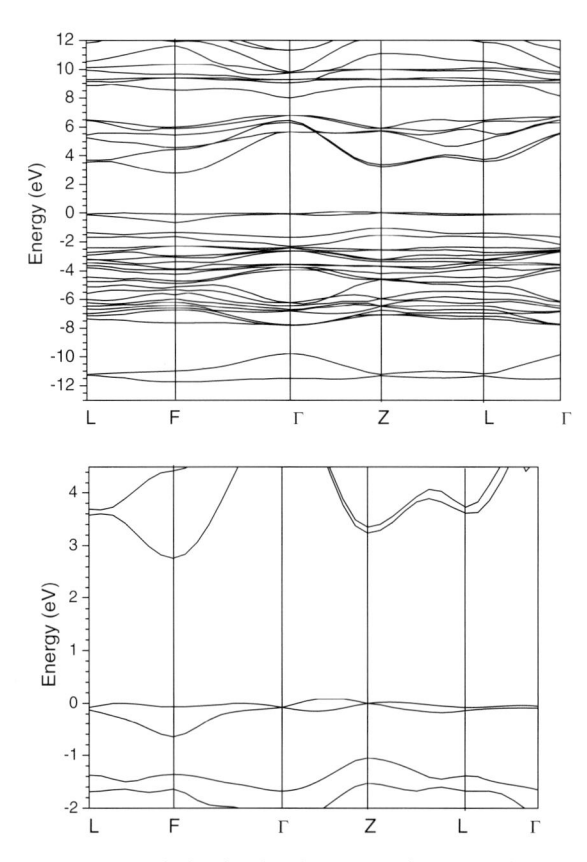

Figure 5.13 Calculated SX band structure of BiFeO$_3$ in the R3c structure, and close up of bands near the gap.

a small distortion of the cubic. It is a semiconductor whose band gap lies in the Fe 3d states of opposite spin (Figure 5.13).

There is no band gap in the LSDA method, even in the distorted structure. The simplest way to create a band gap is to use the LDA + U method, which does produce a band gap, as in the calculation of Neaton et al. [85]. However, the U parameter must be empirically chosen, which is unsatisfactory. We recently calculated the band structure of BFO by the SX method for the experimental structure, with no adjustable parameters [86, 87]. This gave a band gap of 2.7 eV, which was the first available value of the gap. It agrees with subsequent experimental evaluations [88, 89].

5.4
Summary

The SX method has been reviewed as a method to produce band structures of insulators and semiconductors with more accurate band gaps. SX behaves as a hybrid density functional, which can be used for energy minimization. Some examples of the used of SX, particularly in oxides, are given.

Acknowledgements

The authors acknowledge valuable discussions with A. Zunger and S. Lany.

References

1 Sham, L.J. and Schluter, M. (1983) *Phys. Rev. Lett.*, **51**, 1888; Perdew, J.P. and Levy, M. (1983) *Phys. Rev. Lett.*, **51**, 1884; Godby, R.W., Schluter, M., and Sham, L.J. (1986) *Phys. Rev. Lett.*, **56**, 2415.
2 Gygi, F. and Baldereschi, A. (1989) *Phys. Rev. Lett.*, **62**, 2160.
3 Perdew, J.P. and Zunger, A. (1981) *Phys. Rev. B*, **23**, 5048.
4 Filippetti, A. and Spaldin, N.A. (2003) *Phys. Rev. B*, **67**, 125109.
5 Hedin, L. (1965) *Phys. Rev.*, **139A**, 796.
6 Hybertsen, M.S. and Louie, S.G. (1986) *Phys. Rev. B*, **34**, 5390.
7 Aryasetiawan, F. and Gunnarsson, O. (1998) *Rep. Prog. Phys.*, **51**, 2327.
8 Rohlfing, M., Kruger, P., and Pollmann, J. (1993) *Phys. Rev. B*, **48**, 17791.
9 Aulbur, W.G. Jonsson, L., and Wilkins, J. (2000) *Solid State Physics*, edited by Ehrenreich H. and Spaepen, F., Vol. 54 Academic Press, New York, p. 1.
10 van Schilfgaarde, M., Kotani, T., and Faleev, S. (2006) *Phys. Rev. Lett.*, **96**, 226402.
11 Anisimov, V.I., Zaanen, J., and Andersen, O.K. (1991) *Phys. Rev. B*, **44**, 943.
12 Anisimov, V.I., Aryasetiawan, F., and Lichtenstein, A.I. (1997) *J. Phys.: Condens. Matter*, **9**, 767.
13 Janotti, A. and van de Walle, C.G. (2006) *Phys. Rev. B*, **74**, 045202.
14 Becke, A.D. (1993) *J. Chem. Phys.*, **98**, 1372.
15 Perdew, J.P., Ernzerhof, M., and Burke, K. (1996) *J. Chem. Phys.*, **105**, 9982.
16 Muscat, J., Wander, A., and Harrison, N.M. (2001) *Chem. Phys. Lett.*, **342**, 397.
17 Paier, J., Hirschl, R., Marsman, M., and Kresse, G. (2005) *J. Chem. Phys.*, **122**, 234102.
18 Heyd, J. Scuseria, G.E., and Ernzerhof, X.X. (2003) *J. Chem. Phys.*, **118**, 8207.
19 Heyd, J. and Scuseria, G.E. (2004) *J. Chem. Phys.*, **120**, 7274.
20 Heyd, J., Peralta, J.E., Scuseria, E., and Martin, R.L. (2005) *J. Chem. Phys.*, **123**, 174101.
21 Krukau, A.V., Vydrov, O.A., Izmaylov, A.F., and Scuseria, G.E. (2006) *J. Chem. Phys.*, **125**, 224106.
22 Paier, J., Marsman, M., Hummer, K., Kresse, G., Gerber, I.C., and Angyan, J.G. (2006) *J. Chem. Phys.*, **124**, 154709.
23 Bylander, D.M. and Kleinman, L. (1990) *Phys. Rev. B*, **41**, 7868.
24 Seidl, A., Gorling, A., Vogl, P., Majewski, J.A., and Levy, M. (1996) *Phys. Rev. B*, **53**, 3764.
25 Geller, C.B., Wolf, W., Picozzi, S., Continenza, A., Freeman, A.J., and Wimmer, E. (2001) *Appl. Phys. Lett.*, **79**, 368.
26 Asahi, R., Wang, A., Babcock, J.R., Edelman, N.L., Metz, A.W., Lane, M.A., Dravid, V.P., Kannewurf, C.R., Freeman, A.J., and Marks, T.J. (2002) *Thin Solid Films*, **411**, 101.
27 Segall, M.D., Lindan, P.J.D., Probert, M.J., Pickard, C.J., Hasnip, P.J., Clark, S.J., and Payne, M.C. (2002) *J. Phys.: Condens. Matter*, **14**, 2717.
28 Rappe, A.M., Rabe, K.M., and Joannopoulos, J.D. (1996) *Phys. Rev. B*, **41**, 1227.
29 Gibson, M.C., Brand, S., and Clark, S.J. (2006) *Phys. Rev. B*, **73**, 125120.
30 Xiong, K., Robertson, J., Gibson, M.C., and Clark, S.J. (2005) *Appl. Phys. Lett.*, **87**, 183505.
31 Robertson, J., Xiong, K., and Clark, S.J. (2006) *Phys. Status Solidi B*, **243**, 2054; (2006) *Phys. Status Solidi B*, **243**, 2071.

32 Clark, S.J. and Robertson, J. (2007) *Appl. Phys. Lett.*, **90**, 132903; (2009) *Appl. Phys. Lett.*, **94**, 022902. Liu, D. Clark, S.J., and Robertson, J. (2010) *Appl. Phys. Lett.*, **96**, 032905.
33 Vanheusden, K., Warren, W.L., Seager, C.H., Taliant, D.R., Voigt, J.A., and Gnade, B.E. (1996) *J. Appl. Phys.*, **79**, 7983.
34 Ozgur, U. et al., (2005) *J. Appl. Phys.*, **98**, 041301.
35 Look, D.C., Chaflin, B., Allvov, Y.I., and Park, S.J. (2004) *Phys. Status Solidi A*, **201**, 2203.
36 Look, D.C. and Chaflin, B. (2004) *Phys. Status Solidi B*, **241**, 624.
37 Tsukazaki, A. et al. (2005) *Nature Mater*, **4**, 42.
38 Tsukazaki, A., Ohtomo, A., Kita, T., Ohno, Y., Ohno, H., and Kawasaki, M. (2007) *Science*, **315**, 1388.
39 Pearton, S.J., Heo, W.H., Ivill, M., Norton, D.P., and Steiner, T. (2004) *J. Phys.: Condens. Matter*, **19**, R59.
40 Zhang, S.B., Wei, S.H., and Zunger, A. (2001) *Phys. Rev. B*, **63**, 075205. Zhang, S.B., Wei, S.H., and Zunger, A. (1998) *J. Appl. Phys.*, **83**, 3192.
41 Lyons, J.L., Janotti, A., and van de Walle, C.G. (2009) *Appl. Phys. Lett.*, **95**, 252105.
42 Chelikowsky, J.R. (1977) *Solid State Commun.*, **22**, 351.
43 Schroer, P., Kruger, P., and Pollmann, J. (1993) *Phys. Rev. B*, **47**, 6971.
44 Vogel, D., Kruger, P., and Pollmann, J. (1995) *Phys. Rev. B*, **52**, R14316.
45 Usuda, J.M., Hamada, N., Kotani, T., and van Schilfgaarde, M. (2002) *Phys. Rev. B*, **66**, 125101.
46 Fuchs, F., Furthmüller, J., Bechstedt, F., Shishkin, M., and Kresse, G. (2007) *Phys. Rev. B*, **76**, 115109.
47 Preston, A.H.R., Ruck, B.J., Piper, L.F.J., DeMasi, A., Smith, K.E., Schleife, A., Fuchs, F., Bechstedt, F., Chai, J., and Durbin, S.M. (2008) *Phys. Rev. B*, **78**, 155114.
48 Uddin, J. and Scuseria, G.E. (2006) *Phys. Rev. B*, **74**, 245115.
49 Kohan, A.F., Ceder, G. Morgan, D., and van de Walle, C.G. (2000) *Phys. Rev. B*, **61**, 15019.
50 Janotti, A. and van de Walle, C.G. (2005) *Appl. Phys. Lett.*, **87**, 122102.
51 Janotti, A. and Van de Walle, C.G. (2007) *Phys. Rev. B*, **76**, 165202.
52 Oba, F., Togo, A., Tanaka, I., Paier, J., and Kresse, G. (2008) *Phys. Rev. B*, **77**, 245202.
53 Erhart, P., Albe, K., and Klein, A. (2006) *Phys. Rev. B*, **73**, 205203.
54 Patterson, C.H. (2006) *Phys. Rev. B*, **74**, 144432.
55 Paudel, T.R. and Lambretch, W.R.L. (2008) *Phys. Rev. B*, **77**, 205202.
56 Agoston, P., Albe, K., Nieminen, R.M., and Puska, M.J. (2009) *Phys. Rev. Lett.*, **103**, 245501.
57 Lany, S. and Zunger, Av. (2005) *Phys. Rev. B*, **72**, 035215.
58 Lany, S. and Zunger, A. (2007) *Phys. Rev. Lett.*, **98**, 045501.
59 Lany, S. and Zunger, A. (2008) *Phys. Rev. B*, **78**, 235104.
60 Lany, S. and Zunger, A. (2010) *Phys. Rev. B*, **81**, 113201.
61 Clark, S.J., Robertson, J., Lany, S., and Zunger, A. (2010) *Phys. Rev. B*, **81**, 115311.
62 Van de Walle, C.G. (2000) *Phys. Rev. Lett.*, **85**, 1012.
63 Lee, E.C., Kim, Y.S., Jin, Y.G., and Chang, K.J. (2001) *Phys. Rev. B*, **64**, 085120.
64 Girard, R.T., Tjernberg, O., Chiaia, G., Soderholm, S., Jarlsson, U.O., Wigren, C., Nylen, H., and Lindau, I. (1997) *Surf. Sci.*, **373**, 409.
65 Ozawa, K., Sawada, K., Shirotori, Y., and Edamoto, K. (2005) *J. Phys.: Condens. Matter*, **17**, 1271.
66 Probert, M.I.J. and Payne, M.C. (2003) *Phys. Rev. B*, **67**, 075204.
67 Look, D.C., Hemsky, J.W., and Sizelove, J.R. (1999) *Phys. Rev. Lett.*, **82**, 2552.
68 Hagemark, K.I. and Toren, P.E. (1975) *J. Electrochem. Soc.*, **122**, 992. Tuomisto, F. et al., (2003) *Phys. Rev. Lett.*, **91**, 205502.
69 Selin, F.A., Weber, M.H., Solodovnikov, D., and Lynn, K.G. (2007) *Phys. Rev. Lett.*, **99**, 085502.
70 Vlasenko, L.S. and Watkins, G.D. (2005) *Phys. Rev. B*, **71**, 125210.
71 Pacchioni, G., Frigoli, F., Ricci, D., and Weil, J.A. (2000) *Phys. Rev. B*, **63**, 054102. d'Avezac, M., Calandra, M., and Mauri, F. (2005) *Phys. Rev. B*, **71**, 205210.
72 Lany, S. and Zunger, A. (2009) *Phys. Rev. B*, **80**, 085202.

73 Galland, D. and Herve, A. (1970) *Phys. Lett. A*, **33**, 1.
74 Vlasenko, L.S. and Watkins, G.D. (2005) *Phys. Rev. B*, **72**, 035203.
75 Foster, A.S., Sulimov, V.B., Gejo, F.L., Shluger, A.L., and Nieminen, R.N. (2001) *Phys. Rev. B*, **64**, 224108.
76 Shen, C., Li, M.F., Yu, H.Y., Wang, X.P., Yeo, Y.C., Chan, D.S.H.,, and Kwong, D.L. (2005) *Appl. Phys. Lett.*, **86**, 093510.
77 Kerber, A., Cartier, E., Pantisano, L., Degraeve, R., Kauerauf, T., Kim, Y., Groeseneken, G., Maes, H.E., and Schwalke, U. (2003) *IEEE Electron Device Lett.*, **24**, 87.
78 Cartier, E. *et al.* (2006) Tech Digest IEDM.
79 Takeuchi, H., Ha, D., and King, T.J. (2004) *J. Vac. Sci. Technol. A*, **22**, 1337.
80 Nguyen, N.V., Davydov, A.V., Chandler-Horowitz, D., and Frank, M.M.M. (2005) *Appl. Phys. Lett.*, **87**, 192903.
81 Walsh, S., Fang, L., Schaeffer, J.K., Weisbrod, E., and Brillson, L.J. (2007) *Appl. Phys. Lett.*, **90**, 052901.
82 Gavartin, J.L., Ramos, D.M., Shluger, A.L., Bersuker, G., and Lee, B.H. (2006) *Appl. Phys. Lett.*, **89**, 082908.
83 Broqvist, P. and Pasquarello, A. (2006) *Appl. Phys. Lett.*, **89**, 262904.
84 Wang, J., Neaton, J.B., Zheng, H., Nagarajan, V., Ogale, S.B., Liu, B., Viehland, D., Vaithyanathan, V., Schlom, D.G., Waghare, U.V., Spaldin, N.A., Rabe, K.M., Wuttig, M., and Ramesh, R. (2003) *Science*, **299**, 1719.
85 Neaton, J.B., Ederer, C., Waghmare, U.V., Spaldin, N.A., and Rabe, K.M. (2005) *Phys. Rev. B*, **71**, 014113.
86 Clark, S.J., and Robertson, J. (2007) *Appl. Phys. Lett.*, **90**, 132903.
87 Palai, R., Katiyar, R.S., Schmid, H., Tissot, P., Clark, S.J., Robertson, Jv., Redfern, S.A.T., Catalan, G., and Scott, J.F. (2008) *Phys. Rev. B*, **77**, 014110.
88 Ihlefeld, J.F., Podraza, N.J., Liu, Z.K., Rai, R.C., Ramesh, R., and Schlom, D.G. (2008) *Appl. Phys. Lett.*, **92**, 142908.
89 Hauser, A.J., Zhang, J., Meier, L., Ricciardo, R.A., Woodward, P.M., Gustafson, T.L., Brillson, L.J., and Yang, F.Y. (2008) *Appl. Phys. Lett.*, **92**, 222901.

6
Accurate Treatment of Solids with the HSE Screened Hybrid

Thomas M. Henderson, Joachim Paier, and Gustavo E. Scuseria

6.1
Introduction and Basics of Density Functional Theory

Theoretical predictions of electronic properties offer a clear complement to experimental investigations. However, these theoretical predictions are only as useful as they are accurate. Ideally, first-principles calculations of electronic properties in periodic systems would use some high level many-body technique such as coupled-cluster theory; however these methods are horrifically expensive and therefore have restricted applicability in practice. While simpler many-body methods such as GW theory are significantly less expensive, the computational burden they impose nevertheless restricts their scope. Therefore, we often must perforce resort to single-particle descriptions, which are less computationally expensive but of course also less reliable.

One important single-particle reference is the Hartree–Fock (HF) method. The HF density matrix variationally minimizes the HF energy, given by

$$E_{HF} = \langle \Phi | T | \Phi \rangle + \int dr\, n(r) v_{ext}(r) + \frac{1}{2} \int dr_1 dr_2 \frac{n(r_1)n(r_2)}{r_{12}} + E_x^{HF}[\gamma], \quad (6.1)$$

where n is the density, γ the density matrix, $r_{12} = |r_1 - r_2|$, and $E_x^{HF}[\gamma]$ is the nonlocal Fock exchange energy

$$E_x^{HF}[\gamma] = -\frac{1}{2} \int dr_1 dr_2 \frac{\gamma(r_1, r_2) \gamma(r_2, r_1)}{r_{12}}. \quad (6.2)$$

That is, HF takes the usual one-body terms and the Coulomb interaction between electrons and adds a (known) functional of the density matrix to describe exchange. We construct the wave function $|\Phi\rangle$, density and density matrix from HF orbitals, obtained from a reference system of noninteracting electrons in which the external potential is augmented by the Hartree potential

$$v_H(r_1) = \int dr_2 \frac{n(r_2)}{r_{12}}, \quad (6.3)$$

Advanced Calculations for Defects in Materials: Electronic Structure Methods, First Edition.
Edited by Audrius Alkauskas, Peter Deák, Jörg Neugebauer, Alfredo Pasquarello, and Chris G. Van de Walle.
© 2011 Wiley-VCH Verlag GmbH & Co. KGaA. Published 2011 by Wiley-VCH Verlag GmbH & Co. KGaA.

and by an additional nonlocal potential v_x^{HF} which accounts for the effects of the exchange interaction on the orbitals. This nonlocal potential is the functional derivative of the exchange energy E_x^{HF} with respect to the density matrix

$$\frac{\delta E_x^{HF}}{\delta \gamma(r_2, r_1)} = \langle r_1 | v_x^{HF} | r_2 \rangle = -\frac{\gamma(r_1, r_2)}{r_{12}}, \tag{6.4}$$

so that matrix elements of v_x^{HF} between two single-particle functions are

$$\langle \phi | v_x^{HF} | \psi \rangle = \int dr_1 dr_2 \langle \phi | r_1 \rangle \langle r_1 | v_x^{HF} | r_2 \rangle \langle r_2 | \psi \rangle = -\int dr_1 dr_2 \phi^*(r_1) \psi(r_2) \frac{\gamma(r_1, r_2)}{r_{12}}. \tag{6.5}$$

HF has some advantages which should not be forgotten. It is exact for one-electron systems and has no one-electron self-interaction. It also forms a variationally optimal starting point for many-body methods. The orbital energies in HF have a clean interpretation as ionization potentials and electron affinities (i.e., Koopmans' theorem holds), which justifies the use of HF band energy differences to predict the fundamental band gap (the difference between the ionization potential and electron affinity of the system). Numerically, however, the HF band energy difference is a poor predictor of the fundamental gap.

Unfortunately, the advantages of HF are generally outweighed by disadvantages. Because it neglects electron correlation effects entirely, HF is insufficiently accurate for most applications. It has an innate tendency to overly favor electronically localized states, a manifestation of its form of many-electron self-interaction [1–3]. Calculating the nonlocal exchange interaction is computationally demanding in solids. The nonlocal exchange interaction also prevents HF from describing metallic behavior [4].

For all of these reasons, the single-particle method of choice is usually Kohn–Sham (KS) density functional theory (DFT) [5–7], in which the energy is given by

$$E_{KS} = \langle \Phi | T | \Phi \rangle + \int dr n(r) v_{ext}(r) + \frac{1}{2} \int dr_1 dr_2 \frac{n(r_1) n(r_2)}{r_{12}} + E_{xc}^{KS}[n]. \tag{6.6}$$

In other words, KS-DFT includes the usual one-body terms and Coulomb interaction between electrons and describes many-body exchange–correlation effects with a functional $E_{xc}^{KS}[n]$ of the density alone. The density is obtained from a reference system of noninteracting electrons in which the external potential is augmented by the Hartree potential and by an additional local potential v_{xc}^{KS} which accounts for the effects of the exchange and correlation interactions on the orbitals. This local potential is the functional derivative of the exchange–correlation energy E_{xc}^{KS} with respect to the density.

In principle, the exchange component of E_{xc}^{KS} can be taken directly from HF, with the sole distinction being that the density matrix should be constructed from KS orbitals. The functional derivative to build v_x^{KS} would then yield the optimized effective potential (OEP) [8, 9] which is in some sense the best local variant of v_x^{HF}. However, numerical computation of the OEP can be challenging, particularly in a gaussian basis set [10]. Numerous approximations to OEP can solve this problem [11–16], but all remain computationally quite demanding in extended systems because they must construct the nonlocal exchange operator. In practice, HF-type exchange is almost never used in a genuine KS calculation.

The chief difficulty in KS-DFT is the need to approximate E_{xc}^{KS} and v_{xc}^{KS} in practical calculations. Typical density functional approximations (DFAs) include the local density approximation (LDA) and generalized gradient approximations (GGAs) such as the functional of Perdew, Burke, and Ernzerhof (PBE) [17]. These are examples of what we shall term semilocal functionals, in which the exchange–correlation potential and energy density at a point depend only on the density and possibly its derivatives at that point. In effect, they expand the exchange–correlation energy around the homogeneous electron gas result, though they include a rather general dependence on the density gradient which allows them to satisfy known constraints on E_{xc}^{KS} that the second-order gradient expansion violates. Alternatively, one can view GGAs as curing the second-order gradient expansion's violations of known constraints on the exchange–correlation hole [18, 19]. Meta-GGAs such as the TPSS functional of Tao, Perdew, Staroverov, and Scuseria [20] add dependence on the local kinetic energy density, but are also semilocal in character.

For simple solids, these semilocal functionals are reasonably accurate, presumably because their electron densities tend to be slowly varying and the homogeneous electron gas is thus a reasonable starting point. This accuracy, however, does not extend to the description of the band gap, which is severely underestimated. For more complicated systems which may include localized electronic states, magnetic effects, rapidly varying densities, or other complicating factors, semilocal functionals are generally inadequate. Alternative semilocal functionals such as PBEsol [21] and revTPSS [22] can be constructed, but while these yield better lattice parameters and bulk moduli than their parents (respectively, PBE and TPSS), they do not remedy the failures of semilocal DFT in systems with the foregoing features, nor do they yield correct band gaps. Formally, this is largely because semilocal functionals do not include the derivative discontinuity of exact KS (see below) [23].

One possible solution which is very successful in molecular systems is to use a global hybrid functional [24, 25], which mixes a fraction of nonlocal HF-type exchange with conventional semilocal exchange:

$$E_{xc}^{hybrid} = E_{xc}^{DFA} + c_{HF}(E_x^{HF} - E_x^{DFA}). \tag{6.7}$$

Common hybrids include the three-parameter B3LYP hybrid [26] and the nonempirical [27] PBEh global hybrid [28, 29]. It should be emphasized that hybrid calculations are almost always done in what is called the generalized Kohn–Sham (GKS) sense [30], in which the nonlocal exchange *energy* yields a nonlocal exchange *interaction*.

Unfortunately, as already mentioned, nonlocal exchange is problematic in solids. Therefore, Heyd, Scuseria, and Ernzerhof introduced the Heyd–Scuseria–Ernzerhof

(HSE) screened hybrid [31–35]. In HSE, the electron–electron interaction is split into a short-range part and a long-range part, as

$$\frac{1}{r_{12}} = \underbrace{\frac{\text{erfc}(\omega r_{12})}{r_{12}}}_{\text{SR}} + \underbrace{\frac{\text{erf}(\omega r_{12})}{r_{12}}}_{\text{LR}}, \qquad (6.8)$$

with a screening parameter ω which in the latest (HSE06) variant of the functional is numerically $\omega = 0.11 a_0^{-1}$ in terms of the Bohr radius a_0. The short-range part is treated as in the PBEh global hybrid – that is, it uses 25% short-range exact exchange and 75% short-range PBE exchange – while the long-range part is treated purely by PBE. As ω goes to 0, the short-range part dominates and HSE reduces to PBEh; as ω goes to infinity, the short-range part vanishes and HSE reduces to PBE. By screening the electron–electron interaction in this way, one vastly reduces the expense of calculating the exact exchange interaction. Note that while the exchange interaction is short range in nature, the range over which exact exchange is included is approximately $1/\omega \sim 9 a_0$, allowing delocalization to nearest and next-nearest neighboring atoms, which admits some form of nonlocality into the model without including all of it. It is also worth noting that Savin introduced the concept of range-separation in DFT [36–38], though he suggested the use of nonlocal exchange in the long range where HSE includes it instead in the short range.

Remarkably, this simple adjustment to the PBEh global hybrid leads to a computationally affordable functional which is accurate for a wide variety of properties of solids. It is our purpose here to summarize what we believe are the key features of HSE and explain roughly why the functional works; for a more detailed review of the successes and failures of HSE, see Ref. [39]. We begin by discussing what is perhaps the most notable success of HSE (its ability to predict band gaps) in Section 6.2 before stepping back to more carefully consider the physics of screened exchange (SX) in Section 6.3. Section 6.4 discusses several applications of HSE, and we offer a few concluding remarks in Section 6.5.

6.2
Band Gaps

Probably the most important success of HSE is its ability to predict semiconductor band gaps from its band energy differences. Here, we want to rationalize this success and set up the discussion for what follows. Before we begin, however, we had best be precise about what we mean when we say that HSE accurately predicts band gaps.

The fundamental band gap of the system deals with the energy required to add or remove electrons. We define it as

$$\Delta E_{\text{fg}}(N) = E(N+1) + E(N-1) - 2E(N), \qquad (6.9)$$

where $E(N)$ is the energy of the N-electron system. The optical gap instead measures the energy required to excite an electron, and is just the lowest electronic excitation energy of the system. The optical gap is lower than the fundamental gap by the exciton binding energy, which represents the interaction between the excited electron and the hole created upon excitation.

In HF, Koopmans' theorem tells us that the HF fundamental gap is just equal to the HF band energy difference. The HF optical gap can be computed from the time-dependent (TD) linear reponse (generally just known as TD HF) and is lower than the HF fundamental gap; in other words, HF predicts a nonzero exciton binding energy [40, 41]. Due to self-interaction in the unoccupied energy levels, the HF fundamental gap severely overestimates the experimental fundamental gap, at least for semiconductors and typical insulators.

In exact KS, we do not have Koopmans' theorem. However, we can use the Janak–Slater theorem instead [42]. We obtain the well-known result that the exact KS fundamental gap is

$$\Delta E_{fg}^{KS}(N) = \varepsilon_{N+1}^{KS}(N) - \varepsilon_N^{KS}(N) + \Delta_{xc}(N), \tag{6.10}$$

where $\Delta_{xc}(N)$ arises from the discontinuity in v_{xc}^{KS} as a function of N at integer N [43, 44]. Standard semilocal functionals and hybrid functionals in the generalized KS sense do not exhibit this discontinuity [23], so their prediction for the fundamental gap is

$$\Delta E_{fg}^{DFA}(N) = \varepsilon_{N+1}^{DFA}(N) - \varepsilon_N^{DFA}(N), \tag{6.11}$$

where the superscript "DFA" means a conventional density functional approximation.

The appropriate framework for the prediction of the optical gap in exact KS is its TD linear response (i.e., TD KS) [45, 46]. Exact KS must predict the optical gap to be below the fundamental gap, in agreement with experiment, simply by virtue of being a formally exact theory. For semilocal functionals, the optical gap turns out to be [47–50]

$$\Delta E_{og}^{SL}(N) = \varepsilon_{N+1}^{SL}(N) - \varepsilon_N^{SL}(N). \tag{6.12}$$

In other words, semilocal functionals do not describe excitons, and predict the optical gap and fundamental gap to be the same. One must therefore not be dogmatic in claiming that the semilocal band energy difference constitutes a prediction for only one of the optical gap or fundamental gap. Due to self-interaction error, the occupied bands in semilocal DFT are too high in energy, and the semilocal optical gap severely underestimates the experimental optical gap. Thanks to the their nonlocal exchange component, global hybrids predict a nonzero exciton binding energy, and the quality of their prediction for the experimental optical or fundamental gap depends on the amount of nonlocal exchange they incorporate.

Because HSE is done in the generalized KS sense, the HSE band energy difference is the HSE fundamental gap. As a practical matter, the HSE band energy differences correspond better with the experimental optical gap than they do with the experimental fundamental gap, especially as the gap gets larger. Therefore in practice we claim that the HSE band energy difference should be understood as a prediction of the experimental optical gap; because the optical and fundamental gaps are so close in semiconductors, for these systems the HSE band energy difference can also be used as a predictor of the experimental fundamental gap. As the gap gets larger, HSE band energy differences more dramatically underestimate the experimental fundamental gap, and as they get larger still, the HSE band energy difference becomes a poor predictor even for the experimental optical gap.

The success of HSE band energy differences can be rationalized quite straightforwardly, once one knows a few properties of the parent semilocal global hybrid (PBEh) and its underlying GGA (PBE) [51]. As is typically the case for semilocal functionals, PBE band energy differences underestimate the gap. For semiconductors, the PBEh band energy differences overestimate the gap. One can think of HSE as an interpolation between PBE and PBEh; as the range-separation parameter ω varies between 0 and ∞, HSE varies between PBEh and PBE. Thus, there is some nonzero value of ω for which the HSE band energy difference reproduces the experimental gap. Schematically, this is illustrated in Figure 6.1. Note that because HSE includes a fraction of screened exact exchange, it has only a small exciton binding, so the TD HSE optical gap should not differ much from the band energy difference.

While it is reasonably clear that there should exist a value of ω that gives the correct gap for a given system, it is not so clear that this ω should be universal. In fact, it is not (see Ref. [52]). Nonetheless, $\omega = 0.11 a_0^{-1}$ as used in HSE06 seems to be a reasonable system-averaged value across a wide variety of systems. The HSE band energy difference accurately reproduces the optical gap in semiconductors, but severely underestimates the gap in insulators, and the HSE band width in metallic systems is

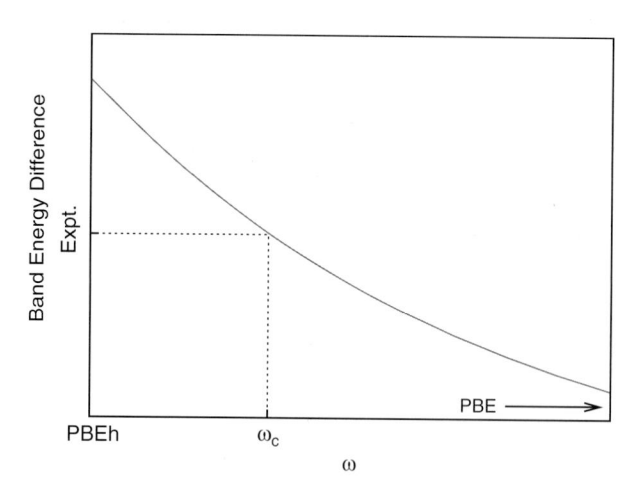

Figure 6.1 (online color at: www.pss-b.com) Schematic illustration of the ω-dependent band energy difference in HSE. The x-axis indicates that $\omega = 0$ corresponds to PBEh and $\omega \to \infty$ corresponds to PBE. The y-axis shows that the experimental gap lies between the PBE and PBEh band energy differences, so that there is some ω_C whose band energy difference reproduces the experimental gap.

generally too large [53, 54]. The failures of HSE for insulators and metals can be traced to the breakdown of the assumption that the PBE gap is too small and the PBEh gap is too large.

6.3 Screened Exchange

In this section we briefly outline a few important relations between the GW approximation, the closely related Coulomb-hole screened exchange (COHSEX) method, and screened hybrid functionals. The pragmatic justification for hybrid functionals is clear (HF and KS-DFT tend to err in opposite directions), as is the pragmatic justification for the use of screened hybrids (they make the calculation less expensive, and are more accurate in solids). There is, however, also a formal justification for screened hybrid DFT: the presence of the other electrons in the systems reduces the full, unscreened exchange used in HF to a certain (material-dependent) extent. We will shortly see that this screening is a direct consequence of the Coulomb correlation between the electrons. For a more rigorous discussion, we refer the reader to Ref. [55].

As explained in Section 6.1, the HF method neglects Coulomb correlation entirely and therefore suffers from several shortcomings. This problem can be overcome by explicitly including correlation corrections. The GW approximation introduced by Hedin [56] in the context of many-body perturbation theory may help. Its name follows from the basic equation defining the frequency dependent self-energy $\Sigma(\omega)$ in terms of the interacting, frequency-dependent Green function $G(\omega)$ and the frequency-dependent *screened* Coulomb interaction $W(\omega)$:

$$\Sigma(\omega) = i\langle GW\rangle(\omega), \tag{6.13}$$

where $\langle GW \rangle$ indicates a convolution over the frequency ω'. Both G and W are nonlocal quantities. The physics is clearly seen by rewriting the screened Coulomb interaction:

$$\begin{aligned}W(\omega) &= \varepsilon^{-1}(\omega)v = (1+v\chi(\omega))v \\ &= v + W_{\text{pol}}(\omega)\end{aligned} \tag{6.14}$$

using the dielectric function $\varepsilon^{-1}(\omega) = (1+v\chi(\omega))$. The dielectric function in turn depends on the polarizability $\chi(\omega)$ and the "bare" (i.e., unscreened) Coulomb kernel $v = 1/|r_1-r_2|$. The screened Coulomb interaction W can therefore be separated into an unscreened term v and the screened or polarizable term $W_{\text{pol}} = v\chi v$, stemming from polarization of the electron density due to a small external potential [48]. Inserting Eq. (6.14) into Eq. (6.13) readily leads to

$$\Sigma(\omega) = iGv + i\langle GW_{\text{pol}}\rangle(\omega) = \Sigma_x + \Sigma_c(\omega), \tag{6.15}$$

demonstrating that the GW approximation goes beyond HF by inclusion of both exchange and correlation effects in the self-energy.

As suggested by Hedin (see Ref. [56]), ignoring the frequency dependence of Σ and taking care of a dependence on virtual orbitals in W_{pol} leads to the so-called static COHSEX approximation. The GW approximation is based on the random phase approximation (RPA) for ε^{-1}, which essentially means a diagrammatic ring or bubble graph expansion for the dielectric function [57]. For the homogeneous electron gas system, the static limit of RPA is the Thomas–Fermi (TF) approximation.[1] In Fourier space, the well known effective TF interaction is

$$W(q, \omega = 0) = \frac{4\pi e^2}{q^2 + k_{\text{TF}}^2}, \qquad k_{\text{TF}}^2 \sim r_s k_F^2, \qquad (6.16)$$

where r_s is known as the Wigner–Seitz radius and the Fermi wave-vector $k_F = (3\pi^2 n)^{1/3}$ (n is the electron density). Note that the TF interaction in real-space corresponds to the Yukawa potential $(e^2/r) \; e^{-k_{\text{TF}} r}$, where the effective screening length is determined by k_{TF}^{-1}. At first glance this can be seen as a route back to motivate the realization of a screened hybrid functional like HSE. Of course, the physics of metals does not directly apply to the physics of semiconductors and insulators. In insulators screening is "weaker" and the effective interaction has a more slowly decaying asymptotic behavior. However, closely related to the spirit of LDA, TF screening is commonly applied in so-called "screened exchange LDA" (SX-LDA) calculations using screened nonlocal exchange and LDA correlation potentials (see Ref. [30] and references therein). For further discussions and comparisons between the aforementioned SX calculations following Bylander and Kleinman [58] and HSE, we refer to Ref. [39].

6.4
Applications

Now that we have discussed some formal aspects of screened hybrid functionals like HSE, let us briefly touch on its performance for a variety of problems.

The success of HSE in the prediction of band gaps is well documented. As an example, Heyd et al. [59] considered a set of 40 solids of which 35 were semiconductors. After removing those solids for which the experimental gaps were either unavailable or larger than 4 eV, 28 semiconductors remain. Table 6.1 shows the predictions of HSE, LDA, PBE, and TPSS for the band gaps in the semiconductors in that set; the HSE numbers here use HSE06 and differ from those in Ref. [59]. While the semilocal functionals considerably underestimate the gap, HSE is quite accurate. Figure 6.2 shows a scatter plot of the HSE band gaps versus the experimental gaps (most of which are fundamental gaps, and which HSE generally underestimates slightly). These results were confirmed in plane-wave basis by Paier et al. [54] who also pointed out that HSE apparently includes too much exact exchange in metals (where it tends to overestimate band widths) [53] and too little in wide gap

---

[1] For small and large wave-vectors q in the high density limit; see, for example, Ref. [6].

Table 6.1 Mean error (ME) and mean absolute error (MAE) in the semiconductor band gaps for the SC 40 test set of Heyd et al. Results reported at the optimized geometry for each functional.

functional	ME (eV)	MAE (eV)
LSDA	−1.01	1.01
PBE	−0.98	0.98
TPSS	−0.83	0.83
HSE	−0.20	0.28

insulators, where it tends to underestimate the gap. As we have discussed, this is because HSE adopts a single, system-independent screening parameter ω which is most accurate for semiconductors. These same studies also demonstrated that HSE accurately predicts semiconductor lattice constants and gives improved estimates of bulk moduli.

Batista et al. [60] carried out a study on silicon, comparing several functionals (including HSE) to diffusion Monte Carlo (DMC), focusing on predictions for the formation energies of interstitial defects and on the relative stabilities of the diamond and β-tin phases. They found that LDA underestimates the defect formation energies by about 1.5 eV relative to the DMC results, while PBE underestimates them by about 1.0 eV. The TPSS meta-GGA is essentially no improvement on PBE. In sharp contrast to these semilocal functionals, HSE underestimates the DMC results by only 0.25 eV. Similar results are found for the relative stabilities of the diamond and β-tin phases; LDA, PBE, and TPSS underestimate the relative stability by roughly a factor of 2, but HSE is quite close to DMC (and in fact the two agree within the uncertainty of the DMC result). Figure 6.3 illustrates these results graphically.

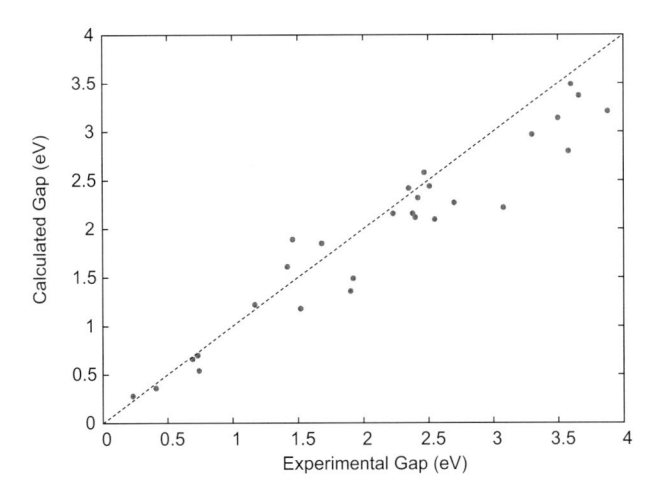

Figure 6.2 (online color at: www.pss-b.com) HSE band gap versus experimental band gap in the SC40 test set.

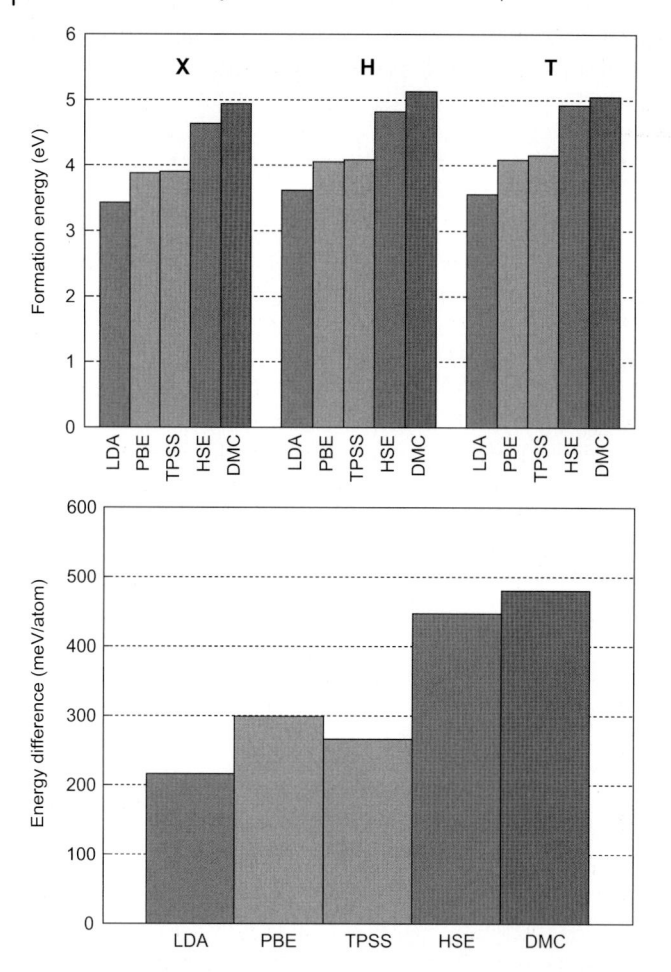

Figure 6.3 (online color at: www.pss-b.com) Comparison between DMC and several DFT functionals for the formation energy of X, H, and T interstitial defects (top panel) and for the relative stability of diamond and β-tin phases (bottom panel) in silicon.

In addition to these accurate results for semiconductors, HSE has properly described Mott insulators. Kasinathan et al. [61] showed that HSE correctly predicts the pressure-dependent Mott transition in MnO.

Lanthanide and actinide systems can be particularly challenging, since the f electrons are sometimes itinerant and other times localized. The balance between these possibilities leads to a multitude of possible ground states and phases. Semilocal functionals are not generally well suited for these systems because they do not properly describe localized states; this causes them to often give qualitatively incorrect results for band gaps and magnetic properties. On the other hand, HSE has had considerable success in describing rare earth systems, since its inclusion of screened exact exchange allows it treat itinerant and localized states in a balanced way.

Prodan et al. [62] studied UO_2, PuO_2, and β-Pu_2O_3. They showed that LDA and PBE incorrectly predict all three systems to be metallic, and TPSS is scarcely better, predicting the antiferromagnetic phases of PuO_2 and β-Pu_2O_3 to have a gap on the order of 0.05 eV. In contrast, HSE is qualitatively correct and provides reasonable gaps. While PBE and TPSS give reasonable lattice constants, HSE predicts essentially the exact result. The semilocal functionals predict that all three systems are ferromagnetic, while HSE correctly predicts that UO_2 and Pu_2O_3 are antiferromagnetic. The magnetic ordering of PuO_2 is controversial, but it is definitely not ferromagnetic and could be, as HSE predicts, antiferromagnetic.

Separate studies by Hay et al. [63], Da Silva et al. [64], and Ganduglia-Pirovano and coworkers [65] showed that HSE is particularly accurate for the structures of CeO_2 and Ce_2O_3, offering significant improvements over semilocal functionals. Semilocal functionals predict the ground state of Ce_2O_3 to be ferromagnetic, while HSE correctly predicts it to be antiferromagnetic. While HSE overestimates the gap for both CeO_2 and Ce_2O_3, it is significantly more accurate than are the semilocal functionals.

Another area where HSE has had considerable success is in the description of carbon nanotubes and graphene nanoribbons. Investigations have shown that HSE accurately predicts the optical excitation spectra of metallic nanotubes [66], and is also quite accurate for optical transitions in semiconducting nanotubes [67]. Additionally, HSE has been used to study the work function of nanotubes [68], as well as their polarizability [69, 70]. The electronic structure of graphene nanoribbons has also been considered [71, 72], and yield predictions that have been experimentally confirmed [73, 74].

Recently, Paier et al. [75] applied HSE and TD HSE to the potential photovoltaic material Cu_2ZnSnS_4. Their HSE results compare very favorably to experimental data for the lattice constants and the band gap, and the HSE band structure has been validated using the more expensive G_0W_0 quasiparticle calculations. Note that both HSE and previous absorption measurements coincide and predict the band gap to be on the order of 1.5 eV. This is in excellent agreement with an independent recent investigation by Chen et al. [76]. Note that HSE is much closer to experiment than PBE, the latter predicting a gap of 0.1 eV [75].

The HSE functional clearly outperforms conventional semilocal functionals for the description of defect transition levels. One important example is localizing acceptor levels in Cu_2O [77]. Chemical intuition suggests a localized $Cu(I) \rightarrow Cu(II)$ oxidation when creating an electron hole in the system. Semilocal functionals artificially delocalize the defect state, whereas HSE gives a very good description and produces defect transition levels in good agreement with experiment.

6.5
Conclusions

SX approximations are very powerful tools for density functional treatments of condensed systems. The HSE screened hybrid functional provides a computationally efficient treatment of lattice parameters, bulk moduli, and band gaps of semiconductors and insulators, while incorporating only a single-empirical parameter. Interestingly, while this parameter was optimized for molecular atomization energies [31], it appears to contain universal information, as the same parameter yields semiconductor band gaps in excellent agreement with experiment. Because it includes a portion of nonlocal exchange and thereby partially alleviates the self-interaction error of semilocal functionals, HSE more correctly describes localized states. This is particularly relevant in the consideration of defects, where semilocal functionals will tend to incorrectly favor delocalized behavior.

The formal aspects of range-separated DFT exchange have been extensively investigated over decades, but several intellectually stimulating questions remain. One is to further elucidate the entanglement between correlation and exchange effects, or in other words, the mechanisms by which many-body correlations screen the exchange interaction. In particular, insights from many-body perturbation theory should help devise more effective screening models which can be applied to DFT. A single range-separation parameter seems sufficient to describe most of the physics we need for semiconductors, but additional flexibility seems to be important if the same functional is also to describe metals or insulators. A straightforward way to add this flexibility is to add additional ranges in what we would term a multirange hybrid [78, 79]. In order to more completely incorporate the physics of SX, one might consider position-dependent fractions of exact exchange [80–87] or position-dependent range separation [88, 89] in successors to HSE. Nonempirical treatments of these quantities are desirable, and should be provided by connections to many-body theory.

Acknowledgements

This work was supported by the National Science Foundation (CHE-0807194), the Department of Energy (DE-FG02-04ER15523 and DE-FG02-09ER16053), a Los Alamos National Labs subcontract (81277-001-10), and the Welch Foundation (C-0036).

References

1 Ruzsinszky, A., Perdew, J.P., Csonka, G.I., Vydrov, O.A., and Scuserua, G.E. (2006) *J. Chem. Phys.*, **125**, 194112.
2 Mori-Sánchez, P. and Cohen, A.J. (2006) *J. Chem. Phys.*, **125**, 201102.
3 Mori-Sánchez, P., Cohen, A.J., and Yang, W. (2008) *Phys. Rev. Lett.*, **100**, 146401.
4 Monkhorst, H.J. (1979) *Phys. Rev. B*, **20**, 1504.
5 Parr, R.G. and Yang, W. 1989 *Density Functional Theory of Atoms and Molecules* (Oxford University Press, New York).
6 Dreizler, R.M. and Gross, E.K.U. (1995) *Density Functional Theory* (Plenum Press, New York).
7 Scuseria, G.E. and Staroverov, V.N. (2005) in: *Theory and Applications of Computational Chemistry: The First 40 Years*, (eds C.E., Dykstra, G., Frenking,

K.S., Kim, and G.E., Scuseria), Elsevier, Amsterdam, The Netherlands, pp. 669–724.
8 Sharp, R.T. and Horton, G.K. (1953) *Phys. Rev.*, **90**, 317.
9 Talman, J.D. and Shadwick, W.F. (1976) *Phys. Rev. A*, **14**, 36.
10 Staroverov, V.N., Scuseria, G.E., and Davidson, E.R. (2006) *J. Chem. Phys.*, **124**, 141103.
11 Sala, F.D. and Görling, A. (2001) *J. Chem. Phys.*, **115**, 5718.
12 Gritsenko, O.V. and Baerends, E.J. (2001) *Phys. Rev. A.*, **64**, 042506.
13 Grüning, M., Gritsenko, O.V., and Baerends, E.J. (2002) *J. Chem. Phys.*, **116**, 6435.
14 Staroverov, V.N., Scuseria, G.E., and Davidson, E.R. (2006) *J. Chem. Phys.*, **125**, 081104.
15 Izmaylov, A.F., Staroverov, V.N., Scuseria, G.E., and Davidson, E.R. (2007) *J. Chem. Phys.*, **127**, 084113.
16 Bulat, F.A. and Levy, M. (2009) *Phys. Rev. A*, **80**, 052510.
17 Perdew, J.P., Burke, K., and Ernzerhof, M. (1996); *Phys. Rev. Lett.*, **77**, 3865; *Phys. Rev. Lett.*, **78**, 1396(E) (1997).
18 Perdew, J.P. and Wang, Y. (1986) *Phys. Rev. B*, **33**, 8800; *Phys. Rev. B*, **40**, 3399(E) (1989).
19 Perdew, J.P., Burke, K., and Wang, Y. (1996) *Phys. Rev. B*, **54**, 16533.
20 Tao, J., Perdew, J.P., Staroverov, V.N., and Scuseria, G.E. (2003) *Phys. Rev. Lett.*, **91**, 146401.
21 Perdew, J.P., Ruzsinszky, A., Csonka, G.I., Vydrov, O.A., Scuseria, G.E., Constantin, L.A., Zhou, X., and Burke, K. (2008) *Phys. Rev. Lett.*, **100**, 136406; Erratum: *Phys. Rev. Lett.*, **102**, 039902 (2009).
22 Perdew, J.P., Ruzsinszky, A., Csonka, G.I., Constantin, L.A., and Sun, J. (2009) *Phys. Rev. Lett.*, **103**, 026403.
23 Cohen, A.J., Mori-Sánchez, P., and Yang, W. (2008) *Phys Rev. B*, **77**, 115123.
24 Becke, A.D. (1993) *J. Chem. Phys.*, **98**, 1372.
25 Becke, A.D. (1993) *J. Chem. Phys.*, **98**, 5648.
26 Stephens, P.J., Devlin, F.J., Chabalowski, C.F., and Frisch, M.J. (1994) *J. Phys. Chem.*, **98**, 11623.
27 Perdew, J.P., Ernzerhof, M., and Burke, K. (1996) *J. Chem. Phys.*, **105**, 9982.
28 Ernzerhof, M. and Scuseria, G.E. (1999) *J. Chem. Phys.*, **110**, 5029.
29 Adamo, C. and Barone, V. (1999) *J. Chem. Phys.*, **110**, 6158.
30 Seidl, A., Görling, A., Vogl, P., Majewski, J.A., and Levy, M. (1996) *Phys. Rev. B*, **53**, 3764.
31 Heyd, J., Scuseria, G.E., and Ernzerhof, M. (2003) *J. Chem. Phys.*, **118**(18), 8207.
32 Heyd, J. and Scuseria, G.E. (2004) *J. Chem. Phys.*, **120**, 7274.
33 Heyd, J. and Scuseria, G.E. (2004) *J. Chem. Phys.*, **121**, 1187.
34 Heyd, J., Scuseria, G.E., and Ernzerhof, M. (2006) *J. Chem. Phys.*, **124**, 219906.
35 Henderson, T.M., Izmaylov, A.F., Scalmani, G., and Scuseria, G.E. (2009) *J. Chem. Phys.*, **131**, 044108.
36 Savin, A. and Flad, H.J. (1995) *Int. J. Quantum Chem.*, **56**, 327.
37 Leininger, T., Stoll, H., Werner, H.J., and Savin, A. (1997) *Chem. Phys. Lett.*, **275**, 151.
38 Savin, A. (1996) in: *Recent Developments and Applications of Modern Density Functional Theory*, (ed. J.M. Seminario), Elsevier, Amsterdam, The Netherlands, pp. 327–357.
39 Janesko, B.G., Henderson, T.M., and Scuseria, G.E. (2009) *Phys. Chem. Chem. Phys.*, **11**, 443.
40 Hanke, W. (1978) *Adv. Phys.*, **27**, 287.
41 Bruneval, F., Sottile, F., Olevano, V., and Reining, L. (2006) *J. Chem. Phys.*, **124**, 144113.
42 Janak, J.F. (1978) *Phys. Rev. B*, **18**, 7165.
43 Perdew, J.P. and Levy, M. (1983) *Phys. Rev. Lett.*, **51**, 1884.
44 Sham, L.J. and Schlüter, M. (1983) *Phys. Rev. Lett.*, **51**, 1888.
45 Gross, E.K.U. and Dreizler, R.M. (eds.), *Density Functional Theory II* (Springer, Heidelberg, 1996).
46 Casida, M.E. (1995) in: *Recent Advances in Density Functional Methods, Part I*, (ed. D.P., Chong), World Scientific, Singapore, p. 155.
47 Hirata, S., Head-Gordon, M., and Bartlett, R.J. (1999) *J. Chem. Phys.*, **111**, 10774.
48 Onida, G., Reining, L., and Rubio, A. (2002) *Rev. Mod. Phys.*, **74**, 601.

49 Botti, S., Schindlmayr, A., Sole, R.D., and Reining, L. (2007) *Rep. Prog. Phys.*, **70**, 357.
50 Izmaylov, A.F. and Scuseria, G.E. (2008) *J. Chem. Phys.*, **129**, 034101.
51 Brothers, E.N., Izmaylov, A.F., Normand, J.O., Barone, V., and Scuseria, G.E. (2008) *J. Chem. Phys.*, **129**, 011102.
52 Paier, J., Marsman, M., and Kresse, G. (2008) *Phys. Rev. B*, **78**, 121201.
53 Stroppa, A., Termentzidis, K., Paier, J., Kresse, G., and Hafner, J. (2007) *Phys. Rev. B*, **76**, 195440.
54 Paier, J., Marsman, M., Hummer, K., Kresse, G., Gerber, I.C., and Ángyán, J.A. (2006) *J. Chem. Phys.*, **124**, 154709; Erratum: *J. Chem. Phys.*, **125**, 249901 (2006).
55 Del-Sole, R., Reining, L., and Godby, R.W. (1994) *Phys. Rev. B*, **49**, 8024.
56 Hedin, L. (1965) *Phys. Rev.*, **139**, A796.
57 Hedin, L. (1999) *J. Phys.: Condens. Matter*, **11**, R489.
58 Bylander, D.M. and Kleinman, L. (1990) *Phys. Rev. B*, **41**, 7868.
59 Heyd, J., Peralta, J.E., Scuseria, G.E., and Martin, R.L. (2005) *J. Chem. Phys.*, **123**, 174101.
60 Batista, E., Heyd, J., Hennig, R.G., Uberuaga, B.P., Margin, R.L., Scuseria, G.E., Umrigar, C.J., and Wilkins, J.W. (2006) *Phys. Rev. B*, **74**, 121102.
61 Kasinathan, D., Kunes, J., Koepernik, K., Diaconu, C.V., Martin, R., Prodan, I.D., Scuseria, G.E., Spaldin, N., Petit, L., Schulthess, T.C., and Pickett, W.E. (2006) *Phys. Rev. B*, **74**, 195110.
62 Prodan, I.D., Scuseria, G.E., and Martin, R.L. (2007) *Phys. Rev. B*, **76**, 033101.
63 Hay, P.J., Martin, R.L., Uddin, J., and Scuseria, G.E. (2006) *J. Chem. Phys.*, **125**, 034712.
64 Da Silva, J. L. F., Ganduglia-Pirovano, M.V., Sauer, J., Bayer, V., and Kresse, G. (2007) *Phys. Rev. B*, **75**, 045121.
65 Ganduglia-Pirovano, M.V., Da Silva, J., and Sauer, J. (2009) *Phys. Rev. Lett.*, **102**, 026101.
66 Barone, V., Peralta, J.E., and Scuseria, G.E. (2005) *Nano Lett.*, **5**, 1830.
67 Barone, V., Peralta, J.E., Wert, M., Heyd, J., and Scuseria, G.E. (2005) *Nano Lett.*, **5**, 1621.
68 Barone, V., Peralta, J.E., Uddin, J., and Scuseria, G.E. (2006) *J. Chem. Phys.*, **124**, 024709.
69 Brothers, E.N., Scuseria, G.E., and Kudin, K.N. (2006) *J. Phys. Chem. B*, **110**, 12860.
70 Brothers, E.N., Izmaylov, A.F., Scuseria, G.E., and Kudin, K.N. (2008) *J. Phys. Chem. C*, **112**, 1396.
71 Ezawa, M. (2006) *Phys. Rev. B*, **73**, 045432.
72 Barone, V., Hod, O., and Scuseria, G.E. (2006) *Nano Lett.*, **6**, 2748.
73 Hand, M.Y., Özyilmaz, B., Zhang, Y., and Lim, P. (2007) *Phys. Rev. Lett.*, **98**, 206805.
74 Chen, Z., Lu, Y.M., Rooks, M.J., and Avouris, P. (2007) *Physica E*, **40**, 228.
75 Paier, J., Asahi, R., Nagoya, A., and Kresse, G. (2009) *Phys. Rev. B*, **79**, 115126.
76 Chen, S., Gong, X.G., Walsh, A., and Wei, S.H. (2009) *Appl. Phys. Lett.*, **94**, 041903.
77 Scanlon, D.O., Morgan, B.J., Watson, G.W., and Walsh, A. (2009) *Phys. Rev. Lett.*, **103**, 096405.
78 Henderson, T.M., Izmaylov, A.F., Scuseria, G.E., and Savin, A. (2007) *J. Chem. Phys.*, **127**, 221103.
79 Henderson, T.M., Izmaylov, A.F., Scuseria, G.E., and Savin, A. (2008) *J. Theor. Comput. Chem.*, **4**, 1254.
80 Burke, K., Cruz, F.G., and Lam, K.C. (1998) *J. Chem. Phys.*, **109**, 8161.
81 Jaramillo, J., Scuseria, G.E., and Ernzerhof, M. (2003) *J. Chem. Phys.*, **118**, 1068.
82 Janesko, B.G. and Scuseria, G.E. (2007) *J. Chem. Phys.*, **127**, 164117.
83 Bahmann, H., Rodenberg, A., Arbuznikov, A.V., and Kaupp, M. (2007) *J. Chem. Phys.*, **126**, 011103.
84 Arbuznikov, A.V. and Kaupp, M. (2007) *Chem. Phys. Lett.*, **440**, 160.
85 Kaupp, M., Bahmann, H., and Arbuznikov, A.V. (2007) *J. Chem. Phys.*, **127**, 194102.
86 Janesko, B.G. and Scuseria, G.E. (2008) *J. Chem. Phys.*, **128**, 084111.
87 Haunschild, R., Janesko, B.G., and Scuseria, G.E. (2009) *J. Chem. Phys.*, **131**, 154112.
88 Krukau, A.V., Scuseria, G.E., Perdew, J.P., and Savin, A. (2008) *J. Chem. Phys.*, **129**, 124103.
89 Henderson, T.M., Janesko, B.G., Scuseria, G.E., and Savin, A. (2009) *J. Quantum Chem.*, **109**, 2023.

7
Defect Levels Through Hybrid Density Functionals: Insights and Applications

Audrius Alkauskas, Peter Broqvist, and Alfredo Pasquarello

7.1
Introduction

Defects strongly affect properties of materials. For example, doping a semiconductor with a small number of impurity atoms leads to a significant change of its conductivity making such materials useful for technological applications. Similarly, optical transitions involving electronic states of defects can induce a coloration of the otherwise transparent solid, a frequently encountered phenomenon in natural crystals. Also, the mechanical properties and the long-term stability of materials are largely controlled by point and line defects.

It is therefore not suprising that the theoretical study of point defects in solid materials has a long history [1]. Many properties of defects are nowadays well understood. These include for instance the nature of hydrogenic impurities in elemental semiconductors and the energy splittings resulting from local crystal fields. However, these kinds of defects represent just a small class of all possible defects.

Technological developments, particularly in the areas associated to energy and information, lead to the consideration of a vast variety of novel and complex materials. A nonexhaustive list of applications includes solar cells [2], novel metal–oxide–semiconductor field-effect transistors [3], longer-serving batteries [4], solid-state light-emitting diodes [5], and solid fuel cells. The behavior of such devices is generally influenced or governed by a myriad of defects that form in the bulk or at the interfaces between the different materials. Those defects are often *deep*, i.e., they are characterized by localized electronic states which bare little resemblance to the electronic states of the host material and possess ionization energies much larger than typical thermal energies. The experimental characterization of such defects is often very difficult, and thus theoretical studies are not only valuable but also essential. However, because of their deep nature, the theoretical description of the associated electronic states is beyond the reach of simple models. This explains the continuous efforts deployed in the theoretical studies of defects [6].

To study localized defects with deep energy levels, it is necessary to treat the atomic and the electronic structure in a self-consistent way. In the last decades, density functional theory (DFT) has been the workhorse for such calculations. Since the exchange–correlation energy, a crucial ingredient in the theory, is not available in an exact form, practical calculations generally rely on approximate expressions, such as the local density approximation (LDA) or the generalized gradient approximation (GGA). Though largely successful, these standard approximations to DFT suffer from several shortcomings. The most serious one for the study of defect levels is the infamous "band-gap problem." Band gaps calculated in the LDA and the GGA are significantly smaller than experimental ones. In some cases, a vanishing band gap is obtained for materials which possess a finite one. In the present paper, we mainly focus on formation energies of defects in different charge states and on the associated electronic transition levels. The defect charge state depends on the electron chemical potential, for which the band gap is the relevant energy scale. Therefore, a correct reproduction of the bulk band gap is imperative for achieving a successful theoretical description [7–15].

There exist many methods which go beyond semilocal approximations to DFT and which alleviate the "band-gap problem." Examples of the application of different theoretical methods, some specific to the defect problem, can be found in Refs. [16–35]. In this review article, we focus on hybrid density functionals. These functionals employ the correlation potential from semilocal approximations, but admix a small fraction of nonlocal exchange to the exchange described within the GGA [36–38]. Since band gaps are underestimated with semilocal density functionals and overestimated with full Hartree–Fock exchange, they are naturally improved through the use of hybrid functionals [39, 40]. We here limit the discussion to hybrid functionals based on bare Coulomb exchange. Recently, a variety of hybrid functionals have been proposed, including *screened* nonlocal exchange [41, 42], spatially varying mixing coefficients, and range-separated functionals. For an overview, we refer the reader to Ref. [43].

The present review article is organized as follows. In Section 7.2 we describe the computational toolbox employed in our calculations, and then discuss the quantities that need to be calculated. The hybrid functionals used in the present work are introduced and their performance is discussed. The comparison between defect energy levels calculated with semilocal functionals on the one hand and with hybrid functionals on the other hand provides some fundamental insight into the properties of deep defects. These are presented and discussed in Section 7.3. In order to improve the description of the band gap in hybrid functionals, it has become common practice to adjust the fraction of admixed nonlocal exchange. In Section 7.4, several arguments supporting this empirical procedure are discussed. In Section 7.5 we apply our theoretical scheme to two defect systems and compare our results with available experimental data. Finally, in Section 7.6 we discuss shortcomings and advantages of the proposed scheme, and draw conclusions.

7.2
Computational Toolbox

The calculations in this work were performed within the plane-wave pseudopotential scheme. Defect systems and interfaces are modeled through large supercells. The ionic cores are described with soft norm-conserving pseudopotentials [44]. Throughout this study, we used a kinetic energy cutoff of 70 Ry, sufficiently high to converge the properties of all the pseudopotentials in this work, including Hf, O, and C. As reference GGA functional, we adopted the functional proposed by Perdew, Burke, and Ernzerhof (PBE) [45]. Pseudopotentials were generated at the PBE level and were used unmodified in calculations with hybrid functionals. Although this is not formally justified, this approach gives a very good description for systems with first-row elements, as can be inferred from comparisons with all-electron calculations [46]. However, for heavier elements this approximation may admittedly involve errors [47]. The Brillouin zone was sampled at the sole Γ-point in most calculations, but denser k-point meshes were used when necessary, for example in the determination of accurate band-edge shifts. Further technical details about specific systems are given below. Structural relaxations were carried out at the PBE level. We used the codes CPMD [48, 49] and Quantum-ESPRESSO [50].

7.2.1
Defect Formation Energies and Charge Transition Levels

The principal quantity that needs to be calculated is the formation energy $E_f^{D,q}$ of the defect D in its charge state q as a function of the electron chemical potential E_F [51]:

$$E_f^{D,q}(E_F) = E_{tot}^{D,q} - E_{tot}^{bulk} - \sum_\alpha n_\alpha \mu_\alpha + q(E_V + E_F). \tag{7.1}$$

In this expression $E_{tot}^{D,q}$ is the total energy of the defect system, E_{tot}^{bulk} the total energy of the unperturbed host, n_α the number of extra atoms of species α needed to create the defect D, and μ_α is the corresponding atomic chemical potential. The electron chemical potential is referred to the valence band maximum (VBM) E_V. It varies between zero and the band-gap E_g. Charge transition levels correspond to specific values of the electron chemical potential for which two charge states have equal formation energies. Let us for example consider charge states q and q'. Equating the expressions of the formation energies defined in Eq. (7.1);, we obtain the value for the charge transition level $\varepsilon(q/q')$:

$$\varepsilon(q/q') = \frac{E_{tot}^{D,q} - E_{tot}^{D,q'}}{q'-q} - E_V. \tag{7.2}$$

For example, the charge transition level $\varepsilon(0/+)$ is given via:

$$\varepsilon(0/+) = E_{tot}^{D,0} - E_{tot}^{D,+} - E_V. \tag{7.3}$$

The total energies of charged systems appearing in Eq. (7.1) need to be corrected to speed up the convergence with respect to the supercell size. First, the total energy is corrected by a term $q\Delta V$, ΔV being the potential difference needed to align the potential far from the *neutral* defect to that of the unperturbed bulk. Second, the total energy is corrected for the spurious electrostatic interaction due to the periodic boundary conditions, for which we use the electrostatic correction of Makov and

Payne [52]. These two corrections are always used unless otherwise stated. While the Makov–Payne correction is known to fail in some specific cases, it is generally quite accurate in the case of extremely localized defects. For instance, for defects in SiO_2, the charge transition levels are already converged for moderate supercell sizes (72 atoms) when this electrostatic correction is included. When accurate quantitative results are needed, it is recommended to use either careful extrapolation schemes [53–55] or more elaborate methods for correcting the electrostatic interactions in the supercell [14, 56]. We note that in this respect the present work is mostly concerned with the comparison between results obtained with different functionals, for which the electrostatic corrections are nearly the same and thus do not represent an issue.

Since the electron chemical potential in experiments is generally referred to the VBM, $\varepsilon(q/q')$ defined in Eq. (7.2) is the relevant physical quantity. However, we find useful in Section 7.3 to consider charge transition levels $\bar{\varepsilon}(q/q')$ referred to an appropriately defined average local potential ϕ in the supercell rather than to the VBM:

$$\bar{\varepsilon}(q/q') = \frac{E_{tot}^{D,q} - E_{tot}^{D,q'}}{q'-q} - \phi. \tag{7.4}$$

7.2.2
Hybrid Density Functionals

While a number of hybrid functionals have been proposed, we focus in this paper only on one-parameter hybrid functionals based on the bare Coulomb exchange, in which a fraction a of nonlocal exact exchange is admixed to the exchange described within the GGA. By nonlocal exact exchange we here refer to the orbital-dependent expression for exchange appearing in the Hartree–Fock theory [57]. This leads to a generalized Kohn–Sham scheme in which the exchange potential is different for each electronic state the non-local part of which is defined as $\hat{V}^i \psi_i = \partial E_x^{exact}/\partial \psi_i$. The exchange energy is thus given by

$$E_x^{hybrid} = a E_x^{exact} + (1-a) E_x^{GGA}. \tag{7.5}$$

The correlation potential is usually taken unmodified from the GGA. When the fraction $a = 0.25$ is used together with the PBE approximation for the semilocal part [38], the hybrid functional is referred to as the PBE hybrid. We here use the notation PBE0, but other notations such as PBEh or PBE1PBE are also in use. The value of $a = 0.25$ has been rationalized in case of molecular systems and is considered to be a good compromise for many systems [38]. However, there is no firm theoretical justification for this choice and the optimal mixing coefficient is admittedly system or even property dependent [38, 58].

For many materials, the PBE0 provides an improved description of the energetic and structural properties when compared to the PBE [40, 59]. Lattice constants, formation energies, and bulk moduli of semiconductors and insulators are generally in a better agreement with experimental data [59]. A similar improvement is also

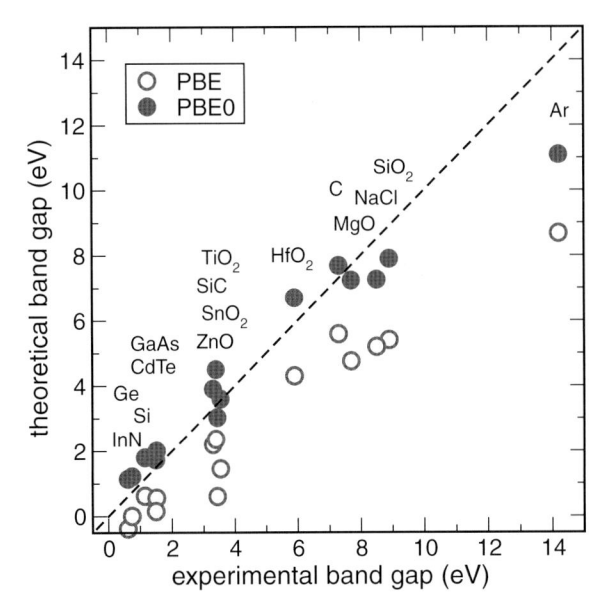

Figure 7.1 (online color at: www.pss-b.com) Calculated versus measured single-particle band gaps for 15 different materials. PBE: open disks, PBE0: filled disks. Results for GaAs, C, MgO, NaCl, and Ar are taken from Ref. [40]; results for InN and ZnO are taken from Ref. [60]; the result for SnO$_2$ is taken from Ref. [33]; the result for TiO$_2$ (anatase form) is taken from Ref. [61]; results for Si, SiC (4H polytype), HfO$_2$, CdTe, and SiO$_2$ are from our calculations.

observed for molecules. However, for metallic systems the use of exact exchange gives rise to unphysical derivative discontinuities at the Fermi level.

Even more importantly than the improvement in structure and energetics, the PBE0 substantially improves the calculated band gaps [40]. This is shown in Figure 7.1. The improvement is especially evident for materials such as Ge or InN which have a vanishing or even negative band gap in semilocal approximations. However, it is also evident that the improvement of the PBE0 over the PBE is not systematic. In the PBE0, band gaps are overestimated for low band-gap materials and underestimated for large band-gap ones. The best agreement is thus for intermediate band-gap materials. It is also clear why the fraction $a = 0.25$ is just a good compromise rather than a universal parameter. In fact for both the PBE and the PBE0 the dependence of the theoretical gap on the experimental one is approximately described by a concave function. For PBE0 ($a = 0.25$), the concave function crosses the diagonal defined by $E_g^{th} = E_g^{expt}$ at about 5 eV. Hence, the theoretical band gaps are overestimated for some materials and underestimated for others. This behavior applies to hybrid functionals with any other reasonable mixing coefficient a.

7.2.2.1 Integrable Divergence

In calculations based on plane-wave basis sets and periodic boundary conditions, exact exchange poses one more challenge due to the long-range nature of the Coulomb interaction. In Fourier space this interaction is proportional to $1/q^2$, and thus diverges at small q. The singularity is integrable, but its straightforward calculation would require very dense k-point meshes. Gygi and Baldereschi [62] proposed a method to treat this singularity. In the matrix element of the exchange operator, they normalized the integrand by substracting an auxiliary function which admits an analytical integration over the Brillouin zone. This method is suitable to be adapted to much sparser k-point samplings, including those limited to the sole Γ-point [46]. The effect of the singularity of the exchange potential can be cast into a correction to the $G = 0$ term of the potential [46, 63]. The Fourier transform $\Phi(G)$ of the exchange interaction is then given by

$$\Phi(G) = \begin{cases} \dfrac{1}{\Omega} \dfrac{4\pi}{G^2} & \text{for } G \neq 0, \\ \chi & \text{for } G = 0, \end{cases} \quad (7.6)$$

where Ω is the volume of the supercell and the singularity correction χ is expressed as

$$\chi = \lim_{\gamma \to 0} \left[\dfrac{1}{\sqrt{\pi\gamma}} - \dfrac{4\pi}{\Omega} \sum_G \dfrac{e^{-\gamma G^2}}{G^2} \right]. \quad (7.7)$$

The total energy of a system of N_{el} electrons is thus corrected by a term $-a\chi N_{el}/2$, where a is the fraction of exact exchange used in the hybrid functional. In Figure 7.2, we give the total energies of Si and SiO$_2$ as a function of the supercell size and/or the density of the k-point mesh for both PBE and PBE0 functionals [46]. In the latter case, the total energies are given with and without the singularity correction. When

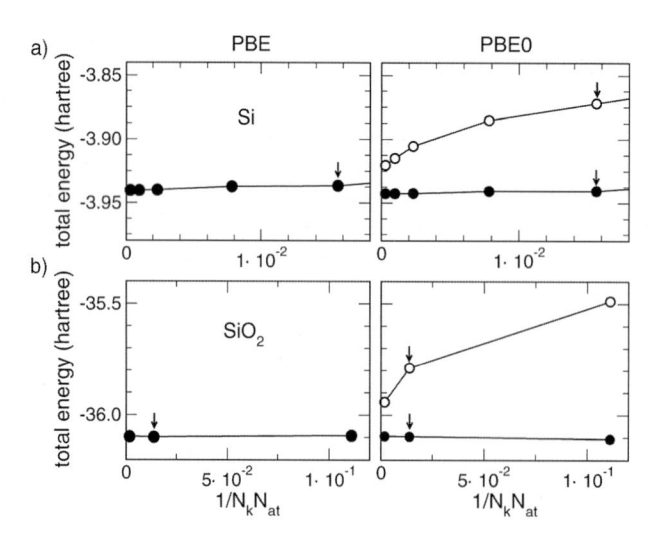

Figure 7.2 Total energies of (a) Si and (b) α-quartz SiO$_2$ per formula unit versus $1/N_k N_{at}$, where N_k is the total number of k-points and N_{at} the total number of atoms in the supercell. Results obtained in the PBE and the PBE0 are reported in left and right panels, respectively. For PBE0, closed and open symbols indicate values obtained with the singularity correction turned on and off, respectively. Arrows show data points which were also obtained with Γ-point sampling.

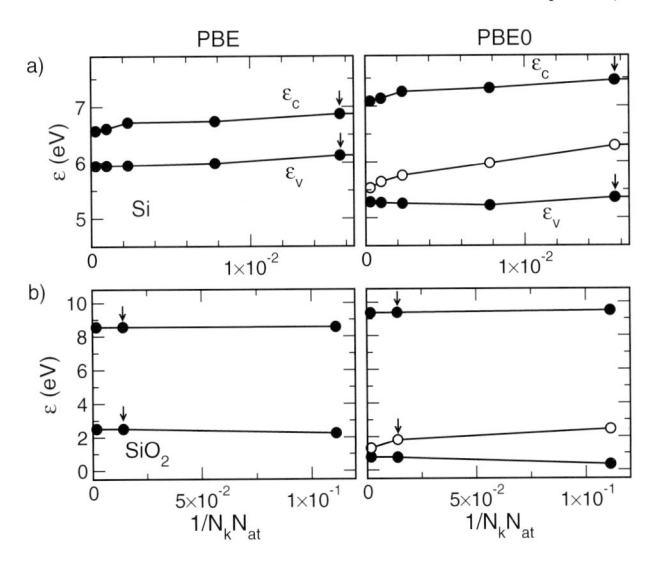

Figure 7.3 Eigenvalues corresponding to the valence (ε_v) and conduction (ε_c) band edges of (a) Si and (b) α-quartz SiO$_2$ versus $1/N_k N_{at}$, where N_k is the total number of k-points and N_{at} the total number of atoms in the supercell. Results obtained in the PBE and the PBE0 are reported in left and right panels, respectively. For PBE0, closed and open symbols indicate values obtained with the singularity correction turned on and off, respectively. Arrows show data points which were also obtained with Γ-point sampling. The energies obtained with the two functionals are aligned through the average electrostatic potential (cf. Ref. [13]).

the correction is included, the convergence properties of PBE0 calculations closely resemble those of PBE calculations. Without the singularity correction the convergence properties clearly deteriorate.

The singularity correction also affects the single-particle eigenvalues. Eigenvalues of unoccupied states remain unchanged, while those of occupied ones shift by $-a\chi$. This is demonstrated in Figure 7.3 for the cases of Si and SiO$_2$. When the singularity correction is included, hybrid-functional calculations converge as fast as those based on semilocal functionals. This equally holds for calculations with k-point samplings restricted to the Γ-point. Singularity corrections apply equivalently to both delocalized bulk-like states and localized defect or molecular states [46].

Furthermore, singularity corrections are particularly useful in the case of elongated supercells, which may otherwise show an unphysical convergence behavior [46]. Even in the case of screened hybrid functionals which do not show any formal singularity, an analogous treatment of the $q = 0$ limit may lead to a speed up of the convergence with k-point sampling [46].

7.3
General Results from Hybrid Functional Calculations

As shown above, hybrid functionals containing a fixed fraction of exact exchange, such as the PBE0 functional, do not bring theoretical band gaps in agreement with experimental ones for all materials. Thus, while their straightforward application to the determination of defect levels is expected to lead to an improvement with respect to semilocal functionals, the comparison with experiment remains ambiguous. Nevertheless, we can gain insight into how calculated and measured defect levels should be compared by performing a comparative study between defect energy levels calculated with semilocal and hybrid density functionals. Such a study is expected to reveal how defect levels shift as the description of the band gap improves [13, 64, 65].

To this end, we find useful to refer charge transition levels calculated with different functionals to a common reference potential ϕ. We denote such charge transition levels by $\bar{\varepsilon}(q/q')$ (cf. Eq. 7.4). In our pseudopotential supercell approach, ϕ is obtained from the supercell average of the sum of the local pseudopotential and of the Hartree potential. We argue in the following that this alignment is a convenient choice for determining energy-level shifts induced by the hybrid functional with respect to a reference semilocal calculation.

7.3.1
Alignment of Bulk Band Structures

In this section, we focus on the alignment of bulk band structures obtained with semilocal and hybrid functionals. To simplify the reasoning, let us assume that the same supercell parameters and the same pseudopotentials are used in the two calculations [66]. In this case, the pseudopotential contribution to ϕ is the same in the two calculations, and the adopted alignment consists in aligning the average electrostatic potential in the two theoretical schemes. This alignment allows one to position band edges in the hybrid calculation with respect to those in the semilocal one, i.e., to determine the shift of the VBM ΔE_V and the conduction band minimum (CBM) ΔE_C on a common energy scale, as shown in Figure 7.4.

To analyze the significance of the adopted alignment scheme, it is convenient to conceptually refer to the band offset at the interface between two materials, A and B. The band offset is a well-defined physical property that can be measured. Following the scheme introduced by Van de Walle and Martin [67], band offsets can be determined by three calculations, namely an interface calculation from which one extracts the line-up of the local average electrostatic potential across the interface and two bulk calculations of materials A and B which allow one to locate the band edges with respect to the respective average electrostatic potential in each material. This procedure can separately be carried out for the semilocal and for the hybrid scheme.

Alternatively but equivalently, the band alignment in the hybrid scheme can also be obtained from the alignment in the semilocal scheme by the consideration of three sources of difference. By comparing the charge densities in the interface calculations performed at the semilocal and hybrid levels, one can extract the difference in line-up of the average electrostatic potential. Such a difference directly results from the dipoles associated to the difference between the charge densities in the semilocal and in the hybrid schemes. The two other sources of difference can be achieved by separately comparing semilocal and hybrid calculations for bulk materials A and B. The required differences correspond precisely to the shifts undergone by the band

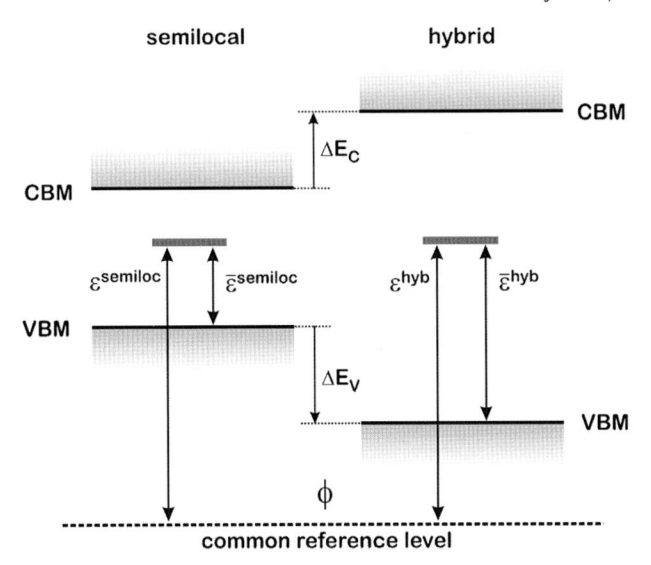

Figure 7.4 (online color at: www.pss-b.com) Charge transition levels calculated within a semilocal and a hybrid functional scheme, aligned to a common reference level φ. ε is the charge transition level referred to the respective VBM (Eq. 7.2), $\bar{\varepsilon}$ is the charge transition level referred to the level φ (Eq. 7.4). ΔE_V and ΔE_C are the shifts of the VBM and of the CBM in the hybrid-functional calculation with respect to the respective edges in the semilocal calculation. Adapted from Ref. [13].

edges when aligned with respect to the average electrostatic potential. This reasoning thus illustrates that the band offsets in the hybrid scheme would be obtained by combining information that can only be extracted from charge density variations as derived from interface calculations with information that can be derived by aligning the energy scales of periodic bulk calculations as proposed.

In the particular case in which material B is the vacuum, we are concerned with a surface system. In this case, it is more natural to adopt the vacuum level at a large distance from the interface as the common reference level for the semilocal and hybrid calculations. From the reasoning in the previous paragraph, it appears clearly that this alignment scheme is not equivalent to the one proposed in this work. Indeed, the alignment to the vacuum level explicitly includes the consideration of a surface and correspondingly charge density differences between the two theoretical schemes might lead to different line-up terms, which would in turn determine a different alignment. Hence, we stress once again that the alignment adopted here is conceptually particularly convenient because it highlights effects of the different theoretical formulations as they result from bulk calculations, without the need for an explicit treatment of interface or surface systems. However, the connection with measurable properties such as work functions and band offsets cannot be made unless such an explicit treatment is considered.

When the charge density is invariant in the two theoretical schemes under comparison, the difference in line-up term vanishes and relative shifts in the presence of an aligned electrostatic potential also acquire direct physical significance. In practical calculations involving semilocal and hybrid calculations, this condition is very close to being satisfied, as demonstrated for both interface [68] and surface systems [69]. In such cases, the relative band-edge shifts determined through an alignment of the average electrostatic potential give the dominant contribution to the variations undergone by the band offsets [68]. In the case of invariant charge densities, the alignment with respect to the average electrostatic potential is fully equivalent to the alignment with respect to a an external vacuum level achieved through the consideration of a surface.

7.3.2
Alignment of Defect Levels

Once the bulk band structures in the two theories are aligned as described in Section 7.3.1, the alignment of charge transition levels is trivial. This is shown in Figure 7.4. Hitherto, the common reference ϕ was taken to be the average electrostatic potential, but it can for convenience be shifted to coincide with the VBM in the hybrid calculation. In this case $\bar{\varepsilon}^{hyb}(q/q') = \varepsilon^{hyb}(q/q')$, and $\bar{\varepsilon}^{semiloc}(q/q') = \varepsilon^{semiloc}(q/q') + \Delta E_V$, ΔE_V being the shift of the valence band in the hybrid calculation with respect to that in the semilocal one.

In Figure 7.5, we compare charge transition levels $\bar{\varepsilon}(q/q')$ calculated with a semilocal functional (PBE) with corresponding ones obtained with a hybrid functional (PBE0) for a large set of deep defects in four different materials [13]. The chosen materials show band gaps covering a wide range of values: Si (with an experimental band gap of 1.17 eV), 4H–SiC (3.3 eV), monoclinic HfO_2 (5.75 eV), and α-quartz SiO_2 (8.9 eV).

Let us first focus on the defects in SiO_2 [64]. Due to the very different band gap of SiO_2 in PBE (5.4 eV) and in PBE0 (7.9 eV), charge transition levels referred to the respective valence band maxima differ significantly. At variance, when the charge transition levels of these defects are referred to the average electrostatic potential, they are very close in the two theoretical schemes. The mean deviation from the ideal alignment is only 0.14 eV. This value is only indicative since it depends on the adopted set of defects. Nevertheless, the alignment of charge transition levels is surprisingly good over a large range of energies. The same alignment property approximately also holds for other materials. For example, in HfO_2 the defect set includes oxygen vacancies and interstitials and the correspondence is similarly very good, with mean deviation of 0.16 eV. The departure from the ideal alignment [$\bar{\varepsilon}^{hyb}(q/q') = \bar{\varepsilon}^{semiloc}(q/q')$] is only slightly larger in SiC and in Si, with a mean deviation of 0.19 eV in both cases.

A detailed inspection reveals that defect levels in the upper part of the band gap tend to shift upwards while those in the lower part tend to shift downwards as the band gap is opened. It was found numerically that the deterioration from the ideal alignment correlates with the increase of the average spread of defect wave functions [13]. In this respect, SiO_2 is an optimal case, because the defect states in this material are characterized by very localized wave functions.

7.3 General Results from Hybrid Functional Calculations

These results can be understood by drawing an analogy between charge transition levels of defect states and ionization potentials or electron affinities of atoms and molecules [13]. The latter quantities can be expressed as total-energy differences and are already well described in semilocal approximations [70, 71], as demonstrated by extensive quantum chemistry calculations [72]. Typical mean deviations of about 0.2 eV are found between calculated and experimental results. Through the use of the Slater–Janak transition-state theory [73, 74], the total energy difference appearing in Eq. (7.4), i.e., $E_{tot}^{D,q} - E_{tot}^{D,q'}$, can be related to a matrix element of the defect state at half occupation, which can then be rationalized to carry the same properties as atomic or molecular states insofar its wave function is sufficiently localized [13, 64]. The ideal alignment $\bar{\varepsilon}^{hyb}(q/q') \approx \bar{\varepsilon}^{semiloc}(q/q')$ is therefore expected to hold best for atomically localized defects and to deteriorate with the extension of the defect wave function.

The correspondence between energy levels in semilocal and hybrid functional schemes does not hold for single-particle eigenvalues of *extended* bulk-like states, as can be inferred from Figure 7.5 for various materials. We stress that this effect should be explained by invoking the delocalized nature of these states rather than a different behavior of eigenvalues and total-energy differences. In fact, the energy of the VBM E_V (and likewise for E_C), appearing in the definition of the charge transition level in

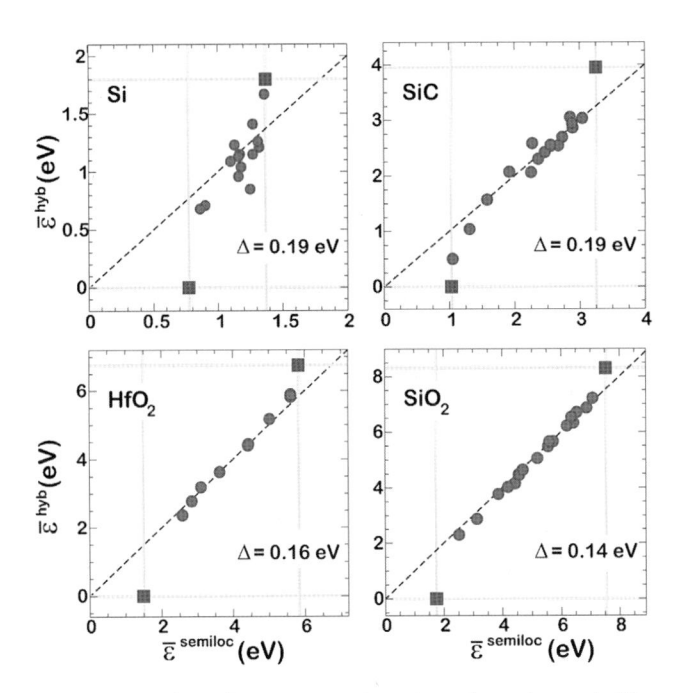

Figure 7.5 (online color at: www.pss-b.com) Comparison between charge transition levels calculated with a semilocal ($\bar{\varepsilon}^{semiloc}$) and a hybrid ($\bar{\varepsilon}^{hyb}$) functional for a variety of defects in Si, SiC, HfO$_2$, and SiO$_2$. The energy levels corresponding to the VBM and CBM are also shown (squares). All energies are referred to a common reference level ϕ (see text), shifted to coincide with the VBM in the hybrid scheme for convenience. For each material, Δ is the r.m.s. error with respect to the ideal alignment (dashed). Adapted from Ref. [13].

Eq. (7.2), can also be expressed through a total-energy difference: $E_V = E_{tot}^{bulk,0} - E_{tot}^{bulk,+}$. However, for delocalized states, in sharp contrast to localized ones, this total-energy difference is subject to large variations when calculated in semilocal and hybrid functional schemes, reflecting the effect of the "band-gap problem" in the same way as single-particle eigenvalues do. This is the main reason why defect charge transition levels in different theoretical schemes differ so much when referred to their respective valence band maxima.

The use of the unknown exact functional [75, 76] would in either case lead to a correct description of the total energies of localized and extended states. The different description of localized and extended states can be related to specific properties of the approximate functional adopted [70, 71, 77, 78]. The success of approximate functionals in describing total energies of localized systems is related to their fulfillment of the sum rule for the exchange–correlation hole [70]. This stringent criterion is fulfilled at integer electron numbers, yielding accurate total energy differences in calculations for atomic and molecular systems [72]. However, such approximate energy functionals fail in reproducing the linear behavior of the exact functional for fractional electron numbers [75, 76]. As has recently been shown by Mori-Sánchez et al. [78], this failure is at the origin of the incorrect description of single-particle eigenvalues and total energies of delocalized systems. Thereby, these theoretical results establish a clear relation between the "band-gap problem" of approximate density functionals and the delocalized or localized nature of electronic states [78]. Over which length scales the transition takes place between localized and delocalized states is at present still a matter of debate. For an interesting discussion on this issue, we refer to Ref. [71].

The results of this section have provided useful indications concerning the way the "band-gap problem" affects energy levels of deep defects. Defects localized on an atomic scale appear to be already well described at the semilocal level when referred to the average electrostatic potential. In particular, this implies that energy separations between such defect levels are accurately described at the semilocal level and barely affected by the "band-gap problem." The calculations indicate that when the defect state becomes more extended this ideal alignment tends to deteriorate. The shortcoming due to the "band-gap problem" only affects delocalized states such as the valence and conduction band edges. While this of course hinders the correct location of defect levels within the band gap, it nevertheless provides significant insight into the way corrections should be made.

7.3.3
Effect of Alignment on Defect Formation Energies

The fact that charge transition levels of deep defects calculated at different levels of theory tend to be aligned when referred to the average electrostatic potential has important implications for the formation energies of charged defects.

In Figure 7.6, we give a diagram which schematically shows the formation energies of a defect in the positive and the neutral charge state as a function of the electron chemical potential, when calculated with a semilocal (dashed lines) or with a hybrid

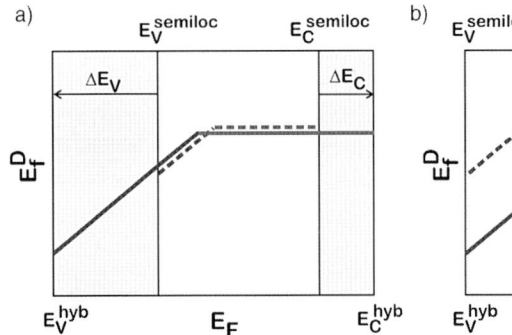

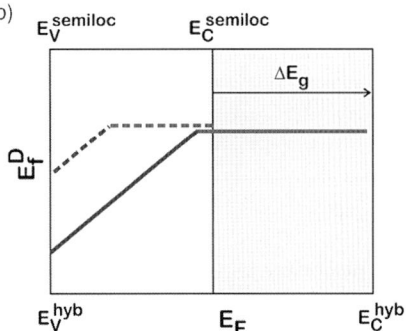

Figure 7.6 (online color at: www.pss-b.com) Formation energies of a point defect as a function of the electron chemical potential E_F calculated with a semilocal (dashed lines) and with a hybrid functional (solid lines). The positive and neutral charge states of the defect are considered. (a) The semilocal and the hybrid calculations are aligned through the average electrostatic potential as in Fig. 7.4; ΔE_V and ΔE_C are the corresponding shifts in the VBM and in the CBM. (b) The VBM of the two calculations are aligned; ΔE_g is the band-gap underestimation in the semilocal calculation with respect to the hybrid one.

(solid lines) functional. The formation energy of a positively charged defect has a positive slope, while that of a neutral defect has a zero slope. The charge transition level corresponds to the value of the electron chemical potential at which the two charge states have equal formation energies. In Figure 7.6a, the transition levels in the two approaches are aligned with respect to the average electrostatic potential as discussed in Section 7.3.1. Let us assume that for the specific defect in Figure 7.6 the charge transition levels in the two theories are indeed very close when referred to the average electrostatic potential. The formation energy of the neutral defect does not depend on the electron chemical potential and we here additionally assume that this energy is quite similar when calculated with semilocal and hybrid functionals. Consequently, this also implies that the formation energies of the positively charged defect are also similar in the two calculations, provided they are taken at the same value of the electron chemical potential referred to the average electrostatic potential. However, since the position of the VBM is different in the two theoretical approaches, the formation energies of the positively charged defect are different when the electron chemical potential is referred to the respective VBM. Thus this clearly illustrates that when the electron chemical potential is found at the VBM as, e.g., for a p-type material, the formation energy of the positively charge defect depends on the location of the VBM relative to the average electrostatic potential. The position of band edges with respect to the average electrostatic potential will be discussed in the next section.

For comparison, we also discuss an alternative alignment scheme which has often been adopted in the literature and which consists in aligning the VBM in the two theoretical approaches, as schematically shown in Figure 7.6b. This alignment scheme assumes that the band-gap problem originates from the wrong placement of the CBM. With this alignment, the charge transition levels are no longer aligned, and the formation energies of the positively charged defect differ for any value of the electron chemical potential. Under the assumption that the VBM are aligned, it appears contradictory that two theoretical approaches that bear similar total energies for the neutral state would instead differ systematically for the positive charge state, especially when such a charge state results from the absence of electrons in the defect state.

7.3.4
"The Band-Edge Problem"

In this section, we elaborate on the "band-gap problem" in relation with the determination of defect levels and argue that it is more appropriate to refer to a "band-edge problem." For this purpose, let us consider two different theories, I and II, which both yield a theoretical band gap in agreement with the experimental one, but different positions of band edges when referred to the average electrostatic potential, as schematically shown in Figure 7.7. For illustration, we consider three kinds of defects. The defect of kind (a) corresponds to an atomically localized defect, for which the energy level does not undergo a significant shift, in accord with our observations in Section 7.3.2. On the other extreme, defect (c) corresponds to a shallow hydrogenic-like impurity which is known to shift with the band edge to which it is tied. We also consider a defect level (b) of intermediate extension, which follows the band edges only to a limited extent.

Figure 7.7 clearly illustrates that reproducing the correct band gap is not a sufficient condition to achieve a correct description of defect levels. When calculated defect levels are compared with experimental ones, the VBM and CBM are natural reference levels. Indeed, the charge transition levels of defects (a) and (b) referred to their respective theoretical VBM are different despite the fact that the two theories correctly reproduce the experimental band gap! This is a direct consequence of the analysis in Section 7.3.2. For the specific case of the oxygen vacancy in ZnO, such considerations explain to a large extent the scatter of calculated charge transition

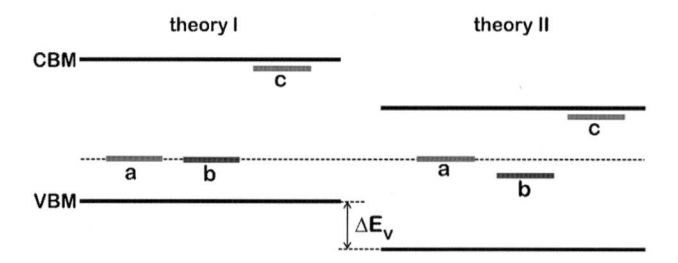

Figure 7.7 (online color at: www.pss-b.com) "The band-edge problem." Comparison of two electronic structure methods, theory I and theory II, for calculations of energy levels of different types of defects: (a) an atomically localized defect; (b) a defect of intermediate extension; (c) a shallow hydrogenic-like defect. The two theories yield the same band gap but different absolute positions of the band edges when referred to the average electrostatic potential. Adapted from Ref. [65].

levels found in the literature [65]. Indeed, different band-gap correction schemes lead to different band-edge positions, while the defect level is generally well defined when referred to the average electrostatic potential [65].

7.4
Hybrid Functionals with Empirically Adjusted Parameters

The comparison of defect charge transition levels calculated with semilocal and hybrid functionals provides insight into the way energy levels of deep defects shift as the description of the band gap improves. The analysis suggests that such defect levels are generally well described at the semilocal level when referred to the average electrostatic potential. Hence, the positioning of the defect levels within the band gap mainly depends on the accuracy by which the adopted functional determines the band edges with respect to this alignment scheme.

As stressed above, a hybrid functional scheme based on the use of a fixed mixing coefficient a does not always yield band gaps in good agreement with experiment. In this section, we address the issue whether band edges determined by hybrid functionals are accurately positioned with respect to the average electrostatic potential, when the mixing coefficient a is tuned to reproduce the experimental band gap. While such an empirical approach is currently in use in the literature, it is ultimately not satisfactory and higher levels of theory will be required to improve the description of the band edges. Nevertheless, the band-gap tuning approach offers a practical scheme in which defect levels are positioned within a band gap of the right value and which can completely be treated within a hybrid functional formulation. It is also based on the well-documented assumption that the structural parameters are generally only moderately affected when the mixing coefficient varies [16, 40, 79], unless the electronic structure itself undergoes important modifications.

As shown in Figure 7.8 for a selected set of materials, semilocal functionals systematically underestimate the experimental band gap. Hybrid functionals generally yield band gaps increasing linearly with the mixing coefficient a [68]. Hence, an optimal mixing coefficient can generally be found for any material:

$$a_{\text{opt}} = \frac{E_g^{\text{expt}} - E_g^{\text{semiloc}}}{\kappa}, \tag{7.8}$$

where $\kappa = dE_g/da$ is the derivative of the band gap with respect to the mixing coefficient. The linear dependence results from the fact that the electron wave functions associated to the band edges do not change significantly with a. Furthermore, this property also signifies that the nature of the band edge states does not change. Indeed, the energy levels of different bands generally behave differently as a varies. Thus, a departure from linearity is expected when the character of the VBM or the CBM changes, as might occur, for example, when sp and d bands cross.

We can provide some support for the tuning of the parameter a by invoking the reasoning of Gygi and Baldereschi [80] in their construction of an approximate GW scheme. Let us consider the nonlocal exchange–correlation potential provided by the

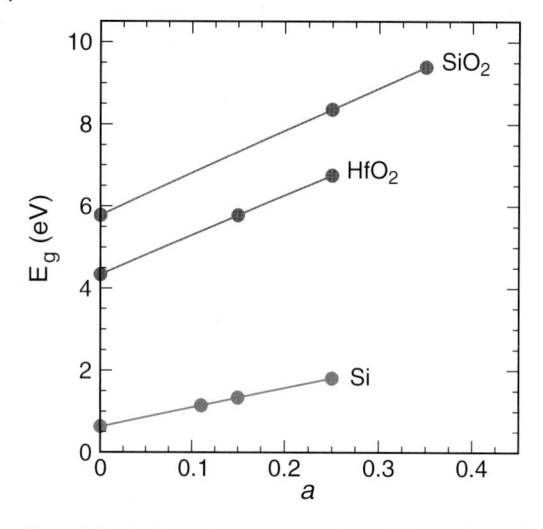

Figure 7.8 (online color at: www.pss-b.com) Dependence of the theoretical band gap on the mixing coefficient a (Eq. 7.5) for several materials. For each material, there is an optimal mixing coefficient a_{opt} for which the hybrid functional reproduces the experimental band gap.

hybrid density functional as a certain approximation to the many-electron exchange–correlation self-energy in the GW approximation, and more particularly to its frequency-independent form, the COHSEX approximation. In this approximation, the long-range interaction is described by screened exchange (SEX), which asymptotically approaches $-1/\varepsilon_\infty(|\mathbf{r}-\mathbf{r}'|)$. In the hybrid functional formulation, the semilocal part of the exchange–correlation is short-ranged [81] and the long-range part is therefore entirely described by the fraction a of exact exchange: $-a/|\mathbf{r}-\mathbf{r}'|$. The assumption that the hybrid functional correctly describes the long-range limit, gives the following relation for the optimal mixing coefficient:

$$a_{opt} \approx \frac{1}{\varepsilon_\infty}. \qquad (7.9)$$

For metals $\varepsilon_\infty = \infty$, and thus $a_{opt} = 0$, which is a correct and intuitive result. In a metal any fraction of exact exchange would produce unphysical derivative discontinuities (e.g., $\partial \varepsilon_k/\partial \mathbf{k}$) at the Fermi level. The present discussion is also fully consistent with the reasoning of Fiorentini and Baldereschi [82], who showed that the error in the semilocal band gap approximately scales like $1/\varepsilon_\infty$. Indeed, the smaller the difference between the semilocal and the experimental band gap, the larger the required fraction a_{opt} of exact exchange in the optimal hybrid functional.

In Figure 7.9a, we show the optimal mixing coefficient a_{opt} versus ε_∞ for various materials. Despite some scatter, a clear correlation between a_{opt} and $1/\varepsilon_\infty$ is indeed apparent. The most evident "out-liers," Ge, GaAs, and ZnO, all possess semicore 3d states. This suggests that the long-range screening is not the only property affecting the band gap, and that there is also an effect associated to the s–d coupling [84], which cannot be captured by tuning a to $1/\varepsilon_\infty$. Other reasons for the data scatter are that

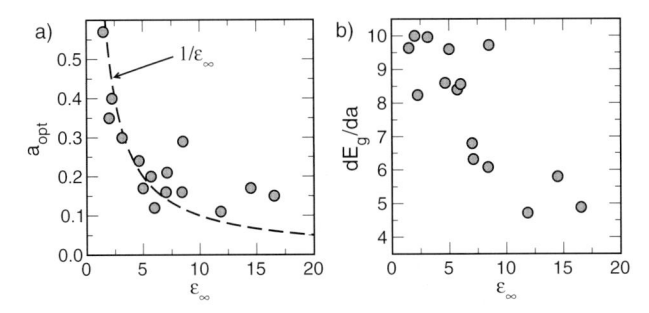

Figure 7.9 (online color at: www.pss-b.com) (a) Optimal mixing coefficients a_{opt} and (b) the derivative of the band gap $\kappa = dE_g/da$ versus the dielectric constant ε_∞ for various materials. The data for the band gaps are taken from the results in Fig. 7.1 and a linear dependence of the band gap on the mixing coefficient a is assumed. For ε_∞, we used experimental data from Ref. [83].

hybrid functionals with $a_{opt} = 1/\varepsilon_\infty$ might inappropriately describe the short-range limit, completely lack the frequency-dependence of the many-body self-energy, and do not account for the anisotropy of the long-range screening when present. Nevertheless, the correlation in Figure 7.9 suggests that the dominant physics is given by the long-range exchange behavior. This is further supported by the good correlation with ε_∞ shown in Figure 7.9b for the quantity $\kappa = dE_g/da$, i.e., the derivative of the band gap with respect to the mixing coefficient. The latter quantity essentially corresponds to $E_g^{HF} - E_g^{semiloc}$, i.e., the difference between the band gap calculated in the Hartree–Fock and in the semilocal scheme.

Once the band gap is tuned to the experimental one, the hybrid functional scheme also provides the shifts of the valence and conduction bands, ΔE_V and ΔE_C. These shifts result from the alignment of the semilocal and hybrid schemes through the average electrostatic potential (cf. Section 7.3.1). They indicate to what extent the conduction and valence bands contribute to the band-gap opening. In a hybrid functional formulation, their relative contributions solely depend on the effect of the nonlocal exact exchange operator. The values of these shifts are critical for a correct placement of defect levels within the band gap. The issue that concerns us here is to what extent exact exchange is reliable for the evaluation of such shifts.

In principal, the accuracy of calculated band-edge shifts can be assessed through the consideration of surface systems. Given a well-defined surface structure, the ionization potential and the electron affinity with respect to the vacuum level could be calculated. However, photoemission data for semiconductor and insulator surfaces might be affected by charging effects and by the occurrence of defects and impurities influencing the electrostatics. Therefore, we here prefer to consider band offsets at semiconductor–oxide interfaces, where such electrostatic effects appear better controlled. For a specific interface system, the band offsets can be achieved through the method of Van de Walle and Martin [67]. The band offsets in the hybrid functional scheme can be derived from those in the semilocal scheme through the consideration of the variation of the electrostatic potential offset and through the application of the

Table 7.1 Valence (ΔE_V) and conduction (ΔE_C) band offsets at the Si/SiO$_2$, SiC/SiO$_2$, and Si/HfO$_2$ interfaces calculated in PBE, PBE0, and the mixed scheme (Ref. [68]), in which the mixing coefficient a is different for the two interface components. Experimental band offsets are from Refs. [89, 90] and [91], respectively.

interface		PBE	PBE0	mixed	expt.
Si/SiO$_2$	ΔE_v	2.5	3.3	4.4	4.4
	ΔE_c	2.3	2.7	3.4	3.4
SiC/SiO$_2$	ΔE_v	1.4	2.0	3.0	2.9
	ΔE_c	1.7	2.0	2.6	2.7
Si/HfO$_2$	ΔE_v	2.3	3.1	2.9	2.9
	ΔE_c	1.5	1.9	1.7	1.7

shifts ΔE_V and ΔE_C [68, 85]. On either side of the interface, the theoretical band-gap matches the experimental one by construction. Such a scheme generally requires the use of different a_{opt} for the two bulk components of the interface and is applicable owing to the weak dependence of the interfacial dipole on the mixing parameter a [68]. The comparison between calculated and measured band offsets then provides a sensitive test for the accuracy of band edges as obtained in hybrid functional schemes. In Table 7.1, we present band offsets calculated for three interface model systems: Si/SiO$_2$ (Refs. [86, 87]), SiC/SiO$_2$ (Ref. [88]), and Si/HfO$_2$ (Ref. [28]). When the hybrid functionals are tuned to match the experimental band gaps of the two interface components (cf. "mixed" in Table 7.1), the calculated band offsets are found to agree with experiment within only 0.1 eV [89–91]. Despite the limited number of studied systems, the good agreement in Table 7.1 is very encouraging and suggests that hybrid functionals may be relied upon for positioning band edges. This would also imply that nonlocal exchange is the primary cause determining the relative size of band-edge shifts.

Another way of validating the shifts of the band edges obtained from the hybrid functional calculation is through comparison with those calculated with a theory of higher level, such as for instance the GW many-body perturbation theory. Recently, Shaltaf et al. [92] performed GW calculations focusing on such shifts. In Table 7.2, we compare shifts in band edges as obtained with a hybrid functional with those obtained by Shaltaf et al. [92] for Si and SiO$_2$. The band gaps in the hybrid functional calculations with a tuned mixing coefficient are by construction exactly equal to the experimental ones, whereas this is not necessarily the case for the various GW approaches. It is therefore more useful to compare relative shifts in the valence and conduction band, i.e., $\Delta E_V/\Delta E_g$ and $\Delta E_C/\Delta E_g$ as obtained in various theoretical schemes. One observes that results obtained with the hybrid functional and with the various GW schemes differ by approximately the same amount as the various GW schemes differ among themselves. This suggests that the quality of the VBM and CBM shifts provided by the hybrid functional scheme is comparable to that achieved with GW methods.

It should be stressed that the study of Shaltaf et al. [92] also showed that in GW schemes these shifts are more difficult to converge than the band gap, requiring

Table 7.2 Relative shifts of the valence band (ΔE_V) and the conduction band ΔE_C with respect to the change in the band gap ΔE_g for Si and SiO$_2$, as determined with a hybrid functional (PBE0), a GW, and a quasiparticle self-consistent GW (QS GW) scheme. The latter two results are taken from Ref. [92].

material	quantity	hybrid	GW	QS GW
Si	$\Delta E_V/\Delta E_g$	−0.54	−0.67	−0.75
	$\Delta E_C/\Delta E_g$	0.46	0.33	0.25
SiO$_2$	$\Delta E_V/\Delta E_g$	−0.69	−0.56	−0.68
	$\Delta E_C/\Delta E_g$	0.31	0.44	0.32

a very high number of empty states in the calculation of the response functions. Furthermore, these shifts are sensitive to various ingredients of the calculation, such as the plasmon pole approximation, the level of self-consistency of the GW approximation, and the vertex corrections [92]. These considerations limit the amount of materials for which such shifts have hitherto been obtained in a reliable way at the GW level.

7.5
Representative Case Studies

In this section, we illustrate the application of hybrid functionals to the study of defects through two case studies.

7.5.1
Si Dangling Bond

The first case study concerns the Si dangling bond. This defect corresponds to the atomic structure of the P_b center, which has clearly been observed at interfaces between silicon and its oxide [93]. The dangling bond was modeled by removing four neighboring atoms in a bulk supercell of 216 silicon atoms [29]. Nine of the ten dangling bonds generated in this way were then passivated with H atoms. The core structure of the model is identical to that used in Ref. [23] for modeling the Ge dangling bond.

The relevant charge states of the dangling bond are the positive, the neutral, and the negative charge states. Charge transition levels were calculated in the PBE (the mixing coefficient $a = 0$), the PBE0 ($a = 0.25$), and with a hybrid functional defined by an intermediate mixing coefficient $a = 0.10$. The evolution of the charge transition levels as well as that of the band edges are shown in Figure 7.10a as function of a [29]. The band structures are aligned through the average electrostatic potential. All displayed levels shift linearly with a. The largest shifts are observed for the band edges. The shifts of the charge transition levels $\varepsilon^{+/0}$ and $\varepsilon^{0/-}$ are more moderate, in agreement with general findings [13]. The charge transition levels reported in

Figure 7.10a include the electrostatic Makov–Payne correction. For the present supercell calculation, the application of this correction yields converged values for the charge transition levels [29].

A hybrid functional calculation with a mixing coefficient $a = 0.11$ precisely reproduces the experimental value of the Si band gap, $E_g = 1.17$ eV. For this value of a, we obtained charge transition levels at $\varepsilon^{+/0} = E_V + 0.2$ eV and $\varepsilon^{0/-} = E_V + 0.8$ eV. We compare these values in Figure 7.10b to the experimental density of interfacial traps at the Si–SiO$_2$ interface as obtained from C/V measurements [93]. The calculated charge transition levels are found to closely correspond to the two experimental peaks at $E_V + 0.26$ eV and $E_V + 0.84$ eV, generally assigned to Si dangling bonds. The good agreement in Figure 7.10b provides support to the practice

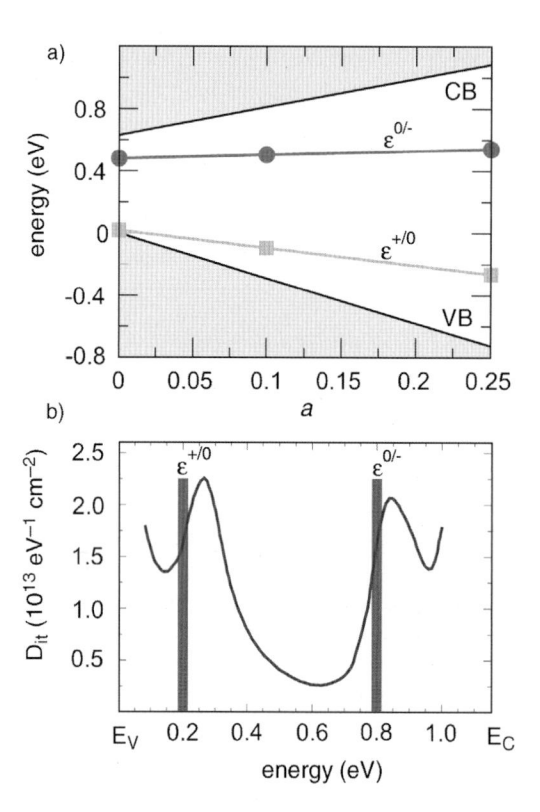

Figure 7.10 (online color at: www.pss-b.com) (a) Dependence of Si band edges and of the charge transition levels $\varepsilon^{+/0}$ and $\varepsilon^{0/-}$ of the Si dangling bond defect on the mixing coefficient a. The vertical energy scale is referred to the VBM in the PBE calculation. Adapted from Ref. [29]. (b) Density of interfacial traps at the Si–SiO$_2$ as measured by the low-frequency C/V technique in Ref. [93] (solid line). The two pronounced peaks at 0.26 and 0.84 eV originate from P_b defects and correspond to the charge transition levels $\varepsilon^{+/0}$ and $\varepsilon^{0/-}$. The charge transition levels obtained through a hybrid functional calculation with $a = 0.11$ are represented by vertical bars.

of using a mixing coefficient a that brings the theoretical band gap obtained within the hybrid functional scheme in accord with the experimental one.

7.5.2
Charge State of O_2 During Silicon Oxidation

The second case study concerns the charge state of the O_2 molecule during silicon oxidation. The silicon oxidation process has attracted considerable interest because of its key role in the manufacturing of Si-based microelectronic devices. Our present understanding relies to a large extent on the oxidation model proposed by Deal and Grove [94]. In this model, the growth of SiO_2 proceeds by (i) the adsorption of the O_2 molecule on the oxide surface, (ii) the diffusion of molecular O_2 through the bulk-like oxide, and (iii) its subsequent reaction at the semiconductor–oxide interface. Simulation techniques based on DFT have been instrumental for achieving an atomic-scale description of the involved processes [95], such as, e.g., the diffusion mechanism of O_2 in amorphous SiO_2 [96], the oxidation reaction [97], etc. However, one aspect that has long been difficult to address is the charge state of the diffusing oxygen molecule. The difficulty of providing a clear answer to this issue stems from the "band-gap problem" of semilocal approximations to DFT [30, 98].

In bulk SiO_2, the oxygen molecule is stable in the neutral and in the negative charge states [96]. The charge state of the O_2 in the vicinity of the Si/SiO_2 interface is determined by the position of the $(0/-)$ charge transition level with respect to silicon band edges. It is assumed that the molecule is close enough to the interface to allow for charge equilibration with the silicon substrate, yet remaining far from the suboxide region where the oxidation reaction takes place. Thus, the $(0/-)$ charge transition level of the O_2 molecule is first determined in a bulk-like amorphous SiO_2 environment and then positioned with respect to Si band edges through the band alignment at the Si/SiO_2 interface. In Figure 7.11, we show the result of such an alignment procedure as obtained within three different theoretical schemes [30]: (i) the semilocal (PBE) functional; (ii) the hybrid (PBE0) functional; (iii) a mixed scheme, in which the fraction of exact exchange is tuned for each interface component, following the prescription for the calculation of band offsets given above [68].

All three theoretical schemes consistently indicate that the $(0/-)$ charge transition level locates above the Si CBM (Figure 7.11), providing convincing evidence that for electron chemical potentials in the Si band gap the neutral charge state is thermodynamically favored. The three schemes show only small quantitative differences. The separation between the $(0/-)$ charge transition level and the Si CBM is 1.1 eV in the PBE and in the mixed scheme, and reduces to 0.8 eV in the PBE0.

To obtain such a level of qualitative agreement between different theoretical schemes, charge transition levels and band offsets were obtained consistently within each scheme. This should be contrasted with the practice of determining transition levels with respect to the oxide band edges within PBE and using the experimental band offsets for alignment with respect to the Si band edges. Such an alignment procedure implicitly takes the erroneous assumption [68, 92] that the band-gap correction is achieved by the sole displacement of the conduction band. In the case of

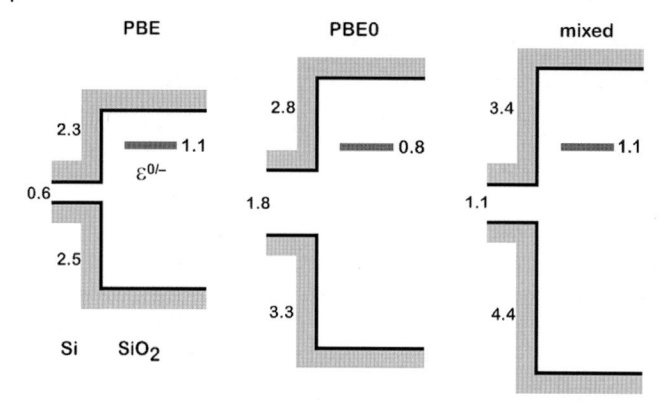

Figure 7.11 (online color at: www.pss-b.com) The alignment of the (0/−) charge transition level of the O$_2$ molecule at the Si/SiO$_2$ interface is obtained within three different theoretical schemes: a semilocal functional (PBE), a hybrid functional (PBE0), and the mixed scheme. The Si band gap, the Si/SiO$_2$ band offsets, and the separation between the defect level and the silicon CBM are given in eV. From Ref. [30].

the O$_2$ molecule in SiO$_2$, this approach results in the opposite conclusion that it is the negative charge state which is thermodynamically favored [98].

7.6 Conclusion

Our investigation indicates that hybrid functional schemes offer a viable theoretical tool for determining the location of energy levels of deep defects with respect to the band edges of the bulk material. The issue is conveniently addressed by separately aligning the defect level and the band edges with respect to the average electrostatic potential. In this way, the determination of the defect level can to a large extent be decoupled from the determination of bulk band edges.

As far as the defects are concerned, it appears that their energy levels with respect to the average electrostatic potential are already well described at the semilocal level and that the hybrid-functional description does not lead to any significant modification. In the case of ionization potentials of molecular systems, a similar agreement is recorded and the comparison with experiment shows that an accurate description is achieved. In analogy with the molecular case, this therefore suggests that the energy separation between the defect level and the average electrostatic potential is in many cases already accurately determined at the semilocal and hybrid functional level.

At variance, the position of the band edges is highly sensitive to the fraction of exact exchange considered in the hybrid functional calculation. The use of any fixed fraction of exact exchange does not lead to a systematic improvement of the band-gap description, thereby hindering the use of a hybrid-functional scheme as a predictive tool. It is therefore necessary to resort to an electronic structure method of higher accuracy to identify the position of the band edges. It should be noted that such a method would be applied in the absence of the defect and could therefore take computational advantage of the full translational symmetry of the host material. This analysis allows us to reformulate the band-gap problem in terms of a band-edge problem, highlighting the importance of a reliable description of bulk bands relative to the average electrostatic potential.

Our work also explored the route of determining band-edge shifts while remaining within the hybrid functional approach. This is done at the cost of empirically adjusting the mixing coefficient a to reproduce the experimental band gap. While such an approach does not offer an ideal solution, several supporting arguments can nevertheless be invoked. In particular, the accuracy of the determined band edges can be assessed either by comparison with electronic-structure theories of higher accuracy or by direct comparison with experimental band offsets. In both cases, available data indicate that the agreement is very encouraging. The present description achieved with hybrid functionals constitutes a noticeable step forward with respect to the case of semilocal functionals in which any conclusion is heavily biased by the band-gap problem.

A generalized use of hybrid functional schemes still requires further work. One important issue is the shift in the band edges with respect to the average electrostatic potential. The validity of shifts determined within hybrid functional schemes should further be investigated by extending the comparison with experiment to a larger set of semiconductor–semiconductor and semiconductor–oxide band offsets. In addition, a more systematic comparison with theoretical schemes of higher accuracy, such as many-body perturbation schemes based on the GW approximation, would be invaluable for further supporting the shifts obtained with hybrid functionals. In particular, such comparisons should also provide insight into whether it is conceptually reasonable to expect that exact exchange dominates the relative value of conduction and valence band shifts.

The use of hybrid functionals with tuned mixing coefficients is clearly unsatisfactory. In the inhomogeneous case of an interface between two materials with very different band gaps, the mixed scheme is advantageous for achieving a good description of band offsets [68]. However, the use of different functionals for the two interface components precludes the study of the transition region and defects therein. Since the optimal mixing coefficients are different on both sides of the interface, a reliable description of the evolution of band edges is not possible with any fixed value of a. Accordingly, the mixing coefficient should depend on the coordinate perpendicular to the interface. This is not practical when plane-wave basis sets are used, but could in principal be achieved in the case of localized basis functions. However, such an approach is conceptually not appealing since the variation across the interface would remain *ad hoc*.

Another problem is related to localized electronic states, such as d and f orbitals, for which self-interaction errors of semilocal density functionals are high. For instance, in many transition metal oxides, such as ZnO, the fraction a required to correctly position the semicore 4s with respect to VBM is not necessarily equal to the value of a reproducing the band gap. This results in a serious deficiency when the role of these d states cannot be neglected. The role of self-interaction errors might be even more significant in the description of very localized defect states affected by strong polaronic effects, such as Al substitutional to Si in SiO_2 [99, 100], Li substitutional to Zn in ZnO [101–103], the self-trapped hole in NaCl [104], defect-trapped and self-trapped electrons and holes in TiO_2 [105], or the Mg vacancy in MgO [106]. Semilocal density functionals yield excessively delocalized electron densities resulting in inaccurate defect geometries. Hybrid functionals generally improve upon this. However, the mixing coefficient a required for reproducing correct defect geometries can be very different than the value optimizing the band gap [103]. This implies that a single value of a cannot concurrently reproduce the bulk band edges and the ground-state geometries of specific defects. Such an *ad hoc* fixation of the hybrid-functional parameters is clearly disturbing.

In conclusion, hybrid functionals certainly represent a powerful tool for the study of defect levels, but problematic aspects persist requiring great caution in their application. In this work, we restricted the discussion to the class of hybrid functionals based on bare nonlocal exchange, in which the mixing coefficient is allowed to vary. Hybrid functionals based on screened exchange might offer greater flexibility, but it is anticipated that even such functionals would not lead to a universally effective tool when used with fixed parameters [107]. In this context, the approach discussed in this work provides useful insights and guidelines for the theoretical determination of defect levels.

Acknowledgements

We acknowledge useful discussions with J.-F. Binder, A. Carvalho, H.-P. Komsa, A. Janotti, P. Rinke, and C. G. Van de Walle. Financial support from the Swiss National Science Foundation (grant no. 200020-119733/1) is acknowledged. The calculations were performed on computational facilities at DIT-EPFL, CSEA-EPFL, and CSCS.

References

1 Stoneham, A.M. (1975) *Theory of Defects in Solids: Electronic Structure of Defects in Insulators and Semiconductors*, Oxford University Press, Oxford.
2 Green, M. (2003) *Third Generation Photovoltaics: Advanced Solar Energy Conversion*, Springer-Verlag, Berling-Heidelberg.
3 Robertson, J. (2006) *Rep. Prog. Phys.*, **69**, 327.
4 Winter, M. and Brodd, R.J. (2004) *Chem. Rev.*, **104**, 4245.
5 Tsao, J.Y. (2004) *IEEE Circuits Devices*, **20**, 28.
6 Van de Walle, C.G. and Neugebauer, J. (2004) *J. Appl. Phys.*, **95**, 3851.
7 Zhang, S.B., Wei, S.-H., and Zunger, A. (2001) *Phys. Rev. B*, **63**, 075205.
8 Janotti, A. and Van de Walle, C.G. (2005) *Appl. Phys. Lett.*, **87**, 122102.
9 Persson, C., Zhao, Y.-J., Lany, S., and Zunger, A. (2005) *Phys. Rev. B*, **72**, 035211.
10 Janotti, A. and Van de Walle, C.G. (2007) *Phys. Rev. B*, **76**, 165202.
11 Lany, S. and Zunger, A. (2007) *Phys. Rev. Lett.*, **98**, 045501.

12 Zhukovskii, Y.F., Kotomin, E.A., Evarestov, R.A., and Ellis, D.E. (2007) *Int. J. Quantum Chem.*, **107**, 2956.
13 Alkauskas, A., Broqvist, P., and Pasquarello, A. (2008) *Phys. Rev. Lett.*, **101**, 046405.
14 Lany, S. and Zunger, A. (2008) *Phys. Rev. B*, **78**, 235104.
15 Rinke, P., Janotti, A., Scheffler, M., and Van de Walle, C.G. (2009) *Phys. Rev. Lett.*, **102**, 026402.
16 Deák, P., Gali, A., Sólyom, A., Buruzs, A., and Frauenheim, Th. (2005) *J. Phys.: Condens. Matter*, **17**, S2141.
17 Knaup, J.M., Deák, P., Frauenheim, Th., Gali, A., Hajnal, Z., and Choyke, W.J. (2005) *Phys. Rev. B*, **72**, 115323.
18 Lany, S. and Zunger, A. (2006) *J. Appl. Phys.*, **100**, 113725.
19 Hedström, M., Schindlmayr, A., Schwarz, G., and Scheffler, M. (2006) *Phys. Rev. Lett.*, **97**, 226401.
20 Broqvist, P. and Pasquarello, A. (2006) *Appl. Phys. Lett.*, **89**, 262904.
21 Gavartin, J.L., Muñoz Ramo, D., Shluger, A.L., Bersuker, G., and Lee, B.H. (2006) *Appl. Phys. Lett.*, **89**, 082908.
22 Li, J., and Wei, S.-H. (2006) *Phys. Rev. B*, **73**, 041201.
23 Weber, J.R., Janotti, A., Rinke, P., and Van de Walle, C.G. (2007) *Appl. Phys. Lett.*, **91**, 142101.
24 Broqvist, P. and Pasquarello, A. (2007) *Appl. Phys. Lett.*, **90**, 082907.
25 Alkauskas, A. and Pasquarello, A. (2007) *Physica B*, **401/402**, 546.
26 Paudel, T.R. and Lambrecht, R.L. (2008) *Phys. Rev. B*, **77**, 205202.
27 Oba, F., Togo, A., Tanaka, I., Paier, J., and Kresse, G. (2008) *Phys. Rev. B*, **77**, 245202.
28 Broqvist, P., Alkauskas, A., and Pasquarello, A. (2008) *Appl. Phys. Lett.*, **92**, 132911.
29 Broqvist, P., Alkauskas, A., and Pasquarello, A. (2008) *Phys. Rev. B*, **78**, 075203.
30 Alkauskas, A., Broqvist, P., and Pasquarello, A. (2008) *Phys. Rev. B*, **78**, 161305.
31 Stroppa, A. and Kresse, G. (2009) *Phys. Rev. B*, **79**, 201201(R).
32 Broqvist, P., Alkauskas, A., Godet, J., and Pasquarello, A. (2009) *J. Appl. Phys.*, **105**, 061603.
33 Ágoston, P., Albe, K., Nieminen, R.M., and Puska, M.J. (2009) *Phys. Rev. Lett.*, **103**, 245501.
34 Lany, S. and Zunger, A. (2010) *Phys. Rev. B*, **81**, 113201.
35 Deák, P., Aradi, B., Frauenheim, Th., Janzén, E., and Gali, A. (2010) *Phys. Rev. B*, **81**, 153203.
36 Becke, A.D. (1993) *J. Chem. Phys.*, **98**, 1372.
37 Becke, A.D. (1993) *J. Chem. Phys.*, **98**, 5648.
38 Perdew, J.P., Burke, K., and Ernzerhof, M. (1996) *J. Chem. Phys.*, **105**, 9982.
39 Muscat, J., Wander, A., and Harrison, N.M. (2001) *Chem. Phys. Lett.*, **342**, 397.
40 Paier, J., Marsman, M., Hummer, K., Kresse, G., Gerber, I.C., and Ángáyan, J.G. (2006) *J. Chem. Phys.*, **124**, 154709; (2006) *J. Chem. Phys.*, **125**, 249901 (E).
41 Heyd, J., Scuseria, G.E., and Enzerhof, M. (2003) *J. Chem. Phys.*, **118**, 8207; (2006) *J. Chem. Phys.*, **124**, 219906 (E).
42 Krukau, A.V., Vydrov, O.A., Izmaylov, A.F., and Scuseria, G.E. (2006) *J. Chem. Phys.*, **125**, 224106.
43 Janesko, B.G., Henderson, T.M., and Scuseria, G.E. (2009) *Phys. Chem. Chem. Phys.*, **11**, 443.
44 Troullier, N. and Martins, J.L. (1991) *Phys. Rev. B*, **43**, 1993.
45 Perdew, J.P., Burke, K., and Ernzerhof, M. (1996) *Phys. Rev. Lett.*, **77**, 3865.
46 Broqvist, P., Alkauskas, A., and Pasquarello, A. (2009) *Phys. Rev. B*, **80**, 085114; (2010) *Phys. Rev. B*, **81**, 039903.
47 Stroppa, A. and Kresse, G. (2008) *New J. Phys.*, **10**, 063020.
48 Hutter, J. and Curioni, A. (2005) *ChemPhysChem*, **6**, 1788; Car, R. and Parrinello, M. (1985) *Phys. Rev. Lett.*, **55**, 2471; CPMD, Copyright IBM Corp. 1990–2006 Copyright MPI für Festkörperforschung Stuttgart 1997–2001.
49 Todorova, T., Seitsonen, A.P., Hutter, J., Kuo, I. F. W., and Mundy, C.J. (2006) *J. Phys. Chem. B*, **110**, 3685.
50 Giannozzi, P., Baroni, S., Bonini, N., Calandra, M., Car, R., Cavazzoni, C., Ceresoli, D., Chiarotti, G.L., Cococcioni, M., Dabo, I., Dal Corso, A., de Gironcoli, S., Fabris, S., Fratesi, G., Gebauer, R.,

50 ...Gerstmann, U., Gougoussis, C., Kokalj, A., Lazzeri, M., Martin-Samos, L., Marzari, N., Mauri, F., Mazzarello, R., Paolini, S., Pasquarello, A., Paulatto, L., Sbraccia, C., Scandolo, S., Sclauzero, G., Seitsonen, A.P., Smogunov, A., Umari, P., and Wentzcovitch, R.M. (2009) *J. Phys.: Condens. Matter*, **21**, 395502, http://www.quantum-espresso.org.
51 Zhang, S.B. and Northrup, J.E. (1991) *Phys. Rev. Lett.*, **67**, 2339.
52 Makov, G. and Payne, M.C. (1995) *Phys. Rev. B*, **51**, 4014.
53 Lento, J., Mozos, J.-L., and Nieminen, R.M. (2002) *J. Phys.: Condens. Matter*, **14**, 2637.
54 Castleton, C.W.M., Höglund, A., and Mirbt, S. (2006) *Phys. Rev. B*, **73**, 035215.
55 Hine, N.D.M., Frensch, K., Foulkes, W.M.C., and Finnis, M.W. (2009) *Phys. Rev. B*, **79**, 024112.
56 Freysoldt, C., Neugebauer, J., and Van de Walle, C.G. (2009) *Phys. Rev. Lett.*, **102**, 016402.
57 Kümmel, S. and Kronik, L. (2008) *Rev. Mod. Phys.*, **80**, 3.
58 Ernzerhof, M., Perdew, J.P., and Burke, K. (1997) *Int. J. Quantum Chem.*, **64**, 285.
59 Paier, J., Hirschl, R., Marsman, M., and Kresse, G. (2005) *J. Chem. Phys.*, **122**, 234102.
60 Fuchs, F., Furthmuller, J., Bechstedt, F., Shishkin, M., and Kresse, G. (2007) *Phys. Rev. B*, **76**, 115109.
61 Labat, F., Baranek, P., Domain, C., Monot, C., and Adamo, C. (2007) *J. Chem. Phys.*, **126**, 154703.
62 Gygi, F. and Baldereschi, A. (1986) *Phys. Rev. B*, **34**, 4405.
63 Carrier, P., Rohra, S., and Görling, A. (2007) *Phys. Rev. B*, **75**, 205126.
64 Alkauskas, A. and Pasquarello, A. (2007) *Physica B*, **401/402**, 670.
65 Alkauskas, A. and Pasquarello, A. unpublished.
66 More complex situations can in principle also be addressed.
67 Van de Walle, C.G. and Martin, R.M. (1987) *Phys. Rev. B*, **35**, 8154.
68 Alkauskas, A., Broqvist, P., Devynck, F., and Pasquarello, A. (2008) *Phys. Rev. Lett.*, **101**, 106802.
69 Lyons, J.L., Janotti, A., and Van de Walle, C.G. (2009) *Phys. Rev. B*, **80**, 205113.
70 Perdew, J.P. and Levy, M. (1997) *Phys. Rev. B*, **56**, 16021.
71 Öğüt, S., Chelikowsky, J.R., and Louie, S.G. (1998) *Phys. Rev. Lett.*, **80**, 3162; Godby, R.W., and White, I.D. (1998) *Phys. Rev. Lett.*, **80**, 3161.
72 Curtiss, L.A., Redfern, P.C., Raghavachari, K., and Pople, J.A. (1997) *J. Chem. Phys.*, **106**, 1063; (1998) *J. Chem. Phys.*, **109**, 42.
73 Slater, J.C. (1972) *Adv. Quantum Chem.*, **6**, 1.
74 Janak, J.F. (1978) *Phys. Rev. B*, **18**, 7165.
75 Perdew, J.P., Parr, R.G., Levy, M., and Balduz, J.L. (1982) *Phys. Rev. Lett.*, **49**, 1691.
76 Perdew, J.P. and Levy, M. (1983) *Phys. Rev. Lett.*, **51**, 1884.
77 Vydrov, O.A., Scuseria, G.E., and Perdew, J.P. (2007) *J. Chem. Phys.*, **126**, 154109.
78 Mori-Sánchez, P., Cohen, A.J., and Yang, W. (2008) *Phys. Rev. Lett.*, **100**, 146401.
79 Heyd, J., Peralta, J.E., Scuseria, G.E., and Martin, R.L. (2005) *J. Chem. Phys.*, **123**, 174101.
80 Gygi, F. and Baldereschi, A. (1989) *Phys. Rev. Lett.*, **62**, 2160.
81 Sham, L.J. and Kohn, W. (1966) *Phys. Rev.*, **145**, 561.
82 Fiorentini, V. and Baldereschi, A. (1995) *Phys. Rev. B*, **51**, 17196.
83 Landolt-Börnstein database, http://www.springermaterials.com.
84 Persson, C. and Zunger, A. (2003) *Phys. Rev. B*, **68**, 073205.
85 Broqvist, P., Binder, J.F., and Pasquarello, A. (2009) *Appl. Phys. Lett.*, **94**, 141911.
86 Sarnthein, J., Pasquarello, A., and Car, R. (1995) *Phys. Rev. Lett.*, **74**, 4682; (1995) *Phys. Rev. B*, **52**, 12690.
87 Giustino, F. and Pasquarello, A. (2005) *Phys. Rev. Lett.*, **95**, 187402; Bongiorno, A., Pasquarello, A., Hybertsen, M.S., and Feldman, L.C. (2003) *Phys. Rev. Lett.*, **90**, 186101; Bongiorno, A., and Pasquarello, A. (2003) *Appl. Phys. Lett.*, **83**, 1417.
88 Devynck, F., Giustino, F., Broqvist, P., and Pasquarello, A. (2007) *Phys. Rev. B*, **76**, 075351.
89 Himpsel, F.J., McFeely, F.R., Teleb-Ibrahimi, A., Yarmoff, J.A., and

Hollinger, G. (1988) *Phys. Rev. B*, **38**, 6084; Keister, J.W., Rowe, J.E., Kolodziej, J.J., Niimi, H., Madey, T.E., and Lucovsky, G. (1999) *J. Vac. Sci. Technol. B*, **17**, 1831.

90 Afanas'ev, V.V., Ciobanu, F., Dimitrijev, S., Pensl, G., and Stesmans, A. (2004) *J. Phys.: Condens. Matter*, **14**, S1839.

91 Oshima, M., Toyoda, S., Okumura, T., Okabayashi, J., Kumigashira, H., Ono, K., Niwa, M., Usuda, K., and Hirashita, N. (2003) *Appl. Phys. Lett.*, **83**, 2172; Renault, O., Barrett, N.T., Samour, D., and Quiais-Marthon, S. (2004) *Surf. Sci.*, **566–568**, 526.

92 Shaltaf, R., Rignanese, G.-M., Gonze, X., Giustino, F., and Pasquarello, A. (2008) *Phys. Rev. Lett.*, **100**, 186401.

93 Poindexter, E.H., Gerardi, G.J., Rueckel, M.-E., Caplan, P.J., Johnson, N.M., and Biegelsen, D.K. (1984) *J. Appl. Phys.*, **56**, 2844.

94 Deal, B.E. and Grove, A.S. (1965) *J. Appl. Phys.*, **36**, 3770.

95 Bongiorno, A. and Pasquarello, A. (2005) *J. Phys.: Condens. Matter*, **17**, S2051.

96 Bongiorno, A. and Pasquarello, A. (2002) *Phys. Rev. Lett.*, **88**, 125901; (2004) *Phys. Rev. B*, **70**, 195312.

97 Bongiorno, A. and Pasquarello, A. (2004) *Phys. Rev. Lett.*, **93**, 086102.

98 Stoneham, A.M., Szymanski, M.A. and Shluger, A.L. (2001) *Phys. Rev. B*, **63**, 241304; Szymanski, M.A., Stoneham, A.M., Shluger, A. (2001) *Solid-State Electron.*, **45**, 1233.

99 Pacchioni, G., Frigoli, F., Ricci, D., and Weil, J.A. (2000) *Phys. Rev. B*, **63**, 054102.

100 Laegsgaard, J. and Stokbro, K. (2001) *Phys. Rev. Lett.*, **86**, 2834.

101 Lany, S. and Zunger, A. (2009) *Phys. Rev. B*, **80**, 085202.

102 Du, M.-H. and Zhang, S.B. (2009) *Phys. Rev. B*, **80**, 115217.

103 Carvalho, A., Alkauskas, A., Pasquarello, A., Tagantsev, A., and Setter, N. (2009) *Phys. Rev. B*, **80**, 195205; (2009) *Physica B*, **23–24**, 4797.

104 Gavartin, J.L., Sushko, P.V., and Shluger, A.L. (2003) *Phys. Rev. B*, **67**, 035108.

105 Morgan, B.J., and Watson, G.W. (2009) *Phys. Rev. B*, **80**, 233102.

106 Droghetti, A., Pemmaraju, C.D., and Sanvito, S. (2010) *Phys. Rev. B*, **81**, 092403.

107 Komsa, H.-P., Broqvist, P., and Pasquarello, A. (2010) *Phys. Rev. B*, **81**, 205118.

8
Accurate Gap Levels and Their Role in the Reliability of Other Calculated Defect Properties

Peter Deák, Adam Gali, Bálint Aradi, and Thomas Frauenheim

8.1
Introduction

Bulk defects and surfaces give rise to characteristic fingerprints in the electrical, optical, and magnetic spectra of non-metallic crystals, critically influencing their functionality in applications. The task of defect theory is to establish the equilibrium concentration of conceivable defects (without or in the presence of other defects) and to calculate their properties for comparison with the experimental spectra. The information gained from the joint efforts of defect theory and spectroscopy serves as database for defect engineering, which has become an integral part of technology design in electronics, optoelectronics, and photovoltaics, but also serves the understanding of surface processes like heterogeneous catalysis.

From the viewpoint of the electrical and optical properties of the bulk, as well as of the chemical reactivity of surfaces, band gap states are of paramount importance in non-metallic solids. Their position with respect to the band edges determines the electrical and optical spectra but contributes also dominantly to the formation energy. Calculating defect level positions has been the toughest challenge for defect theory, because its "work horse" in the past decades [1], density functional theory (DFT) in its standard implementations – the local density approximation (LDA) and the semi-local generalized gradient approximation (GGA) – leads to a serious underestimation of the band gap (sometimes to no gap at all), and to big uncertainties in the defect level positions in it. For a long time, this deficiency has been regarded as a relatively minor problem, hampering only the comparison of the calculated spectra with the experimental ones but, in fact, it has serious implications for the formation energy [2], and so for the relative stability of different defect configurations. In this paper we will demonstrate this and review our experience with different correction schemes for calculating defect properties free of the gap error.

Before doing so, however, let us consider the ways of calculating electrical and optical spectra. Usually, what is being measured is the energy promoting an electron from the valence band maximum (VBM) to a defect level (acceptors), or to the conduction band minimum (CBM) from a defect level (donors). Of course, internal

---

Advanced Calculations for Defects in Materials: Electronic Structure Methods, First Edition.
Edited by Audrius Alkauskas, Peter Deák, Jörg Neugebauer, Alfredo Pasquarello, and Chris G. Van de Walle.
© 2011 Wiley-VCH Verlag GmbH & Co. KGaA. Published 2011 by Wiley-VCH Verlag GmbH & Co. KGaA.

excitation of the defect is also possible, and the reverse (recombination) processes can also be measured. In some experiments, the measured energy absorption or emission corresponds to the change in the electronic energy alone, while in others to the change in the total energy. With reference to a Franck–Condon diagram, the former are called vertical transitions, while the latter, where the ions have time to relax, are called adiabatic. Since (except for internal excitations) the charge state of the defect itself changes, these transition energies are often referred to as "optical" or "thermal" charge transition levels of the defect, respectively. In principle, an excitation energy should be calculated as the total energy difference between the ground and excited states, but DFT can be applied only to the former. However, excitations energies can also be deduced by comparing the ground state energies of different charge states. Using charge transition energies of the perfect and the defective crystal, as shown in Figure 8.1, one can obtain the energies of the required electronic transitions from/to the band edges to/from the defect level. For vertical (optical) transitions, the total energy differences have to be taken at the geometry of the initial state, while for adiabatic (thermal) transitions at the respective equilibria of the final and initial states. Actually, the charge transition energies with respect to the vacuum level correspond to the ionization energies and electron affinities of the systems. (N.B.: the electron affinity is the ionization energy of the negative state.) In DFT, if the applied functional was exact, the negative of the highest occupied Kohn–Sham (KS) level would be exactly equal with the ionization energy [3, 4]. In supercell calculations the vacuum level is not defined but the common reference level of Figure 8.1 does not appear in the required ionization energy differences. So, e.g., for the infinite system, $I_C - A_C = \varepsilon_{CB} - \varepsilon_{VB} = E_g$ [2]. The standard local and semi-local approximations of the exchange functional introduce a spurious electron self-interaction which leads to the "band gap error," irrespective of which way one calculates it (assuming that proper band-filling corrections were applied in the total energy calculated at special k-points) [2]. Also, while the total energy should be a linear function of the occupation number (between integer values), in the standard

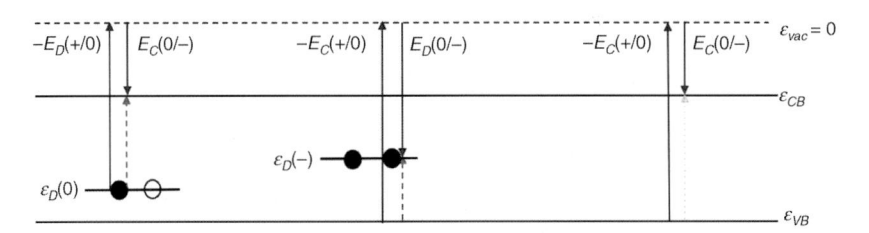

Figure 8.1 (online color at: www.pss-b.com) Transition energies as ground state energy differences. Lines denoted with the Greek letter ε represent (Kohn–Sham) one-electron levels, arrows denoted with Latin E correspond to transition energies between the charge states given in parentheses for the perfect (C) and the defective (D) crystal. I and A are the corresponding ionization energies and electron affinities, respectively. Arrows with dashed lines (red) are the donor and acceptor transitions, while the dotted (green) one is the fundamental absorption (E_g) of the bulk.

approximations this function has a positive curvature. This leads to the improper placement of the KS levels of the defect with respect to the band edges [4].

"Historically," the first attempts to remedy the gap problem were aimed at correcting the KS levels of the standard approximations. In Section 2 we will consider the most common methods and conclude that, save for many-body calculations on the defect containing solid, the applicability of them all are restricted to special defects even within one host material.

The "gap error," however, does not only concern the calculation of the defect level positions – it also appears in the formation energy of the defect, influencing the correct prediction of the ground state and the activation energies for diffusion or reactions. In Section 3 we will demonstrate this on a few examples. In Section 4 we will also show that total energy corrections based on the KS level corrections can only be applied in special cases.

Obviously, the whole "gap problem" could be solved by applying GW or Quantum Monte Carlo methods. The difficulty is, that – while simplified (G_0W_0) techniques can be applied to large supercells [5] to provide *a posteriori* quasiparticle (QP) corrections – self-consistent GW calculations [6] are still rather costly, restricting the size of the supercell presently, e.g., to 64–72 atoms for tetrahedral semiconductors (at the very limit), and calculation of the total energy is as yet not possible even for these. Cost factors limit Quantum Monte Carlo calculations even more, mostly forbidding even to take into account relaxation effects [7]. Therefore, in the time being, we think that the best way of dealing with the gap problem is to turn to a generalized Kohn–Sham scheme, based on approximate non-local exchange functionals [8]. There are many possibilities available (see, e.g., Refs. [9–14]). In Section 5 we will investigate the performance of the screened hybrid exchange functional of Heyd, Scuseria, and Ernzerhof [15, 16] for a set of well chosen defects. Our conclusion is that carefully tested semi-empirical, range-separated hybrids are presently the best tools for economically feasible studies of both defect spectra and energetics, with transferability within a class of hosts of similar bonding and irrespective of the nature of the defect.

8.2
Empirical Correction Schemes for the KS Levels

First attempts for correcting the consequences of the gap error for defects have concentrated on the gap levels. Since calculated effective masses were acceptably accurate, the assumption was that (semi)local approximations to DFT described both bands accurately, just the energy difference between the VBM and the CBM was too small [17]. This has led to the idea of taking the VBM as reference, open up the gap to its experimental value, and scale the energy difference of the defect level to the VBM accordingly. Needless to say that this procedure has no justification whatsoever. A more intelligent approach was to consider whether the defect state was VB or CB related and – as a first approximation – apply the same shift (w.r. to the VBM) as for the CBM in the latter case, or no shift at all in the former. Mostly, however, no clear-cut decision is possible. This was taken into account by the "scissor operator," introduced by Baraff and Schlüter [18] for the case of a vacancy. Since the wave function of the defect can be expanded on the basis of the perfect crystalline states, they assumed that the amount of necessary shifting for a defect level can be determined by the weight of the conduction-band states in the expansion. Therefore, the shift necessary for the CBM to reproduce the experimental gap is to be scaled by the sum of overlaps between the defect wave function and all CB states of the perfect crystal. In principle, the scissor operator can be applied self-consistently, but most often it was applied *a posteriori*, which made sense only if the too low-lying CBM in the LDA or GGA calculation did not mask the localized defect level [2]. (Otherwise the calculation would lead – incorrectly – to an effective mass like state and a full shift with the CBM.)

In our experience, the scissor operator has worked reasonably well for substitutional defects and split-interstitials but for interstitial defects in the low electron density region of the crystal the results seemed to be problematic. Therefore, we carried out a case study using hydrogen as a probe in silicon and silicon carbide [19]. The hydrogen interstitial in Si has established configurations both in the high and low electron density regions of the crystal [20]. In the neutral charge state, it intercepts a Si–Si bond. At this, so-called bond-center (BC) site the hydrogen is a donor. There exists, however, a metastable site behind a Si–Si bond, in the so-called antibonding (AB) position (near to the tetrahedral interstitial site T). Here the hydrogen acts as an acceptor. This provides a unique opportunity to check how the validity of the scissor correction depends on the position of the defect. In SiC the interstitial H is always an acceptor at the AB site, behind a silicon atom. Comparing that to the case of H_{AB} in Si can show how the scissor correction works in materials with very different gaps. In our study we have used standard local (spin) density approximation, and compared the scissor correction to the QP-correction of a G_0W_0 calculation. (Details are described in Ref. [19].)

The results are shown in Table 8.1. The scissor correction is close to the G_0W_0 result for H_{BC}. This corroborates the experience obtained with substitutional defects that the scissor operator works well in the high electron density region of the crystal.

Table 8.1 Comparison of the scissor- and the G_0W_0 QP-corrections [in (eV)] to the KS levels obtained by LDA for interstitial H in Si and SiC. (Also corrections predicted by a one-parameter hybrid functional are given).

level[a]	scissor	G_0W_0	hybrid
Si: CBM	0.61	0.66	0.61
Si : H_{BC}^0	0.53	0.44	0.47
Si : H_{AB}^-	0.44	0.17	0.12
SiC:CBM	1.08	1.17	1.12
SiC : H_{AB}^-	0.90	0.12	0.04

a) The position of the CBM has been set according to the experimental gap in the case of the scissor operator. The mixing parameter of the hybrid functional has been fitted to reproduce the lattice parameter, cohesive energy, bulk modulus, and the band gap.

In contrast the scissor yields a gross overcorrection for H_{AB} in both materials. (The error increases in 4H–SiC, relative to Si, by about the same amount as the increase in the gap correction.) The explanation lies in the different nature of the defect state at BC and AB. At BC in Si it is essentially an antibonding combination of the sp^3 hybrids on the Si neighbors, i.e., clearly conduction band derived. Therefore, it can very well be described by a linear combination of the CB states of the perfect crystal. The defect state of H at the AB site (i.e., near the tetrahedral interstitial site T) is a good example for an $1s$ effective-mass state. However, in this interstitial "hole" of the tetrahedrally bonded semiconductors, the electron density is small, so only CB states can be involved in the expansion of even such an essentially valence state [21]. As a result, the scissor gives a correction almost as big as that of the CBM, being increasingly wrong with increasing gap. Obviously, the basic assumption of the scissor operator works only in the high electron density region of the perfect crystal, and not if the defect resides in a low electron density region where the VB states of the perfect crystal do not provide an adequate basis for expanding a strongly localized defect state.

Another way of dealing with the gap problem is to correct the host band structure of a standard (semi)local DFT approximation. One possibility for that is the application of an *a posteriori* alignment scheme [22, 23]. It has been observed that the QP-corrections work like a symmetric scissor, pushing the VBM down and the CBM up, and – at least for some defects – the position of the gap levels, with respect to a suitable chosen "external" reference, hardly changes [23]. This would allow to correct the positions of the VBM and the CBM, based on – say GW – calculations done on the small unit cell of the perfect system, which is much less costly to carry out than for the defective super cell. Unfortunately, however, this idea works only for defects with well localized mid-gap levels [23], and provides no way of correcting the total energy (see next section). A self-consistent but empirical way of correcting the host band structure is offered by the use of the LDA + U (or GGA + U) methods, and/or by applying non-local empirical pseudopotentials (NLEP) for correction, as described in Ref. [4]. One problem with this procedure is that the adjustment of the KS levels is often accompanied by the deterioration in the ground state properties of the system (e.g., ionicity and lattice constant). The other problem is that a reasonably justified choice of the empirical parameters can only be made for the host system but not for an impurity. Our attempt, in the latter case, to choose parameters by enforcing the equality of the ionization energies, calculated either as total energy difference or as the negative of the highest occupied KS level, has failed.

8.3
The Role of the Gap Level Positions in the Relative Energies of Various Defect Configurations

While the methods mentioned in the previous Section can be used – at least for some defects – to obtain spectra comparable to experiment, the consequences of the gap error in the total (and formation) energy still remain. Here we would like to recall two examples in silicon [24], which demonstrate how the error in the gap level position directly influences the relative energy of different configurations. We compare here calculations with pure GGA exchange to those with a one-parameter hybrid functional. The mixing parameter of Hartree–Fock and GGA exchange in the latter have been chosen to optimize the lattice constant, the cohesive energy, the bulk modulus, and its derivative, as well as the widths of the VB and the fundamental gap [25]. With this optimal value, also higher bulk excitations are well reproduced and, as shown in Table 8.1, the level positions of H in Si are very near to the ones obtained by GW. Later, in Section 8.5, we will show that hybrid functionals can really be used as reference, providing not only correct gap level positions but also total energy differences free of the gap error.

The first example is interstitial oxygen in silicon, in its ground state O_i, as puckered BC interstitial, and in the so called O_Y configuration, which is the saddle point along the diffusion path (Figure 8.2). The activation energy for diffusion is well established experimentally: 2.53 eV in the 270–700 °C range [26]. In a theoretical calculation at 0 K, one should expect a somewhat higher value. Based on the observed Si–O–Si stretch frequency of ~ 1100 cm^{-1}, the zero-point energy in the ground state can be estimated to be 0.07 eV, so the 0 K theoretical barrier should be above 2.6 eV. It is known, that well converged LDA or GGA calculations underestimate this activation energy: our GGA calculation in a 64 atom supercell resulted in 2.37 eV. In contrast, the hybrid functional gave 2.69 eV, which is well in line with experiment. For both functionals, the increase of the total energy follows the emergence of a gap level from the VB, when going from the electrically inactive O_i configuration toward the saddle point at O_Y, where the central Si atom has a p-type dangling bond, doubly occupied due to electron transfer from the trivalent oxygen atom. Considering that the gap level is doubly occupied, the 0.14 eV difference in the level positions between the hybrid and the pure GGA calculation at the saddle point seems to explain most of the deviation of the activation energy, 0.32 eV. Although the numerical agreement is somewhat accidental (as shown in the next Section), this finding indicates that the error of GGA in predicting the activation energy is related to the gap error.

The other example is the complex of substitutional boron with a self-interstitial, $B_{Si} + Si_i$. This defect gives rise to the charge transition level shown in Table 8.2, is paramagnetic in the neutral charge state and, according to the observed hyperfine interactions, it has C_{1h} symmetry [27]. LDA and GGA studies result in a metastable configuration with such a symmetry but give also a more stable one with C_{3v}

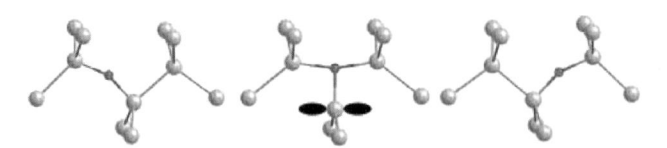

Figure 8.2 (online color at: www.pss-b.com) The diffusion path of interstitial oxygen in silicon (O_i) from ground state to ground state through the O_Y transition state. The undercoordinated Si in the latter is indicated by the lone pair p state. This state gives rise to a level in the gap. The oxygen atom is the dark (red in color) sphere.

8.3 The Role of the Gap Level Positions in the Relative Energies of Various Defect Configurations

Table 8.2 (+/0) charge transition levels in silicon [with respect to the perfect crystal VBM in (eV)], calculated by an LDA functional, and after correcting the total energies according to Eq. (8.1), based on a calculation by G_0W_0 or by a one-parameter hybrid functional.

E(+/0) w.r. VBM		with corrections based on level positions in		
	LDA	G_0W_0	hybrid	exptl.
H_{BC}	0.54	0.98	0.98	0.94[a]
$B_{Si} + Si_i$	0.66		0.94	0.99[b]

a) Ref. [29],
b) Ref. [27]

symmetry, as shown in Figure 8.3 (see Ref. [25] and references therein). This has always been suspected to be a consequence of the gap problem [28]. Calculation with the hybrid functional proves that, resulting the C_{1h} configuration lower in energy than the C_{3v} one. The reason is that in the C_{1h} configuration, which is nearly a [110] split-interstitial, a different kind of orbital is occupied than in the C_{3v} configuration, where practically the boron is an on-center substitutional and Si_i is near the tetrahedral interstitial site. The gap level position is shifted up between the GGA and the hybrid calculation much more in the C_{3v} case, than in the C_{1h}. This difference brings about a larger increase of the total energy for the C_{3v} configuration than for the C_{1h}, leading to a reversal in the stability sequence. This is clearly a case where LDA and GGA both fail to predict the correct ground state of a system because of the gap error.

These examples show, that the error in the gap level position directly influences the relative energies of defects and, without appropriate correction, this can lead to serious quantitative and qualitative errors in the predictions based on (semi)local approximations of DFT. Note, that the error of the gap level positions increases with the width of the gap (see Table 8.1), and can easily cause catastrophic problems for defects in wide band gap materials.

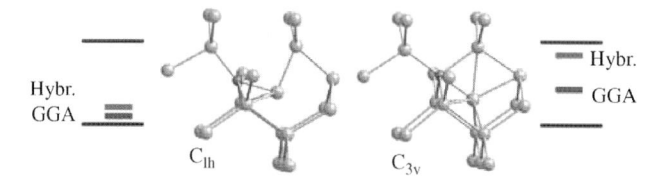

Figure 8.3 (online color at: www.pss-b.com) The C_{1h} (left) and C_{3v} (right) configurations of the $B_{Si} + Si_i$ complex in silicon. The boron atom is the dark (green in color) sphere. Also shown are the defect level positions in the gap, as obtained by a pure GGA and a hybrid functional.

8.4
Correction of the Total Energy Based on the Corrected Gap Level Positions

The examples described in the previous section suggest a possibility of correcting the total energy. The latter can always be given as the sum of the band energy and of double-counting terms, and the band energy can be split into the contribution of the occupied defect level in the gap and that of the valence band:

$$E_{tot} = E_{BE} + E_{dc} = n_D \varepsilon_D + \sum_i^{VB} n_i \varepsilon_i + E_{dc}(\varrho) \tag{8.1}$$

where ε_D, ε_i denote KS levels in the gap and in the VB, respectively, and n_D, n_i are the corresponding occupation numbers. Equation (8.1) clearly shows that an error in the first term of the right-hand side will influence the total energy, since the double-counting term depends only on the electron density $\varrho(r)$, which – according to GW calculations – are quite well described by the (semi)local approximations, and there is no reason to expect compensation by the second term. Based on Eq. (8.1), we have used the following correction scheme for the total energy:

$$E_{tot}^{corr.} = E_{tot} + n_D \Delta \varepsilon_D, \tag{8.2}$$

$$\Delta \varepsilon_D = [\varepsilon_D^{corr.} - \varepsilon_{VBM}^{corr.}] - [\varepsilon_D^{uncorr.} - \varepsilon_{VBM}^{uncorr.}] \tag{8.3}$$

Table 8.2 shows the results for the two donors discussed above in silicon, using the gap level position obtained in the hybrid calculation for correction [24]. In the first case the correction is also given based on a G_0W_0 calculation [19]. As can be seen the corrected results compare favorably to experiment. This correction scheme has recently been proposed again using GW data in Eq. (8.3) [30, 31].

Actually, we have been using this correction scheme for quite some time, initially with the scissor correction for the gap levels [32], later with the level positions obtained from one-parameter hybrid functional calculations [33]. Our experience about its success was, however, somewhat mixed, so we have examined [24] the working of Eqs. (8.2) and (8.3) on the energy difference (i) between the BC and AB sites (in the same charge state) for interstitial Si:H_i, (ii) between the ground state and saddle point configurations of Si:O_i, and (iii) between two possible configurations of a silicon vacancy in SiC, as shown in Figure 8.4. (The obvious V_{Si} configuration can transform into a $V_C + C_{Si}$ one [34]). Assuming that the hybrid calculation provides both gap level positions and self-consistent total energies free of the gap error, the corrections (with respect to a pure GGA calculation) to each term in Eq. (8.1) can be calculated separately. (The corrections to E_{tot} and E_{BE} can be calculated directly, while that of E_{dc} is their difference.) These are compared in Table 8.3 with the approximate correction based on Eqs. (8.2) and (8.3). As can be seen, the agreement between the first and last row is good in only one case. Looking at the contributions from E_{BE} and E_{dc}, however, it is clear that even this agreement is the result of a lucky compensation effect, which does not occur in the other two. How can this be understood in light of the success with the charge transition levels in Table 8.2?

8.4 Correction of the Total Energy Based on the Corrected Gap Level Positions

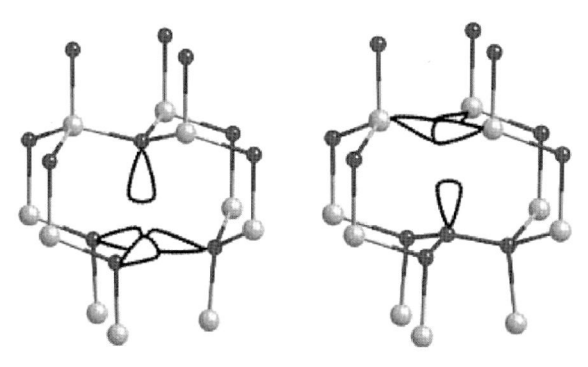

Figure 8.4 (online color at: www.pss-b.com) V_{Si} (left) and its isomer $V_C + C_{Si}$ (right) in SiC. The carbon atoms are the dark (blue) spheres. The dangling bonds are schematically indicated.

In each of the three cases examined in Table 8.3, there is considerable change in the bonding. Hydrogen at the BC site in silicon has a singly occupied gap level near the CB edge, while at the AB site it has a doubly occupied resonance just below the VB edge. No correction for the latter has been taken into account in the last row, but is included in the first. (Apparently the two corrections happen to cancel each other accidentally to a high degree.) The oxygen atom in its interstitial position in silicon is divalent, with two tetravalent Si neighbors. At the saddle point of its motion it becomes trivalent and one Si neighbor becomes undercoordinated. Obviously, very different VB resonances correspond to the two cases and the change in E_{BE} is considerably more than just the correction coming from the appearance of the gap level. (Still, the error in the latter does influence the total energy significantly.) Between the two different configurations the electron density, and so E_{dc} changes as well. For H in Si this seems to compensate the change in E_{BE}, for O only the VB part of the latter. In the case of the vacancy in SiC there is no compensation at all. The approximate correction covers the change from three dangling bonds on Si to three on C, but at the same time three Si–C bonds are replaced by three C–C bonds, so a non-self-consistent correction is obviously meaningless.

Table 8.3 The difference between a hybrid and a GGA calculation [24] in the relative energy of two defect configurations, a and b, split according to the various terms of Eq. (8.1). The last row gives the approximate total energy corrections based on Eqs. (8.2) and (8.3).

$a - b$	Si		SiC
	$H_{BC} - H_{AB}$	$O_Y - O_i$	$V_{Si} - V_C + C_{Si}$
$\Delta(E_{tot}^a - E_{tot}^b)$	+0.01	+0.32	−0.75
$\Delta(E_{BE}^a - E_{BE}^b)$	−0.04	+0.51	+1.08
$\Delta(E_{dc}^a - E_{dc}^b)$	+0.05	−0.19	−1.83
$n_D^a(\varepsilon_D^a + \Delta\varepsilon_D^a) - n_D^b(\varepsilon_D^b + \Delta\varepsilon_D^b)$	+0.19	+0.28	−1.89

In contrast to these cases, the simple change in the occupation of a gap level may shift it a little, but VB resonances will hardly be affected and little change will occur in E_{dc} – at least in the cases mentioned here. However, in case of bistable defects, the charge transition may induce a substantial relaxation of the nuclei and a very different bonding. Therefore, the *a posteriori* corrections scheme of Eqs. (8.2) and (8.3) should be used with utter care.

8.5
Accurate Gap Levels and Total Energy Differences by Screened Hybrid Functionals

Considering all the problems with correcting the results of the (semi)local exchange functionals on the one hand, and the unfeasibility of applying many-body theories to large supercells on the other, generalized KS (gKS) schemes with approximate non-local exchange functionals seem to offer the solution in the time being. Such approximations are the hybrid functionals. Based on the adiabatic connection formula Becke has suggested [35, 36] an approximation to the exact DFT exchange energy by mixing GGA and Hartree–Fock (HF) exchange. The mixing parameter of these hybrid functionals were chosen semi-empirically to optimize thermochemical data of molecules. The optimal choice of 25% HF-exchange can also be justified theoretically [37]. It has been observed early on that hybrid functionals systematically improve the gKS gap of semiconductors [38]. Encouraged by that, we have determined materials-specific mixing parameters for crystalline Si [25], SiC [33] and SiO_2 [39], by fitting ground state properties as well as the gap to experimental values, as mentioned earlier. Table 8.4 shows the band gaps obtained. Although these hybrids have proved themselves in several applications (see, e.g., Refs. [33, 39–42]), their lack of transferability from one material to another is a severe restriction, especially for interface studies.

More recently, a new class of hybrid functionals has been introduced [15], where the mixing is done only for the short range part of the electron–electron interaction. This corresponds to screened non-local exchange (screened hybrids). The screening parameter introduces an additional degree of freedom, and optimizing these can give excellent gaps for a wide range of semiconductors (but not for all). The first version

Table 8.4 Band gaps obtained by a one-parameter hybrid exchange functional. Values in Italics in each column were obtained by the mixing parameter optimized for ground state properties and the gap of the given material. The experimental values are shown in parentheses. All values are in (eV).

HF-part	fundamental gap (eV)			
	SiO_2 (9.5)	Si (1.17)	3C-SiC (2.36)	C (5.48)
12%		*1.17*		
20%	9.0	1.44	*2.42*	5.12
28%	*9.5*			

Table 8.5 Comparison of PBE and HSE06 results [46] for the lattice constant (a, c), cohesive energy (E_{coh}), bulk modulus (B_0), fundamental (indirect) gap (E_g), first allowed direct transition at the Brillouin-zone center ($\Gamma_{25}' \rightarrow \Gamma_2'$), and valence band width (VB) with experimental data [47, 48] in case of the Group IV semiconductors.

	method	a (Å)	c (Å)	E_{coh} (eV)	B_0 (GPa)	E_g (eV)	$\Gamma_{25}' \rightarrow \Gamma_2'$ (eV)	VB (eV)
diamond	PBE	3.574		7.85	425	4.21	13.3	21.5
	exptl.	3.567		7.37	443	5.48	15.3	24.2
	HSE06	3.544		7.58	464	5.42	15.7	23.8
4H-SiC	PBE	3.091	10.116	6.51		2.22		
	exptl.	3.073	10.053			3.23		
	HSE06	3.069	10.045	6.37		3.21		
3C-SiC	PBE	4.375		6.51	209	1.37	6.1	15.3
	exptl.	4.360		6.34	224	2.36	7.4	17.
	HSE06	4.345		6.37	230	2.25	7.7	17.1
silicon	PBE	5.468		4.62	88	0.61	3.14	11.8
	exptl.	5.429		4.63	99	1.17	4.15	12.5
	HSE06	5.434		4.54	98	1.17	4.33	13.3
germanium	PBE					0.00		
	exptl.	5.658		3.88		0.74	0.90	13.0
	HSE06	5.670		3.66		0.84	0.88	13.9

(HSE03) of this screened hybrid by Heyd–Scuseria–Ernzerhof [15] has shown substantial improvement of the band gap over a one-parameter hybrid with 25% HF-part (termed PBE0 in Ref. 9) and, above all, the same quality in similarly bonded materials. It is important to emphasize that at the same time also the reproduction of the basic ground state properties (lattice constant, heat of formation, and bulk modulus) have also improved [43].

In Table 8.5 we show this for the Group IV semiconductors, diamond, SiC, Si, and Ge, using the revised HSE06 version [16]. (The screening parameter is set to 0.2 Å^{-1}, keeping the 25% admixture of HF-exchange to 75% GGA exchange calculated by the Perdew, Burke, and Ernzerhof – or PBE – functional [44].) We would like to point out that details of the band structure, like width of the VB or the first allowed direct transition at Γ, which have not been included into the fitting procedure, agree also well with experiment. The band gaps are reproduced in all these materials with the same high accuracy. This is even true for Ge (taking into account the lack of spin-orbit coupling in our calculation, which would lower the fundamental gap), for which the PBE calculation gives no gap at all. The transferability of HSE06 in materials with similar bonds is encouraging, provided it pertains also to defect levels. In a recent paper we have found that HSE06 works extremely well for the inner excitation of a defect in diamond [45]. We have, therefore, also investigated the transition energies between the band edges and defect states in a systematic manner [46].

From the viewpoint of defects, the approximate non-local exchange functional in the gKS scheme is expected to remedy both the gap problem and the inappropriate dependence of the total energy on the occupation number, or in other words: to provide both total energy and gKS levels of the defect free of the gap error. To check this, we have compared the band ↔ defect transitions computed (cf. Figure 8.1) as differences of self-consistent total energies (ΔSCF method) to values obtained as differences between highest occupied gKS levels (ΔKS method). The average potentials between the perfect crystal and the defective supercell have been aligned using the method suggested in Ref. [1]. Charged supercells were calculated assuming a jellium charge of opposite sign. This leads to an error, dependent on the supercell size, both in the total energy and in the gKS levels [31]. In general, the error also depends on the nature of the defect state, requiring non-trivial correction procedures [49]. Therefore we have chosen fairly large, 512-atom supercells (in the Γ approximation) to minimize all size effects, and applied 65% of the monopole correction for charged supercells, as suggested in Ref. [2]. In case of acceptors both the ΔKS and ΔSCF transitions need correction, and we assumed them to be approximately equal for a single negative charge.

In Table 8.6 we compare the ΔKS and ΔSCF vertical transitions for a series of donors and acceptors in diamond, silicon, and germanium. For Si, the defects have been chosen to scan the whole width of the gap by their gap levels. With one exception, the agreement is within 0.1 eV, irrespective of the gap width of the host, or the shallow or deep nature of the defect.

Table 8.7 shows the comparison of the calculated adiabatic transition energies to the experimental ones. The adiabatic "ΔKS" values have been obtained by adding the relaxation energy of the charged state (with respect to the neutral one) to the vertical ΔKS transition. The values obtained this way are in stunning agreement with experiment. (N.B. the defects in this study have been chosen by the criterion of having accurate experimental values beyond doubt.) The same is true for the ΔSCF results, except for the case of the iron interstitial in silicon (Si:Fe$_i$), the "odd guy out" also in Table 8.6. Comparison of the two Tables show that the error is in the calculated ΔSCF

Table 8.6 Vertical transition energies [in (eV)] calculated by comparing highest occupied levels (ΔKS) or total energies (ΔSCF) according to Figure 8.1. The E(+/0) transition levels of donors are given with respect to the VBM, the E(+/0) transitions levels of acceptors with respect to the CBM.

	donors			acceptors		
E(+/0) w.r. CBM	ΔKS	ΔSCF	E(0/−) w.r. VBM	ΔKS	ΔSCF
C_{512}:P$_C$	−0.6	−0.6	C_{512}:B$_C$	+0.3	+0.4
Si_{512}:S$_{Si}$	−0.3	−0.3	Si_{512}:In$_{Si}$	+0.2	+0.2
Si_{512} : S$_{Si}^+$	−0.6	−0.6	Si_{512}:O$_{Si}$	+0.9	+1.0
Si_{512}:Fe$_i$	−1.0	−0.7			
Si_{512}:Au$_{Si}$	−0.9	−0.8			
Si_{512}:C$_i$	−1.0	−0.9	Si_{512}:C$_i$	+1.0	+1.0
Ge_{512}:S$_{Ge}$	−0.4	−0.3	Ge_{512}:O$_{Ge}$	+0.4	+0.4

Table 8.7 Adiabatic $\Delta KS^{a)}$ and ΔSCF transition energies compared to experiment. Experimental values are from Ref. [50].

	donors				acceptors		
$E(+/0)$ w.r. CBM	ΔKS	ΔSCF	exptl.	$E(0/-)$ w.r. VBM	ΔKS	ΔSCF	exptl.
$C_{512}:P_C$	−0.5	−0.5	−0.6	$C_{512}:B_C$	+0.3	+0.4	+0.4
$Si_{512}:S_{Si}$	−0.3	−0.3	−0.3	$Si_{512}:In_{Si}$	+0.2	+0.2	+0.2
$Si_{512}:S_{Si}^+$	−0.5	−0.5	−0.6	$Si_{512}:O_{Si}$	+0.8	+0.9	+0.9
$Si_{512}:Fe_i$	−0.8	−0.5	−0.8				
$Si_{512}:Au_{Si}$	−0.8	−0.8	−0.8				
$Si_{512}:C_i$	−0.9	−0.8	−0.9	$Si_{512}:C_i$	+0.9	+1.0	+1.0
$Ge_{512}:S_{Ge}$	−0.3	−0.3	−0.3	$Ge_{512}:O_{Ge}$	+0.3	+0.3	+0.3

a) The relaxation energy of the charged state (with respect to the neutral one) was added to the vertical ΔKS transition.

vertical transition. This may be partly an error of the simplified charge correction, but definitely not entirely. (The error is bigger than the whole monopole correction.) Si:Fe$_i$ has a gap state which is highly localized on a Fe 3d orbital, i.e., compared to the other defects, has the least contribution form host derived states. (In all other cases the gap state is either effective mass like or a combination of host dangling bonds. The donor state of Si:C$_i$ is a pure p orbital on C.) Therefore, a possible explanation for the error might be that the screened hybrid can mimic the accurate non-local exchange functional in every respect only for states characteristic to the class of hosts for which the parameters have been chosen. For other states the dependence of the total energy on the occupation number is still seemingly correct, but not the absolute value.

If the above analysis is correct, the defect energetics obtained by a screened hybrid can only be fully trusted if the defect state is dominantly host-related. Still, the position of the gap level seem to be supplied by high accuracy in every case. This has three advantages: (i) the experimental spectrum can be predicted, (ii) comparison of the ΔKS and ΔSCF vertical transition energies provides a convenient way of assessing the reliability of the energetics, and (iii) the transition energies can be calculated without the need for a charge correction.

8.6 Summary

We have considered, how the "gap error" of the standard (semi)local DFT approximations influences both the spectra and relative energies of defects, and investigated several ways of correction. We conclude that from the point of view of the spectrum,

i) the scissor operator is a reasonably accurate and very convenient method of correcting the gap level position, if the defect is in the high electron density region of the perfect crystal, but fails outside of that.

ii) Methods of correcting the KS levels of the host work only for a certain class of defects even in one material.
iii) Gap level positions in semi-empirical screened hybrid calculations are just as accurate as the gap for which they have been parameterized, independent of the nature of the defect. The parameters are transferable within a class of materials with similar bonding.

From the view point of calculating the relative energy of different defect configurations (or charge states) we conclude that

iv) Different kind of gap states and VB resonances in the two configurations lead to different errors in the band energy, and so to a substantial error in the relative energy. The latter increases with the gap error ($\sim$gap width) and can lead to reversal of the stability ordering. Therefore, in such cases the LDA or GGA total energy has to be corrected for the "gap error" as well.
v) Correcting the band energy for the gap level alone is only sufficient if the relaxation upon changing the defect state is small.
vi) The total energy supplied by semi-empirical screened hybrids seems to be largely free of the consequences of the "gap error" for defect states with the character of the orbitals in those materials for which the parameters have been optimized.

It appears that, in the time being, well parameterized semi-empirical screened hybrids are the preferred method for calculating relative energies and electronic transitions for defects. Although much less expensive than GW or Quantum Monte Carlo calculations, their computational cost is still about an order of magnitude higher than for LDA or GGA. Therefore, a careful check of (iv) is recommended.

Acknowledgements

The authors are grateful for fruitful discussions with S. Lany, G. Kresse, and G. E. Scuseria. Support of the Supercomputer Center of Northern Germany (HLRN grant no. hbc00001), as well as that of the German–Hungarian bilateral research fund (436 UNG 113/167/0-1) are appreciated. AG acknowledges the support of the Hungarian grants OTKA (no. K67886) and NKTH (Nr. NKFP-07-A2-ICMET-07), as well as the János Bolyai program from the Hungarian Academy of Sciences.

References

1 Van de Walle, C.G. and Neugebauer, J. (2004) *J. Appl. Phys.*, **95**, 3851.
2 Lany, S. and Zunger, A. (2008) *Phys. Rev. B*, **78**, 235104.
3 Janak, J.F. (1978) *Phys. Rev. B*, **18**, 7165; Ambladh, C.-O. and von Barth, U. (1985) *Phys. Rev. B*, **31**, 3231.
4 Lany, S. and Zunger, A. (2009) *Phys. Rev. B*, **80**, 085202. (Chapter 11).
5 Furthmüller, J., Cappellini, G., Weissker, H.C., and Bechstedt, F. (2002) *Phys. Rev. B*, **66**, 045110.
6 Shishkin, M. and Kresse, G. (2007) *Phys. Rev B*, **75**, 235102.

7 Batista, E.R., Heyd, J., Hennig, R.G., Uberuaga, B.P., Martin, R.L., Scuseria, G.E., Umrigar, C.J., and Wilkins, J.W. (2006) *Phys. Rev. B*, **74**, 121102(R).

8 Fuchs, F., Furthüller, J., Bechstedt, F., Shishkin, M., and Kresse, G. (2007) *Phys. Rev. B*, **76**, 115109.

9 Adamo, C. and Barone, V. (1999) *J. Chem. Phys.*, **110**, 6158.

10 Robertson, J., Xiong, K., and Clark, S.J. (2006) *Phys. Status Solidi B*, **243**, 2054. (Chapter 5).

11 Janotti, A., Segev, D., and Van deWalle, C.G. (2006) *Phys. Rev. B*, **74**, 045202.

12 Baumeier, B., Krüger, P., and Pollmann, J. (2007) *Phys. Rev. B*, **76**, 085407.

13 Broqvist, P., Alkauskas, A., and Pasquarello, A. (2009) *Phys. Rev. B*, **80**, 085114.

14 Wu, X., Selloni, A., and Car, R. (2009) *Phys. Rev. B*, **79**, 085102.

15 Heyd, J., Scuseria, G.E., and Ernzerhof, M. (2003) *J. Chem. Phys.*, **118**, 8207; Heyd, J. and Scuseria, G.E. (2004) *J. Chem. Phys.*, **121**, 1187. (Chapter 6).

16 Krukau, A.V., Vydrov, O.A., Izmaylov, A.F., and Scuseria, G.E. (2006) *J. Chem. Phys.*, **125**, 224106.

17 In fact, the total width of the bands is always underestimated. E.g., the indirect band gap in silicon is underestimated by only 0.5 eV, but the direct transition at Γ from the VBM to the second subband in the CB by 1.0 eV, indicating a "compressed" CB. The width of the VB is also too small by 6%.

18 Baraff, G.A. and Schlüter, M. (1984) *Phys. Rev. B*, **30**, 1853.

19 Deák, P., Frauenheim, T., and Gali, A. (2007) *Phys. Rev. B*, **75**, 153204.

20 Estreicher, S.K. (1995) *Mater. Sci. Eng. R*, **14**, 319; Ammerlaan, C.A.J. (2004) *Silicon. Evolution and Future of a Technology*, vol. 217 (eds P. Siffert and E. Krimmel), Springer-Verlag, Berlin, p. 261.

21 This can be seen in the almost as high a sum of overlaps with the CB states (0.72) as at the BC site (0.87).

22 Schultz, P.A. (2206) *Phys. Rev. Lett.*, **96**, 246401.

23 Alkauskas, A., Broqvist, P., and Pasquarello, A. (2008) *Phys. Rev. Lett.*, **101**, 046405.

24 Deák, P., Aradi, B., Frauenheim, T., and Gali, A. (2008) *Mater. Sci. Eng. B*, **154/155**, 187.

25 Deák, P., Gali, A., Sólyom, A., Buruzs, A., and Frauenheim, T. (2005) *J. Phys.: Condens. Matter*, **17**, S2141.

26 Stavola, M., Patel, J.R., Kimerling, L.C., and Freeland, P.E. (1983) *Appl. Phys. Lett.*, **42**, 73; Takeno, H., Hayamizu, Y., and Miki, K. (1998) *J. Appl. Phys.*, **84**, 3113.

27 Watkins, G.D. (1975) *Phys. Rev. B*, **12**, 5824; Harris, R.D., Newton, J.L., and Watkins, G.D. (1987) *Phys. Rev. B*, **36**, 1094. (Chapter 7).

28 Hakala, M., Puska, M.J., and Nieminen, R.M. (2000) *Phys. Rev. B*, **61**, 8155.

29 Nielsen, K.B., Dobaczewski, S., Sogård, S., and Nielsen, B.B. (2002) *Phys. Rev. B.*, **65**, 075205.

30 Rinke, P., Janotti, A., Scheffler, M., and Van de Walle, C.G. (2009) *Phys. Rev. Lett.*, **102**, 026402.

31 Lany, S. and Zunger, A. (2010) *Phys. Rev. B*, **81**, 113201.

32 Aradi, B., Gali, A., Deák, P., Lowther, J.E., Son, N.T., Janzén, E., and Choyke, W.J. (2001) *Phys. Rev. B*, **63**, 245202; Aradi, B., Deák, P., Son, N.T., Janzén, E., Devaty, R.P., and Choyke, W.J. (2001) *Appl. Phys. Lett.*, **79**, 2746; Szűcs, B., Gali, A., Hajnal, Z., Deák, P., and Van de Walle, Ch. G. (2003) *Phys. Rev. B*, **68**, 085202.

33 Gali, A., Deák, P., Ordejón, P., Son, N.T., Janzén, E., and Choyke, J.W. (2003) *Phys. Rev. B*, **68**, 125201; Gali, A., Hornos, T., Deák, P., Son, N.T., Janzén, E., and Choyke, W.J. (2005) *Appl. Phys. Lett.*, **86**, 102108; Knaup, J.M., Deák, P., Gali, A., Hajnal, Z., Frauenheim, Th., and Choyke, W.J. (2005) *Phys. Rev. B*, **71**, 235321.

34 Bockstedte, M., Mattausch, A., and Pankratov, O. (2004) *Phys. Rev. B*, **69**, 235202.

35 Becke, A.D. (1993) *J. Chem. Phys.*, **98**, 5648.

36 Becke, A.D. (1997) *J. Chem. Phys.*, **107**, 8554.

37 Perdew, J.P., Ernzerhof, M., and Burke, K. (1996) *J. Chem. Phys.*, **105**, 9982.

38 Bredow, T. and Gerson, A.R. (2000) *Phys. Rev. B*, **61**, 5194; Muscat, J., Wander, A.,

and Harrison, N.M. (2001) *Chem. Phys. Lett.*, **342**, 397.

39 Knaup, J.M., Deák, P., Gali, A., Hajnal, Z., Frauenheim, Th., and Choyke, W.J. (2005) *Phys. Rev. B*, **71**, 235321; Knaup, J.M., Deák, P., Gali, A., Hajnal, Z., Frauenheim, Th., and Choyke, W.J. (2005) *Phys. Rev. B*, **72**, 115323.

40 Deák, P., Knaup, J.M., Hornos, T., Thill, Ch., Gali, A., and Frauenheim, Th. (2007) *J. Phys. D*, **40**, 6242–6253. Corrigendum: **41**, 049801 (2008).

41 Deák, P., Buruzs, A., Gali, A., and Frauenheim, Th. (2006) *Phys. Rev. Lett.*, **96**, 236803.

42 Aradi, B., Ramos, L.E., Deák, P., Köhler, Th., Bechstedt, F., Zhang, R.Q., and Frauenheim, Th. (2007) *Phys. Rev. B*, **76**, 035305.

43 Marsman, M., Paier, J., Stroppa, A., and Kresse, G. (2008) *J. Phys.: Condens. Matter*, **20**, 064201.

44 Perdew, J.P., Burke, K., and Ernzerhof, M. (1996) *Phys. Rev. Lett.*, **77**, 3865.

45 Gali, A., Janzén, E., Deák, P., Kresse, G., and Kaxiras, E. (2009) *Phys. Rev. Lett.*, **103**, 186404.

46 Deák, P., Aradi, B., Frauenheim, T., and Gali, A. (2010) *Phys. Rev. B*, **81**, 153203.

47 Madelung, O. (ed.) (1991) Semiconductors. Group IV Elements and II-V Compounds, in: *Data in Science and Technology*, Springer, Berlin.

48 Levinshtein, M.E., Rumyantsev, S.L., and Shur, M.S. (eds) (2001) *Properties of Advanced Semiconductor Materials: GaN, AlN, InN, BN, SiC, and SiGe*, John Wiley and Sons, New York.

49 Freysoldt, Ch., Neugebauer, J., and Van de Walle, C.G. (2009) *Phys. Rev. Lett.*, **102**, 016402. (Chapter 14).

50 Schulz, M., Dalibor, T., Martienssen, W., Landolt, H., and Börnstein, R. (2003) *Impurities and Defects in Group IV Elements, IV-IV and III-V Compounds*, Landolt-Börnstein, New Series, Group III, vol. 41, Pt. ß Springer, Berlin, Subvol. a,2.

9
LDA + U and Hybrid Functional Calculations for Defects in ZnO, SnO$_2$, and TiO$_2$

Anderson Janotti and Chris G. Van de Walle

9.1
Introduction

Defects and impurities greatly influence the optical and electrical properties of semiconductors. Control of their concentration and their effects is essential for enabling the utilization of a semiconductor material for electronic and optoelectronic device applications [1–3]. As examples, ZnO, SnO$_2$, and TiO$_2$ are promising materials for light emitters, transparent contacts, and photocatalysis; nevertheless, their use in devices has been hindered by the inability to control their electrical and optical properties, which are strongly affected by the presence of native defects and background impurities [4–8]. ZnO has a direct band gap of 3.4 eV and shows excellent luminescence and carrier-transport properties, but the lack of p-type conductivity prevents the use of ZnO in LEDs and lasers. Better control over the n-type conductivity would also improve its prospects as a transparent electrical contact [4–6]. SnO$_2$ has a band gap of 3.6 eV and has been considered as an alternative to the transparent conducting oxide Sn-doped In$_2$O$_3$ (ITO) [7]. Similar to ZnO, SnO$_2$ would greatly benefit from an improved control over doping and defect concentrations. TiO$_2$ has a band gap of 3.0 eV and its expanded use in photocatalysis and photoelectrolysis depends on engineering its band gap for extending its activity to the visible spectrum, as well as on controlling carrier transport and unwanted carrier recombination [8]. These properties are again strongly influenced by impurities and native point defects.

Calculations based on the density functional theory (DFT) within the local density approximation (LDA) or its semilocal extensions, such as the generalized gradient approximation (GGA), form the standard approach for studying defects in semiconductors and insulators [10]. The problem with the DFT-LDA or GGA is the severe underestimation of band gaps [9], which impart typically large errors in the calculated formation energies and the position of transition levels [10, 11]. Empirical corrections, such as applying a scissors operator, have been proposed over the years, with conclusions varying qualitatively from one research group to another [10, 12]. Recently, the use of LDA + U [13] and the development of screened hybrid functionals [14] have led to significant progress toward a quantitative description of

Advanced Calculations for Defects in Materials: Electronic Structure Methods, First Edition.
Edited by Audrius Alkauskas, Peter Deák, Jörg Neugebauer, Alfredo Pasquarello, and Chris G. Van de Walle.
© 2011 Wiley-VCH Verlag GmbH & Co. KGaA. Published 2011 by Wiley-VCH Verlag GmbH & Co. KGaA.

defects in semiconductors. In particular, a systematic approach based on LDA + U has been proposed and applied for defects in ZnO [11]. The extra Coulomb potential U has been added to improve the description of the Zn semi-core d states in a justified physical manner. As a consequence, the interaction between the Zn d states and O p states that compose the upper valence band states, and the position of the Zn 4s states are also affected, leading to a partial correction of the band gap in ZnO [15, 16]. An extrapolation based on LDA and LDA + U results was then performed and corrected transition levels and formation energies for all native point defects have been obtained [11, 17, 18]. As a main result, it has been predicted that oxygen vacancies are not responsible for the unintentional n-type conductivity in ZnO since it is a deep donor. These results have been favorably compared to recent experimental measurements on high quality ZnO single crystals [19, 20].

The use of LDA + U for studying defects is limited to materials with semicore d states such as the d-bands in ZnO, SnO_2, or InN [11, 21, 22]. In our opinion, the use of LDA + U for states that are more appropriately described as delocalized or itinerant bands is unwarranted and may lead to spurious results. For instance, applying LDA + U to the Ti d states of TiO_2 and related materials, or to the O p states in any of these oxides is not physically justified, since these states clearly lead to extended states in the band structure. The advent of screened hybrid functionals [14] and its implementation with periodic boundary conditions has allowed overcoming this limitation [23]. Mixing a fraction of non-local Hartree–Fock exchange with the GGA exchange potential [24] and imposing a screening length leads to an improved description of the electronic properties of a wide range of materials. By adjusting the mixing parameter it is possible to accurately describe band gaps. The imposition of a screening length is essential for describing semiconductors and metals on the same footing [25], which is necessary for determining formation energies in which metallic phases enter as references for reservoir energies. Based on Heyd, Scuseria, and Ernzerhof (HSE) is has been possible to describe the different charge states of the oxygen vacancy in TiO_2 [26].

In the present work, we discuss the results for native defects in ZnO and SnO_2 using the LDA/LDA + U extrapolation and the HSE hybrid functional. We address the advantages and limitations of these two methods, and draw comparisons with experimental data where available. We also present results for oxygen vacancies in TiO_2 based on HSE, shedding light on the differences among the wide range of results and conclusions reported in the literature.

9.2
Methods

Formal definitions of formation energies and transition levels are given in the paper by Janotti and Van de Walle in this volume [27] and will not be repeated here. Instead, we focus on the uncertainties introduced by the use of DFT. The standard approach based on DFT-LDA or GGA for calculating defects in semiconductors fails to provide quantitative predictions of transition levels and formation energies. This failure can be largely attributed to the band-gap error in semiconductors and insulators. LDA and GGA underestimate band gaps by more than 50%, and the errors are usually in the position of both valence and conduction bands. As a consequence, calculated transition levels that describe transitions between charge states of the defect carry uncertainties that can be as large as the band-gap error [10, 11]. Transition levels are defined as total energy differences between adding an electron and/or removing an electron from defect-induced gap states. This is analogous to the definition of the band gap as an energy difference between adding an electron to the conduction band and removing an electron from the valence band. Thus, transition levels suffer from the same error as band gaps in the LDA or GGA [9].

An often-overlooked problem is the error in formation energies due to the band-gap underestimation in the LDA/GGA [11]. It is often assumed that the formation energy of a neutral defect is a ground-state property that is well described within DFT-LDA or GGA. However, if the defect induces single-particle states in the band gap that are occupied with electrons, the error in the energetic position of these levels will also affect the formation energy. As the band gap is corrected by going beyond the LDA/GGA approximations, the defect-induced states shift with respect to the valence-band maximum, resulting in changes in the defect formation energy. As an extreme example, in the case of shallow acceptors, the defect-induced states are expected to shift with the valence band and the correction in the formation energy of the acceptor when the Fermi level is at the valence-band maximum is directly related to the correction in the position of the valence-band maximum on an absolute energy scale [12]. In the case of deep acceptors (or donors), the correction to the formation energies results from both the shift in the occupied single-particle states in the gap and the correction in the valence-band maximum (or conduction-band minimum) [11].

Various approaches to correct defect formation energies have recently been developed, including the LDA + U, hybrid functionals, and GW [11, 26, 28, 29]. In this paper we discuss the first two. The LDA + U has been applied to defects in materials with semicore d states such as ZnO, SnO_2, and InN [11, 21, 22]. An external Coulomb potential is added to the semicore d electrons of the metal atom, leading to a downshift of the d bands, which become more localized and narrower, and indirectly affects both the valence-band and conduction-band edges. It affects the states at the top of the valence band through the coupling with the O p states. As the ionic cores are more screened by the localization of the d states, it shifts the s states of the metal atoms, which compose the conduction-band minimum in these materials, upward in energy. These two effects lead to an opening of the band gap [15, 16]. Note that LDA + U provides only a partial correction to the band gap, through the correction of the semicore d states, since the LDA/GGA problem associated with the discontinuity of the exchange-correlation potential as a function of the number of electrons still persists [9]. Since the LDA + U improvements affect the position of the band edges, defect states which are derived from valence- and conduction-band states are also affected. Hence, one can perform calculations based on LDA and LDA + U, and inspect how defect transition levels change in response to a partial band-gap correction. Based on this information, one can extrapolate transition levels and correct formation energies, as described in Refs. [11, 18]. The extrapolation scheme has a physics basis, since the states of the host crystal form a complete basis set for expressing the defect-related states. Therefore, by going from LDA to LDA + U, the defect-related single-particle states in the gap change according to their conduction- versus valence-band character.

The advent of hybrid functionals, in particular the screened form proposed by HSE represents a significant improvement in the predictive power of defect calculations in semiconductors and insulators [14]. By adding a fraction of Hartree-Fock exchange to the GGA exchange only within a fixed radius (screening length), the HSE has been successful in describing the structure and electronic properties of many materials [14, 23, 25]. However, there are now two adjustable parameters, namely the fraction of Hartree–Fock exchange and the screening length. No rigorous *ab initio* procedures exist to determine the choice of these parameters, although a screening length of 10 Å and a mixing parameter of 0.25 are frequently assumed. It is to be expected that the screening length and the amount of Hartree–Fock exchange may vary from material to material. A common approach has been to fix the screening length at 10 Å and to vary the fraction of Hartree–Fock exchange in order to reproduce the band gap of a given material. This is acceptable in the absence of rigorous prescriptions, but prudence dictates that the sensitivity of the results to the value of the mixing parameter be examined. In our own work, we always ensure that the qualititative conclusions of our studies are independent of the precise value of the mixing parameter.

In the following we will discuss the results of LDA + U and HSE applied to the study of selected defects in ZnO, SnO_2, TiO_2. Formation energies as a function of chemical potentials and Fermi level position are calculated as described in Refs. [10, 11].

9.2.1
ZnO

With a direct band gap of 3.4 eV, an exciton binding energy of 60 meV, and being available as large single crystals, ZnO is a promising material for light emitting diodes, laser diodes, and high-power transistors. Since optical transitions from the lowest conduction band to the next available conduction-band states involve photons with energies in the UV range, ZnO has also been considered as transparent electrode. However, the development of ZnO for these applications has been hindered by a lack of understanding and difficulties in controlling the electrical conductivity [4–6]. ZnO in bulk and thin-film forms is almost always n-type, the cause of which has been highly debated. p-Type ZnO has been reported by many authors, but reliability and reproducibility are questionable [4–6].

The unintentional n-type conductivity in ZnO has long been attributed to the presence of native point defects such as oxygen vacancies or zinc interstitials [4]. However, the identification of such defects in as-grown (as opposed to irradiated) material has been elusive, and the evidence of their relation to the observed conductivity has always been indirect, *e.g.*, based on the variation of conductivity with O_2 partial pressure in the annealing environment. In the absence of reliable

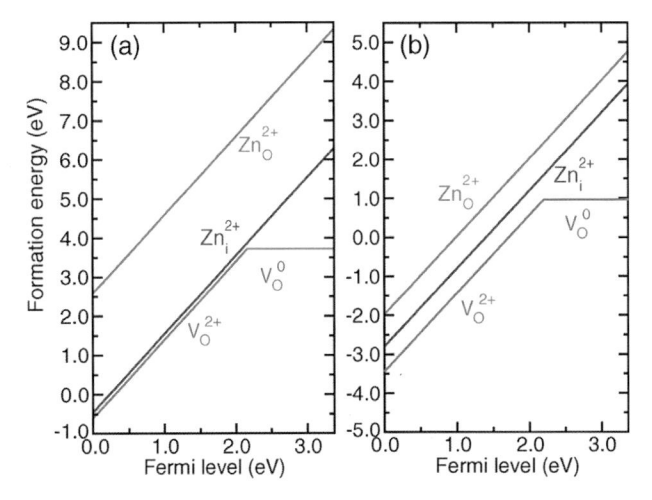

Figure 9.1 (online colour at: www.pss-b.com) Formation energy as a function of Fermi level for donor-type native point defects in ZnO: oxygen vacancies (V_O), zinc interstitials (Zn_i), and zinc antisites (Zn_O). (a) Energies according to the LDA/LDA + U method [11]. (b) Energies according to the HSE approach, after Oba et al. [28]. Both plots represent Zn-rich conditions. The Fermi level is referenced to the valence-band maximum.

experiments, first-principles calculations can provide direct insight in the role played by native point defects. The conclusion is that neither O vacancies nor Zn interstitials can explain the observed n-type conductivity in ZnO [11]. Recent experiments on high-quality bulk single crystals indeed agree with the conclusions based on first-principles calculations [19, 20].

In Figure 9.1(a) we show the formation energy *versus* Fermi-level position for donor-type native point defects in ZnO, in the Zn-rich limit. These results were based on an extrapolation of LDA and LDA + U calculations as described in Refs. [11, 18]. As a main result, it has been found that the oxygen vacancy is a deep donor with a transition level (2+/0) at about 1 eV below the conduction band. Therefore, V_O cannot explain the observed n-type conductivity in ZnO. The zinc interstitial is a shallow donor, but it is unstable. With a migration barrier of only 0.6 eV [11], Zn interstitials are mobile even below room temperature. Zinc antisites (Zn_O) are also shallow donors, stable in the 2+ charge state for Fermi-level positions near the conduction band. The large off-site displacement of the Zn atom indicates that Zn_O^{2+} is actually a complex of V_O^0 and Zn_i^{2+}. The high formation energy in n-type ZnO indicates that Zn_O^{2+} is unlikely to play a role in the observed unintentional conductivity in as-grown or annealed materials, unless Zn_O^{2+} is created by non-equilibrium processes such as irradiation. The transition levels related to higher charge states, Zn_O^{3+} and Zn_O^{4+}, are not shown in Figure 9.1.

Note that LDA + U applied to the Zn d states only results in a partial correction to the band gap (1.5 eV for $U = 4.7$ eV vs. 0.8 eV from LDA). Further opening of the band gap in order to recover the experimental value of 3.4 eV can in principle be obtained by applying very large values of U ($U_s = 43.5$ eV) to the Zn s states [30]. Such large values of U lead to unphysical effects, e.g., an artificially increased electron effective mass. It also explains the observed downward shift in defect transition levels of the oxygen vacancy [30]. The vacancy-related states in the gap are composed of Zn dangling bonds, which have s and p character. Contrary to the case in which U is only applied to the Zn d states, U_s therefore acts directly on the defect states themselves. If these defect states are occupied (such as in the case of the neutral charge state of an oxygen vacancy), the state will shift downwards, resulting in a lowering of the $(2+/0)$ transition level. In our opinion it is unclear whether this effect of applying U_s reflects the correct physics.

In Figure 9.1(b) we show the results of HSE hybrid functional calculations for donor defects in ZnO [28]. These HSE calculations were performed with an adjusted mixing parameter of 37.5% in order to reproduce the experimental value of the band gap of ZnO. We note that the positions of the transition levels with respect to the band edges are in remarkable agreement with the results obtained with extrapolation of LDA and LDA + U results [Figure 9.1(a)]. However, the formation energies themselves for these donor defects are lower in the HSE approach, although the main conclusions regarding their relation to the unintentional n-type conductivity in ZnO are unchanged. That is, the oxygen vacancy is a deep donor, and zinc interstitials and zinc antisites are shallow donors but have very high formation energies under n-type conditions and are hence unlikely to be responsible for the observed n-type conductivity.

It is interesting to note that for V_O^{2+} and Zn_i^{2+} the difference in formation energies in Figures. 9.1(a) and (b) is roughly equal to twice the valence-band offset of 1.4 eV between LDA + U and HSE estimated from Ref. [15, 31]. That is, we can attribute the formation-energy difference largely to a downward shift of the valence-band maximum on an absolute energy scale. In the LDA/LDA + U approach, no further correction was assumed for the valence-band-maximum beyond the effects of U on the Zn d states. It has now become clear that, in fact, further corrections to the valence-band positions are necessary. Such corrections are included in the HSE [31].

It is important to note that the results for V_O using LDA or GGA are qualitatively different from those using the LDA/LDA + U approach and the HSE. In the LDA/GGA the $(2+/0)$ transition level is within 0.2 eV from the conduction-band minimum [11], implying that V_O could be a source of conductivity in ZnO. In contrast, according to the LDA/LDA + U or HSE results the $(2+/0)$ level is ~1 eV below the conduction-band minimum, ruling out the possibility of V_O contributing electrons to the conduction band by thermal ionization.

9.2.2
SnO$_2$

Tin dioxide is a wide-band-gap semiconductor of high interest for transparent electrodes [7]. It crystallizes in the rutile structure and has a band gap of 3.6 eV [32]. The ease of making it n-type, its highly dispersive conduction band (small effective mass), and the large energy difference between the conduction-band minimum and the next-higher conduction band at Γ contribute to SnO$_2$ supporting high carrier concentrations while still maintaining a high degree of optical transparency [7].

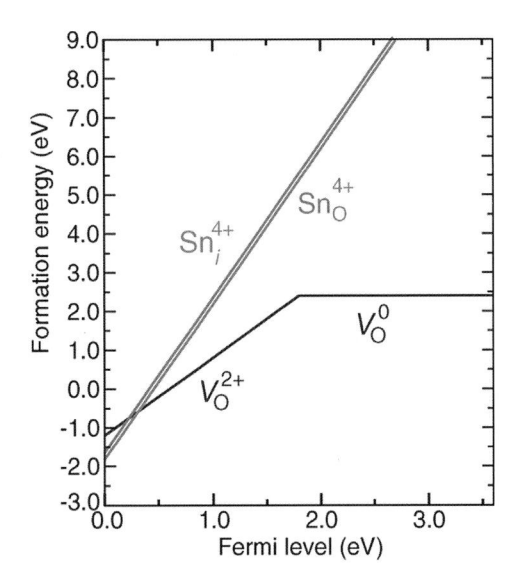

Figure 9.2 (online colour at: www.pss-b.com) Formation energy as a function of Fermi level for donor-type native defects in SnO_2 obtained by the LDA/LDA + U approach [21]. For Fermi-level positions near the conduction band V_O is stable in the neutral charge state whereas Sn_i and Sn_O are stable in the 4+ charge state.

SnO_2 can be made n-type by adding impurities such as Sb or F, which incorporate on Sn and O sites, respectively. In addition, it has been widely believed that oxygen vacancies are also a source on n-type conductivity. In analogy to ZnO, the evidence for oxygen vacancies has been based on the correlation between electron concentrations and oxygen partial pressure in annealing experiments: increasing the oxygen partial pressure leads to lower conductivities [7]. However, the attribution of conductivity to oxygen vacancies is not supported by recent first-principles calculations [21].

In Figure 9.2 we show the calculated formation energies of donor native point defects in SnO_2. These results were obtained from a combination of LDA and LDA + U calculations as described in Ref. [21]. Similarly to ZnO, the oxygen vacancy is a deep donor, and the Sn interstitial is unstable with very high formation energy if the Fermi level is positioned near the conduction-band minimum. The Sn antisite has even higher formation energy and is also an unlikely source of conductivity in SnO_2. Therefore, the unintentional n-type conductivity is probably caused by the presence of impurities. For example, hydrogen in either the interstitial form or substituting for oxygen has been predicted to act as a shallow donor in SnO_2 [21, 33].

9.2.3
TiO_2

Titania is most stable in the rutile crystal structure, with a band gap of 3.1 eV [32]. The upper part of the valence band is composed of O 2p states, and the lower part of the

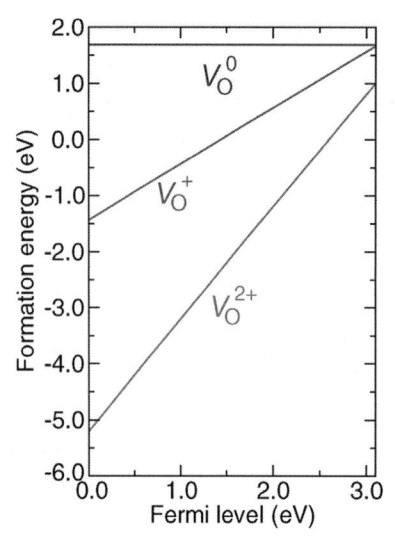

Figure 9.3 (online colour at: www.pss-b.com) Formation energy as a function of Fermi level for the oxygen vacancy (V_O) in TiO_2 in the Ti-rich limit using the HSE hybrid functional. V_O^{2+} is lower in energy than V_O^+ and V_O^0 even for the Fermi level positioned at the conduction-band minimum.

conduction band of Ti 3d states [26]. TiO_2 can be made n-type by incorporation of shallow donor impurities (e.g., Nb, F, and H), and by annealing in reducing environments [8]. Because its conductivity varies with O_2 partial pressure, it is often argued that oxygen vacancies and/or titanium interstitials are sources of conductivity in TiO_2 [8].

In Figure 9.3 we show the calculated formation energies for oxygen vacancies in TiO_2 according to the HSE hybrid functional [26]. These results were corrected for the effects of using a finite-size supercell by performing GGA calculations for V_O^{2+} and V_O^+ using supercells of 72, 216, and 576 atoms and extrapolating to the dilute limit. We conclude that oxygen vacancies are shallow donors, with V_O^+ and V_O^0 higher in energy than V_O^{2+} for any value of the Fermi level within the band gap [26].

The formation energy of V_O^{2+} in the extreme Ti-rich limit is relatively low even when the Fermi level is positioned near the conduction-band-minimum. This might lead us to conclude that oxygen vacancies are the cause of conductivity in vacuum-annealed TiO_2. However, care should be taken, since the extreme Ti-rich limit is probably not experimentally accessible since it corresponds to very low oxygen partial pressures. We also need to keep in mind that impurities that act as shallow donors, such as hydrogen, also likely contribute to the observed conductivity [34].

The use of the HSE hybrid functional is essential for describing the neutral and positive charge states of V_O in TiO_2. In the LDA and GGA the single-particle state induced by V_O^0 and V_O^+ is above the conduction-band minimum, so that these charge states cannot be stabilized [26, 35, 36], prohibiting drawing reliable conclusions about the relative energetic stability of the various charge states. In HSE, the neutral and positive charge states can be explicitly calculated and their energy compared with that of the 2 + charge state [26].

9.3 Summary

We have discussed the results of calculations that go beyond the LDA and GGA approximations to describe defects in oxide semiconductors. The LDA/GGA deficiency in describing band gaps leads to large errors in transition levels and formation energies, and corrections or methods that overcome the band gap problems are necessary for quantitative predictions. We argue that an extrapolation of LDA and LDA + U calculations for systems with semicore d states (such as ZnO and SnO_2) is a reliable method for predicting transition levels, while formation energies depend on how well LDA + U describes the absolute position of the valence-band maximum. The HSE hybrid functional approach is more general but also much more computationally demanding. It has been shown to be promising for describing the structural and electronic properties of defects in semiconductors and insulators. HSE can describe all possible charge states of the oxygen vacancy in TiO_2, resulting in a physical picture that is much closer to what is expected experimentally than that provided by the LDA and GGA.

Acknowledgements

We acknowledge fruitful collaborations and discussions with J. Lyons, J. Varley, A. K. Singh, P. Rinke, N. Umezawa, and G. Kresse. This work was supported by the NSF MRSEC Program under Award No. DMR05-20415, by the UCSB Solid State Lighting and Energy Center, and by the MURI program of the Army Research Office under Grant No. W911-NF-09-1-0398. It made use of the CNSI Computing Facility under NSF grant No. CHE-0321368 and Teragrid.

References

1 Lannoo, M. and Bourgoin, J. (1981) *Point Defects in Semiconductors I: Theoretical Aspects* (Springer-Verlag, Berlin 1983); *Point Defects in Semiconductors II: Experimental Aspects* (Springer-Verlag, Berlin 1983).
2 Pandelides, S.T. (ed.) (1992) *Deep Centers in Semiconductors: A State-of-the-Art Approach*, second ed. (Gordon and Breach Science, Yverdon.
3 Stavola, M. (ed.) (1999) *Identification of Defects in Semiconductors, Semiconductors and Semimetals*, vol. 51A, 51B (Academic, San Diego.
4 Look, D.C. (2001) *Mater. Sci. Eng. B*, **80**, 383.
5 Janotti, A. and C.G. Van de Walle (2009) *Rep. Prog. Phys.*, **72**, 126501.
6 McCluskey, M.D. and Jokela, S.J. (2009) *J. Appl. Phys.*, **106**, 071101.
7 Dawar, A.L., Jain, A.K., Jagadish, C., and Hartnagel, H.L. 1995) *Semiconducting Transparent Thin Films* (Institute of Physics, London.

8. Linsebigler, A.L., Lu, G., and Yates, J.T., Jr. (1995) *Chem. Rev.*, **95**, 735.
9. Perdew, J.P. and Levy, M. (1983) *Phys. Rev. Lett.*, **51**, 1884.
10. C.G. Van de Walle and Neugebauer, J. (2004) *J. Appl. Phys.*, **95**, 3851.
11. Janotti, A. and C.G. Van de Walle (2007) *Phys. Rev. B*, **76**, 165202.
12. Zhang, S.B., Wei, S.H., and Zunger, A. (2001) *Phys. Rev. B*, **63**, 075205.
13. Anisimov, V.I., Aryasetiawan, F., and Liechtenstein, A.I. (1997) *J. Phys.: Condens. Matter*, **9**, 767.
14. Heyd, J., Scuseria, G.E., and Ernzerhof, M. (2003) *J. Chem. Phys.*, **118**, 8207; *J. Chem. Phys.*, **124**, 219906 (2006).
15. Janotti, A., Segev, D., and Van de Walle, C.G. (2006) *Phys. Rev. B*, **74**, 045202.
16. Janotti, A. and Van de Walle, C.G. (2007) *Phys. Rev. B*, **75**, 121201.
17. Janotti, A. and Van de Walle, C.G. (2005) *Appl. Phys. Lett.*, **87**, 122102.
18. Janotti, A. and Van de Walle, C.G. (2006) *J. Cryst. Growth*, **287**, 58.
19. Vlasenko, L.S. and Watkins, G.D. (2005) *Phys. Rev. B*, **71**, 125210.
20. Wang, X.J., Vlasenko, L.S., Pearton, S.J., Chen, W.M., and Buyanova, I.A. (2009) *J. Phys. D, Appl. Phys.*, **42**, 175411.
21. Singh, A.K., Janotti, A., Scheffler, M., and Van de Walle, C.G. (2008) *Phys. Rev. Lett.*, **101**, 055502.
22. Janotti, A. and Van de Walle, C.G. (2008) *Appl. Phys. Lett.*, **92**, 032104.
23. Paier, J., Marsman, M., Hummer, K., Kresse, G., Gerber, I.C., and Àngyàn, J.G., (2006) *J. Chem. Phys.*, **124**, 154709.
24. Perdew, J.P., Burke, K., and Ernzerhof, M. (1996) *Phys. Rev. Lett.*, **77**, 3865.
25. Marsman, M., Paier, J., Stroppa, A., and Kresse, G. (2008) *J. Phys.: Condens. Matter*, **20**, 064201.
26. Janotti, A., Varley, J.B., Rinke, P., Umezawa, N., Kresse, G., and Van de Walle, C.G. (2010) *Phys. Rev. B*, **81**, 085212.
27. Van de Walle C.G., and Janotti, A. (2010) *Phys. Status Solidi B*, published online, doi: 10.1002/pssb.201046290
28. Oba, F., Togo, A., Tanaka, I., Paier, J., and Kresse, G. (2008) *Phys. Rev. B*, **77**, 245202.
29. Rinke, P., Janotti, A., Scheffler, M., and Van de Walle C.G. (2009) *Phys. Rev. Lett.*, **102**, 026402.
30. Paudel, T.R., and W.R.L. Lambrecht (2008) *Phys. Rev. B*, **77**, 205202.
31. Lyons, J.L., Janotti, A., and Van de Walle, C.G. (2009) *Phys. Rev. B*, **80**, 205113.
32. Dean, J.A. (ed.) 1992 *Lange's Handbook of Chemistry*, fourteenth ed. (McGraw-Hill, Inc. New York.
33. Varley, J.B., Janotti, A., Singh, A.K., and Van de Walle, C.G. (2009) *Phys. Rev. B*, **79**, 245206.
34. DeFord, J.W. and Johnson, O.W. (1983) *J. Appl. Phys.*, **54**, 889.
35. Sullivan, J.M. and Erwin, E.C. (2003) *Phys. Rev. B*, **67**, 144415.
36. S. Na-Phattalung, Smith, M.F., Kim, K., Du, M.H., Wei, S.H., Zhang, S.B., and Limpijumnong, S. (2006) *Phys. Rev. B*, **73**, 125205.

10
Critical Evaluation of the LDA + U Approach for Band Gap Corrections in Point Defect Calculations: The Oxygen Vacancy in ZnO Case Study

Adisak Boonchun and Walter R. L. Lambrecht

10.1
Introduction

The local density approximation (LDA) is well known to underestimate band gaps in semiconductors. In a recent paper, Paudel and Lambrecht [1] (PL), discussed this problem for the oxygen vacancy V_O in ZnO. The problem appears to be rather dramatic in this case as different authors do not even agree on whether the relevant defect level, the $2+/0$ transition level lies in the upper or the lower half of the band gap. Discrepancies also exist between different authors on the magnitude of the energy of formation of the defect and on the positions of the one-electron levels. At the time that paper was written, previous work had addressed the gap corrections mostly in *a posteriori* fashion [2–7]. The point of the PL paper was to use an adjusted Hamiltonian and total energy functional that gave the correct band gap for the host and then apply it to the defect transition levels. In particular, they used the LDA + U approach with U Coulomb interactions not only for the d states of Zn but also the s orbitals. The idea behind this unorthodox application of LDA + U is explained below. Since that work and even before, the problem has been investigated by several others by a variety of approaches, hybrid functionals [8–10], the GW method [9], and screened exchange (Chapter 5, [11, 12]). Here, we critically re-examine the results of PL and explore the LDA + U model further with additional U_i parameters, which overcome some of the shortcomings of the previous $U_d + U_s$ model. We compare the results from various groups for this benchmark case and look for what consensus can be reached and what remain open questions.

We first briefly remind the reader of the nature of the LDA + U method. Then we discuss some side issues, such as the potential alignment and image charge corrections. We then present the results of a new LDA + U potential and end with an overview of the various results on the oxygen vacancy in ZnO.

Advanced Calculations for Defects in Materials: Electronic Structure Methods, First Edition.
Edited by Audrius Alkauskas, Peter Deák, Jörg Neugebauer, Alfredo Pasquarello, and Chris G. Van de Walle.
© 2011 Wiley-VCH Verlag GmbH & Co. KGaA. Published 2011 by Wiley-VCH Verlag GmbH & Co. KGaA.

10.2
LDA + U Basics

The LDA + U approach was originally introduced to deal with electrons in localized orbitals for which the standard density functional approach [13, 14] in the local (spin) density approximation (L(S)DA) is not sufficient. The main emphasis was on open-shell systems such as transition metal compounds and rare-earth metals and compounds. The original versions were strictly LDA + U rather than LSDA + U with all the magnetic effects arising from the Hubbard like on-site Coulomb terms that were added to the LDA Hamiltonian [15, 16]. A key aspect of LDA + U is that since LDA already contains some exchange and correlation in an orbital-independent way, a double counting correction is required when explicitly adding the Coulomb and exchange effects for these localized orbitals. While the first version [15] used an *around mean-field* approach in which the LDA is supposed to give the right answer for equal occupation of all the d orbitals, the more often used version is the so-called "fully localized limit," (FLL). In this initial discussion, we refer to the orbitals for which U effects are added as the d orbitals although we later will generalize this. In the FLL we assume that the LDA gives the correct total energy for the atomic limit of integer occupations of the specific d_i orbitals. They are either fully occupied or empty. What the LDA + U framework provides in that case, is how starting from such an atomic limit the interaction with the other bands, modifies these occupation numbers and may lead to orbital ordering. A key aspect of the LDA + U model is that the total energy is treated as a functional of the electron density as well as separately as a function of the occupation numbers of d orbitals, $E[n(\mathbf{r}), n_i]$. In its simplest form [16] it is given by

$$E_{LDA+U} = E_{LDA} - UN(N-1)/2 + U\sum_{i,j} n_i n_j. \tag{10.1}$$

The potential for the i-th orbital $V_i = \delta E/\delta n_i(\mathbf{r})$ with $n_i(\mathbf{r}) = |\psi_i(\mathbf{r})|^2$ then becomes

$$V_i = V_{LDA} + U\left(\frac{1}{2} - n_i\right). \tag{10.2}$$

Thus the one-electron levels

$$\varepsilon_i = \partial E/\partial n_i = \varepsilon_i^{LDA} + U\left(\frac{1}{2} - n_i\right) \tag{10.3}$$

are shifted from the LDA even though the LDA + U total energy in the limit of integer n_i remains in principle the same as the LDA total energy.

The version we presently use starts from LSDA and besides U, contains in general non-spherical Coulomb terms written in terms of the Slater F_k integrals and Clebsch–Gordan coefficients, specific to which spherical harmonic d_i orbitals are occupied. Also since those orbitals depend on the coordinate system, the method is formulated in a rotationally invariant form in terms of density matrices n_{ij}. This form of the LSDA + U method is described by Liechtenstein et al. [17]. It comes down to a

Hartree–Fock like treatment of the localized orbitals with a parametrized treatment of their Coulomb and exchange interactions, in particular using a screened Coulomb interaction U. Hence, one can expect some similarities between the results of this approach and hybrid functionals.

Our calculations are all carried out within a full-potential linearized muffin–tin orbital (FP-LMTO) method [18, 19]. This method uses smoothed Hankel functions as envelope functions, which are augmented inside muffin–tin spheres (in terms of ϕ and $\dot\phi = d\phi/dE$ functions, with ϕ solutions of the radial Schrödinger equation), as usual in linear methods. We use a double-κ basis set with optimized smoothing radii and radial decay constants κ including spdf functions in the basis set for the first κ and sp for the second one. In addition, we treat Zn-3d orbitals as valence bands and add 4d-local orbitals for a better description of the conduction bands. For the augmentation we use an angular momentum cut-off of $l_{max} = 4$. We use a 127 atom supercell for which Γ-point sampling of the Brillouin zone is adequate. While the basis set corresponds to a muffin–tin potential, it should be emphasized that the full non-spherical potential inside the spheres and non-constant potential in the interstitial region is treated. Forces are calculated analytically and allow us to optimize the structure.

For the present purpose, dealing with ZnO, the semicore 3d states are completely filled, so the LDA + U treatment simply results in a downward shift of those orbitals by $U_d/2$. At the same time, we note that if U_s is applied to the s orbitals of Zn, which primarily constitute the conduction band minimum and are thus almost empty, they will shift up approximately by $U_s/2$. So, in a strictly empirical manner, we can adjust U_d so as to shift the d-levels down to where they are found by photoemission and U_s so as to open the correct gap at Γ.

It should be kept in mind that the justification for adding U_s is different than for the d orbitals. The physics of the d-states, is indeed that they are strongly localized and have strong Coulomb interactions. This in part opens the gap because of the resulting reduced p–d hybridization with the O-2p orbitals constituting the valence band maximum (VBM). On the other hand, the remainder of the gap is not due to localized atomic effects. Quite to the contrary, analysis of the GW approximation (see below) shows that the long-range or at least medium range $1/r$ behavior of the (dynamically screened) exchange is crucial [20]. The same could also be concluded from hybrid functionals (chapter 6, [21]). Nonetheless, in a GW approach, the difference between quasiparticle and Kohn–Sham eigenvalues is the expectation value of $\langle\psi_c|\Sigma_{xc}^{GW} - v_{xc}^{LDA}|\psi_c\rangle$ with Σ_{xc}^{GW} the self-energy operator and ψ_c the conduction band minimum wave function. To the extent that the conduction band wave function ψ_c is dominated by the Zn-s orbitals, it amounts to a shift of the latter in the Hamiltonian. Thus in a strictly pragmatic sense, the U_s shift mimics this effect. Now, in practice, one should realize that a rather large and seemingly unphysical U_s is required, both because the occupation of the Zn-s orbitals is not zero and because the conduction band minimum is not purely Zn-s like. Finally, we note that the two shifts are not independent of each other. A shift in Zn-s through the self-consistently leads to a more ionic bond and this in turn affects the d-states, but ultimately, the PL model leads to a band gap and a d-band position that agree with experiment.

We also note that while the original idea behind LDA + U is that the added U terms should not destroy the already good agreement of LDA for total energies but merely adjust the one-electron levels, this does not imply that the LDA part of the energy would not change. The LDA term in the energy is indirectly affected because changing the Hamiltonian modifies the wave functions, charge density, etc. through the self-consistency. In fact, we find that we cannot directly use the LDA + U total energy functional for defect formation energies when applied to non-localized orbitals. In particular, we cannot apply the same LDA + U total energy functional to the free atoms or the reference systems that enter in defining the formation energies. In fact, this would not make sense because U is supposed to contain system specific screening and is not transferable from one system to another. While the U_d may be more or less transferable between different solid state environments, U_s in PL is designed to adjust the band gap specifically of ZnO and thus of course has nothing to do with the position of the Zn-s levels in for example metallic Zn. PL thus calculated the energy formation of the neutral charge state in LDA and only used the full LDA + U functional for the difference between different charge states. In the present work with even more U parameters, we decided to use only the LDA-part of the functional, without the added Hubbard-U and double counting terms. The only way in which LDA + U then enters is through the modified one-electron potential. This may seem strange and may seem to break the consistency of our one-electron levels and the total energy functional, e.g., by invalidating Janak's theorem [22]. However, it should be kept in mind that the Hubbard-U and double counting terms of the FLL LDA + U are primarily designed to deal with the open shell orbital ordering within the set of localized orbitals. For closed shell systems, these terms should vanish.

To complete our discussion of LDA + U we note that the actual operator entering the calculations is a non-local projection potential $|\phi_i\rangle V_i \langle\phi_i|$, in which $\phi_i(r)$ is a local partial wave in the muffin–tin sphere at the LMTO linearization energy ε_ν. As such it depends on the sphere radius. This is not important for well-localized wave functions like d states, but when applied to s orbitals, as we will do here, it is sphere radius dependent.

10.3
LDA + U Band Structures Compared to GW

The band structures of ZnO obtained with the LDA, LDA + U_d, and LDA + U_d + U_s potential were shown in PL. To further scrutinize them, we here compare the band structure of that model with a GW calculation [23, 24] in Figure 10.1.

More precisely, because the latter usually overestimates the band gap slightly we compare with a 20% LDA, 80% quasiparticle self-consistent GW (QSGW) band structure, which almost exactly reproduces the band gap of 3.4 eV at room temperature. Strictly, speaking we should use a zero temperature gap corrected for spin–orbit splitting of the VBM, zero-point motion corrections, and exciton effects [23] of 3.6 eV but for easy comparison to experiment, we here prefer a gap

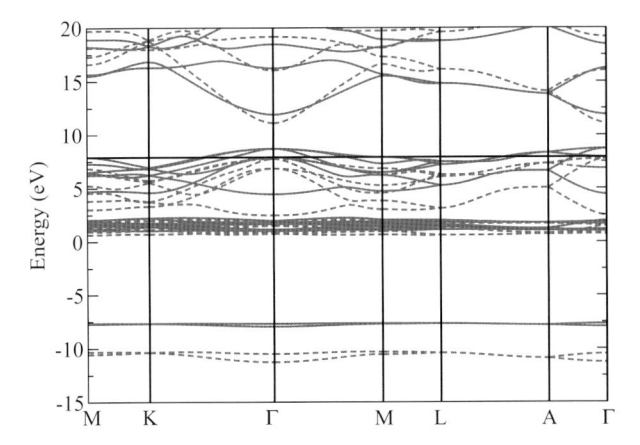

Figure 10.1 (online color at: www.pss-b.com) Band structure of ZnO in "0.8 QSGW" approximation (red dashed line) compared to the PL LDA + U_d + U_s model (blue solid line).

of 3.4 eV. In the QSGW approach, the independent particle Kohn–Sham equations from which the GW self-energy, schematically $\Sigma = iGW$, with G the one-electron Green's function and W the screened Coulomb interaction, is calculated, contains a non-local exchange correlation potential

$$V_{xc}^{QSGW} = \frac{1}{2} \sum_{nm} |\psi_m\rangle \Re\{\Sigma_{mn}(\varepsilon_m) + \Sigma_{mn}(\varepsilon_n)\} \langle\psi_n|, \qquad (10.4)$$

adjusted self-consistently in terms of the self-energy Σ. Here, $\Re$ means taking the Hermitean part, and ψ_n are the eigenstates of the Hamiltonian with this potential. In other words, it forms the best starting point for a single-shot perturbation theory calculation of Σ and leads to Kohn–Sham equations that equal the quasiparticle excitation energies.

In retrospect, one might raise three criticisms of the PL LDA + U_d + U_s model. First, the band gap shift induced by the U_s occurs mainly at the Γ-point only. Thus, instead of a rigid shift, the gaps at K, M, and L are not raised as much and this leads to an overall wrong curvature of the lowest conduction band, with an overestimated effective mass (EM). In particular, since the V_O defect levels are deep, one might expect that their wave function contains contributions from several host band conduction band states at different k-points when a decomposition of the defect state in host states is attempted. One might expect that thus the defect level is not sufficiently raised along with the gap correction.

Second, Alkauskas et al. [25, 26] recently made the observation that localized defect levels with respect to the average electrostatic potential are much less sensitive to computational model than with respect to the band-edges. Thus, to place the defect levels correctly with respect to the band-edges requires additional care in calculating the proper band edges relative to the electrostatic potential. It thus appears important that not only the band gap but also the individual band-edges agree between GW and the LDA + U model. Thus, unlike the usual practice of plotting the bands relative

Table 10.1 Structural and total energy properties of wurtzite in various LDA + U models, in-plane lattice constant a (Å), c/a ratio and internal parameter u, energy difference between rocksalt and wurtzite structure in eV/pair. PL [1], BL (this work) defined in Table 10.2.

	LDA	PL	BL	expt.
a (Å)	3.20	3.31	3.30	3.25
c/a	1.603	1.598	1.57	1.6018
u	0.3811	0.3811	0.3872	0.382
$\Delta E_{\text{RS-WZ}}$	0.223	0.062	0.026	> 0

to the VBM, we here plot the bands relative to the average electrostatic potential as zero. We can see that the PL LDA + U_d + U_s significantly overestimates both the VBM and CBM.

Third, one might worry whether the LDA + U_d + U_s model provides correct total energies. This point for example was raised by Lany and Zunger [27] who found that LDA + U potentials including U_s may lead to too ionic bonding which leads to the h-MgO structure becoming more stable. In that structure, c/a is notably reduced and $u \approx 0.5$ and the system becomes effectively five-fold instead of four-fold coordinated [28]. The latter is closely related to the rocksalt structure. Thus, we here also examine the rocksalt to wurtzite energy difference. From the results in Table 10.1 we can see that indeed the PL model leads to an increase in lattice constant and a reduction of c/a compared to the LDA results. However, the deviation from the experimental wurtzite structure is only minor. Wurtzite stays lower in energy than rocksalt although their energy difference is reduced significantly.

10.4
Improved LDA + U Model

In order to remedy the first two of these problems, we constructed a new LDA + U potential, including additional parameters U on Zn-p, O-s, and O-p. Our goal is to obtain as close agreement as possible with the above GW band structure and then to explore how these potentials behave for the defect states. The role of the U_{Op}, U_{Os} which are mostly occupied is to shift the corresponding upper and lower VBM down. The U_{Znp} further allows us to adjust the lowest conduction band at K, M and L. The old and new LDA + U model parameters are summarized in Table 10.2. We note that compared to the previous model U_{Zns} is significantly reduced. We also show the actual shift potentials V_i [according to Eq. (10.2)] that result from them for bulk ZnO with the self-consistently determined density matrices as well as the average occupation numbers (averaged over the different p and d orbitals). These correspond to more reasonable shift potentials than the U_i values, perhaps with the exception of Zn-s in the PL model. For example, we see that because O-p orbitals have an occupation close to 0.5 one needs a large U to achieve a reasonable shift. One still needs to keep in mind that all actual bands have mixed atomic orbital character by

Table 10.2 Parameters U_i of the LDA + U models, resulting self-consistent occupation numbers n_i and shift potential V_i in eV. PL [1], BL (present work).

i	U_i		n_i		V_i	
	PL	BL	PL	BL	PL	BL
Znd	3.4	4.9	0.947	0.960	−1.56	−2.18
Zns	43.5	13.60	0.039	0.097	19.75	5.48
Znp	0	27.21		0.023		12.98
Os	0	21.77		0.795		−6.47
Op	0	39.45		0.642		−5.65

forming bonding and antibonding states and that the above occupation numbers are sphere size dependent. Note also that the shifts may vary near the defects when their occupation numbers change. This is in fact what distinguishes LDA + U from non-local external potentials.

The band structure of our new LDA + U model compared to GW is shown in Figure 10.2. The new model can be seen to adjust the conduction bands not only at Γ but also its EM at Γ and dispersion all the way to K and M. The position of the VBM relative to the average electrostatic potential is also improved. The model still somewhat underestimates the band width of the O-2p like valence bands and overestimates the valence band EM. Higher conduction bands at Γ are still slightly off and the low lying O-2s like valence band has too low binding energy.

Unfortunately, the newer model gives slightly worse wurtzite structural properties. Still the lattice constant is only 1.5% overestimated, c/a only 2% underestimated and u stays far from 0.5. The total energy difference between rocksalt and wurtzite obtained in both LDA + U models is significantly lower than in LDA. Our LDA result is close to the results by Schleife et al. [29] of 0.29 eV. At least, we can be reassured that

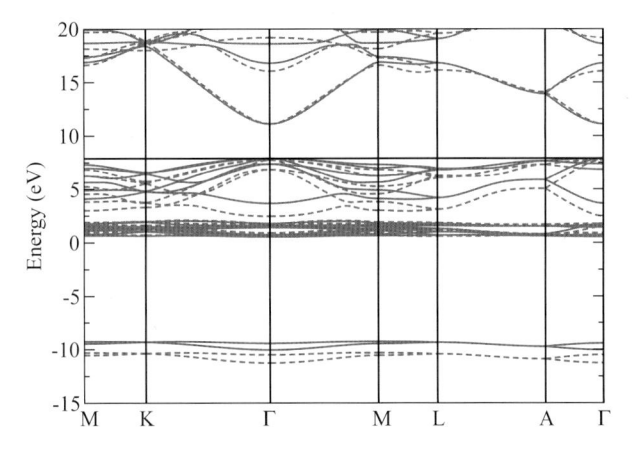

Figure 10.2 (online color at: www.pss-b.com) Band structure of ZnO in "0.8 QSGW" approximation (red dashed line) compared to the present LDA + U model (blue solid line).

the wurtzite structure remains the lower energy structure and we are nowhere near h-MgO-like c/a and u.

We note that this adjustment of the LDA + U potentials is by no means unique. One might also contemplate adding empty sphere shift potentials to adjust the potential in the interstitial region. This might actually more readily mimic a shift of the delocalized conduction band states than Zn-p but has not yet been attempted here.

10.5
Finite Size Corrections

Before proceeding to the results of the LDA + U models for the defect, we need to address two side issues which influence the results. The first of those is the finite size correction. PL examined the size convergence in different size supercells but concluded that no clear $1/L$ behavior was seen. Furthermore, the expected behavior of the image charge correction [30, 31] $q^2/L\varepsilon$ to be proportional to q^2 was not observed. They therefore did not include the image charge corrections, but instead used scaling versus $1/V$ which led to a rather small extrapolation from the largest supercells used (192 atoms). In retrospect, this failure to obtain the $1/L$ behavior is probably in part due to the use of relaxed structures, mixing up the $1/V$ elastic effects as well as possibly problems in sufficiently accurately determining the alignment potential and mostly the limited range of supercell sizes investigated which makes it difficult distinguishing between $1/L$ and $1/L^3$ behavior. Such effects can dominate the result especially for relatively small cells. Only for cells larger than say 200 atoms, the purely electrostatic terms q^2/L become dominant. This can for instance be seen in Figure 7 of Lany and Zunger [27].

We now believe that even for relatively delocalized defect electron densities, the image point charge correction is important because the latter contains always point like contributions from the ionic charge change introduced by the defect. When we add the image point charge correction, or rather 2/3 of it as recommended by Lany and Zunger [27] to mimic the additional quadrupole term, to the results of PL for the largest cell, we find that the $\varepsilon(2+/0)$ transition level shifts becomes 0.64 eV above the VBM the $\varepsilon(+/0)$ level 0.72 eV, and the $\varepsilon(2+/+)$ level 0.57 eV. The defect formation energies as function of Fermi level are shown in Figure 10.3.

The quadrupole background interaction term $\propto Qq/\varepsilon L^3$ identified by Makov and Payne [31] should strictly speaking not include the screening charge density. It is the latter that leads to an effective $Q \propto L^2$ behavior which turns the $1/L^3$ into $1/L$ behavior and allows one to combine it with the point charge term [27]. But since the screening charge is due at least in part to the background density itself, it is not clear one should include this correction. Without the factor 2/3, the levels would shift down even further to 0.53, 0.66, and 0.40 eV, respectively.

The defect then becomes a positive U_{eff} defect rather than negative U type. We note that PL already found much less negative U_{eff} behavior in other words a smaller $|U_{\text{eff}}|$ than the LDA calculations or LDA + U_d only. Here, U_{eff} should not be confused with

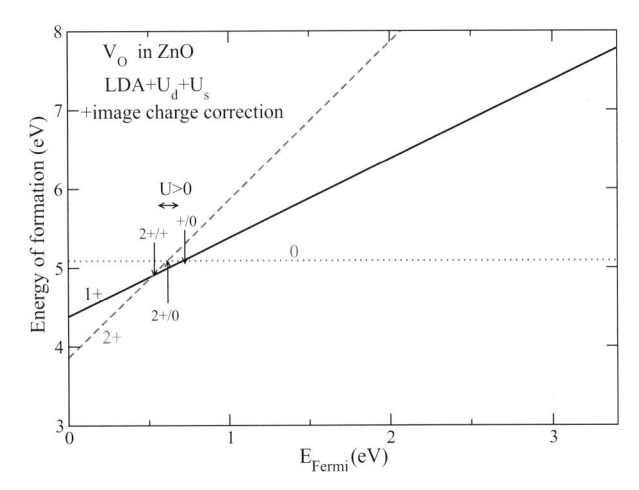

Figure 10.3 (online color at: www.pss-b.com) Formation energies of the oxygen vacancy in ZnO with the PL LDA + U_d + U_s model for 192 atoms including image charge correction as function of Fermi level.

the LDA + U parameters but is defined as $U_{eff} = \varepsilon(+/0) - \varepsilon(2+/+)$ This result in fact is consistent with Lany and Zunger's slightly different LDA + U_d + U_s model [27]. To be sure, these authors did not propose this to be their favorite approach for dealing with the gap correction. They used it for illustrative purposes and compared a U_s-only, $U_s + U_d$, and U_d-only model. In their $U_d + U_s$ model they obtain $2+/0$ at 0.61 and $+/0$ at 0.79 eV and $2+/+$ at 0.43 eV, very close to ours when including the factor 2/3 in the image point charge correction.

10.6
The Alignment Issue

Another important issue is how to align the VBM of the perfect crystal with that of the defect cell. The defect formation Gibbs free energy (at zero temperature) is defined as

$$\Delta G_f(D, q) = E(D, q) - E(X) - \sum_i \mu_i \Delta n_i + q\mu_e, \qquad (10.5)$$

where $E(D, q)$ is the total energy of the supercell with the defect in the charge state q, compensated by a neutralizing homogeneous background charge density, $E(X)$ is the total energy of the perfect crystal calculated in the same size supercell to avoid k-point convergence issues, μ_i is the chemical potential of the elements whose occupation changes by the defect, and μ_e is the chemical potential of the electrons. The latter represents the energy of the electrons in the perfect crystal reservoir $\mu_e = \varepsilon_{vbm} + \varepsilon_F$ with the Fermi level ε_F measured relative to the VBM and ε_{vbm} determined relative to the average electrostatic potential. To determine ε_{vbm} we use a local reference potential V_{loc} mark on an atom far away from the defect where the potential presumably becomes bulk like, and add $\varepsilon_{vbm}^{bulk} - V_{loc}^{bulk}$ where the latter is calculated in the perfect crystal primitive cell.

To do this accurately, one needs to average over a few atoms far away from the defect and make sure that the cell is large enough that the local potential marker indeed becomes constant over the region far away from the defect. In PL this alignment potential was determined for the neutral charge state only and then used for the other charge states. This avoids possible long-range contributions of the defect potential for the charged states.

In Figure 10.4 we plot the potential at the muffin–tin radii as function of distance from the oxygen vacancy for all three charge states. We see that sufficiently far from the defect, this potential becomes indeed constant apart from some small oscillations and is nearly the same for the three charge states.

10.7
Results for New LDA + U

Here, we used supercells of 128 atoms in the wurzite structure. We summarize the results of our new LDA + U potentials compared with those of PL for the same size cell in Table 10.3. The image point charge correction added was taken as $(9/10)q^2/\varepsilon R$ with R the radius of a sphere with the same volume as the supercell. This image charge correction amounts to 0.18 eV for $q = 1$ using a dielectric constant $\varepsilon = 10$ and a four times larger value for $q = 2$. This does not include the correction factor of 2/3 for the quadrupole term for the reasons explained earlier. Note that for the 1+ charge state, the one-electron levels are spin-polarized. The lower one is occupied, the higher one of minority spin is

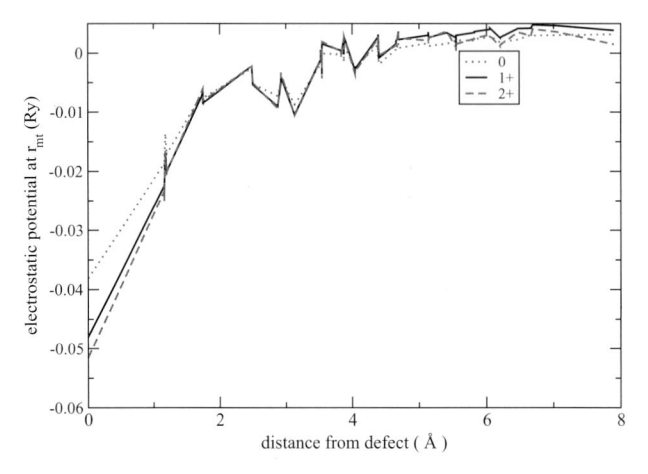

Figure 10.4 (online color at: www.pss-b.com) Potential at the muffin–tin radius as function of distance from the defect for different charge states.

Table 10.3 One-electron levels relative the VBM for different charge states (and spin σ in the $q = +1$ case) $\varepsilon_{q\sigma}$, Gibbs free energies of formation in different charge states $\Delta G_f(q)$ for $\varepsilon_F = 0$, transition level $\varepsilon(q,q')$, and U_{eff}, all in eV.

	ε_0	$\varepsilon_{+\downarrow}$	$\varepsilon_{+\uparrow}$	ε_{2+}	$\Delta G_f(0)$	$\Delta G_f(1+)$	$\Delta G_f(2+)$	$\varepsilon(2+/0)$	$\varepsilon(+/0)$	$\varepsilon(2+/+)$	U_{eff}
PL[a]	1.0	0.6	1.1	1.8	5.06	4.41	3.92	0.57	0.65	0.49	0.16
PL[b]					4.70	4.48	3.62	0.54	0.22	0.86	−0.64
BL[c]	1.50	1.22	1.86	2.46	3.24	0.93	−2.41	2.82	2.31	3.34	−1.0
JV[d]					6.7	4.6	1.8	2.42	1.94	2.90	−0.96
LZ[e]					4.2	3.3	1.0	1.60	0.94	2.24	−1.32
Oba[f]	1.0			$> \varepsilon_{cbm}$	4.1			2.20			< 0
SX[g]					3.95			2.20			< 0
GGA+U+GW[h]	0.8	0.51	2.5	$> \varepsilon_{cbm}$	4.55	3.3	1.8	1.36	1.26	1.46	−0.2
HSE+GW[h]								1.66			
P[i]	0.34	0.64	2.24	2.74				0.34	0.04	0.64	−0.6

a) Based on PL [1]; 128 atom cell but adding image charge corrections.
b) Using LDA + U_d + U_s as in PL but using LDA part of functional only.
c) This work's new LDA + U model.
d) Janotti and Van de Walle [2]: extrapolated LDA + U_d.
e) Lany and Zunger [3]: LDA + U_d correction of VBM.
f) HSE [32] hybrid functional with Hartree-Fock mixing fraction $\alpha = 0.375$ (Oba et al. [8]).
g) Screened exchange (Clark et al. [12]).
h) GGA + U + GW and HSE($\alpha = 0.25$) + GW (Lany and Zunger [9]).
i) B3LYP functional (Patterson [10]).

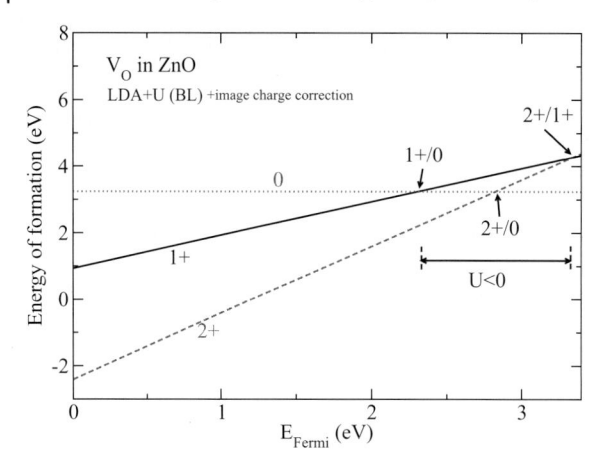

Figure 10.5 (online color at: www.pss-b.com) Energies of formation of V_O in different charge states as function of Fermi level in the oxygen rich limit calculated in present LDA + U model.

empty. The results for the energies of formation as function of Fermi level position are shown in Figure 10.5 In Table 10.3 the second row uses the same LDA + U_d + U_s model but instead of using the full LDA + U functional as PL did, we use our present approach of only using the LDA part of the total energy, using effectively only the V_i shifts. Image point charge corrections are the same as before. Note that this leads to about the same 2 + /0 level but changes the U_{eff} value to become negative.

10.8
Comparison with Other Results

In Table 10.3 we have added selected information from the literature, focusing on the latest results. Unfortunately, not all authors give values for all quantities shown here. Some are estimated from figures to the best of our ability or energies of formation were converted from Zn-rich to O-rich using an energy of formation of ZnO of −3.1 eV.

We can see that in the present LDA + U model, the one-electron level in the neutral charge state lies somewhat higher above the VBM than in PL. The one electron levels in the 1+ and 2+ charge states also lie significantly higher in the gap but still below the CBM even for the 2+ charge state. Clearly, this is as expected by the fact that in our new LDA + U model the VBM drops down relative to the electrostatic potential and the gap is opened not only at Γ but throughout the lowest conduction band at other k-points.

Lany and Zunger [9] recently pointed out that one-electron levels require a finite-size correction for the effects of image point charges and the background charge density. We estimate this effect as follows. If we approximate the cells by spheres, neighboring cells do not give any contribution and the correction amounts to the potential due the constant background charge density. This is readily calculated to be

$$\phi(r)-\phi(R) = \frac{q}{2\varepsilon}\left(\frac{1}{R} - \frac{r^2}{R^3}\right). \tag{10.6}$$

The question is now how localized the defect wavefunction is. If it is a δ-function at the origin, the upward shift is $q/(2\varepsilon R)$. If it is spread uniformly over the whole cell, we need to average the above potential over the sphere, which gives $(1/5)q/(\varepsilon R)$. For a 128 atom cell ZnO cell, these amount to 0.1 and 0.04 eV providing upper and lower bounds. Averaging, the correction is estimated to be a downward shift of order 0.07 eV for $q = 1$ and proportional to q. Our results in Table 10.3 do not include this negligible correction.

The defect formation energies here correspond to the oxygen rich limit. Our value of about 3.24 eV for the neutral charge state is somewhat lower than Paudel's result and Lany and Zunger [3], Oba et al. [8], and Clark et al. [12] but significantly smaller than that of Janotti and Van de Walle [2]. In the Zn-rich limit, the energies are lowered by 3.1 eV and the $\Delta G_f(0)$ becomes less than 1 eV, supporting the idea that this could be an abundant defect. It will, however, not be a large source of free electrons because it is a deep donor.

The transition levels move to significantly higher values, in fact even higher than Janotti and Van de Walle's and we find back a relatively strongly negative $U_{eff} = -1.0$ eV. This is mainly a result of the different relaxations in the 2 + (outward by 15.6%), 1+ (inward by 0.3%) and neutral (inward by 4.4%) charge states.

We note that the results by Janotti and Van de Walle [2] using an extrapolated U_d correction as well as the Lany and Zunger results from 2005 [3] without extrapolation are both based on LDA results with *a posteriori* corrections only. The VBM was simply shifted down by the LDA + U_d shift in pure ZnO. In that case, the position of the 1+ one-electron level is above the CBM and this leads to erroneous occupation of the CBM instead of the defect level. In that sense the position of the 1+ level is not well defined and that makes the U_{eff} untrustworthy. While most of the discrepancy between Janotti and Van de Walle [2] and Lany and Zunger [3] arises from their choice of extrapolating or not extrapolating the shifts induced by LDA + U_d, part also arises from the fact that Janotti and Van de Walle did not apply the image charge correction. GGA and LDA also slightly differ in lattice constants and this also contributes to the confusion [6, 7]. Actual GGA + U_d or LDA + U_d calculations applied to the defect place the transition levels closer to the VBM than simply adding a LDA + U_d induced downward shift of the VBM *a posteriori* to the LDA results. For example, Lany and Zunger [9] show that GGA + U gives $\varepsilon(2+/0) = 0.98$ eV, while their 2005 result [3] just shifting the VBM down *a posteriori* gave 1.60 eV. Erhart et al. [6, 7] used a different way of extrapolating to infinite cells and used GGA + U instead of LDA + U and reported the results either with or without the Janotti–Van de Walle type of extrapolation [6, 7] leading to somewhat intermediate results. The work by Zhang et al. [4] used smaller cells and other ways to estimate the gap correction effects.

Recently, several hybrid functional calculations have been carried out. The oldest one is the B3LYP calculation of Patterson [10]. It finds transition levels even closer to the VBM but also finds defect levels in the gap for all charge states, including the 2+ charge state. Oba et al. [8] used the HSE [32] hybrid functional but with the fraction of Hartree–Fock adjusted so as to get the correct band gap. That gives a $\varepsilon(2+/0)$ at 2.2 eV similar to Janotti and Van de Walle [2]. Unfortunately, Oba et al. [8] do not mention the $+/0$ or U_{eff} values nor do they show the band structure in the 1+ charge state. Their band structure in the neutral charge state finds a defect level in the usual place, about 1 eV above the VBM, while the 2+ charge state shows no level in the gap.

A similar high energy location of the $\varepsilon(2+/0)$ is obtained using screened exchange by Clark et al. [12] at 2.2 eV. Again, they do not mention the position of the single plus charge state levels. Lany and Zunger [9, 12] recently also applied the HSE functional but with the standard Hartree–Fock mixing fraction of 1/4 and that calculation places the defect level at 1.67 eV. They also used HSE and GGA + U as starting point for a GW calculation. They use an approach originally introduced by Rinke et al. [33] in which the vertical transition energies between charge states in a Franck–Condon coordination diagram are calculated as quasiparticle excitations. In other words, the transition from the neutral to the 1+ charge state is considered to be a transition of an electron from the defect level to the CBM at frozen geometry of the neutral charge state and this energy difference is first calculated as a difference between quasiparticle levels in GW. Afterward, the relaxation energy in each charge state is added as calculated either in HSE or in GGA + U. The comparison is a bit complicated because these authors use a smaller zincblende cell and hence find it necessary to correct the one-electron levels for finite size corrections as already mentioned above. They find a $\varepsilon(2+/0)$ at 1.36 eV in the GGA + U + GW approach and at 1.66 eV in the HSE + GW approach, still significantly lower than the other results. Their $|U_{\text{eff}}|$ is also significantly smaller than the LDA values or our present value.

10.9
Discussion of Experimental Results

Finally, we should discuss the connection of all these results to experiment. Two experiments are particularly relevant to the present discussion. The first is the electron paramagnetic resonance (EPR) experiment by Evans et al. [34]. They observe the appearance of the V_O^+ charge state from the neutral start state under excitation of light with $h\nu > 2.1$ eV. Interpreting this as an optical transition from the neutral V_O^0 one-electron level to the conduction band, and neglecting excitonic effects, this places the neutral defect's one-electron level at 1.3 eV above the VBM. In fact, this could be viewed as an upper limit taking in to account the exciton binding energy. This appears consistent with most calculations placing this level in the lower half of the band gap given the uncertainties on the experimental determination. Our present value for this one-electron level at 1.5 eV seems a bit high.

The second experiment is by Vlasenko and Watkins [35]. Using optically detected EPR (ODEPR) they provide evidence for a process of the type

$$V_O^+ + EM^0 \rightarrow V_O^0 + EM^+, \tag{10.7}$$

in which an electron is transferred from an EM type donor to the oxygen vacancy in the single positive charge state, thereby quenching its EPR signal. Since the EM and ODEPR signal (L3) associated with the V_O^+ are positive in a photoluminescence band with estimated zero phonon line at 2.48 eV, this could indicate that the empty (minority spin) one-electron eigenvalue of the V_O^+ charge state lies 2.48 eV below the EM level or roughly the same amount below the CBM or at about 1 eV above the VBM if this photoluminescence is directly a result of the electron capture. There is still considerable uncertainty on this experimental value. The peak of this photoluminescence occurs at 600 nm or closer to 2.0 eV. On the other hand, an alternative explanation of this process is that the positive photoluminescence results from subsequent recombining of the (radiatonlessly) captured electron with a hole from the VBM. If we assume that this happens before the defect has time to relax to the neutral ground state and still reflects the geometry of the single positive charge state, then it means that the empty one-electron level lies at about 2.0–2.5 eV above the VBM. This result is qualitatively consistent with Lany and Zunger's GW calculation [9] who find that the occupied level of the V_O^+ charge state shifts along with the VBM while the empty state shifts along with the CBM when the GW self-energy shifts are applied. The values for these levels in Table 10.3 are estimated from their figure. We find similarly that the $1+$ charge state minority spin level lies in the upper part of the gap at 1.86 eV, supporting the second interpretation of the Vlasenko Watkins experiment. Our value is a bit on the low side. This was the interpretation proposed by Janotti and Van de Walle [2] although they based it on the position of the $+/0$ transition state in the gap.

We conclude that neither of the two experiments provides direct evidence for the location of the thermodynamic transition levels. Instead they provide information on the one-electron eigenvalues in the neutral and $1+$ charge state, respectively.

10.10
Conclusions

In summary, the LDA + U approach as applied to band gap correction for defects was reviewed. Shortcomings in an earlier application of the approach by PL [1] were identified and corrected. The LDA + U approach was applied to Zn-s, p, d and O-s, p orbitals with U_i values adjusted so as to reproduce as closely as possible the band structure of ZnO in the QSGW approach. Not only band gaps but also dispersions and the position of the levels relative to the electrostatic reference potential were adjusted. This new LDA + U model leads to transition levels close to (but slightly higher than) the recent hybrid functional and screened exchange calculations. In addition, the position of the empty minority spin level in the

1+ charge state is argued to support an explanation of the ODEPR data by Vlasenko and Watkins in terms of a two step capture + recombination with valence band hole model. The position of the one-electron level in the neutral charge state is consistent with EPR optical activation of the V_O^+ signal. The energy of formation of the oxygen vacancy is found to be relatively low, supporting the notion that this could be an abundant defect in Zn-rich material.

Acknowledgements

This work was supported by the Army Research Office under grant no. W911NF-06-1-0476 and the National Science Foundation under grant No. DMR 0710485. A. Boonchun also thanks the Commission on Higher education of Thailand for support. Computations were carried out at the Ohio Supercomputercenter under project number PDS-0145 and the HPC center at CWRU. We thank Tula Paudel for useful discussions and Mark Van Schilfgaarde for the FP-LMTO and QSGW programs.

References

1 Paudel, T.R. and Lambrecht, W.R.L. (2008) *Phys. Rev. B*, **77** (20), 205202.
2 Janotti, A. and Van de Walle, C.G. (2005) *Appl. Phys. Lett.*, **87** (12), 122102.
3 Lany, S. and Zunger, A. (2005) *Phys. Rev. B*, **72** (3), 035215.
4 Zhang, S.B., Wei, S.H., and Zunger, A. (2001) *Phys. Rev. B*, **63** (7), 075205.
5 Kohan, A.F., Ceder, G., Morgan, D., and Van de Walle, C.G. (2000) *Phys. Rev. B*, **61** (22), 15019–15027.
6 Erhart, P., Klein, A., and Albe, K. (2005) *Phys. Rev. B*, **72** (8), 085213.
7 Erhart, P., Albe, K., and Klein, A. (2006) *Phys. Rev. B*, **73** (20), 205203.
8 Oba, F., Togo, A., Tanaka, I., Paier, J., and Kresse, G. (2008) *Phys. Rev. B*, **77** (24), 245202.
9 Lany, S. and Zunger, A. (2010) *Phys. Rev. B*, **81** (11), 113201.
10 Patterson, C.H. (2006) *Phys. Rev. B*, **74** (14), 144432.
11 Clark, S.J. and Robertson, J. (2010) *Phys. Status Solidi B*, doi: 10.1002/pssb.201046110 (chapter 5 in this book).
12 Clark, S.J., Robertson, J., Lany, S., and Zunger, A. (2010) *Phys. Rev. B*, **81** (11), 115311.
13 Hohenberg, P. and Kohn, W. (1964) *Phys. Rev.*, **136** (3B), B864–B871.
14 Kohn, W. and Sham, L.J. (1965) *Phys. Rev.*, **140** (4A), A1133–A1138.
15 Anisimov, V.I., Zaanen, J., and Andersen, O.K. (1991) *Phys. Rev. B*, **44**, (3) 943–954.
16 Anisimov, V.I., Solovyev, I.V., Korotin, M.A., M.T. Czyżyk, and Sawatzky, G.A. (1993) *Phys. Rev. B*, **48** (23), 16929–16934.
17 Liechtenstein, A.I., Anisimov, V.I., and Zaanen, J. (1995) *Phys. Rev. B*, **52** (8), R5467–R5470.
18 Methfessel, M., van Schilfgaarde, M., Casali, R.A. 2000) A full-potential LMTO method based on smooth Hankel functions, in: *Electronic Structure and Physical Properties of Solids. The Use of the LMTO Method* (ed. H. Dreyssé), Lecture Notes in Physics, vol. 535 Springer Verlag, Berlin, p. 114.
19 Kotani, T. and van Schilfgaarde, M. (2010) *Phys. Rev. B*, **81** (12), 125117.
20 Maksimov, E.G., Mazin, I.I., Savrasov, S.Y., and Uspenski, Y.A. (1993) *J. Phys.: Condens. Matter*, **1**, 2493.
21 Henderson, T.M., Paier, J., and Scuseria, G.E. (2011) *Phys. Status Solidi B*, doi: 10.1002/pssb.201046303

References

22 Janak, J.F. (1978) *Phys. Rev. B*, **18** (12), 7165–7168.

23 Kotani, T., van Schilfgaarde, M., and Faleev, S.V. (2007) *Phys. Rev. B*, **76** (16), 165106.

24 van Schilfgaarde M., Kotani, T., and Faleev, S.V. (2006) *Phys. Rev. B*, **74** (24), 245125.

25 Broqvist, P., Alkauskas, A., and Pasquarello, A. (2009) *Phys. Rev. B*, **80** (8), 085114.

26 Alkauskas, A., Broqvist, P., and Pasquarello, A. (2010) *Phys. Status Solidi B*, doi: 10.1002/pssb.201046195 (chapter 7 in this book).

27 Lany, S. and Zunger, A. (2008) *Phys. Rev. B*, **78** (23), 235104.

28 Limpijumnong, S. and Lambrecht, W.R.L. (2001) *Phys. Rev. B*, **63** (10), 104103.

29 Schleife, A., Fuchs, F., Furthmüller, J., and Bechstedt, F. (2006) *Phys. Rev. B*, **73** (24), 245212.

30 Leslie, M. and Gillan, M.J. (1985) *J. Phys. C, Solid State Phys.*, **18**, 973.

31 Makov, G. and Payne, M.C. (1995) *Phys. Rev. B*, **51** (7), 4014–4022.

32 Heyd, J., Scuseria, G.E., and Ernzerhof, M. (2006) *J. Chem. Phys.*, **124** (21), 219906.

33 Rinke, P., Janotti, A., Scheffler, M., and C.G. Van de Walle (2009) *Phys. Rev. Lett.*, **102** (2), 026402.

34 Evans, S.M., Giles, N.C., Halliburton, L.E., and Kappers, L.A. (2008) *J. Appl. Phys.*, **103** (4), 043710.

35 Vlasenko, L.S. and Watkins, G.D. (2005) *Phys. Rev. B*, **71** (12), 125210.

11
Predicting Polaronic Defect States by Means of Generalized Koopmans Density Functional Calculations

Stephan Lany

11.1
Introduction

In semiconductors and insulators, non-isovalent atomic substitution critically controls the electrical behavior by introduces carriers (electrons or holes), and the utilization of such "doping" [1] lies at the heart of modern semiconductor technology. The dopants are generally classified into two categories, "shallow" and "deep" [2]: Shallow donor or acceptor states, respectively, can be thermally excited into the conduction band minimum (CBM) or the valence band maximum (VBM) thereby releasing the carriers that give rise to n- or p-type conductivity. Deep states, in contrast, are often undesired, since they can cause carrier trapping and recombination. In order to theoretically model a doped semiconductor it is, therefore, indispensable to be able to predict whether an impurity or defect acts as a shallow or as a deep center and to predict accurately the energy levels relative to the respective band edges (CBM or VBM). (In the following, we will use the term "semiconductor" in the wider sense as comprising also wide-gap materials and insulators). In this paper, we review recent work on a particular class of deep defects, *i.e.*, the impurity- or defect-bound small polarons [3], which are atomically localized and strongly bound defect states that create large lattice distortions.

The modeling of isolated point defects in semiconductors requires to treat in the order of 100 atoms, *e.g.*, in a supercell method, which necessitates rather efficient electronic structure methods. Thus, most total-energy calculations for defects in semiconductors have so far been performed using density functional theory (DFT) [4, 5] in its standard local density approximation (LDA) [6–8] for exchange and correlation, or gradient corrected versions thereof (GGA) [9–11]. However, in many cases, these density functional approximations (DFA) fail even qualitatively in the prediction of defect states with localized wavefunctions. For example, experiment has shown that acceptor-bound holes in many oxides are deep centers having wavefunctions that are centered at single oxygen atoms, *e.g.*, SiO$_2$:Al [12], ZnO:Li [13], or the singly charged Zn vacancy (V_{Zn}^-) in ZnO [14, 15] and ZnSe [16]. In contrast, DFA predicts in all these cases that the hole-wavefunction is distributed over the equivalent

Advanced Calculations for Defects in Materials: Electronic Structure Methods, First Edition.
Edited by Audrius Alkauskas, Peter Deák, Jörg Neugebauer, Alfredo Pasquarello, and Chris G. Van de Walle.
© 2011 Wiley-VCH Verlag GmbH & Co. KGaA. Published 2011 by Wiley-VCH Verlag GmbH & Co. KGaA.

O atoms neighboring the defect [17–21]. All these cases are characterized by an open-shell electronic configuration of the bound state, e.g., the $a_1^2 t_2^5$ configuration of Li_{Zn} in ZnO (using the notation of the approximate T_d point group symmetry of Li_{Zn}). The wrong wave-function localization can be understood as resulting from the residual self-interaction error of DFA, leading to an insufficient energy splitting between occupied and unoccupied states [18, 19]. The wave-function localization and resulting structural properties can be corrected by a range of theoretical methods, such as Hartree–Fock (HF) [17, 22], hybrid functionals [21, 23–25], screened exchange [26], DFA + U applied to O-p orbitals [27–30], self-interaction correction [31, 32], or our recently introduced hole-state potential [19, 20] which, although related to DFA + U, is constructed such to avoid the rather uncontrolled modification of the defect-free host-bandstructure when DFA + U is applied to the anion p-states [19].

For illustration, we compare in Figure 11.1 the calculated spin–density of the Li_{Zn}^0 center in ZnO in DFA and after applying the correction of Ref. [19]. (Note that the spin–density isosurface shown in Figure 11.1 closely resembles the wave-function-square of the unoccupied acceptor state. We prefer to show the spin–density, because this quantity is probed in magnetic resonance experiments [13].) We see that in DFA the acceptor state is not only delocalized over the neighboring oxygen atoms, but spreads over the entire supercell. This behavior is clearly that of a shallow state, similar to what one would expect from effective-mass theory [33, 34]. Accordingly, the acceptor ionization energy is relatively small in DFA, around 0.1–0.2 eV [35, 36]. However, both the delocalization over many atomic sites (Figure 11.1a) and the shallow acceptor level are inconsistent with experiment, which shows localization on a single O atom [13] and an much deeper acceptor state around 0.8 eV [37, 38]. Applying the hole-state potential of Ref. [19], the acceptor state becomes localized on a single O atom leading to a local magnetic moment at this O site, and strong structural relaxations occur which break the (approximate) tetrahedral (T_d) local symmetry around the Li impurity (see Figure 11.1). Alternatively, the mixing of Fock exchange into the DFA Hamiltonian, as done in hybrid-functionals [39–41] has very

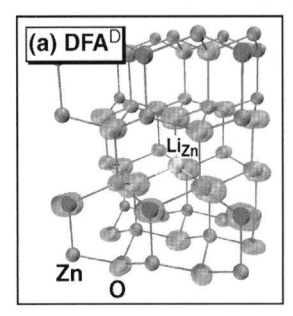

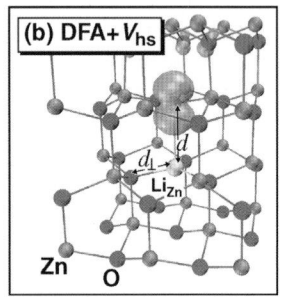

Figure 11.1 (online colour at: www.pss-b.com) Spin–density isosurface (green) of the Li_{Zn} acceptor in ZnO. In standard DFA (a), the acceptor wavefunction is effective-mass like and the structure is symmetric ($d_{Li-O} = 2.02$ Å). (b) After correction by the onsite potential V_{hs} the acceptor state is localized on a single O atom, and the structure is symmetry broken ($d_\perp = 1.91$ Å; $d_\parallel = 2.71$ Å) (Ref. [19]).

similar effects on the defect geometry and the localization of the acceptor state [21, 24] of Li_{Zn}.

By phasing in the on-site correction for O-p orbitals, we found in Ref. [19] that the geometry, the wavefunction localization, and the local magnetic moment exhibit an almost "digital" behavior, i.e., the change from the situation of Figure 11.1a to that of Figure 11.1b occurs abruptly above a critical value of the on-site potential and then changes very little when further increasing the potential strength parameter. Similarly, hybrid-functional calculations of the Al_{Si} center in SiO_2 using the B3LYP functional [39] with 20% Fock exchange did not restore the localization of the hole on a single oxygen site [17, 18], but the localization kicks in when the fraction of Fock exchange is increased [23]. Therefore, a guiding principle is desired that helps to determine appropriate parameters for such methods. Whereas the correct description of the structural and magnetic properties mostly require that the parameterized DFA correction (U, V_{hs}, Fock-exchange, etc.) is sufficient to stabilize the localized solution above the critical threshold, an accurate determination of these parameters is even more important when one is interested in energy differences between the localized and delocalized states to determine, for instance, acceptor binding energies in oxides, because these change continuously with the strength of the parameterized correction, e.g., the on-site potential V_{hs} [19]. Indeed, different parameterizations of hybrid-functionals have also led to rather different ionization energies for Li in ZnO [21, 24]. We now formulate a generalized Koopmans condition [19] that can serve as such a guiding principle to determine appropriate parameters for DFA corrections.

11.2
The Generalized Koopmans Condition

The Hohenberg–Kohn theorem [4] of DFT can be extended to fractional electron numbers N, describing a separated open system with fluctuating electrons [42, 43]. The *exact* total energy is then a piecewise linear function $E(N)$ with a discontinuous slope at integer N. In DFA, however, $E(N)$ is generally a convex function [43, 44], due to the approximate nature of the local density formalism. In order to relate the curvature of $E(N)$ to the behavior of Kohn–Sham (KS) single particle energy e_i when changing the occupation $0 \leq n_i < 1$ of the state i, we employ Janak's theorem [45, 46],

$$dE(n_i)/dn_i = e_i, \tag{11.1}$$

and find that the convexity of $E(N)$ is caused by a shift of e_i to higher energies during the occupation of state i in DFA, i.e.,

$$d^2 E(n_i)/dn_i^2 > 0, \quad \text{or}$$
$$de_i(n_i)/dn_i > 0, \tag{11.2}$$

(Note that we assume that the density functional does not have an explicit discontinuity [47, 48], which is the case for all methods considered here).

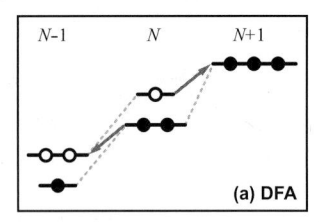

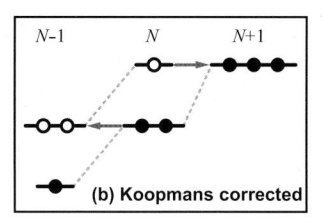

Figure 11.2 (online colour at: www.pss-b.com) Schematic illustration of the single particle energy shifts upon electron addition or removal in DFA (a) and after enforcing the generalized Koopmans condition (b). In (b), the state whose occupancy is changed (red arrows) maintains a constant energy.

For illustration, Figure 11.2 shows the single particle energy scheme for electron removal from or electron addition into a partially occupied state. This situation occurs, e.g., in case of the p^5 configuration of the isolated F atom [43, 49], where the three F-p (say, p_x, p_y, and p_z) orbitals of the spin-down channel are occupied by only two electrons. As illustrated in Figure 11.2a, the energy gap between the occupied and the unoccupied orbitals is usually rather small (or even vanishes) in DFA. For example, we obtained a gap of only 0.7 eV for the F-atom in its non-spherical, symmetry-broken DFA ground state [49]. When an electron is added, the energy of all three states increases, and the gap closes due to energetic degeneracy when all states are occupied (Figure 11.2a). Conversely, when an electron is removed, the energy of all states is lowered. Whereas the change of the single particle energy of *one state* due to the electron addition into *another state* reflects simply the increased Coulomb repulsion, the energy change of the highest state i following the change of its *own* occupation reflects a spurious self-interaction effect of DFA, which gives rise to erroneous convexity of $E(N)$, cf. Eq. (11.2). Indeed, the correct situation that leads to the linearity of $E(N)$,

$$d^2 E(n_i)/dn_i^2 = 0, \quad \text{or}$$
$$de_i(n_i)/dn_i = 0, \tag{11.3}$$

requires that the energy of the state i (i.e., the one whose occupation changes) remains constant during electron addition or removal, as shown in Figure 11.2b. If the DFA is corrected such to fulfill this requirement, we obtain for the electron addition energy (negative of the electron affinity A)

$$E(N+1) - E(N) = e_i(N), \tag{11.4}$$

by integration of Janak's theorem, or, equivalently,

$$E(N-1) - E(N) = -e_i(N), \tag{11.5}$$

for the electron removal energy (ionization potential I). In this case, the KS eigenvalue e_i of the state i acquires the meaning of a quasi-particle energy. Since, the index i refers to the state whose occupancy changes, $e_i(N)$ is either the lowest unoccupied state of the N electron system in case of electron addition [Eq. (11.4)], or it is the highest occupied state system of the N electron system in case of electron removal [Eq. (11.5)] (see Figure 11.2). Thus, if the conditions (4) and (5) are met, the single-particle gap equals the quasi-particle gap $I - A$, which, e.g., in case of the above mentioned example of the F-atom is 14 eV, much larger than the 0.7 eV single-particle energy gap in DFA [49] (*cf.* Figures 11.2a and b).

Equation (11.4) [and the equivalent Eq. (11.5)] resembles the Koopmans theorem which states an *approximate equality* in HF theory [50]. We emphasize, however, that here it has instead the meaning of a *condition* that has to be made fulfilled for parameterized corrections of DFA, such as the on-site potentials defined in Ref. [19], or the appropriate fraction of Fock-exchange in hybrid-functionals (see below). To clarify the relation between Eq. (11.4) and the Koopmans theorem, we consider that the electron addition energy – for a fixed structural geometry – can be expressed as [45, 51]

$$E(N+1) - E(N) = e_i(N) + \Pi_i + \Sigma_i. \tag{11.6}$$

Here, Π_i is the SI energy after electron addition to the orbital i under the constraint of the wave-functions being fixed at the initial-state, and Σ_i is the energy contribution arising due to wave-function relaxation. The original Koopmans theorem [50] was formulated for HF theory, where $\Pi_i \equiv 0$ holds rigorously, as an *approximation* which is good only when relaxation effects are small. In solids, however, the (negative) relaxation energy $\Sigma_i < 0$ is usually not negligible, in particular because dielectric screening leads to a significant charge rearrangement (requiring wave-function relaxation) following the electron addition into the state e_i. Indeed, by comparing Eqs. (11.4) and (11.6) we see that due to $\Sigma_i < 0$, the HF eigenenergy $e_i(N)$ of the initially unoccupied state is higher than the electron addition energy, just opposite to the situation in DFA. Accordingly, HF calculations exhibit the well-known [43, 44] concave behavior $d^2 E(n_i)/(n_i) d n_i^2 < 0$, opposite to the convex behavior of DFA. The correct linearity of $E(N)$ [Eq. (11.3)] is obtained in between the DFA and HF limits, when the SI energy Π_i and the relaxation energy Σ_i cancel each other, i.e., $\Pi_i + \Sigma_i = 0$.

11.3
Adjusting the Koopmans Condition using Parameterized On-Site Functionals

By avoiding the necessity to evaluate linearity of the function $E(N)$ explicitly, the generalized Koopmans condition, Eqs. (11.4) and (11.5) serve as a convenient tool to restore the correct behavior of the functional upon variation of the occupation. In order to compensate for the convex shape of $E(N)$ in DFA, one needs a suitable, parameterized perturbation of the DFA Hamiltonian that allows to make Eqs. (11.4) or (11.5) satisfied by adjustment of the parameter. Based on the observation that DFA and HF theory show opposite curvatures of $E(N)$, one obvious possibility is mix DFA and the Fock exchange in hybrid-functionals, so to balance the two opposite behaviors. A computationally less expensive method is DFA + U [52], which has indeed been successful in restoring the correct localization of the Al_{Si} defect in

SiO$_2$ [27]. However, the application of DFA + U to anion-p states, as needed for the treatment of, e.g., O-localized holes (cf. Figure 11.1b), is somewhat problematic: the DFA + U potential, e.g., in its simplified form of Ref. [53],

$$V_U = (U-J)\left(\frac{1}{2}-n_{m,\sigma}\right), \qquad (11.7)$$

depends on the atomic orbital projected occupancy $n_{m,\sigma}$ for the m-sublevels of spin σ. On the other hand, the anion-p states are generally much less localized than d-states on which DFA + U is typically applied, and the respective occupancy, e.g., of an O-site in defect free environment of a pure oxide host is therefore considerably smaller than the nominal full occupancy $n_{m,\sigma}=1$ expected for O(−II) anions, and it depends on the integration radius used for DFA + U. For example, we found [19] for the O-p occupancy in pure, defect-free ZnO, $n_{m,\sigma}=0.4$–0.7 depending on the size of the integration radius associated with different pseudopotentials. Considering the form of the DFA + U potential, Eq. (11.7), we see that DFA + U for O-p has a rather uncontrolled effect on the O-p host states, creating either an attractive (if $n_{m,\sigma}>0.5$) or a repulsive (if $n_{m,\sigma}<0.5$) potential causing a significant and uncontrolled distortion of the band structure of the pure oxide, even in the absence of any defect or impurity. For example, application of DFT + U to the defect-free oxide would decrease or increase the band gap (by shifting the O-p states down or up) depending on the choice of the pseudopotential.

In order to avoid the uncontrolled side effects of DFA + U, we defined in Ref. [49, 54] a "hole-state potential" of the form

$$V_{hs} = \lambda_{hs}(1-n_{m,\sigma}/n_{host}), \qquad (11.8)$$

which can be created by superposition of the occupation dependent DFA + U potential, Eq. (11.7), and the occupation-independent non-local external potential of Ref. [55]. Here, the reference occupation n_{host} is taken as the occupancy in the defect-free oxide host, so that the V_{hs} vanishes for all normally occupied O-p orbitals in the pure host. The parameter λ_{hs} controls the strength of the hole-state potential and will be adjusted so to match the Koopmans condition. If now a hole polaron is trapped at an O-site, this will cause a much lower occupancy $n_{m,\sigma}$ for the sublevel hosting the hole (e.g., the O-p_z orbital shown in Figure 11.1b), creating a repulsive potential for this level, and therefore stabilizing the localized hole. The effect of V_{hs} is illustrated schematically in Figure 11.3, showing for the Li acceptor in ZnO the O-p orbital energies (minority-spin, σ = ↓) for the O neighbor that has the hole trapped. Since these O-p orbitals occur as resonant states centered at energies below the VBM, the small splitting between the occupied and unoccupied sub-levels in DFA (cf. Figure 11.2) is not enough to lift the unoccupied level into the gap. Consequently, the hole relaxes to the VBM, and occupies the shallow effective-mass like level, as shown in Figure 11.1a. The increased splitting between the occupied and unoccupied sublevels due to the hole-state potential V_{hs} moves the localized hole state into the gap, thereby creating an acceptor state that is localized on a single O-atom (Figure 11.1b).

From the level diagram shown in Figure 11.3, one can expect that a minimum strength of V_{hs} [controlled via the strength parameter λ_{hs}, see Eq. (11.8)] is needed to

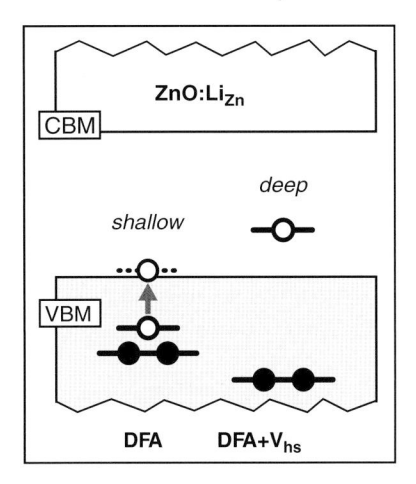

Figure 11.3 (online colour at: www.pss-b.com) Schematic illustration of the occupied and unoccupied single particle energies for the oxygen hole due to Li$_{Zn}$ in ZnO. In DFA (left), the localized hole at the O-site is unstable and relaxes into the shallow effective-mass state just above the VBM. Applying the hole-state potential V_{hs} (right) increases the splitting which stabilizes the localization of the hole in one O-p sub-orbital (cf. Figure 11.1).

lift the unoccupied O-p state into the gap and to stabilize the polaronic hole state. Indeed, when we phase-in V_{hs}, we observe that beyond a critical value $\lambda_{hs} > \lambda_{hs}^{cr}$ of the hole state potential, the symmetry breaking occurs and a strong local magnetic moment develops at the O-site at which the hole is localized, as shown in Figure 11.4a. In this calculation, in which we used the exchange-correlation functional of Ref. [11] for the underlying DFA, the condition Eq. (11.4) is fulfilled for $\lambda_{hs}^{lin} = 4.3$ eV [19] (see Figure 11.4b), at which point the correct linear behavior [cf. Eq. (11.3)] is recovered. Since, λ_{hs}^{lin} lies well above the critical value λ_{hs}^{cr} required to stabilize the polaronic state (see Figure 11.4b), the polaron state is predicted to be the physically correct state. Note that when Eqs. (11.4) or (11.5) are employed to determine the appropriate value for the parameterized functional (e.g., λ_{hs} for the on-site potential V_{hs}), one has to correct for supercell finite-size effects that affect both total energies $E(N)$ (see Refs. [49, 54]) and single-particle energies $e(N)$ [56] in case the electron number N corresponds to a charged defect state.

11.4
Koopmans Behavior in Hybrid-functionals: The Nitrogen Acceptor in ZnO

While HF theory was successful in describing qualitatively correctly the localization of holes on single oxygen sites, e.g., for the Al$_{Si}$ defect in SiO$_2$ (smoky quartz) [17, 18], or Li$_{Mg}$ in MgO [22], it does not provide a quantitative description: e.g., it predicts much too large band gaps and exceedingly large hole binding energies, e.g., the hole state bound at an O-neighbor of Li$_{Mg}$ in MgO was found roughly 10 eV above the valence band in Ref. [22]. Accordingly HF predicts often polaronic carrier trapping

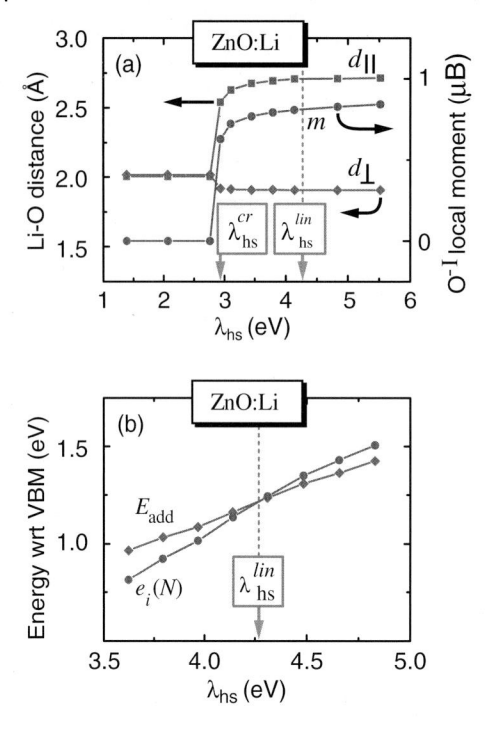

Figure 11.4 (online colour at: www.pss-b.com) (from Ref. [19]). (a) Structural and magnetic properties of the Li_{Zn} impurity in ZnO, as a function of the hole-state potential strength λ_{hs}. The polaronic state is stable above a critical value $\lambda_{hs} > \lambda_{hs}^{cr}$. The distance $d_{\parallel}$ between Li and the O atom with the trapped hole becomes larger than the distance $d_{\perp}$ between Li and the O atoms in the basal plane. (cf. Figure 11.1b), and a strong local magnetic moment m occurs. (b) The electron addition energy $E_{add} = E(N+1) - E(N)$ and the energy eigenvalue $e_i(N)$ of the initially unoccupied acceptor state of Li. λ_{hs}^{lin} marks the value of λ_{hs} for which Eq. (11.4) is satisfied.

even in cases where it should not [57]. However, a reasonable compromise may be achieved by mixing only a fraction of the non-local Fock exchange into the DFA Hamiltonian. The non-local exchange potential in such hybrid-functionals has the general form

$$V_x^{nl}(\mathbf{r}, \mathbf{r}') = -\alpha \sum_i \frac{\psi_i^*(\mathbf{r}')\psi_i(\mathbf{r})}{|\mathbf{r}-\mathbf{r}'|} f(|\mathbf{r}-\mathbf{r}'|), \tag{11.9}$$

where the parameter α and the attenuation function f vary among different formulations of hybrid-functionals, e.g., B3LYP [39] ($\alpha = 0.2, f = 1$), PBEh [40] ($\alpha = 0.25, f = 1$), HSE [41] [$\alpha = 0.25, f = \text{erfc}(\mu|\mathbf{r}-\mathbf{r}'|)$], or screened exchange [58–60] [$\alpha = 1, f = \exp(-k_{TF}|\mathbf{r}-\mathbf{r}'|)$]. For suitable parameters, such hybrid-functional calculations give reasonable band gaps, and therefore are increasingly applied for the prediction of defects in semiconductors [61–64]. (Note that the mentioned functionals further differ in the amount of semi-local gradient corrections for exchange and correlation, which has, however, only minor effects on the band-structure properties). Hybrid-functionals have also been used to describe anion-localized hole states for defects in various oxides, *i.e.*, those cases where standard DFA fails even qualitatively, like Al_{Si} in SiO_2 [23], Al_{Ti} in TiO_2 [65], and Li_{Zn} in ZnO [21, 24].

Since, as discussed above, HF theory exhibits the opposite $E(N)$ non-linearity (concave) of DFA (convex), the mixture of DFA and Fock exchange can in principle also be used to cancel the non-linearity of $E(N)$, *i.e.*, to make the generalized Koopmans condition, Eq. (11.4), fulfilled. Typically, however, hybrid-functional parameters are either taken from the pre-defined standards of the respective hybrid-functional formulation [21, 66] or are adjusted to match the experimental band gap [24, 64], and neither choice guarantees that the cancellation of non-linearity is complete. Indeed, some previous hybrid-functional calculations showed deviation from experimentally established facts, either quantitatively (ZnO:Li, Ref. [21]) or even qualitatively (SiO_2:Al, Refs. [17, 18]). The ability of hybrid-functionals to match the generalized Koopmans condition was recently addressed for defects in elemental semiconductors [67], and for the case of the N_O acceptor in ZnO [68].

Acceptor-doping of ZnO with nitrogen is subject of a controversy in the experimental literature [69]. While substitutional N_O dopants are often considered as being shallow acceptors, magnetic resonance experiments found a strongly localized hole-wavefunction [70, 71] that is inconsistent with the picture of a shallow effective-mass acceptor.

As shown in Figure 11.5a, the N_O acceptor state is already at the DFA level more localized than an effective-mass state, in contrast to Li_{Zn} (Figure 11.1). In DFA, the hole-state has p_{xy} character (p-orbitals perpendicular to the crystal c-axis), stemming from a half-occupied e_g symmetric state. As seen in Table 11.1, the all four N–Zn nearest neighbor distances are almost identical. When applying the on-site potential V_{hs} to N-p orbitals (in addition to V_{hs} for the O-p orbitals as above), using a parameter λ_{hs} such to satisfy the Koopmans condition, Eq. (11.4), [72] the hole becomes largely localized within a single N-p_z orbital, stemming from an unoccupied a_1 symmetric state. The nearest neighbor distances are now strongly anisotropic, the Zn atom along the c-axis having an $\sim$0.2 Å larger distance from N than the Zn atoms in the basal plane (Table 11.1). Thus, in Koopmans-corrected DFT the partial occupancy is lifted, which leads to a Jahn–Teller relaxation, in accord with experimental interpretations [70, 71]. Comparing the effect of non-local Fock exchange with that of the on-site potential V_{hs}, we see that both methods predict very similar acceptor wave-functions (Figure 11.5) and defect geometries (Table 11.1).

Whereas the structural properties and the wavefunction localization of ZnO:Li showed an almost digital switching between the symmetric delocalized and the symmetry-broken localized configurations with variation of the potential strength parameter λ_{hs} (Figure 11.4a) the vertical acceptor ionization energy showed a more continuous variation with λ_{hs} (Figure 11.4b). A similar sensitivity on the details of the parameterized functional can be expected for the thermal (relaxed) acceptor ionization energy. Therefore, we examined the relation between the Koopmans behavior and the depth of the N_O acceptor level [68]: standard DFA calculations predicted the acceptor level 0.4 eV above the VBM [36]. When we apply DFA + U to account for

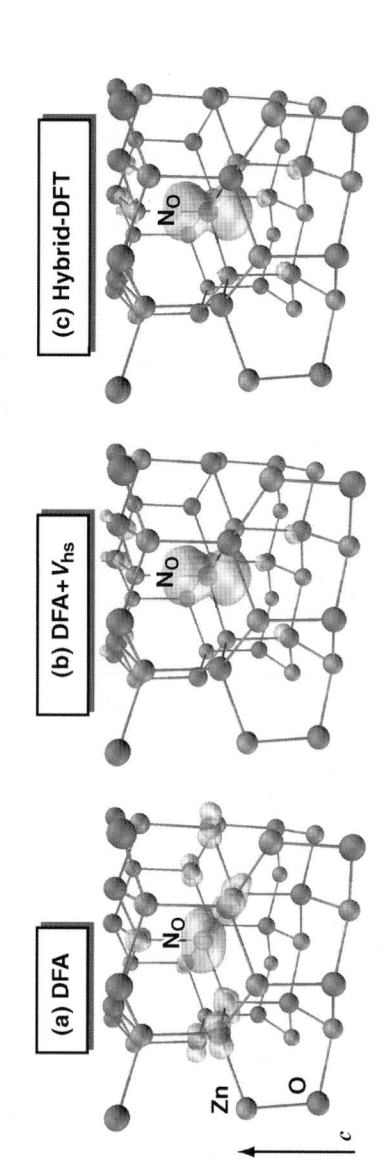

Figure 11.5 (online colour at: www.pss-b.com) (modified from Ref. [68]). Calculated spin–density (green: isosurface of 0.03 $\mu_B/Å^3$) of the neutral N_O^0 acceptor state in (a) standard DFA, (b) in Koopmans-corrected DFA with the onsite potential V_{hs}, and (c) in the HSE hybrid-functional. The arrow indicates the c-axis of the Wurtzite crystal.

Table 11.1 Properties of the neutral N_O acceptor in ZnO in different methods: the nearest neighbor N–Zn distances $d_\parallel$ and $d_\perp$ (cf. Figure 11.1), the acceptor level $\varepsilon(0/1-)$, and the non-Koopmans energy Δ_{nK}.

	$d_\parallel / d_\perp$ (Å)	$\varepsilon(0/1-)$ (eV)	Δ_{nK} (eV)
DFA[a]	1.93/1.95	$E_V + 0.74$	+0.62
DFA[a] + V_{hs}	2.18/1.94	$E_V + 1.62$	0
HSE ($\alpha = 0.25$)	2.16/1.96	$E_V + 1.40$	−0.05
HSE ($\alpha = 0.38$)	2.16/1.96	$E_V + 2.05$	−0.40

a) see Ref. [73].

the too high Zn-d orbital energy and the resulting exaggerated p–d repulsion [49], we get already a quite deep acceptor level at 0.7 eV above the VBM (see Table 11.1). This, however does not yet address the Koopmans behavior of the N–p like hole state. Indeed, when we calculate the non-Koopmans energy $\Delta_{nK} = E(N+1) - E(N) - e_i(N)$ [cf. Eq. (11.4)], we find a large positive value $\Delta_{nK} = +0.6$ eV (Table 11.1) [72] originating from the convex $E(N)$ behavior of DFA. When the generalized Koopmans condition $\Delta_{nK} = 0$ is restored by means of the on-site potential V_{hs}, the acceptor level lies even much deeper at 1.6 eV above the VBM.

We further tested the Koopmans behavior of N_O in the HSE hybrid functional, comparing two different values for the parameter α [see Eq. (11.9)], i.e., the "standard" value $\alpha = 0.25$ [40, 41], and an increased fraction of Fock exchange $\alpha = 0.38$, chosen such to reproduce the experimental band gap of ZnO [63]. We find that for $\alpha = 0.25$ the Koopmans condition is quite well fulfilled, although the band gap is still underestimated by about 1 eV. The acceptor level at 1.4 eV is close to the prediction with the on-site potential V_{hs}. A similar acceptor level was also found in a recent hybrid-functional study [64], although for a rather different parameter $\alpha = 0.36$. For the gap-corrected value $\alpha = 0.38$, we find a negative value $\Delta_{nK} = -0.4$ eV (Table 11.1), indicating concave $E(N)$ behavior, i.e., overcorrection relative to the underlying DFA. Therefore, the corresponding acceptor level at 2.1 eV is most likely unrealistically deep. From the cancellation of the $E(N)$ non-linearity in different functional, as summarized in Table 11.1, we can conclude on theoretical grounds that shallow acceptor states that have been reported in ZnO [74–76] cannot originate from substitutional N_O impurities, and must have other causes. One recent suggestion is that the shallow levels are related to stacking faults, possibly decorated with additional defects or impurities [77].

11.5
The Balance Between Localization and Delocalization

In Ref. [78], we described two fundamentally different behaviors an electrically active defect (i.e., a donor or an acceptor) can assume: (i) the primary defect-localized state (DLS), which results from the atomic orbital interaction between the defect atom and its ligands, forms a resonance inside the continuum of host bands. In this case of a shallow defect, the carriers (electrons or holes) occupy a secondary perturbed host state (PHS) with a delocalized, band-like wavefunction and an energy close to the band edge. (ii) The DLS lies inside the band gap. This is the signature of a deep defect state, and the wavefunction is usually localized at the site of the defect and its ligands.

Even though Li is clearly a deep acceptor in ZnO on account of its large ionization energy and the localized nature of the bound hole [19], it is an interesting observation that the charged Li_{Zn}^- acceptor does not show a quasi-particle energy state inside the band gap, as shown in Figure 11.6a, and therefore shows the signature of the case (i) of a shallow state. In its equilibrium structure, the ionized Li acceptor exhibits no symmetry breaking, and all nearest neighbor are practically equal ($d_{Li-O} = 2.0$ Å) as expected from the approximate local T_d symmetry in the wurtzite lattice [79]. The large anisotropy in the NN-distances (cf. Figure 11.4a) occurs only after a hole is bound on one of the four initially equivalent O neighbors. One can, therefore, raise the "Chicken or egg" like question whether the hole localization causes the symmetry breaking of the atomic structure, or whether the symmetry breaking drives the hole localization. The answer to this question depends on whether the hole localizes on a single O-site even in the absence of the lattice distortion, or, in other words, whether there exists an energy barrier in the configuration coordinate diagram that causes

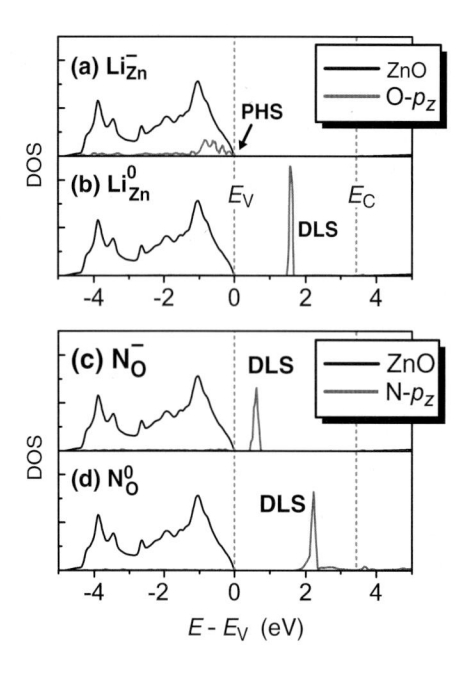

Figure 11.6 (online colour at: www.pss-b.com) Density of states (DOS) for the ionized and charge-neutral Li_{Zn} and N_O acceptor states in ZnO, calculated in DFA + V_{hs} (see Ref. [73]). The local DOS is projected on the O-p_z and N-p_z orbitals which host the bound hole in case of the charge-neutral acceptors (cf. Figures 11.1b and 11.5b, respectively).

a local minimum for the symmetric structure. Indeed, in the special situation that one defect, depending on its charge state, can assume both behaviors (i) and (ii) above, there exists generally an energy barrier between the two structural configurations. Such a barrier in the configuration coordinate diagram leads often to a range of experimentally observable metastability effects [78, 80, 81]. As seen in Figures 11.6a and b, the Li acceptor in ZnO indeed exhibits a change between shallow (i) and deep (ii) behavior being associated with a change of the charge state, which hints toward the presence of metastability effects.

Due to the energy of the Li-induced DLS below the VBM (Figure 11.6a), a free a hole can become bound at the Li acceptor in a effective-mass like (VBM-like) state without occupying the localized defect state. This shallow state of Li is, however only a transient state [82], because the energy can be lowered by the activated lattice relaxation and ensuing localization of the hole (Figure 11.1b) in a deep gap state (Figure 11.6b). Even though the transient shallow state is not suited to produce p-type conductivity, it could be observed for a short time after photo-excitation, thereby explaining the experimental observation of both a shallow and a deep state of the Li acceptor in photoluminescence [83, 84]. We recently [82] found a similar duality also for the metal-site acceptors in GaN, where Mg-doping has led to the observation of two distinct acceptor states in optical experiments [85] and of both effective-mass like and non-effective mass like hole wavefunctions in magnetic resonance experiments [86, 87]. We found that the ground state of the divalent acceptors Be, Mg, and Zn in GaN has always a localized hole wavefunction, akin to that of Li in ZnO (Figure 11.1b), which is indicative of a deep acceptor. However, Mg_{Ga} represents the unique case where the ionization energy of the deep state exceeds only slightly (by 0.03 eV) that of the ideal effective mass state, and is therefore still small enough for effective p-type doping. This explains the exceptional success of Mg-acceptor doping in GaN [88].

More generally, in regard of the balance between localization and delocalization, and the existence of an energy barrier, one can distinguish a total of four different cases, as illustrated in Figure 11.7. We now describe each case briefly with a specific example:

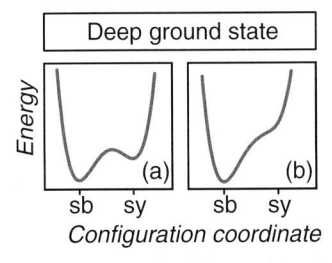

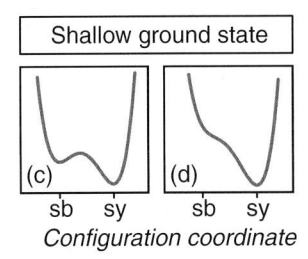

Figure 11.7 (online colour at: www.pss-b.com) Schematic configuration coordinate diagrams for acceptor states in semiconductors, illustrating the four different cases resulting from the energy ordering of the symmetry-broken (sb) and symmetric (sy) configurations, and from the existence or non-existence of an energy barrier.

i) *Deep ground state with barrier* (Figure 11.7a, e.g., ZnO:Li). In the symmetric structure of the ionized Li$_{Zn}$ acceptor, there is no defect induced quasi-particle state inside the band gap (Figure 11.6a). Thus, the neutral Li acceptor has a locally stable symmetric configuration with a delocalized effective-mass like wavefunction (PHS). Only after an activated symmetry breaking and large lattice relaxation, the localized O-p_z like hole state (cf. Figure 11.1b) occurs as a deep quasi-particle state (DLS) inside the band gap (Figure 11.6b). Examples include ZnO:Li$_{Zn}$, GaN:Mg$_{Ga}$ [82], and ZnTe:V$_{Zn}$ [20].

ii) *Deep ground state without barrier* (Figure 11.7b, e.g., ZnO:N). The symmetric, ionized N$_O$ acceptor (cf. Table 11.1) has its quasi-particle defect state already deep inside the band gap (Figure 11.6c). When forming the neutral acceptor state by removing an electron, the resulting hole immediately occupies the deep defect state (DLS), leading to the relaxation into the symmetry-broken configuration (cf. Table 11.1) without barrier. During relaxation, the DLS moves deeper into the gap (Figure 11.6c). Examples include ZnO:N$_O$ [68] and ZnO:V$_{Zn}$ [20].

iii) *Shallow ground state with barrier* (Figure 11.7c, e.g., ZnTe:Li). So far, we have considered only acceptor states whose (charge neutral) ground states are symmetry broken and have a localized hole state. Of course, there exist also acceptors in semiconductors where the ground state is symmetric with a band like effective-mass wavefunction. Considering Li$_{Zn}$ in ZnTe, we can utilize an initial lattice distortion to obtain a symmetry-broken state where the hole is located at only one of the four equivalent Te ligands, akin to the state shown in Figure 11.1b for ZnO. The parameter $\lambda_{hs} = 3.1$ eV for Te-p is then calculated analogous to the case of ZnO (Figure 11.4). However, we find that the T_d symmetric ground state with a delocalized effective-mass like hole wavefunction (PHS) lies 0.3 eV lower in energy than the symmetry-broken configuration. Thus, the generalized Koopmans formalism correctly predicts the well established effective-mass behavior of Li$_{Zn}$ in ZnTe, and the calculated Li acceptor ionization energy of 0.08 eV reflects the shallow effective-mass acceptor level (experiment: 0.06 eV [89]).

iv) *Shallow ground state without barrier* (Figure 11.7d, e.g., GaAs:Mg$_{Ga}$). For the Mg acceptor in GaAs, we find that the symmetry broken configuration cannot be stabilized even for large values of λ_{hs} for As-p (we can estimate $\lambda_{hs} = 2.7$ eV by evaluating the Koopmans condition for a constrained lattice distortion). Thus, the Mg$_{Ga}$ acceptor in GaAs has only one energy minimum, i.e., the T_d symmetric, shallow effective-mass state.

11.6
Conclusions

The physical condition of the piecewise linearity of the total energy $E(N)$ as a function of the fractional electron number plays an important role for the prediction of the structural configuration, the wave-function localization, and in particular, the ionization energies of acceptors in wide-gap semiconductors. Based on DFT this condition may be achieved via on-site potentials or by mixing of non-local Fock exchange. When the bias of standard DFT toward symmetrical and delocalized solutions is overcome in such Koopmans corrected calculations, a symmetry broken solution often emerges as the ground state, usually leading to a deep non-conductive state (with the notable exception of GaN:Mg). The symmetry breaking of the defect wavefunction can either be the result of an initial breaking of the structural symmetry, or be purely electronically driven, corresponding to the existence or non-existence, respectively, of an energy barrier in the configuration coordinate diagram. In smaller-gap semiconductors with heavier anions the tendency toward hole localization is reduced, leading to shallow effective mass like states of substitutional acceptors.

Acknowledgements

This research was supported by the U.S. Department of Energy under Contract No. DE-AC36-08GO28308 and used high performance computing resources of the National Energy Research Scientific Computing Center.

References

1 Woodyard, J.R. (1944) US Patent No. 2,530,110 filed, June 2, 1944, awarded Nov. 14, 1950.
2 Pantelides, S.T. (ed) (1986) *Deep Centers in Semiconductors: A State of the Art Approach*, Gordon and Breach, Yverdon.
3 Schirmer, O.F. (2006) *J. Phys.: Condens. Matter*, **18**, R667.
4 Hohenberg, P. and Kohn, W. (1964) *Phys. Rev.*, **136**, B864.
5 Kohn, W. and Sham, L.J. (1965) *Phys. Rev.*, **140**, A1133.
6 von Barth, U. and Hedin, L. (1972) *J. Phys. C*, **5**, 1629.
7 Ceperley, D.M. and Alder, B.J. (1980) *Phys. Rev. Lett.*, **45**, 566.
8 Perdew, J.P. and Zunger, A. (1981) *Phys. Rev. B*, **23**, 5048.
9 Lee, C., Yang, W., and Parr, R.G. (1988) *Phys. Rev. B*, **37**, 785.
10 Perdew, J.P. and Wang, Y. (1992) *Phys. Rev. B*, **45**, 13244.
11 Perdew, J.P., Burke, K., and Ernzerhof, M. (1996) *Phys. Rev. Lett.*, **77**, 3865.
12 O'Brien, M.C.M. (1955) *Proc. R. Soc. Lond. A*, **231**, 404.
13 Schneider, J. and Schirmer, O.F. (1963) *Z. Naturf.*, **18a**, 20.
14 Vlasenko, L.S. and Watkins, G.D. (2005) *Phys. Rev. B*, **72**, 035203.
15 Wang, X.J., Vlasenko, L.S., Pearton, S.J., Chen, W.M., and Buyanova, I.A. (2009) *J. Phys. D, Appl. Phys.*, **42**, 175411.
16 Jeon, D.Y., Gislason, H.P., and Watkins, G.D. (1993) *Phys. Rev. B*, **48**, 7872.
17 Pacchioni, G., Frigoli, F., Ricci, D., and Weil, J.A. (2000) *Phys. Rev. B*, **63**, 054102.
18 Lægsgaard, J. and Stokbro, K. (2001) *Phys. Rev. Lett.*, **86**, 2834.
19 Lany, S. and Zunger, A. (2009) *Phys. Rev. B*, **80**, 085202.
20 Chan, J.A., Lany, S., and Zunger, A. (2009) *Phys. Rev. Lett.*, **103**, 016404.
21 Du, M.H. and Zhang, S.B. (2009) *Phys. Rev. B*, **80**, 115217.
22 Lichanot, A., Larrieu, C., Zicovich-Wilson, C., Roetti, C., Orlando, R., and Dovesi, R. (1998) *J. Phys. Chem. Solids*, **59**, 1119.
23 To, J., Sokol, A.A., French, S.A., Kaltsoyannis, N., and Catlow, C. R.A. (2005) *J. Chem. Phys.*, **122**, 144704.

24 Carvalho, A., Alkauskas, A., Pasquarello, A., Tagantsev, A.K., and Setter, N. (2009) *Phys. Rev. B*, **80**, 195205.

25 Gallino, F., Di Valentin, C., Pacchioni, G., Chiesa, M., and Giamello, E. (2010) *J. Mater. Chem.*, **20**, 689.

26 Clark, S.J., Robertson, J., Lany, S., and Zunger, A. (2010) *Phys. Rev. B*, **81**, 115311.

27 Nolan, M. and Watson, G.W. (2006) *J. Chem. Phys.*, **125**, 144701.

28 Elfimov, I.S., Rusydi, A., Csiszar, S.I., Hu, Z., Hsieh, H.H., Lin, H.-J., Chen, C.T., Liang, R., and Sawatzky, G.A. (2007) *Phys. Rev. Lett.*, **98**, 137202.

29 Morgan, B.J. and Watson, G.W. (2009) *Phys. Rev. B*, **80**, 233102.

30 Scanlon, D.O., Walsh, A., Morgan, B.J., Nolan, M., Fearon, J., and Watson, G.W. (2007) *J. Phys. Chem. C*, **111**, 7971.

31 d'Avezac, M., Calandra, M., and Mauri, F. (2005) *Phys. Rev. B*, **71**, 205210.

32 Droghetti, A., Pemmaraju, C.D., and Sanvito, S. (2008) *Phys. Rev. B*, **78**, 140404(R).

33 Kittel, C. and Mitchell, A.H. (1954) *Phys. Rev.*, **96**, 1488.

34 Luttinger, J.M. and Kohn, W. (1955) *Phys. Rev.*, **97**, 869.

35 Wardle, M.G., Goss, J.P., and Briddon, P.R. (2005) *Phys. Rev. B*, **71**, 155205.

36 Park, C.H., Zhang, S.B., and Wei, S.H. (2002) *Phys. Rev. B*, **66**, 073202.

37 Schirmer, O.F. and Zwingel, D. (1970) *Solid State Commun.*, **8**, 1559.

38 Meyer, B.K., Alves, H., Hofmann, D.M., Kriegseis, W., Forster, D., Bertram, F., Christen, J., Hoffmann, A., Straßburg, M., Dworzak, M., Haboeck, U., and Rodina, A.V. (2004) *Phys. Status Solidi B*, **241**, 231.

39 Becke, A.D. (1993) *J. Chem. Phys.*, **98**, 1372; (1993) *J. Chem. Phys.*, **98**, 5648.

40 Perdew, J.P., Ernzerhof, M., and Burke, K. (1996) *J. Chem. Phys.*, **105**, 9982.

41 Heyd, J., Scuseria, G.E., and Ernzerhof, M. (2003) *J. Chem. Phys.*, **118**, 8207; Krukau, A.V., Vydrov, O.A., Izmaylov, A.F., Scuseria, G.E. (2006) *J. Chem. Phys.*, **125**, 224106.

42 Perdew, J.P., Parr, R.G., Levy, M., and Balduz, J.L., Jr. (1982) *Phys. Rev. Lett.*, **49**, 1691.

43 Perdew, J.P., Ruzsinszky, A., Csonka, G.I., Vydrov, O.A., Scuseria, G.E., Staroverov, V.N., and Tao, J. (2007) *Phys. Rev. A*, **76**, 040501(R).

44 Mori-Sánchez, P., Cohen, A.J., and Yang, W. (2008) *Phys. Rev. Lett.*, **100**, 146401.

45 Slater, J.C., *Quantum Theory of Molecules and Solids*, Vol. 4 (1974) McGraw-Hill, New York.

46 Janak, J.F. (1978) *Phys. Rev. B*, **18**, 7165.

47 Perdew, J.P. and Levy, M. (1983) *Phys. Rev. Lett.*, **51**, 1884.

48 Sham, L.J. and Schlüter, M. (1983) *Phys. Rev. Lett.*, **51**, 1888.

49 Lany, S. and Zunger, A. (2008) *Phys. Rev. B*, **78**, 235104.

50 Koopmans, T.C. (1934) *Physica*, **1**, 104.

51 Zunger, A. and Freeman, A.J. (1977) *Phys. Rev. B*, **16**, 2901.

52 Anisimov, V.I., Solovyev, I.V., Korotin, M.A., Czyzyk, M.T., and Sawatzky, G.A. (1993) *Phys. Rev. B*, **48**, 16929.

53 Dudarev, S.L., Botton, G.A., Savrasov, S.Y., Humphreys, C.J., and Sutton, A.P. (1998) *Phys. Rev. B*, **57**, 1505.

54 Lany, S. and Zunger, A. (2009) *Modelling Simul. Mater. Sci. Eng.*, **17**, 084002.

55 Lany, S., Raebiger, H., and Zunger, A. (2008) *Phys. Rev. B*, **77**, 241201(R).

56 Lany, S. and Zunger, A. (2010) *Phys. Rev. B*, **81**, 113201.

57 Stoneham, A.M., Gavartin, J., Shluger, A.L., Kimmel, A.V., Muñoz Ramo, D., Rønnow, H.M., Aeppli, G., and Renner, C. (2007) *J. Phys.: Condens. Matter*, **19**, 255208.

58 Bylander, B.M. and Kleinman, L. (1990) *Phys. Rev. B*, **41**, 7868.

59 Asahi, R., Mannstadt, W., and Freeman, A.J. (1999) *Phys. Rev. B*, **59**, 7486.

60 Robertson, J., Xiong, K., and Clark, S.J. (2006) *Thin Solid Films*, **496**, 1.

61 Patterson, C.H. (2006) *Phys. Rev. B*, **74**, 144432.

62 Gali, A., Hornos, T., Son, N.T., Janzén, E., and Choyke, W.J. (2007) *Phys. Rev. B*, **75**, 045211.

63 Oba, F., Togo, A., Tanaka, I., Paier, J., and Kresse, G. (2008) *Phys. Rev. B*, **77**, 245202.

64 Lyons, J.L., Janotti, A., and Van de Walle, C.G. (2009) *Appl. Phys. Lett.*, **95**, 252105.

65 Islam, M.M., Bredow, T., and Gerson, A. (2007) *Phys. Rev. B*, **76**, 045217.
66 Stroppa, A. and Kresse, G. (2009) *Phys. Rev. B*, **79**, 201201(R).
67 Deák, P., Aradi, B., Frauenheim, T., Janzén, E., and Gali, A. (2010) *Phys. Rev. B*, **81**, 153203.
68 Lany, S. and Zunger, A. (2010) *Phys. Rev. B*, **81**, 205209.
69 McCluskey, M.D. and Jokela, S.J. (2009) *J. Appl. Phys.*, **106**, 071101.
70 Carlos, W.E., Glaser, E.R., and Look, D.C. (2001) *Phys. B*, **308–310**, 976.
71 Brant, A.T., Yang, S., Evans, S.M., Halliburton, L.E., and Giles, N.C. (2009) Mater. Res. Soc. Fall meeting, abstract H10.46
72 Due to the small DFA band gap, the N_O state hybridizes with the conduction band when $\lambda_{hs} > 3$ eV, which prevents the accurate calculation of $E(N + 1)$. Therefore, we determined $\lambda_{hs} = 4.3$ eV by linear extrapolation of the difference $E(N + 1) - E(N) - e_i$ between $1\,\text{eV} \leq \lambda_{hs} \leq 3\,\text{eV}$. Also, since a minimum threshold exists for λ_{hs} to produce the correct wave-function symmetry (see Figure 5), we determined the non-Koopmans energy Δ_{nK} for DFA (see Table 1) by back-extrapolation of the same linear function to $\lambda_{hs} = 0$.
73 For the underlying DFA we used the gradient corrected exchange correlation potential of Ref. [11] and an on-site Coulomb interaction [53] for Zn-d with $U = 6$ eV, see Ref. [49]. The conduction band edge was shifted to reflect the experimental band gap.
74 Zeuner, A., Alves, H., Hofmann, D.M., Meyer, B.K., Hoffmann, A., Haboeck, U., Strassburg, M., and Dworzak, M. (2002) *Phys. Status Solidi B*, **234**, R7.
75 Pan, M., Nause, J., Rengarajan, V., Rondon, R., Park, E.H., and Ferguson, I.T. (2007) *J. Electron. Mater.*, **36**, 457.
76 Liu, W., Gu, S.L., Ye, J.D., Zhu, S.M., Wu, Y.X., Shan, Z.P., Zhang, R., Zheng, Y.D., Choy, S.F., Lo, G.Q., and Sun, X.W. (2008) *J. Cryst. Growth*, **310**, 3448.
77 Schirra, M., Schneider, R., Reiser, A., Prinz, G.M., Feneberg, M., Biskupek, J., Kaiser, U., Krill, C.E., Thonke, K., and Sauer, R. (2008) *Phys. Rev. B*, **77**, 125215.
78 Lany, S. and Zunger, A. (2005) *Phys. Rev. B*, **72**, 035215.
79 The fact that the C3v point group symmetry of the wurtzite lattice sites is already lower than the Td symmetry plays practically no role for the present discussion. Analogous effects are observed also in the zincblende modification of ZnO [20].
80 Lany, S. and Zunger, A. (2004) *Phys. Rev. Lett.*, **93**, 156404.
81 Lany, S. and Zunger, A. (2006) *J. Appl. Phys.*, **100**, 113725.
82 Lany, S. and Zunger, A. (2010) *Appl. Phys. Lett.*, **96**, 142114.
83 Meyer, B.K., Volbers, N., Zeuner, A., Sann, J., Hofmann, A., and Haboeck, U. (2006) *Mater. Res. Soc. Symp. Proc.*, **891**, 491.
84 Meyer, B.K., Stehr, J., Hofstaetter, A., Volbers, N., Zeuner, A., and Sann, J. (2007) *Appl. Phys. A*, **88**, 119.
85 Monemar, B., Paskov, P.P., Pozina, G., Hemmingsson, C., Bergman, J.P., Kawashima, T., Amano, H., Akasaki, I., Paskova, T., Figge, S., Hommel, D., and Usui, A. (2009) *Phys. Rev. Lett.*, **102**, 235501.
86 Glaser, E.R., Carlos, W.E., Braga, G.C.B., Freitas, J.A., Jr., Moore, W.J., Shanabrook, B.V., Henry, R.L., Wickenden, A.E., Koleske, D.D., Obloh, H., Kozodoy, P., DenBaars, S.P., and Mishra, U.K. (2002) *Phys. Rev. B*, **65**, 085312.
87 Glaser, E.R., Murthy, M., Freitas, J.A., Jr., Storm, D.F., Zhou, L., and Smith, D.J. (2007) *Physica B*, **401-402**, 327.
88 Nakamura, S., Mukai, T., Senoh, M., and Iwasa, N. (1992) *Jpn. J. Appl. Phys.*, **31**, L139.
89 Magnea, N., Bensahel, D., Pautrat, J.L., and Pfister, J.C. (1979) *Phys. Status Solidi B*, **94**, 627.

12
SiO$_2$ in Density Functional Theory and Beyond
L. Martin-Samos, G. Bussi, A. Ruini, E. Molinari, and M.J. Caldas

12.1
Introduction

From a theoretical point of view, amorphous silica is considered one of the most typical strong glasses belonging to the category of disordered tetrahedral networks. From a technological point of view, silica is widely used in different fields such as microelectronic industry (for metal-oxide-semiconductor devices), optical fiber technologies, and nanoimprint lithography. During the manufacturing process, a large variety of defects may be generated in the samples, which modifies the performances of the silica-based devices. Furthermore, when used in harsh environments, these pre-existing defects can act as precursor sites for the generation of new defects, or new defects can directly arise through ionization or knock-on processes making the defect assortment even wider and the device performances unpredictable. In the recent past, with the advent of nanodevices, the reduced dimensions of the oxide layers and the required abruptness of the interface demand for an atomic-scale understanding of the microscopic processes governing electronic performances, such as the carrier mobility and energy levels.

At the atomic scale, defects influence the electrical and optical properties of materials by adding localized states into their band structure. The main impact of such defect states occur when they are located in the band gap. Therefore, it would be of fundamental relevance to provide a reliable description of the gap surroundings. However, even for crystalline phases, the *ab initio* evaluation of the band gap is a hard task. In fact, the most celebrated approach for the calculation of the electronic properties for the ground-state is based on the density functional theory (DFT), but it is in principle unqualified for the description of excited state properties, at least in its usual formulation, and consequently for a correct prediction of the band gap; on the other hand, the application of appropriate and sophisticated approaches based on the many-body-perturbation theory is often hindered by the huge computational effort that is required for realistic systems. Moreover, the analysis of the electronic structure of amorphous systems implies a deep understanding of the impact of disorder – in its different possible manifestations – on the electronic properties of the system.

Advanced Calculations for Defects in Materials: Electronic Structure Methods, First Edition.
Edited by Audrius Alkauskas, Peter Deák, Jörg Neugebauer, Alfredo Pasquarello, and Chris G. Van de Walle.
© 2011 Wiley-VCH Verlag GmbH & Co. KGaA. Published 2011 by Wiley-VCH Verlag GmbH & Co. KGaA.

In this chapter, we discuss the possibility of defining a consistent criterium to univocally define the electronic gap of amorphous systems, and we calculate it from first principles for silica using different theoretical schemes. The results of our systematic investigations allow us to trace back the electronic structure of the system to the specific kinds of disorder present in the target system. Our findings are also discussed in relation to the Anderson model, which predicts the formation of localized band tails at the band edges of disordered systems.

12.2
The Band Gap Problem

Let us first recall that the energy difference between photo-emission and inverse photo-emission signals is the difference between the $N + 1$ excitation (-electron affinity) and the $N - 1$ excitation (-ionization potential), where N is the number of electrons in the ground state. The most common first-principles theoretical approach applied to study the electronic structure of solids is the density DFT. In DFT [1, 2], the exact ground-state density and energy for a given system are obtained by minimization of an energy functional which depends on the external potential and some "Universal Functional" of the density. As such "Universal Functional" is unknown, one has to face its practical realization: the building of a functional is not unique, which may lead to uncontrolled approximations and transferability issues. In the Kohn–Sham formalism the functional becomes the exchange–correlation functional. The oldest and most widely used approximation is the local density approximation (LDA) and the classical method to correct LDA is the explicit inclusion of inhomogeneity effects via gradient expansions, i.e., generalized gradient approximation (GGA). It is well known that DFT, declined in any of its two flavors (LDA or GGA), has reached a great success for modeling material properties such as equilibrium cell parameters, phonon spectra, etc. However, experimental band gap values are underestimated by about 30–40%. Even if in principle every excited state energy can be considered as a functional of the ground-state density, there remains the question on how such functionals can be realized in more explicit terms. The main lacks of LDA and GGA are their locality, energy independence, continuity with the adding or removing of electrons, and wrong long-wavelength limit which does not cancel the self-interaction (SI) part in the Hartree potential. Recently, functionals with a fraction of Hartree–Fock exchange (hybrid-functionals) have been build [3–6], in order to compensate for part of the above mentioned lacks (i.e., SI and discontinuity issue).

The GW approximation offers a valuable parameter-free alternative to such *ad hoc* building of new exchange–correlation functionals. GW has its roots in the work by Hedin and Lundqvist [7, 8]. In its first formulation, an approximated form for the electron self-energy (exchange–correlation self-energy) was obtained through an expansion in terms of the screened Coulomb potential. This procedure can be interpreted as a generalization of the Hartree–Fock method, where the expansion is based on the screened rather than the bare Coulomb potential. In more recent times, see among others [9–16], the *GW* method has been used on top of DFT calculations, often within the LDA for the exchange–correlation functional. This combined approach (called one-shot *GW* or G^0W^0) allows for great improvement on the agreement of electronic band structures with experimental results, and is presently becoming the state-of-the-art for the *ab initio* prediction of electronic properties in extended systems. The main ingredient that makes *GW* so successful is related to the fact that it contains most of the SI cancellation, and that the self-energy is non-local and energy dependent. Moreover, in the *GW* framework, as in Hartree–Fock through Koopman's Theorem, the HOMO (and each occupied band) is by definition the $N-1$ electronic excitation (where N is the number of electrons) and the LUMO (and each unoccupied band) is the $N+1$ excitations, i.e., the self-energy is discontinuous.

Let us concentrate on the self-energy. Within the Hartree–Fock approximation the self-energy can be written as:

$$\Sigma \equiv \Sigma_x(12) = iv(1+2)G(12), \tag{12.1}$$

where v is the bare Coulomb potential and G is the Green function. Indexes 1 and 2 are generalized coordinates plus spin and time ($1 \equiv x_1, t_1; 2 \equiv x_2, t_2$). As the poles of the Green function are the $N+1$ and $N-1$ excitations, the Hartree–Fock self-energy contains the discontinuity. In the long wavelength limit it is easy to prove that the $q=0$ term compensates the SI coming from the Hartree potential. However, as v depends only on $|r-r'|$, it is still a local and energy independent function.

In the *GW* approximation the self-energy takes the form of a "dressed" Hartree–Fock:

$$\Sigma(12) = iW(1+2)G(12), \tag{12.2}$$

where *W* is the screened Coulomb interaction that is calculated through a Dyson-like equation with a random-phase approximation (RPA) irreducible polarizability. The *W* term contains the response of the electronic system when non-interacting electron–hole pairs are created, which is a function of $(r-r')$, $(r''-r''')$ and the energy. The polarizability is also a function of $(r-r')$ and $(r''-r''')$, and its calculation usually needs a summation on all possible transitions (from occupied to unoccupied states). It is easy to understand why even if the equations are known since the sixties, only recently, with some additional approximations, we manage to computationally afford systems with tens of atoms. One of the approximations of great success has been the Coulomb hole and screened exchange approximation [17] (COHSEX). The COHSEX formula is obtained from the general *GW* self-energy by imposing a static *W* before the energy convolution. The COHSEX approximation has the advantage of being independent of the unoccupied states summation in the self-energy. The first term (SEX) is a screened Hartree–Fock, where the screening is the static *W*, while the second (COH) is just a static local potential.

After COHSEX, different multipole expansions have been proposed to include, without explicitly performing the energy convolution, the energy dependence in *W*. The most popular is the plasmon pole approximation [17–19], where the energy dependence in *W* is replaced by an interpolation through a single pole (at some plasmon energy).

Table 12.1 Fundamental/Homo–Lumo gap (eV) for alpha-quartz and amorphous SiO$_2$ within different approximations, DFT–LDA, one-shot HF, one-shot COHSEX and one-shot GW on top of DFT–LDA.

	DFT–LDA	HF	COHSEX	GW
alpha-quartz	5.9	16.9	10.1	9.4
a-SiO$_2$	5.6	16.2	10.1	9.3

In order to highlight the effect of inclusion of off-diagonal elements, also called local-field effects (with explicit dependency on $(r - r')$ and $(r'' - r''')$) on the description of the gap, we have performed four calculations on alpha-quartz and on an a-SiO$_2$ model (for details on the generation of the amorphous sample we refer the reader to Ref. [20]) within the four different approximations DFT–LDA, HF, COHSEX (from LDA), and G^0W^0 (from LDA), see Table 12.1. For more details on the GW calculation see Ref. [21, 22].

As expected, DFT underestimates the gap, HF overestimates it, while COHSEX and, in particular, G^0W^0 give the closest values to experimental outcomes – that range from 8.8 to 11.5 eV [23–25]. From a careful comparison between the gap values for quartz and a-SiO$_2$ (see Table 12.1) we can note that the gap size for quartz is larger than for the a-SiO$_2$ phase within DFT and HF, while the experiments suggest that, if a difference exists, it is smaller than the experimental accuracy. Local-field effects taken into account by W in the COHSEX and GW self-energy enhances the opening of the gap due to the disorder, bringing the a-SiO$_2$ gap closer to that of quartz. This enhancement is not related to the SI and discontinuity issues, already corrected (or not inserted) in HF. This behavior implies that, for disordered materials, it is fundamental to treat correlations through non-diagonal objects. Indeed, it is unlikely that any *ad hoc* choice of local functionals, whatever fraction of Hartree–Fock exchange it contains, would be able to reproduce the whole effect of disorder on the electronic structure.

12.3
Which Gap?

In a crystalline semiconductor/insulator the fundamental gap/band gap is a well defined quantity, i.e., the energy difference between the bottom of the conduction band and the top of the valence band. When one deals with a disordered material, the definition of "gap" is not straightforward. Indeed, following the ideas presented in the pioneering works of Anderson, Mott and Cohen [26–29], disorder induces the formation of localized band tails (flat bands) at each band edge, creating boundary regions (mobility edges) and opening the so-called "mobility gap", see Figure 12.1. Conceptually, the mobility gap for a disordered system is equivalent to the band gap for a crystal. In addition to the mobility gap, one could also define a "HOMO–LUMO" gap (following simply an occupied/unoccupied criterium), which is smaller, by definition, than the mobility gap.

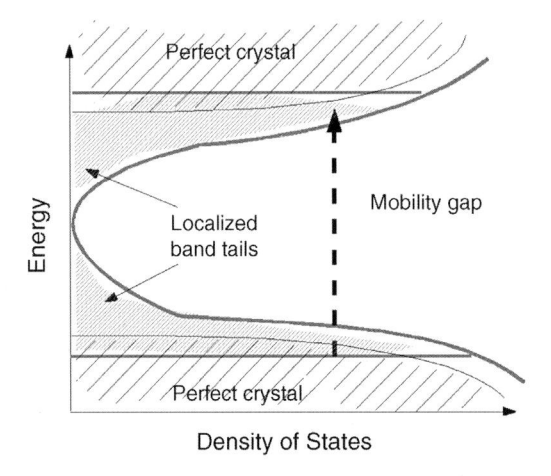

Figure 12.1 Schematic representation of the density of states close to the top of the valence band and the bottom of the conduction band, in a disordered semiconductor/insulator compared to the representative crystal.

For semiconductors such as a-Si/H, the density of states in the mobility gap region has been intensively investigated [30, 31]. However, in the SiO_2 case, there is no experimental evidence of localized band tails between valence and conduction bands. Moreover, experiments do not see any marked difference between the quartz and the a-SiO_2 electronic structure. The effect of disorder has been theoretically studied mostly by addressing model systems through simplified Hamiltonians, that contain parametrized terms to account for the disorder contribution (see, e.g., in Ref. [32]). With respect to the band gap of a perfectly ordered system, a low degree of disorder produces a closure of the mobility gap, while strong disorder opens it. The electronic structure of amorphous SiO_2 obtained within *ab initio* DFT is usually compared to the electronic properties of crystalline alpha quartz, and suggest a small closure of the mobility gap [33]. Amorphous SiO_2 and other low pressure crystalline phases of SiO_2, such as quartz and cristobalite, are constituted by "well-connected" SiO_4 networks: at the short range scale differences between the crystals and the amorphous samples come from bond-angle and bond-length fluctuations, while the medium-range structure is governed by the connectivity of the SiO_4 network itself [34] which is different in all these phases. Given the fact that a low degree of disorder acts by producing bond angle and bond length variation while a topology change is expected to be produced by strong disorder, the reason why DFT results in a small gap closure of the amorphous versus the crystalline system, instead of an opening, is to be clarified. This could be traced back to several reasons, such as to the specific amorphous model used in the simulations, or to the specific theoretical approach, or to the absence "strong" disorder for a-SiO_2, or to the fact that the mobility gap has not been compared to the correct reference crystal. Indeed, the density of a-SiO_2 is around 2.2 g/cm^3 while the density of quartz is around 2.6 g/cm^3. It is well known that the gap increases with the density in tetrahedral networks. Therefore the

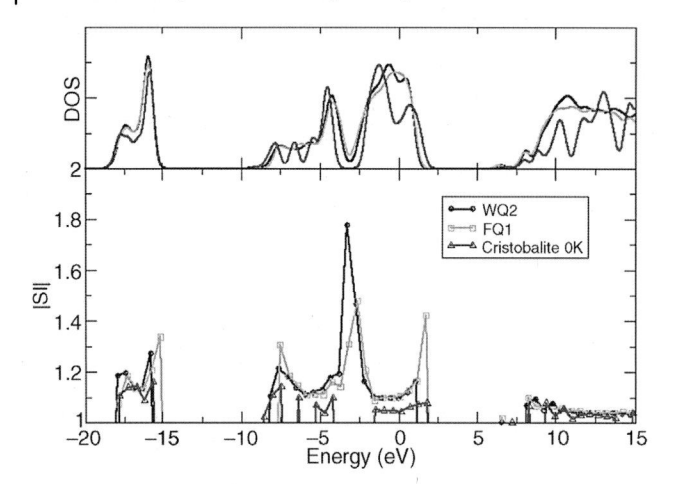

Figure 12.2 (online color at: www.pss-b.com) Upper panel: DFT–LDA energy density of states for two a-SiO$_2$ models WQ2 and FQ1, obtained by quenching from a melt with two different quench rates, 2.6×10^{13} and 1.1×10^{15} K/s, respectively, and a perfectly ordered cristobalite (Cristobalite 0 K). Lower panel: states localization by means of their |SI|. Each |SI| point has been calculated by averaging within an energy interval of 0.09 eV. The DOS have been aligned maximizing the overlap between deep valence levels.

electronic structure of a-SiO$_2$ has to be compared with the crystal phase that exhibits the most similar density, i.e., cristobalite.

We have performed DFT calculations on two 108 atoms amorphous SiO$_2$ models, WQ2 and FQ1 generated at two different quench rate 2.6×10^{13} K/s and 1.1×10^{15} K/s, respectively (where WQ and FQ stay for "well-quenched" and "fast-quenched" to distinguish between the two quench rates, see Ref. [20] for further details), and on two 192 atoms cristobalite models (Fd3-m), one at $T = 0$ K (C0) and one at $T = 300$ K (C300). Here, we use the temperature as a way to add to the system a small stochastic disorder. Calculations have been performed at the Γ point, with a wave function cut-off of 70 Ry and norm-conserving pseudopotentials. Hundred and eight atoms supercell is big enough for having a system-size convergency within 0.1 eV at the self-energy level. In Figure 12.2, we have plotted the density of states and the localization of each state as a function of the energy for the two amorphous models as well as for the perfectly ordered cristobalite, which we use as crystal reference. The localization is described by means of the normalized SI, obtained by dividing the SI, i.e., the Coulomb interaction between an electronic state and itself, as generally defined by the following equation:

$$\text{SI} = \frac{1}{V} \int \int \frac{\varphi_s^*(r)\varphi_s^*(r)\varphi_s^*(r')\varphi_s^*(r')}{||r-r'||} d^3r \ d^3r', \qquad (12.3)$$

by the SI of a plane wave normalized in the corresponding cell (which generalizes the SI tool, extensively used in the quantum chemical community for finding maximally localized basis sets, for the description of extended systems that are modeled through periodic boundary conditions [21]): normalizing the SI allows for a quantitative, energy-independent estimate of localization for systems with different unit-cell volumes (the boundaries being $|SI|=1$ for a fully delocalized state, i.e., a plane wave, and divergent $|SI|$ for a completely localized state, i.e., a Delta function)]. We observe signatures of localization almost exclusively at valence edges, while the bottom of the conduction band exhibits a perfectly delocalized character. As expected, the increase of the quench speed increases the localization, and we see at the neighborhood of the valence band top for the FQ1 model states highly localized, that can be considered as localized band tail states. It is also easy to see that the cristobalite gap is smaller than the *mobility* gap of the disordered systems. If we compare the fundamental gap of the perfectly ordered cristobalite, 5.4 eV, with the system perturbed just by thermal disorder, 5.3 eV [21], it is clear that a small degree of disorder induces a small closure of the gap. Therefore, we recover completely the Anderson, Mott, and Cohen picture [26–29], i.e., small disorder degree close the gap while a strong disorder degree widen it.

The widening of the gap due to strong disorder is particularly enhanced when moving to GW, as it includes non-diagonal screening, as explained above. Indeed, the GW gap for cristobalite is 8.9 eV while the mobility gap for a-SiO$_2$ is 9.4–9.2 eV for WQ2 and FQ1, respectively (for a more extended discussion see Ref. [21]).

12.4
Deep Defect States

Defect levels (donor or acceptor) and formation energies of charged defects are difficult to describe quantitatively within DFT [35]. Indeed, they are related to ionization potentials ($N-1$ excitation)/electron affinities ($N+1$ excitation) of the defect state [36, 37]. Some attempts have been made to try to circumvent the need of going beyond DFT [38, 39]. However, it is unlikely that one can find semi-empirical rules that work in a general case, and for defects in the already disordered structure of an amorphous system this is even more critical. To illustrate the complexity of the problem, we can compare results for the well-quenched model we showed above, to the results of a model with connectivity defects.

We have performed DFT–LDA and G^0W^0 (on top of LDA) on an amorphous SiO$_2$ model with two point defects: a non-bridging oxygen NBO (coordination 1) and a tri-coordinated silicon (the simulation parameters are the same as in Ref. [21]). These two atomic defects have produced five strongly localized defect states, a dangling bond (Figure 12.3a), two occupied oxygen 2p non-bonding orbitals (Figure 12.3b and c), and two unoccupied silicon non-bonding orbitals (Figure 12.3d and e).

The corresponding defect levels are located inside the gap (single lines in Figure 12.4). Comparing the level alignment in the neighborhood of the gap, DFT, and GW, it is evident that neither an *ad hoc* band stretching nor a rigid energy shift can locate correctly the states from their DFT–LDA relative position, even if one tries to treat separately the occupied and unoccupied states. In the case of the highest localized states, such as the O dangling bond of the NBO, the many-body corrections even alter their position relative to the band tails, precluding scissor shift approximations. Even if

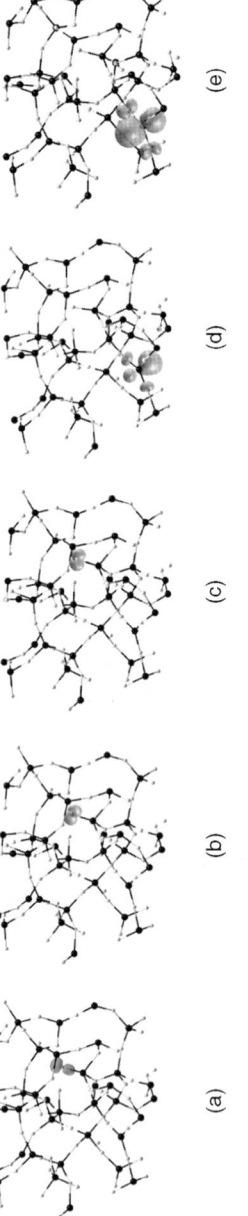

Figure 12.3 (online color at: www.pss-b.com) Localized defect states due to a non-bridging oxygen [(a), (b), and (c)] and a tri-coordinated silicon [(d) and (e)] in an a-SiO$_2$ model (in white oxygen atoms and in black silicon atoms).

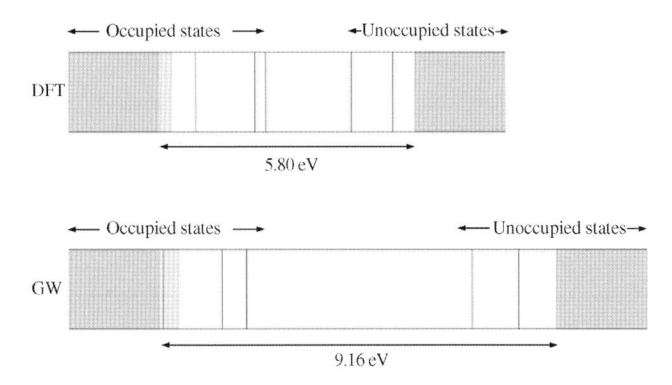

Figure 12.4 (online color at: www.pss-b.com) Electronic structure of a 108 atoms a-SiO$_2$ system with defects introduced by one non-bridging oxygen and one tri-coordinated silicon, within DFT (upper panel) and G^0W^0 (lower panel). Shaded blue and red bands represent, respectively, delocalized states and localized tails due to disorder. Calculation for the defect system was also performed through norm-conserving pseudopotentials and Γ-point sampling; an energy cutoff of 70 Ry was used for the wave-functions and Fock operator. Single gray lines stay for the strongly localized defect states (occupied oxygen non-bonding and unoccupied silicon non-bonding states). The single red line shows the position of the oxygen dangling bond level. Energies have been aligned to the top of the valence mobility edge.

this kind of approximation has been successfully applied to ideal crystalline systems, it is fundamental to include the exact many-body correction for quantitative studies when treating states with markedly different characters. A qualitatively correct ordering of localized states with a different nature is already provided at the Hartree–Fock level, indicating that the inclusion of exact-exchange is mandatory for a sound description of dangling bonds. However, the weight of the exchange part with respect to the correlation part is different from state to state and no semi-empirical rule can be extrapolated from the results. It is also interesting to note the increase in the energy difference between the two 2p non-bonding orbitals, which are seen almost degenerate within DFT–LDA. The sensitivity to local field effects in GW enhances the energy difference, which is due to the different orientation of the orbitals. As for the two unoccupied (and more delocalized) defect states, they maintain almost the same relative position, and are just deeper in the gap, more detached from the mobility edge. We also observe that the sensitivity of defect states to a proper treatment of many-body effects is in agreement with a very recent paper [40], where the application of the GW scheme to the analysis of a carbon vacancy in 4H-SiC turned out to be decisive to correctly account for electron–electron correlations.

12.5 Conclusions

We have shown that the inclusion of *local field effects* may be relevant even just for a correct quantitative evaluation of the gap size. The example of the deep defects in amorphous silica shows how it could be difficult to try to extract general semi-empirical models to circumvent the needs of going beyond DFT. Unfortunately, the way to a bi-univocal modeling of the electronic properties in the gap neighborhood is still under an active debate [36–38]. Indeed, the computational effort needed by a full GW calculation, or GW including vertex-corrections, free from pseudo-potential effects, precludes its application to realistic systems [41, 42], forcing the research community to find shortcuts, that need to be more extensively tested: pseudo-potentials, approximations to the screening, one-shot GW starting from different wave-functions and energies [42]. It is of fundamental relevance to systematically explore all this shortcuts and give complete benchmarks.

References

1. Hohenberg, P. and Kohn, W. (1964) *Phys. Rev.*, **136**, B864.
2. Kohn, W. and Sham, L.J. (1965) *Phys. Rev.*, **140**, A1133.
3. Perdew, J.P., Ernzerhof, M., and Burke, K. (1996) *J. Chem. Phys.*, **105**, 9982.
4. Heyd, J., Scuseria, G.E., and Ernzerhof, M. (2003) *J. Chem. Phys.*, **118**, 8207.
5. Heyd, J., Scuseria, G.E., and Ernzerhof, M. (2006) *J. Chem. Phys.*, **124**, 219906.
6. Stephens, P.J., Devlin, F.J., Chabalowski, C.F., and Frisch, M.J. (1994) *J. Phys. Chem.*, **98**, 11623.
7. Hedin, L. (1965) *Phys. Rev.*, **139**, A796.
8. Hedin, L. and Lundqvist, S. (1969) *Solid State Phys.*, **23**, 1.
9. Hybertsen, M.S. and Louie, S.G. (1984) *Phys. Rev. B*, **30**, 5777.
10. Hybertsen, M.S. and Louie, S.G. (1985) *Phys. Rev. Lett.*, **55**, 1418.
11. Godby, R.W., Schluter, M., and Sham, L.J. (1988) *Phys. Rev. B*, **37**, 10159.
12. Farid, B., Daling, R., Lenstra, D., and van Haeringen, W. (1988) *Phys. Rev. B*, **38**, 7530.
13. Engel, G.E., Farid, B., Nex, C.M.M., and March, N.H. (1991) *Phys. Rev. B*, **44**, 13356.
14. Shirley, E.L. and Martin, R.M. (1993) *Phys. Rev. B*, **47**, 15404.
15. Rojas, H.N., Godby, R.W., and Needs, R.J. (1995) *Phys. Rev. Lett.*, **74**, 1827.
16. Rohlfing, M., Kruger, P., and Pollmann, J. (1995) *Phys. Rev. B*, **52**, 1905.
17. Hybertsen, M.S. and Louie, S.G. (1986) *Phys. Rev. B*, **34**, 5390.
18. Godby, R.W. and Needs, R.J. (1989) *Phys. Rev. Lett.*, **62**, 1169.
19. Engel, G.E. and Farid, B. (1993) *Phys. Rev. B*, **47**, 15931.
20. Martin-Samos, L., Limoge, Y., Crocombette, J.P., and Roma, G. (2005) *Phys. Rev. B*, **71**, 014116.
21. Martin-Samos, L., Bussi, G., Ruini, A., Caldas, M.J., and Molinari, E. (2010) *Phys. Rev. B*, **81**, 081202(R).
22. Martin-Samos, L. and Bussi, G. (2009) *Comput. Phys. Commmun.*, **180**, 1416.
23. van den Keybus, P. and Grevendonk, W. (1986) *Phys. Rev. B*, **33**, 8540.
24. Evrard, R. and Trukhin, A.N. (1982) *Phys. Rev. B*, **25**, 4102.
25. Weinberg, Z.A., Rubloff, G.W., and Bassous, E. (1979) *Phys. Rev. B*, **19**, 3107.
26. Anderson, P.W. (1958) *Phys. Rev.*, **109**, 1492.
27. Cohen, M.H., Fritzsche, H., and Ovshinsky, S.R. (1969) *Phys. Rev. Lett.*, **22**, 1065.
28. Anderson, P. (1975) *Phys. Rev. Lett.*, **34**, 953.
29. Mott, N.F. (1978) *Rev. Mod. Phys.*, **50**, 203.
30. Wronski, C.R., Lee, S., Hicks, M., and Kumar, S. (1989) *Phys. Rev. Lett.*, **63**, 1420.
31. Biryukov, A.V. et al., (2005) *Semiconductors*, **39**, 351.

32 Fazileh, F., Chen, X., Gooding, R.J., and Tabunshchyk, K. (2006) *Phys. Rev. B*, **73**, 035124.
33 Sarnthein, J., Pasquarello, A., and Car, R. (1995) *Phys. Rev. Lett.*, **74**, 4682.
34 King, S.V. (1967) *Nature (London)*, **47**, 3053.
35 Fazzio, A., Caldas, M., and Zunger, A. (1984) *Phys. Rev. B*, **29**, 5999.
36 Rinke, P., Janotti, A., Scheffler, M., and Van de Walle, C.G. (2009) *Phys. Rev. Lett.*, **102**, 026402.
37 Martin-Samos, L., Roma, G., Rinke, P., and Limoge, Y. (2010) *Phys. Rev. Lett.*, **104**, 075502.
38 Xiong, K., Robertson, J., Gibson, M.C., and Clark, S.J. (2005) *Appl. Phys. Lett.*, **87**, 183505.
39 Alkauskas, A., Broqvist, P., and Pasquarello, A. (2008) *Phys. Rev. Lett.*, **101**, 046405.
40 Bockstedte, M., Marini, A., Pankratov, O., and Rubio, A. (2010) *Phys. Rev. Lett.*, **105**, 026401.
41 Bruneval, F. (2009) *Phys. Rev. Lett.*, **103**, 176403.
42 Rostgaard, C., Jacobsen, K.W., and Thygesen, K.S. (2010) *Phys. Rev. B*, **81**, 085103.

13
Overcoming Bipolar Doping Difficulty in Wide Gap Semiconductors

Su-Huai Wei and Yanfa Yan

13.1
Introduction

Application of semiconductors as electric and optoelectronic devices depends critically on their dopability. Failure to dope a material, i.e., to produce enough free charge carriers beyond a certain limit at working temperature, is often the single most important bottleneck for advancing semiconductor-based high technology. Wide band gap (WBG) semiconductors, such as diamond, AlN, GaN, MgO, and ZnO, have unique physical properties that are suitable for applications in short-wavelength and transparent optoelectronic devices [1–9]. To realize these applications, the formation of a high-quality homo p–n junction is essential. In other words, high-quality bipolar (p- and n-type) doping are required for the same material. Unfortunately, most WBG semiconductors experience a serious doping asymmetry problem, i.e., they can easily be doped p- or n-type, but not both [10]. For example, diamond can be doped relatively easily p-type, but not n-type [1–3]. On the other hand, ZnO can easily be made high-quality n-type, but not p-type [7–9]. For materials with very large band gap such as AlN and MgO, both p- and n-type doping are difficult [11]. The doping difficulty also exists in nanostructure semiconductors where the band gaps increase due to the quantum confinement [12, 13]. These doping problems have hindered the potential applications of many WBG and nanostructure materials.

Extensive research has been done to understand the origin of the bipolar doping difficulties in WBG and nanostructure semiconductors and to find possible solutions to overcome the doping difficulty. In the past, we have proposed various approaches to overcome the bipolar doping difficulty in WBG semiconductors. These approaches have been tested by the systematical calculation of defect formation energies and transition energy levels of intrinsic and extrinsic defects in various WBG and nanostructure semiconductors using first-principles density-functional theory [14–23]. In this paper, we review the origins of the bipolar doping difficulty and describe our approaches for overcoming the doping bottleneck in WBG semiconductors. The paper is organized as follows. Section 13.2 discusses the salient features of calculating defect properties, in which the issues related to the finite size of the supercell and the band gap errors are discussed. Section 13.3 analyzes symmetry and occupation of defect levels. Section 13.4 describes what causes the bipolar doping difficulty in WBG semiconductors and the origin of the doping limit rules. Section 13.5 describes our proposed approaches to overcome the bipolar doping difficulties, focusing mostly on ZnO and other WBG semiconductors. Section 13.6 briefly summarizes the paper and provides an outlook for future research in this field.

13.2
Method of Calculation

We performed the band structure and total energy calculations using the first-principles density-functional theory with local density approximation (LDA) or general gradient approximation (GGA) [24, 25]. We used the supercell approach in which the defects or defect complexes are put at the center of a supercell and the periodic boundary condition is applied. For quantum dots (QDs) the surface are passivated by hydrogen or pseudohydrogen [12, 13]. In all calculations, all the atoms are allowed to relax until the Hellman–Feynman forces acting on them become negligible. For charged defects, a uniform background charge is added to keep the global charge neutrality of the supercells [19].

To determine the defect formation energy and defect transition energy levels, one needs to calculate the total energy $E(\alpha, q)$ for a supercell containing defect α in charge state q, the total energy $E(\text{host})$ of the same supercell without the defect, and the total energies of the involved elemental solids or gases at their stable phases. It is important to realize that the defect formation energy depends on the atomic chemical potentials μ_i and the electron Fermi energy E_F. From these quantities, the defect formation energy, $\Delta H_f(\alpha, q)$, can be obtained by:

$$\Delta H_f(\alpha, q) = \Delta E(\alpha, q) + \sum n_i \mu_i + q E_F, \tag{13.1}$$

where $\Delta E(\alpha, q) = E(\alpha, q) - E(\text{host}) + \sum n_i E(i) + q \varepsilon_{\text{VBM}}(\text{host})$. E_F is referenced to the valence band maximum (VBM) of the host. μ_i is the chemical potential of constituent i referenced to elemental solid/gas with energy $E(i)$; n_i is the number of elements and q is the number of electrons transferred from the supercell to the reservoirs in forming the defect cell. The transition energy $\varepsilon_\alpha(q/q')$ is the Fermi energy at which the formation energy of defect α at charge state q is equal to that at charge state q'. Using Eq. (13.1), the transition energy level with respect to the VBM can be obtained by

$$\varepsilon_\alpha(q/q') = \frac{\Delta E(\alpha, q) - \Delta E(\alpha, q')}{q' - q} - \varepsilon_{\text{VBM}}(\text{host}). \tag{13.2}$$

Typically, for finite supercell, the Brillouin zone integration for the charge density and total energy calculations is performed using special k-points or equivalent k-points in the superstructures. This approach gives better convergence on the calculated charged density and total energy. However, it usually gives a poor description on the symmetry and energy levels of the defect state, as well as the

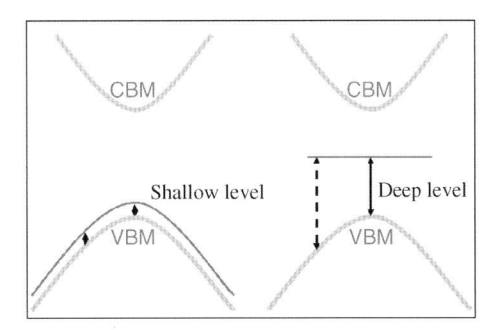

Figure 13.1 (online colour at: www.pss-b.com) Schematic plot of the defect level with respect to the band edge. It is shown that in a supercell calculation, the energy level determined at the Γ-point is correct for both shallow and deep levels.

VBM and conduction band minimum (CBM) states, and the results could be sensitive to the k-points sampling because the band edges are determined by the k-point sampling. To avoid this problem, a Γ-point-only calculation is often used to determine the total energy and transition energy level because the symmetry of the defect and the band edge states are well defined at Γ, and both the shallow and deep levels are correctly described (see Figure 13.1). However, for small supercell, Γ-point-only approach may give poor total energy convergence. Here, we propose to use a hybrid scheme to combine the advantages of both special k-points and Γ-point-only approaches [14, 19]. In this scheme, we first calculate the transition energy level *with respect to VBM*, which is given by

$$\varepsilon(0/q) = \varepsilon_D^\Gamma(0) - \varepsilon_{VBM}^\Gamma(\text{host}) + \frac{E(\alpha, q) - (E(\alpha, 0) - q\varepsilon_D^k(0))}{-q}. \tag{13.3}$$

For donor level ($q > 0$), it is usually more convenient to reference the ionization energy level to the CBM, i.e., we can rewrite Eq. (13.3) as

$$\varepsilon_g^\Gamma(\text{host}) - \varepsilon(0/q) = \varepsilon_{CBM}^\Gamma(\text{host}) - \varepsilon_D^\Gamma(0) + \frac{E(\alpha, q) - (E(\alpha, 0) - q\varepsilon_D^k(0))}{q}, \tag{13.4}$$

where a positive number calculated from Eq. (13.4) indicates the distance of the transition energy level below the CBM. In Eqs. (13.3) and (13.4) $\varepsilon_D^k(0)$ and $\varepsilon_D^\Gamma(0)$ are the defect levels at the special k-points (weight averaged) and at the Γ-point, respectively; and $\varepsilon_{VBM}^\Gamma(\text{host})$ and $\varepsilon_{CBM}^\Gamma(\text{host})$ are the VBM and CBM energies, respectively, of the host at the Γ-point. $\varepsilon_g^\Gamma(\text{host})$ is the band gap at the Γ-point. The first term on the right-hand side of Eqs. (13.3) or (13.4) give the single-electron defect level at the Γ-point. The second term determines the relaxation energy U (including both the Coulomb contribution and the atomic relaxation contribution) of the charged defects calculated at the special k-points, which is the extra cost of energy after moving $(-q)$ charge to the neutral defect level with $E = \varepsilon_D^k(0)$. Afterwards, the formation energy of defect α at charge state q can be obtained by

$$\Delta H_f(\alpha, q) = \Delta H_f(\alpha, 0) - q\varepsilon(0/q) + qE_F, \tag{13.5}$$

where $\Delta H_f(\alpha,0)$ is the formation energy of the charge-neutral defect and E_F is the Fermi level with respect to the VBM. This new approach has been used successfully for studying defects in various semiconductors.

We want to point out that in writing down Eqs. (13.1)–(13.4), we assumed that the common reference energy level is used in the calculation of defect α in different charge states. Because in a periodic supercell calculation the zero potential energy is not well defined, therefore, we have to lineup the potential using a common reference. This is usually done by lineup core level of an atom far away from the defect center. In the case where core level is not available, average potential around the atom can also be used.

In the formula above, we also assumed that the VBM energy position is given by its eigenvalue $\varepsilon_{VBM}(\text{host})$. For small supercell, however, the Koopman's theorem may not hold that is

$$\delta E = \varepsilon_{VBM}(\text{host}) - [E(\text{host}, N) - E(\text{host}, N-1)] \tag{13.6}$$

is not zero, especially for VBM with localized electron states. Similar situation may also exist for the CBM energy position. In this case, the correction term δE should be added to determine the VBM or CBM energy level.

It the supercell calculation, the periodic boundary conditions introduces a spurious Coulomb interaction between the charged defects in different cells. To estimate the magnitude of this interaction, point charges immersed in neutralizing jellium are usually assumed. Attempts to including higher-order terms are difficult, because the higher-order multipoles of the defect charge are not uniquely defined in the supercell approach [26]. However, in reality, the charged defect does not have a delta-function-like distribution, especially for shallow defects, which have a relatively uniform charge distribution. Therefore, direct application of the Makov and Payne [26] correction using only point charge often overestimates the effect [19]. Thus, for shallow low charge state defects we usually assume the Makov and Payne correction are not important. However, this correction term could be large if the defect level is localized or the defect is in a small QD.

Another issue that leads to uncertainty in defect calculation is caused by the fact that LDA or GGA calculations underestimate the band gap of a semiconductor. One way to correct this error is to project the defect level to the CBM and VBM states, and shift the defect level accordingly when the band edges are shifted to correct the band gap error [27]. Another way to correct this error is using high level DFT calculation such as GW approach [28]. However, currently, this type of full scale calculation is still formidable. Recently, more empirical hybrid density-functional method [29, 30] which mixes certain amount of Hartree–Fock potential with the GGA potential is used for defect calculation. Although this kind of approach can correct the band gap in some empirical way, the symmetry breaking caused by the Hartree–Fock potential and the accuracy of the calculated defect levels, in our opinion, still needs experimental verification.

13.3
Symmetry and Occupation of Defect Levels

It is often very useful to know the symmetry and character of the single-particle defect level before we start the calculation, because it can help us identify the defect level and because defects with different symmetry and character will behave differently. Moreover, to modify defect states or correct the band gap error, it is also important to know the symmetry of the state. For example, although both anion vacancy and cation interstitial have the same a_1 defect levels in II–VI semiconductors [23], anion vacancy has the a_1^v character derived from the valence band or cation dangling bond states, whereas cation interstitial has instead the a_1^c character derived from the conduction band. Thus, the energy level of the cation interstitial is expected to follow closely with the CBM, whereas the energy level of anion vacancy state will not.

For simple extrinsic impurities, one can predict in principle whether a dopant is a donor with a single-particle energy level close to the CBM or an acceptor with a single-particle energy level close to the VBM by simply counting the number of the valence electrons of the dopants and the host elements. For example, in CdTe, one can expect that group-I elements substituting on the Cd site, X_{Cd}^I create acceptors, whereas group-VII elements substituting on the Te site, Y_{Te}^{VII} creates donors. Generally speaking, to produce a shallow acceptor, it is advantageous to use a more electronegative dopant, whereas to produce a shallow donor, it is advantageous to use a less electronegative dopant.

For intrinsic defects, the situation is more complicated. Figure 13.2 shows the single-particle energy levels of tetrahedrally coordinated charge-neutral defects in CdTe [23]. Generally speaking, when a high-valence atom is replaced by a low-valence

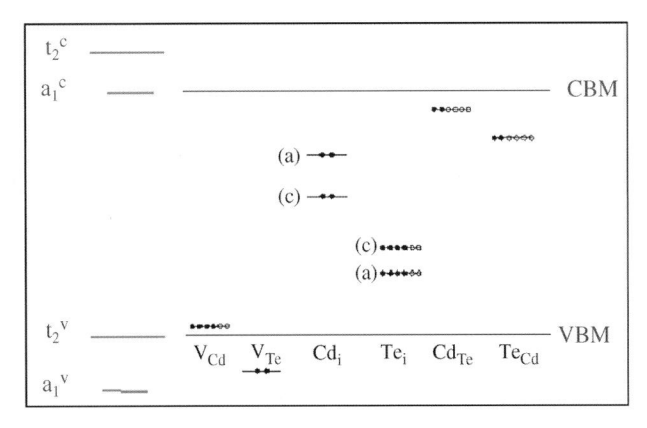

Figure 13.2 (online colour at: www.pss-b.com) Single-particle defect levels for the tetrahedrally coordinated neutral intrinsic defects in CdTe. The solid (open) dots indicate the state is occupied (unoccupied).

atom (e.g., Cd_{Te}) or by a vacancy V_{Cd} and V_{Te}, defect states are created from the host valence (v) band states that move upward in energy. The defect states consist of a low-lying singlet a_1^v state and a high-lying threefold-degenerate t_2^v state. Depending on the potential, both a_1^v and t_2^v can be above the VBM. These states are occupied by the nominal valence electrons of the defect plus the valence electrons contributed from the neighboring atoms (e.g., in CdTe, six electrons if the defect is surrounded by four Te atoms or two electrons if it is surrounded by four Cd atoms). For example, for charge-neutral V_{Cd}, the defect center has a total of $0 + 6 = 6$ electrons. Two of them will occupy the a_1^v state and the remaining four will occupy the t_2^v states just above the VBM, so V_{Cd} is an acceptor. On the other hand, if a low-valence atom is replaced by a high-valence atom (e.g., Te_{Cd}), or if a dopant goes to an interstitial site (e.g., Cd_i and Te_i), the a_1^v and t_2^v are pulled down and will remain inside the valence band. Instead, the defect states a_1^c and t_2^c are created from the host conduction band states that move down in energy. Depending on the potential, both the a_1^c and t_2^c states can be in the gap. For example, for charge-neutral Te_{Cd}, $6 + 6 = 12$ electrons are associated with this defect center. Eight of them will occupy the bonding a_1^v and t_2^v states, two will occupy the a_1^c state, and the remaining two will occupy the t_2^c state. Since the partially occupied t_2^c state is close to the CBM, Te_{Cd} is also a donor. For the interstitial defect, Cd_i has two electrons that will fully occupy the a_1^c state and is thus expected to be a donor. The Te_i defect center has six electrons. Two will occupy the a_1^c state and the remaining four will occupy the t_2^c states. Since the partially occupied t_2^c states are closer to the VBM, Te_i is expected to be a deep acceptor.

13.4
Origins of Doping Difficulty and the Doping Limit Rule

In general, there are three main factors that could cause the doping limit in a semiconductor material [10, 14, 19, 22, 23]: (i) the desirable dopants have limited solubility; (ii) the desirable dopants have sufficient solubility, but they produce deep levels, which are not ionized at working temperatures; and (iii) there is spontaneous formation of compensating defects. The first factor depends highly on the selected dopants and growth conditions. The second factor only depends on the selected dopants. Thus, these two factors can sometimes be suppressed by carefully selecting appropriate dopants and controlling the growth conditions. The third factor is an intrinsic problem for semiconductors; thus, it is the most difficult problem to overcome, especially for WBG semiconductors. This is because the formation energy of charged compensating defects depends linearly on the position of the Fermi level, E_F [see Eq. (13.1)]. When a semiconductor is doped, the Fermi level shifts, which can lead to spontaneous formation of the compensating charged defects. For example, when a semiconductor is doped p-type, E_F moves close to the VBM. In this case, the formation energy of the charged donor defects decreases because they will donate their electrons into the Fermi reservoir (Figure 13.3). In WBG semiconductors with low VBM, the formation energy decrease of donor defects can be so large that at some Fermi energy $E_F = \varepsilon_{pin}^{(p)}$ the formation energy of certain donor defect becomes zero,

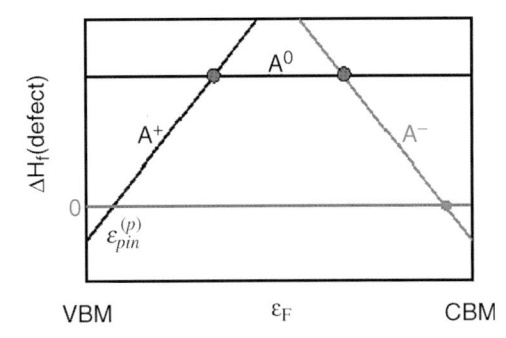

Figure 13.3 (online colour at: www.pss-b.com) Schematic plot of the dependence of the formation energy of charged defects on the Fermi energy position. The p-type pinning energy $\varepsilon_{pin}^{(p)}$ is the Fermi energy E_F at which the formation energy of donor A has zero formation energy.

i.e., it can form spontaneously, so further shift of the Fermi energy is not possible. Moreover, low VBM also leads to high ionization energy. Therefore, *a semiconductor with low VBM is difficult to be doped p-type*. The trend for n-type doping is similar, i.e., *a semiconductor with high CBM is difficult to be doped n-type*. This doping limit rule explains why a semiconductor with large band gap usually cannot be doped one type or even both types under equilibrium thermodynamic growth conditions. It also provides a general guideline about whether a material can be doped p- or n-type if we know the band alignment between different compounds. For example, Figure 13.4

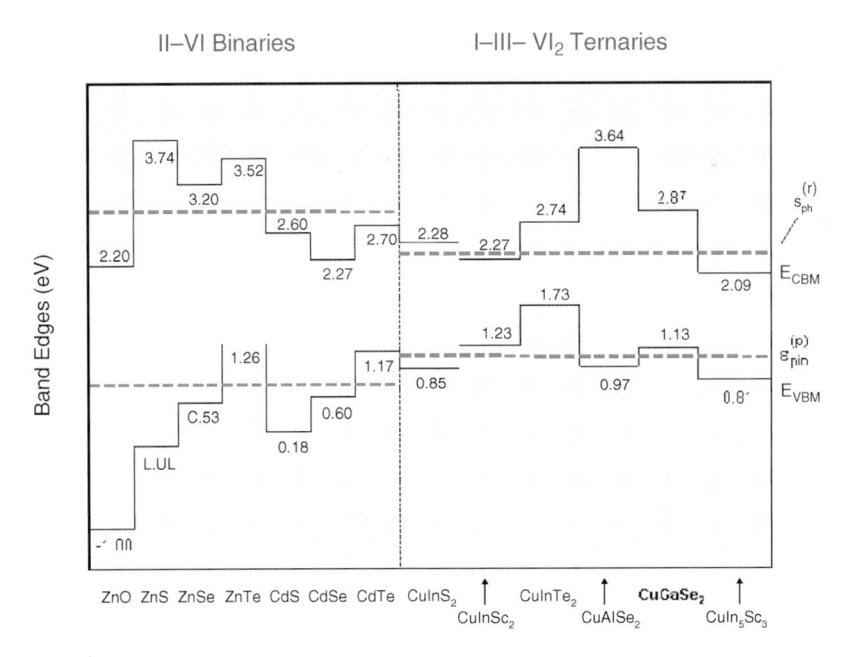

Figure 13.4 (online colour at: www.pss-b.com) Band alignment and n- and p-type pinning energy of II–VI and I–III–VI semiconductors (Ref. [10]).

shows the calculated band alignment for II–VI and I–III–VI semiconductors [10]. We see that ZnO has very low CBM and VBM, so it can be easily doped n-type, but not p-type. On the other hand, ZnTe with high VBM energy can be easily doped p-type, but not n-type. For I–III–VI compounds, CuInSe$_2$ can be doped both p- and n-type, but for CuGaSe$_2$, n-type doping will be difficult.

Based on the above understanding, we will search for corresponding solutions to overcome the doping limit. We will focus on the following approaches: (i) increase defect solubility by "defeating" bulk defect thermodynamics using non-equilibrium growth methods such as extending the achievable chemical potential through molecular doping or raising the host energy using surfactant; (ii) reduce defect ionization energy level by designing shallow dopants or dopant complexes; and (iii) reduce defect compensation and ionization level by modifying the host band structure near the band edges. As examples, we will discuss doping in some representative WBG semiconductors such as n-type doping in ZnTe and diamond and p-type doping in ZnO. The principles discussed here are general and are applicable to other WBG semiconductors.

13.5
Approaches to Overcome the Doping Limit

13.5.1
Optimization of Chemical Potentials

13.5.1.1 Chemical Potential of Host Elements

As Eq. (13.1) indicates, the formation energy of a defect, which determines the solubility of dopants, depends sensitively on the atomic chemical potentials of both the host elements and the dopants [14, 19]. Thus, optimization of the growth conditions and dopant source is critical to enhance the doping ability. So far, computational results and analysis have focused on the dependence of formation energies on host-element chemical potentials [27, 31]. For example, N substituting O (N_O) is expected to be a p-type dopant for ZnO. The formation energy of N_O depends on the chemical potentials of Zn, O, and N. Under thermal equilibrium growth conditions, there are some thermodynamic limits on the achievable values of the chemical potentials. First, to avoid precipitation of the elemental dopant and host elements, the chemical potentials are limited by

$$\mu_{Zn} \leq \mu(\text{Zn metal}) = 0, \tag{13.7}$$

$$\mu_O \leq \mu(O_2 \text{ gas}) = 0, \tag{13.8}$$

$$\mu_N \leq \mu(N_2 \text{ gas}) = 0. \tag{13.9}$$

Second, μ_{Zn} and μ_O are limited to the value of maintaining ZnO. Therefore,

$$\mu_{Zn} + \mu_O = \Delta H_f(\text{ZnO}). \tag{13.10}$$

13.5 Approaches to Overcome the Doping Limit

[Figure 13.5 diagram content:]

$\mu_O = -3.5$ μ_O → $\mu_O = 0$
$\mu_N = 0$

$\mu_N = -0.6$

$\mu_O < 0$
$\mu_{Zn} < 0$
$\mu_N < 0$

$\mu_{Zn} + \mu_O = \Delta H(ZnO) = -3.5$ eV

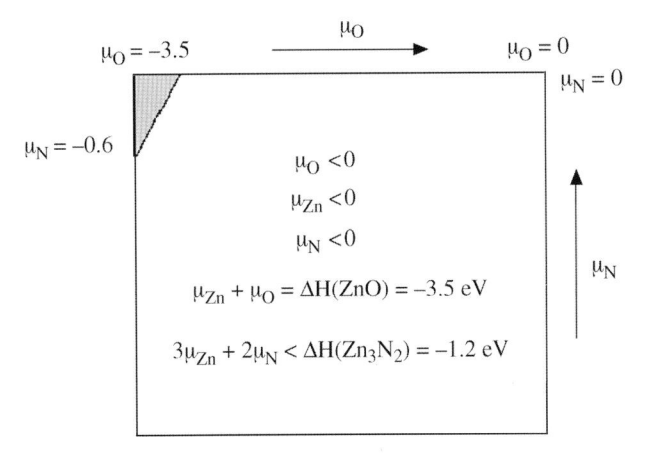

$3\mu_{Zn} + 2\mu_N < \Delta H(Zn_3N_2) = -1.2$ eV

μ_N

Figure 13.5 The achievable chemical potential region for N doped ZnO under equilibrium growth condition.

Here, $\Delta H_f(ZnO)$ is the formation energy of bulk ZnO. The calculated value is about -3.5 eV. Finally, to avoid the formation of the Zn_3N_2 secondary phase, μ_N is also limited by

$$3\mu_{ZN} + 2\mu_N \leq \Delta H_f(Zn_3N_2). \tag{13.11}$$

The calculated formation energy of Zn_3N_2 is about -1.2 eV. Using the equations above, the achievable chemical potential region is shown in Figure 13.5.

Figure 13.6 shows the calculated formation energies of charge neutral defects as a function of the O chemical potential. Here, μ_N is derived from N_2 gas. It is seen that the formation energy of N_O is lower at the O-poor condition, but higher at the O-rich

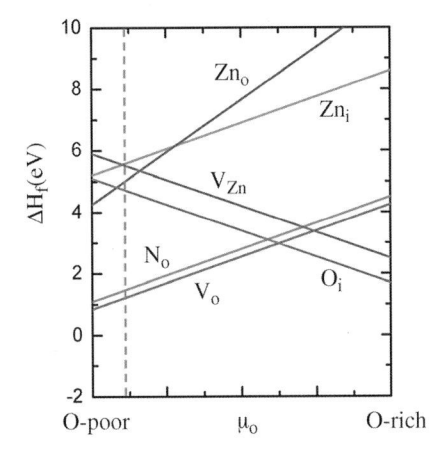

Figure 13.6 (online colour at: www.pss-b.com) Calculated formation energies of charge-neutral defects as a function of O chemical potential. The dashed line indicates the growth condition at which (left region) Zn_3N_2 will precipitate.

condition. Thus, to enhance the solubility of N, ZnO films should be synthesized at O-poor conditions. It should be noted that the formation energies of other intrinsic defects also depend on the growth conditions. At O-poor conditions, the formation energies for "acceptor-killer" defects, such as Zn interstitials (Zn_i) and O vacancies (V_O), are decreased. Thus, there is an intrinsic problem for enhancing the p-type doping using N as the dopant. In Section 13.5.4, we will discuss how this dilemma may be eliminated by selecting appropriate dopants.

The left region of the dashed line indicates the growth condition at which Zn_3N_2 will precipitate under equilibrium growth. This indicates that the O chemical potential should not go into this region so as to avoid the precipitation of Zn_3N_2. Thus, the precipitation of a secondary phase can limit the solubility of dopants. To overcome this, it has been shown that the precipitation of a secondary phase can be suppressed through epitaxial growth [32]. For example, the calculated thermodynamic solubility of N in bulk GaAs is only $[N] < 10^{14}$ cm^{-3} at $T = 650\,°C$ due to the formation of a fully relaxed, secondary GaN phase. However, single-phase epitaxial films grown at $T = 400\text{--}650\,°C$ with $[N]$ as high as $\sim 10\%$ have been reported. Zhang and Wei [32] found that if coherent surface strain is considered, the formation of the secondary GaN phase could be suppressed during epitaxial growth. As a result, the solubility of N can be enhanced significantly. A similar approach could be used to avoid forming Zn_3N_2 in ZnO:N.

Avoiding the formation of secondary phase can also lead to some unexpected consequences. For example, substituting Zn by Al for n-type doping in ZnO, the formation energy of Al_{Zn} depends on $(\mu_{Zn} - \mu_{Al})$, thus one may expect that the formation energy of Al_{Zn} should reach minimum under Zn-poor condition. However, to avoid the formation of Al_2O_3, we need to satisfy the following condition:

$$2\mu_{Al} + 3\mu_O \leq \Delta H_f(Al_2O_3). \tag{13.12}$$

Combine Eqs. (13.10) and (13.12), we have

$$(\mu_{Zn} - \mu_{Al}) \geq \frac{\mu_O + 2\Delta H_f(ZnO) - \Delta H_f(Al_2O_3)}{2}. \tag{13.13}$$

That is, in the achievable chemical potential region, Al_{Zn} has the lowest formation under O-poor or Zn-rich condition.

13.5.1.2 Chemical Potential of Dopant Sources

Although the dependence of doping efficiency on the host element's chemical potential has been studied extensively, the dependence of doping efficiency on *dopant* chemical potential has not attracted much attention because normally there are no significant alternative dopant sources. However, for the case of N doping of ZnO (or other *oxides*), there is a unique and unusual opportunity. There are at least four different gases, namely N_2, NO, NO_2, and N_2O that can be used as the dopant source. If these molecules arrive intact at the growing surface, their chemical potentials will determine the doping efficiency. We found that the N solubility can be enhanced significantly when metastable NO or NO_2 gases are used as the dopant sources [15]. A key feature that underlies our idea of doping with NO or NO_2 is that these molecules can supply single N atoms by breaking only the weak N–O bonds, whereas one has to break the strong N–N bonds to obtain desirable single-N defects when N_2 and N_2O are used. For example, when the NO, N_2O, or NO_2 molecules arrive *intact* at the growing surface, the formation energy of a charge-neutral N_O defect is given by

$$\Delta H_f(N_O, 0) = E(N_O, 0) - E(\text{host}) + 2\mu_O - \mu_{NO} \tag{13.14}$$

or

$$\Delta H_f(N_O, 0) = E(N_O, 0) - E(\text{host}) + 1.5\mu_O - \frac{\mu_{N_2O}}{2}, \tag{13.15}$$

$$\Delta H_f(N_O, 0) = E(N_O, 0) - E(\text{host}) + 3\mu_O - \mu_{NO_2}. \tag{13.16}$$

Our calculated formation energy is about 0.9, 0.2, and −0.2 eV for NO, N_2O, and NO_2 molecules, respectively. The NO and NO_2 molecules may simply be supplied as such or they may be produced by a reaction in the gas phase. For example, NO molecules can be created by $N_2O \Leftrightarrow NO + N$. In this case, $\mu_{NO} = \mu_{N_2O} - \mu_N$. Figure 13.7 shows the calculated formation energy of charge-neutral N_O for four different gases. The difference between N_2/N_2O and NO/NO_2 is very clear, i.e., the use of NO/NO_2 leads to significantly reduced formation energies for N_O because it does not entail any energy to break the N–N bonds. The negative formation energies of N_O at Zn-rich conditions indicate that NO or NO_2 molecules can be incorporated spontaneously to form N_O defects, if these molecules are intact before they are incorporated into ZnO.

However, in practical growth conditions, to avoid the precipitation of the secondary phases such as Zn_3N_2, besides satisfying

$$\mu_O + \mu_N = \mu_{NO} \quad \text{for NO gas as dopant,} \tag{13.17}$$

$$2\mu_O + \mu_N = \mu_{NO_2} \quad \text{for } NO_2 \text{ gas as dopant,} \tag{13.18}$$

$$\mu_O + 2\mu_N = \mu_{N_2O} \quad \text{for } N_2O \text{ gas as dopant,} \tag{13.19}$$

we also need to satisfy Eqs. (13.10) and (13.11), which set a low limit for the achievable O chemical potential. In Figure 13.7, the left, middle, and right dashed vertical lines indicate the low limits for O chemical potentials for N_2O, NO, and NO_2 molecules, respectively.

13.5.2
H-Assisted Doping

As we discussed above, the solubility of both the dopants and compensating defects depend sensitively on the position of the Fermi level. If we can control the Fermi level at a desirable position, then we may enhance the solubility of dopants and suppress the formation of compensating defects. Electron or hole injection could be a method to control the position of the Fermi level during film growth. Another popular

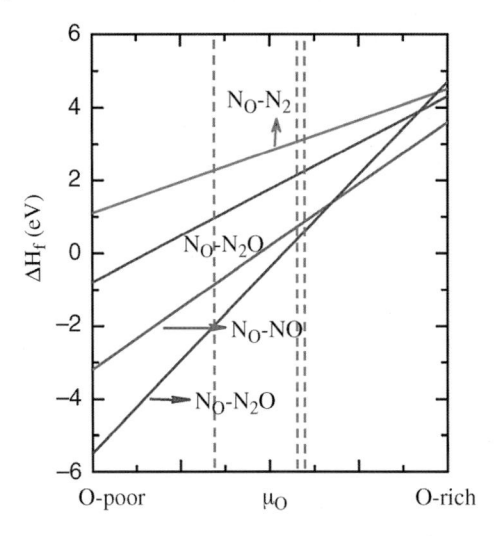

Figure 13.7 (online colour at: www.pss-b.com) Calculated formation energy of charge-neutral N_O with four different dopant sources: N_2, N_2O, NO, and NO_2. The left, middle, and right dashed lines indicate the low limits for achievable O chemical potentials for N_2O, NO, and NO_2 molecule doping.

approach is passivating the dopants by H atoms. For example, in Mg-doped GaN, the introduction of H can prevent such a shift. As a result, the concentration of Mg can be enhanced [34]. After film growth, H can be annealed out to achieve p-type conductivity.

For N-doping in ZnO (also other oxides), the co-existence of H can, in addition to preventing the Fermi level shift, directly passivate N dopants, forming a molecular NH complex on O site [$(NH)_O$]. The binding energy for $(NH)_O$ is 2.9 eV. The $(NH)_O$ complexes electronically mimic O atoms and cause smaller lattice distortion than N_O. Thus, the concentration of $(NH)_O$ in ZnO can be much higher than N_O [14]. Figure 13.8 shows the calculated formation energy for $(NH)_O$ as a function of O chemical potential. For comparison, the formation energies of N_O and some hole-killer defects are also shown. It is seen that the formation energy of $(NH)_O$ is lower than any other defects in the O-poor condition. In addition, the existence of H also pins the Fermi energy level, so the formation of compensating defects enhanced by a shifted Fermi level is also suppressed. Therefore, p-type doping could be achieved after subsequently driving out the hydrogen atoms from the sample by thermal annealing.

13.5.3
Surfactant Enhanced Doping

To lower the defect formation energy, which is the total energy difference between the final doped state and the initial state, we can either increase the initial dopant energy, as discussed in the previous section, or increase the energy of the host. Recently, we have shown that this can be done by introducing an appropriate surfactant during

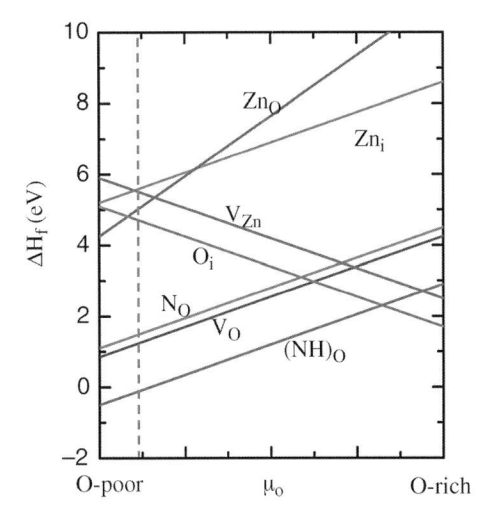

Figure 13.8 (online colour at: www.pss-b.com) Calculated formation energy of $(NH)_O$ in ZnO.

epitaxial growth [35]. The general concept for enhancing dopant solubility via epitaxial surfactant growth is schematically described in Figure 13.9. It is known experimentally that the surfactants in epitaxial growth float on the top surface of the growth front. The enhancement of dopant solubility initiates in the sublayers below the surface. For p-type doping, dopants introduce acceptor levels with holes near the VBM of the host system. On the other hand the surfactants on the growth surface will introduce surfactant levels. If the surfactant levels are higher in energy than the acceptor levels and have electrons available, the surfactants will donate the electrons to the acceptors and consequently leads to a Coulomb binding between the surfactant and the dopant. Such charge transfer reduces the energy of the system and consequently leads to effective reduction on the formation energy of dopant incorporation in the host. The formation energy reduction is large if the energy level

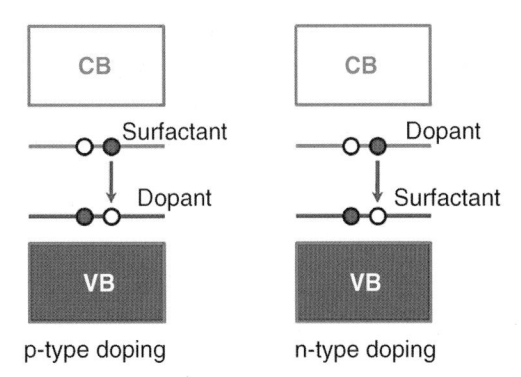

Figure 13.9 (online colour at: www.pss-b.com) Schematic plot of the mechanism of surfactant enhanced doping during epitaxial growth.

difference between the surfactant and acceptor levels is large. The same principle holds for n-type doping, except that the charge transfer is from dopant level to empty surfactant level. In this case, the positions of dopant levels are close to the CBM and the surfactant levels must be lower in energy than the dopant levels. Thus, the key for this concept is how to ensure that the surfactant levels are higher (lower) in energy than the dopant levels and there are indeed electrons (holes) in the surfactant levels in p-type (n-type) doping. We have calculated the formation energy for substitutional Ag_{Zn} in the sublayer of ZnO $(000\bar{1})$ surface and found that the formation energy is lowered by 2.3 eV with S as surfactant as compared to that without surfactant.

13.5.4
Appropriate Selection of Dopants

There are two general rules for choosing an appropriate dopant to produce shallow defect levels. First, an appropriate dopant should favor the growth conditions that will suppress the formation of compensating defects. As we discussed in Section 13.2, the solubility of dopants and concentration of intrinsic defects depend sensitively on the growth conditions. It is highly desirable to have a growth condition that enhances the dopant solubility and suppresses the formation of intrinsic compensating defects. This may be achieved by identifying the compensating defects and choosing suitable dopants. For example, the major defects that compensate acceptors in ZnO are Zn_i and V_O. Figure 13.6 shows that to suppress the formation of these hole-killer defects, an O-rich growth condition is preferred. Of course, this condition is not preferred for N incorporation. However, such a growth condition is preferred for doping at cation sites.

Second, dopants at cation sites in compound semiconductors generally produce shallower acceptor levels than dopants at anion sites. This is because for most cation–anion compound semiconductors, the valance bands are derived mainly from the anion atoms. Dopant substituting at cation site would, in general, cause smaller perturbation than dopants at anion sites on the anion-derived valance band. Thus, theoretical studies have found that Group-I elements such as Li and Na have low acceptor levels, whereas Group-V elements such as N, P, As, and Sb have deep acceptor levels in ZnO [33]. The large ionization energy for Group-V acceptors in ZnO can be understood as follows: The acceptor level, especially the shallow acceptor level, has a wave-function character similar to that of the VBM state, which consists mostly of anion p, and small amounts of cation p and cation d orbitals. Therefore, to have a shallow acceptor level, the dopant should be as electronegative as possible, that is, it should have low p orbital energy. For example, because the atomic p orbital energy level of N is the lowest (Figure 13.10), i.e., most electronegative, among the Group-V elements, N has been the preferred acceptor dopant for II–VI semiconductors because it produces the lowest acceptor levels compared to the other Group-V dopants. However, due to the low VBM of the oxides, the level of N_O in ZnO is still relatively deep [19, 33] at about 0.4 eV above the VBM, making acceptor ionization difficult. The other Group-V elements are less electronegative than the N atom; therefore, they have much larger ionization energy than N_O [33]. This explains why it

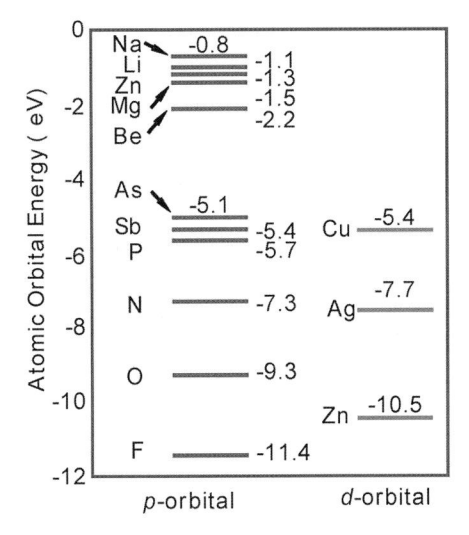

Figure 13.10 (online colour at: www.pss-b.com) LDA-calculated valence p and d energy levels of neutral atom to show the general chemical trends.

is difficult to achieve anion site shallow acceptors for the oxides. Recent experiments, however, have demonstrated good p-type conductivity for As, as well as p-doped ZnO. However, theoretical studies revealed that As impurities actually occupy Zn antisites, forming $As_{Zn} + 2V_{Zn}$ complexes [36]. The real contribution to the p-type conductivity is from V_{Zn}. The ionization energy is reduced due to the interaction between V_{Zn} and As_{Zn}.

According to the above discussion, Group-Ia (Li, Na) and Group-Ib (Cu, Ag) elements may be better choices for producing p-type ZnO. So far, only doping with Group-Ia elements has been studied extensively. There are very few studies on doping of ZnO with Group-Ib elements. However, very few p-type ZnO films have been achieved using Group-Ia elements as dopant. Theoretical studies have revealed the possible reasons for the difficulty. Substitutional Group-Ia elements (Li and Na) at T_d sites are indeed shallow acceptors [33]. However, when the Fermi energy is close to the VBM, Group-Ia elements prefer to occupy the interstitial sites in ZnO, which are electron donors. As a result, Group-Ia elements fail to dope ZnO p-type. The reason why Li and Na prefer the interstitial sites rather than substitutional sites is largely due to the low ionization energies of the valence s electron and large size mismatch of ions of the Group-Ia elements. Such mismatches are much less for Group-Ib elements. Thus, Group-Ib elements may be better candidates than Group-Ia elements for p-type ZnO doping. Therefore, we have studied the doping effect with Group-Ib elements in ZnO.

Our electronic structure calculations have revealed that Cu, Ag, or Au occupying a Zn site creates a single-acceptor state above the VBM of ZnO. Our calculated GGA transition energies $\varepsilon(0/-)$ are at about 0.7, 0.4, and 0.5 eV above the VBM for Cu_{Zn}, Ag_{Zn}, and Au_{Zn}, respectively [20]. These results indicate that (i) the acceptor level created by Ag_{Zn} is shallower than the acceptor levels created by Cu and Au and (ii) the transition energies for the substitutional Group-Ib elements are much deeper than that of the substitutional Group-Ia elements. The reason for (ii) can be understood as the following: The substitutional elements induced acceptor level is derived mostly from the VBM state, which has the anion p and cation d characters. For Group-Ib elements, their occupied d orbital energies are near the oxygen p level. Because both the O, p and the Group-Ib d orbitals have the same t_2 symmetry in the tetrahedral environment, there is strong p–d repulsion between the two levels, pushing the acceptor levels higher. On the other hand, Group-Ia elements have no active valence d orbitals, so their defect levels are shallower than the Group-Ib substitutional defects. Among the three Group-Ib elements, Ag has the largest size and lowest atomic d orbital energy, so the p–d repulsion is the weakest. This explains why Ag_{Zn} has the lowest transition energy level among the three Group-Ib elements.

As we discussed above, although the Group-Ia substitutional acceptor levels are shallower, the Group-Ia elements prefer to occupy interstitial sites in p-type ZnO samples, forming shallow donors. In this case, p-type ZnO cannot be realized due to strong self-compensation. Thus, we have also calculated the formation energy of the Group-Ib dopants at interstitial sites. Figure 13.11 shows the defect formation energies as a function of the Fermi level calculated under the oxygen-rich condition for Group-Ib elements at different sites. The solid dots indicate the transition energy level positions for substitutional Cu, Ag, and Au. We find that the self-compensation is very small with Group-Ib elements in ZnO. This is because Group-Ib elements do not prefer to occupy the interstitial sites, even when the Fermi level is close to the

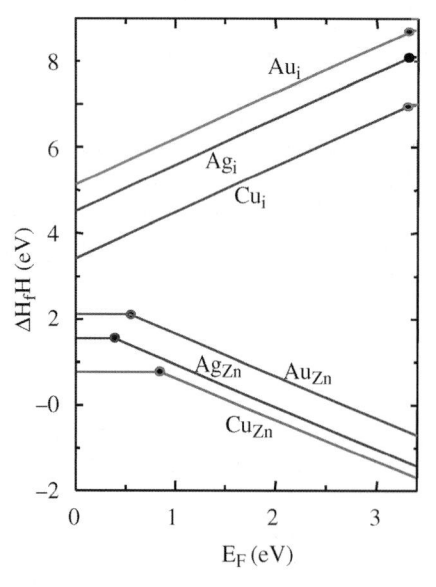

Figure 13.11 (online colour at: www.pss-b.com) Calculated formation energies as a function of Fermi level for Group-Ib elements in ZnO.

VBM. Our calculations reveal that Group-Ib elements may be better candidates than the Group-Ia elements for p-type doping of ZnO.

Our calculations revealed that the acceptor levels created by Group-Ib elements are not very shallow. However, because the formation energies of the substitutional Group-Ib elements in ZnO are very low at the O-rich conditions, a high concentration of dopants can be easily achieved. At this growth condition, the compensation by intrinsic donor defects can be effectively suppressed. Therefore, p-type doping in ZnO could still be achieved with these elements, especially for Ag doping. It is important to point out that the calculated $(0/-)$ transition energy level for Ag_{Zn} is comparable to the calculated $(0/-)$ transition energy for N_O, which is currently one of the most favorable dopants for p-type doping in ZnO. For N doping, ZnO thin films should be grown under an O-poor condition to incorporate N efficiently at O sites. This growth condition also promotes the formation of hole-killer defects, such as O vacancies and Zn and N interstitials. On the other hand, for incorporating Ag at Zn sites, the growth should be done at an O-rich condition, which suppresses the formation of major hole-killer defects. In addition, self-compensation can also be avoided for Ag doping. Therefore, our results suggest that Ag may be a better dopant than N for p-type doping in ZnO, especially when it is combined with passivating donors to form defect complexes (see Section 13.5.5 below). Our conclusion is supported by recent experiment results on p-type ZnO thin films with Ag and Cu dopants [37].

13.5.5
Reduction of Transition Energy Levels

To reduce the acceptor transition energy level in ZnO, co-doping or cluster doping has been proposed [38]. In conventional co-doping, two single acceptors (e.g., N_O) are combined with a single donor (e.g., Ga_{Zn}) to form an acceptor defect complex. It is expected that through donor–acceptor level repulsion, shallow acceptor levels can be created. However, detailed theoretical analyses show that for direct-band gap semiconductors such as ZnO, the reduction of this type of conventional co-doping on the ionization energy is rather small. This is because the donor and acceptor levels usually have different symmetries and wave-function characters: the donor state has the s-like a_1 character, whereas the acceptor has the p-like t_2 character. Furthermore, in the case of two acceptors plus one donor (e.g., $2N_O + Ga_{Zn}$), because the two acceptors are forced to be fcc (or hcp in the wurtzite structure) nearest neighbors, the acceptor–acceptor level repulsion can even raise the ionization energy [23].

To avoid the problem discussed above, we have explored a different and novel idea in which a fully occupied deep donor is used to attract a second partially occupied donor to lower its ionization energy [16]. In particular, we studied a double donor (either Si, Ge, or Sn on the Zn site) paired with a single donor (either F, Cl, Br, or I on the Te site) in ZnTe. Different from the Coulomb binding that exists in charged donor–acceptor complexes in the co-doping approach the binding between the two donors results from the level repulsion between the two donor states (Figure 13.12). The level repulsion significantly reduces the energy of the fully occupied lower level,

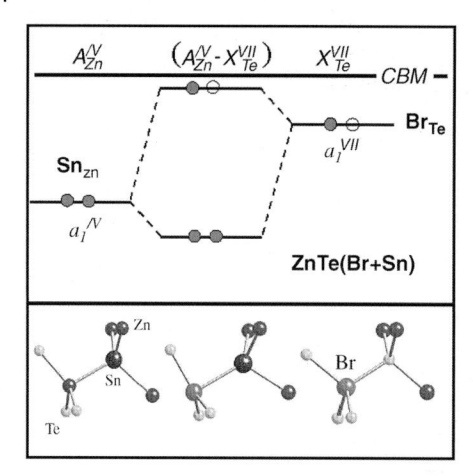

Figure 13.12 (online colour at: www.pss-b.com) Illustration of the interaction between the single donor and the double donor states associated with Br_{Te} and Sn_{Zn} in the formation of complexes Br_{Te}–Sn_{Zn} in ZnTe.

stabilizing the donor–donor pair, while it increases the energy of the partially occupied upper level, thus reducing the ionization energy. Notice that because the doubly occupied a_1^{IV}-derived state is *charge neutral*, there is no Coulomb repulsion between the two nominally donor impurities. Furthermore, because the two donor states have the same symmetry and atomic character the level repulsion is very efficient. For example, we find that the formation of a Br_{Te}–Sn_{Zn} pair in ZnTe is exothermic with a binding energy of 0.9 eV. It lowers the electron ionization energy of Br_{Te} by a factor of more than three from 240 to 70 meV, resulting in an effective shallow donor. Similar idea has also been proposed to enhance n-type doping in diamond [17]. Recently Kim and Park [18] have also suggested that the same idea can be applied to explain oxygen vacancy assisted n-type doping in ZnO by forming Zn_i–V_O paire to lower the formation energy and transition energy levels of Zn_i in ZnO.

We have also proposed two approaches to reduce the ionization energy in p-type doping of ZnO [21]. The proposals are based on the following considerations: (i) as discussed in the previous section, to lower the ionization level, one should find a dopant with low valence p orbital energy (more electronegative), preferably at the anion site. Because the wave function of the V_{Zn} has a large distribution on the neighboring O atomic sites (Figure 13.13a), replacing one of the neighboring O atoms by the more electronegative F (the F 2p level is 2.1 eV lower in energy than the O 2p level, see Figure 13.10) is expected to reduce the energy level of V_{Zn}. The binding energy between the F_O single donor and the V_{Zn} double acceptor is also expected to be large. Furthermore, this defect complex pair $V_{Zn} + F_O$ contains only one acceptor, so there will be no acceptor–acceptor repulsion to raise the ionization level; and (ii) we notice that one of the reasons that the N_O defect level is deep in ZnO is because the N 2p level strongly couples to the nearest-neighbor Zn 3d orbitals (Figure 13.13c),

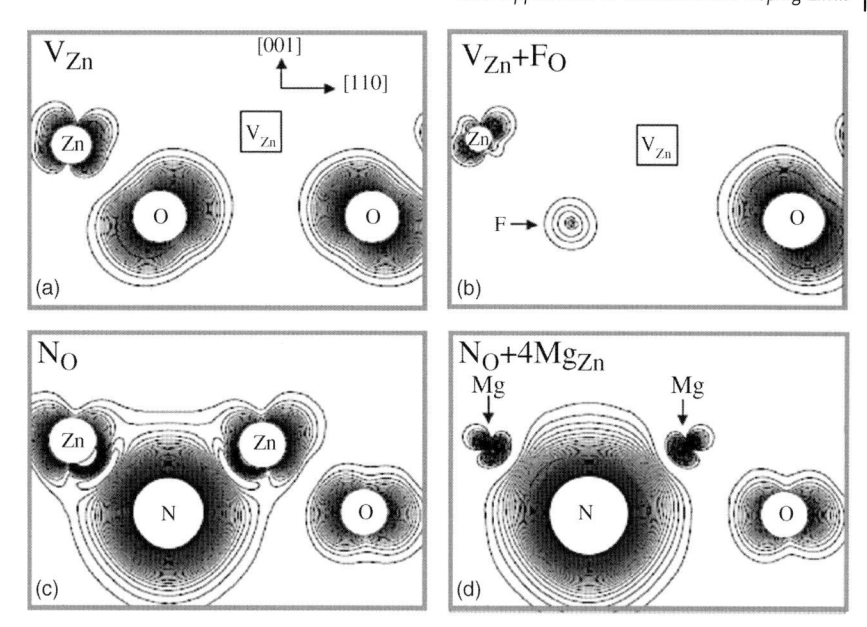

Figure 13.13 (online colour at: www.pss-b.com) Charge density plot of defect levels in ZnO: (a) V_{Zn}, (b) $V_{Zn} + F_O$, (c) N_O, and (d) $N_O + 4Mg_{Zn}$.

which both have t_2 symmetry in this tetrahedron environment. If we can replace the Zn atom by an isovalent Mg atom that has a similar atomic size as Zn but no occupied d orbital, the defect transition energy level of $N_O + nMg_{Zn}$ should be lower than that of N_O in ZnO. The effect should be most efficient for $n = 4$, when the tetrahedral environment around N_O is preserved and no level splitting occurs.

Figure 13.13b shows the charge density plot of the $V_{Zn} + F_O$ defect level. When F is introduced, it creates defect levels inside the valence band, removing one of the oxygen dangling-bond contributions to the acceptor level and making the transition energy lower. The calculated $(0/-)$ transition energy level of $V_{Zn} + F_O$ is 0.16 eV, which is much smaller than the corresponding $(-/2-)$ transition energy level of V_{Zn} at 0.34 eV. It is also lower in energy than the $(0/-)$ transition energy level of V_{Zn} at 0.18 eV. The calculated $V_{Zn} + F_O$ binding energy is -2.3 eV, indicating that the defect pair is very stable with respect to the isolated defects. This large binding energy can be understood by noticing that to form the defect complex, F_O donates one of its electrons to V_{Zn}, which results in a large Coulomb interaction between V_{Zn}^- and F_O^+. Based on this study, we believe that adding a small amount of F in ZnO to form a $V_{Zn} + F_O$ defect pair is beneficial to p-type doping in ZnO. However, we also want to point out that F_O itself is a donor, so too much F_O (more than the amount of V_{Zn}) in the sample can over compensate the acceptors.

Figure 13.13d shows the defect level charge density of $N_O + 4Mg_{Zn}$. Compared to $N_O + 4Zn_{Zn}$, we see that the cation d character is removed and the defect level is more localized on the N atomic site. The calculated $(0/-)$ transition energies are 0.29 eV for $N_O + Mg_{Zn}$ and 0.23 eV for $N_O + 4Mg_{Zn}$ eV, shallower than that for N_O.

However, the calculated binding energy for $N_O + Mg_{Zn}$ is positive at 0.3 eV, indicating that N does not like to bind with Mg in ZnO. This is because the N–Zn bond is stronger than the N–Mg bond. Our calculations show that both N–Zn and Mg–O bonds are shorter than the Zn–O bond, but the N–Mg bond length is longer than the Zn–O bond length. However, for ZnMgO alloys with relatively high Mg concentrations, the opportunity to form $N_O + nMg_{Zn}$ is reasonably high due to the entropy contribution. Furthermore, the VBM of the ZnMgO alloys is similar to that of ZnO, because the wave function is more localized on the ZnO region. This may explain why some ZnMgO alloys can be doped p-type [39]. Further lowering of the acceptor transition energy level is expected if we replace Mg by Be, because the Be 2p orbital energy is much lower than the 3p orbital of Mg (Figure 13.10). Indeed, we find that the (0/−) transition energy levels of $N_O + Be_{Zn}$ and $N_O + 4Be_{Zn}$ are at 0.22 and 0.12 eV, respectively.

Other successful co-doping schemes include the one demonstrated by Limpijumnong et al. [36] who show that As_{Zn}–$2V_{Zn}$ in ZnO creates relatively shallow acceptor levels. In this complex, the shallow acceptor level is realized because the two V_{Zn} acceptors are connected by the As_{Zn} (or P_{Zn}) antisite donor through a cation sublattice; so the separation between the two V_{Zn} is large and the level repulsion between them is weak.

13.5.6
Universal Approaches Through Impurity-Band Doping

We recently proposed a universal approach to overcome the long-standing doping polarity problem for WBG semiconductors [22]. The approach is to reduce the ionization energies of dopants and the spontaneous compensation from intrinsic defects by creating a passivated impurity band, which can be achieved by introducing passivated donor–acceptor complexes or isovalent impurities. In this case, the ionization energy is reduced by shifting the band edge through the impurity band, which is higher than the VBM or lower than the CBM, rather than through the shifting of defect energy levels. When the same element is used to create the impurity band and as dopant, the ionization energy is always small. Furthermore, due to a smaller Fermi level shift, charge compensation is also reduced. Our density-functional theory calculations demonstrate that this approach provides excellent explanations for the available experimental data of n-type doping of diamond and p-type doping of ZnO, which could not be understood by previous theories. In principle, this universal approach can be applied to any WBG semiconductors, and therefore, it will open a broad vista for the use of these materials. Our concept agrees well with the observation by Kalish et al. [40], who suggested that impurity bands could play a role in co-doped diamond.

We first demonstrate our approach for n-type doping in diamond. It is known that n-type doping of diamond is extremely difficult because the donor levels are usually 0.6 eV or deeper below the CBM for most dopants such as N and P [41, 42]. Some n-type diamonds have been reported by using N and P as dopants and the mechanism has been studied theoretically. However, the most exciting n-type doping of diamond

13.5 Approaches to Overcome the Doping Limit

in the last few years is the co-doping of B with deuterium. It is reported that through this co-doping, n-type diamond has been realized with an activation energy of about 0.2–0.3 eV [3].

We now explain how and why our new concept can explain the experimental results of n-type doping by deuteration of B-doped diamonds. It is reported that the deuteration of B-doped diamond undergoes two clear steps: (i) the passivation of B acceptors by deuterium and (ii) the excess deuterium doping that leads to the formation of shallow donors. The experiments suggest strongly that (B, D) complexes are responsible for the shallow donors; here, D indicates deuterium. In our calculation, we use H for deuterium. Our calculation shows that the ionization energy level for an isolated H in diamond is about 2.8 eV below the CBM, which is consistent with the calculated results reported by others [43]. Isolated B + 2H complexes in diamond have also been found theoretically to be deep donors [44]. Our calculations reveal that the passivated (B + H) complexes generate fully unoccupied impurity bands, which lie about 1.0 eV below the host CBM. An isolated H atom in diamond has two low-energy sites: bond center (C–H–C) or anti-bond (C–C–H) sites. When B atoms are available in diamond, H atoms preferentially bond to B atoms, because in their mutual presence, B atoms are negatively charged and H atoms are positively charged. The energy of the bond-center configuration is lower than the anti-bond configuration because an H^+ ion prefers to sit at a high electron-density site. Figure 13.14 shows the calculated total density of states (DOS) for pure diamond host (green curve) and a supercell containing a (B + H) complex (red curve), with the B–H–C configuration. It reveals clearly that the formation of a passivated (B + H) complex does not change the basic electronic structure, but only generates an unoccupied impurity band below the CBM. Our results, therefore,

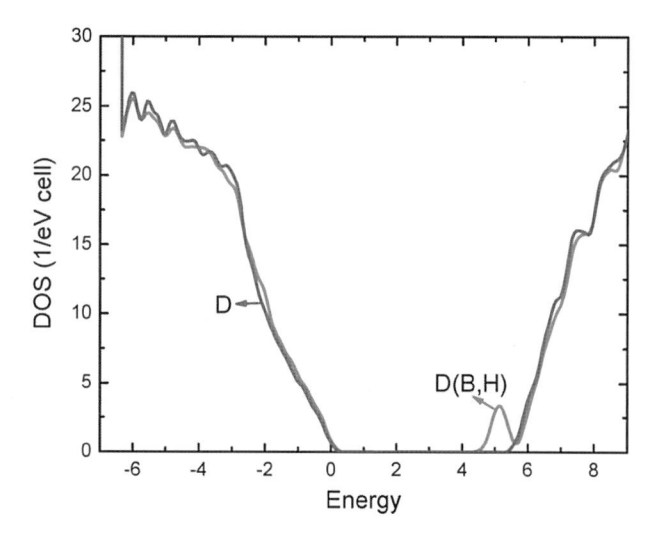

Figure 13.14 (online colour at: www.pss-b.com) Calculated DOS for pure diamond host (green curve) and a supercell containing a (B + H) complex (red curve), with the B–H–C configuration.

suggest that the first step of the deuteration of B-doped diamonds is to passivate the B acceptors, and create the fully unoccupied impurity bands below the CBM.

When excess deuterium/H atoms are available after the first step, they will start to dope the passivated system, i.e., they effectively dope the new host with the unoccupied impurity band, rather than the original conduction band. Thus, in calculating the ionization energy, the term $\varepsilon_{CBM}^{\Gamma}$(host) in Eq. (13.4) should now be replaced by the impurity-band minimum (IBM), $\varepsilon_{IBM}^{\Gamma}$. In other words, the transition now occurs between the H defect levels and the unoccupied impurity bands, rather than the original conduction bands. As a result, the transition energy can be reduced dramatically.

For H doping in the (B + H)-passivated diamonds, the excess H atoms bind to the (B + H) complexes, forming (H–B–H) triplets. For charge-neutral H atoms, the lowest energy configuration is shown in Figure 13.15a, where the excess H is at the B antibonding site. We call this configuration (H–B–H)–AB. When the excess H atom is positively charged ($q = +1$), the fully relaxed structure is shown in Figure 13.15b. We see in Figure 13.15b that the H^+ ion at the antibonding site becomes energetically unstable, and it moves to a bond-center site with high electron density to lower the Coulomb energy. This atomic displacement results in significant bond rearrangements and a large energy lowering of the charged defect (-1.8 eV), which leads to significant reduction of the ionization energy [see Eq. (13.4)]. The calculated $\varepsilon(0/+)$ transition energy level is 0.3 eV below the unoccupied impurity-band edge. We also studied a metastable (H–B–H)–BC triplet defect, where both H atoms are at the puckered B–C bond-center sites. The atomic configurations for neutral and charged defect complexes are shown in Figure 13.15c and d, respectively. This configuration is about 0.6 eV higher in energy than the (H–B–H)–AB complex due to strong H^+–H^+ Coulomb repulsion; but the calculated transition energy level is 0.2 eV, which is 0.1 eV lower than that for the (H–B–H)–AB complex due to less crystal-field splitting.

The calculated transition energies agree very well with the experimentally measured ionization energies, suggesting that the second step of deuteration of B-doped diamond is to effectively dope the (B + H) impurity bands. This new concept, therefore, explains why (B, H) co-doping can create shallow donors in diamonds. It should be noted that to form the impurity bands and have reasonable transport properties, a critical concentration threshold is needed. Furthermore, the edge of the impurity band depends on the concentration of B atoms. The higher B concentration results in a more-broadened (B + H) impurity band. Consequently, the ionization energy will be reduced. This explains another experimental observation, i.e., diamonds with a higher B concentration exhibit shallower donor levels.

Our approach can also be applied to explain p-type doping of ZnO. As discussed above, p-type doping of ZnO is difficult. However, Ga and N co-doping has produced good p-type ZnO [8, 9]. The doping mechanism is not well understood. Most reliable theoretical calculations predicted that the ionization energy for N acceptors in ZnO is about 0.4 ± 0.1 eV above the VBM [21, 33, 42]. But the experimentally measured N acceptor ionization energy in p-type ZnO is much shallower, only 0.1–0.2 eV above the VBM [6, 7]. The conventional co-doping concept cannot explain the discrepancy

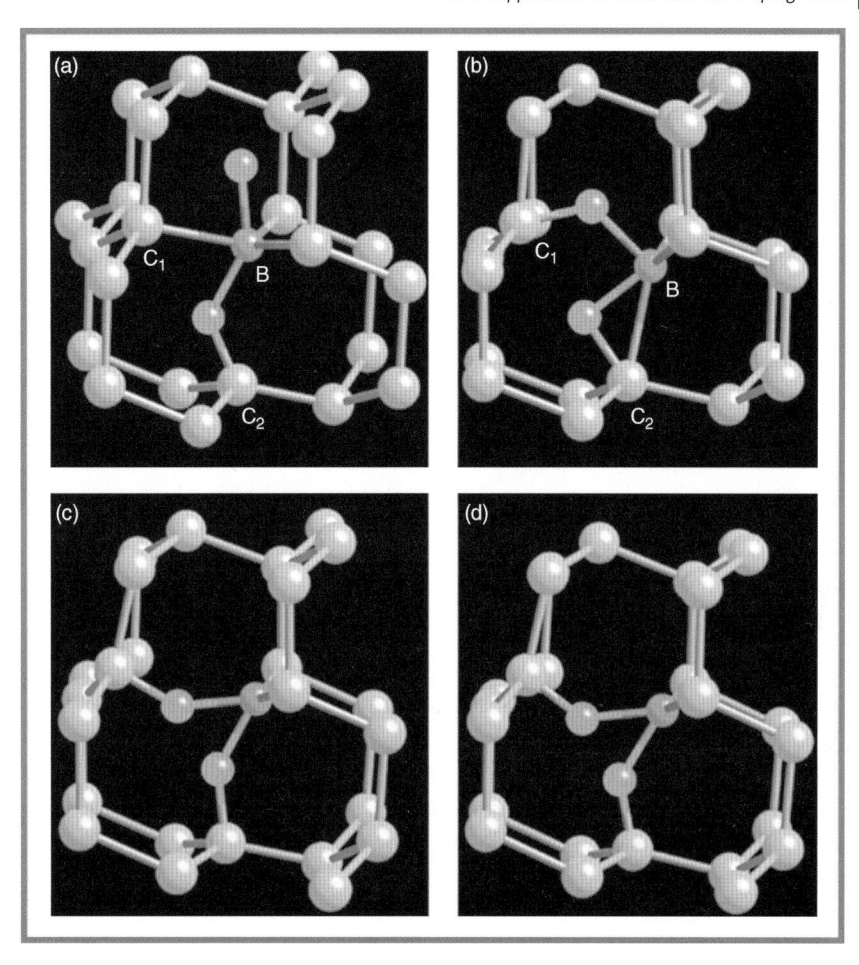

Figure 13.15 (online colour at: www.pss-b.com) Relaxed structures for B + 2H complexes in diamond with charge-neutral and +1 charged states. The blue balls are C atoms. The red balls are B atoms. The green balls are H atoms. (a) Neutral state for complex (H–B–H)–AB, (b) +1 charged state for complex (H–B–H)–AB, (c) neutral state for complex (H–B–H)–BC, and (d) +1 charged state for complex (H–B–H)–BC.

because the calculated ionization level of an isolated Ga + 2N impurity is still deep, at about 0.4 eV.

Here, we show that to successfully use Ga and N co-doping to obtain p-type ZnO, the first step is to form passivated stoichiometric (Ga + N) complexes, and create a fully occupied impurity band above the VBM of ZnO. Ga and N bind together strongly in ZnO because they passivate each other. Figure 13.16 shows the calculated total DOS for pure ZnO host (blue curve) and a system containing a (Ga + N) complex (red curve). It reveals clearly that the formation of a passivated (Ga + N) complex does not change the basic electronic structure, but only generates an additional fully occupied band above the VBM. When excess N atoms are available,

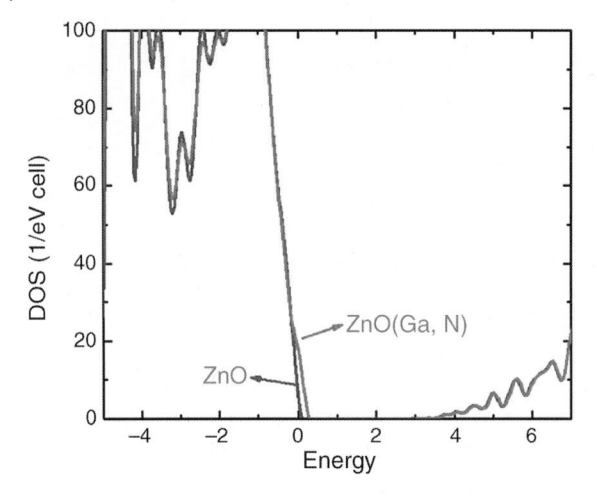

Figure 13.16 (online colour at: www.pss-b.com) Calculated DOS for pure ZnO (green curve) and a supercell containing a (Ga, N) complex.

they will dope the passivated system. The transition will occur between the N defect levels and the fully occupied impurity bands, rather than the original valence bands. Thus, the term $\varepsilon_{VBM}^{\Gamma}$(host) in Eq. (13.3) should now be replaced by the impurity-band maximum, $\varepsilon_{IBM}^{\Gamma}$.

Previous calculations suggested that for the Ga + 2N complexes, the first N occupies the first nearest-neighboring O site of the Ga, which occupies a Zn site [34]. The second N occupies the second nearest-neighboring O site. This N atom does not bind directly to the Ga atom. We call this configuration (N–Ga–N)–A. However, our calculations reveal that the excess N atoms bind to the (Ga + N) sites, forming

Figure 13.17 (online colour at: www.pss-b.com) Relaxed structures for (a) configuration A and (b) configuration B.

a (N–Ga–N)–B complex with both N atoms occupying the first nearest-neighboring O sites of the Ga atom. The relaxed structures for A and B configurations are shown in Figure 13.17a and b, respectively. The B configuration is about 0.5 eV lower in energy than the A configuration. We have calculated the acceptor ionization energies for both configurations, considering effective doping of the passivated (Ga + N) impurity bands. The calculated ionization energies are 0.2 and 0.1 eV for configurations A and B, respectively. Our results, therefore, are able to explain the puzzling experimentally measured ionization energies for N acceptors. Again, we want to point out that to form (Ga + N) impurity bands and have reasonable transport properties, critical Ga and N concentrations are needed. The transition energy is also expected to be reduced in the ZnO with higher Ga concentration.

With our approach, we are able to explain experimentally observed B and H co-doped n-type diamonds and Ga and N co-doped p-type ZnO, which could not be understood by previous theories. The physical principle behind this new concept is clear; that is, we can first create a fully passivated impurity band and then dope the impurity band. This approach can be applied, in principle, to any WBG semiconductors to overcome the doping polarity problems found in these materials. It should be pointed out that to be successful, the concentration of the defects inducing the impurity band must exceed a certain percolation limit, so that reasonable transport properties can be achieved. The small band gap reduction caused by forming an impurity band can also be easily adjusted by alloying with other elements. For example, adding a small amount of Mg or Be in ZnO can easily open the band gap without changing the doping property [45, 46].

13.6
Summary

We have reviewed three main origins for the doping limit in WBG semiconductors, i.e., (i) low dopant solubility; (ii) deep ionization energy levels; and (iii) spontaneous formation of compensating defects. We have also proposed solutions to overcome the doping bottlenecks, which include (i) increase defect solubility by defeating bulk defect thermodynamics using non-equilibrium growth methods such as extending the achievable chemical potential through molecular doping or rasing the host energy using surfactant; (ii) reduce defect ionization energy level by designing shallow dopants or dopant complexes; and (iii) reduce defect compensation and ionization level by modifying the host band structure near the band edges. The issues related to the defect calculations are discussed. We believed that the band gap correction is an important issue in computational defect physics and more studies are needed.

Acknowledgement

We would like to thank A. Janotti, M.M. Al-Jassim, J. Li, S.-S. Li, S. Limpijumnong, C.-H. Park, D. Segev, J.-B. Xia, L. Zhang, and S.B. Zhang for their contributions and helpful discussions in this work. This work is supported by the U.S. Department of Energy under Contract No. DE-AC36-08GO28308.

References

1 Koizumi, S., Watanabe, K., Hasegawa, M., and Kanda, H. (2001) *Science*, **292**, 1899.
2 Isberg, J., Hammersberg, J., Johansson, E., Wikstrom T., Twitchen D.J., Whitehead, A.J., Coe S.E., and Scarsbrook, G.A. (2002) *Science*, **297**, 1670.
3 Teukam, Z., Chevallier, J., Saguy C., Kalish, R., Ballutaud, D., Barbé, M., Jomard, F., Tromson-Carli, A., Cytermann, C., Butler, J.E., Bernard, M., Baron, C., and Deneuville, A. (2003) *Nature Mater.*, **2**, 482.
4 Taniyasu, Y., Kasu, M., and Makimoto, T. (2006) *Nature*, **441**, 325.
5 Huang, M.H., Mao, S., Feick, H., Yan, H., Wu, Y., Kind, H., Weber, E., Russo, R., and Yang, P. (2001) *Science*, **292**, 1897.
6 Tsukazaki, A., Ohtomo, A., Onuma, T., Ohtani, M., Makino, T., Sumiya, M., Ohtani, K., Chichibu, S.F., Fuke, S., Segawa, Y., Ohno, H., Koinuma, H., and Kawasaki, M. (2005) *Nature Mater.*, **4**, 42.
7 Tsukazaki, B., Kubota, M., Ohtomo, A., Onuma, T., Ohtani, K., Ohno, H., Chichibu, S.F., and Kawasaki, M. (2005) *Jpn. J. Appl. Phys.*, **44**, L643.
8 Joseph, M., Tabata, H., and Kawai, T. (1999) *Jpn. J. Appl. Phys.*, **38**, L1205.
9 Look, D.C., Claflin, B., Alivov, Y.I., and Park, S.J. (2004) *Phys. Status Solidi A*, **201**, 2203.
10 Zhang, S.B., Wei, S-H., and Zunger, A. (1998) *J. Appl. Phys.*, **83**, 3192.
11 Neumark, G.F. (1997) *Mater. Sci. Eng.*, **R21**, 1.
12 Li, J., Wei, S.-H., and Wang, L.-W. (2005) *Phys. Rev. Lett.*, **94**, 185501.
13 Li, J., Wei, S.-H., Li, S.-S., and Xia, J.-B. (2008) *Phys. Rev. B*, **77**, 113304.
14 Yan, Y. and Wei, S.-H. (2008) *Phys. Status Solidi B*, **245**, 641.
15 Yan, Y., Zhang, S.B., and Pantelides, S.T. (2001) *Phys. Rev. Lett.*, **86**, 5723.
16 Janotti, A., Wei, S.-H., and Zhang, S.B. (2003) *Appl. Phys. Lett.*, **83**, 3522.
17 Segev, D. and Wei, S.-H. (2003) *Phys. Rev. Lett.*, **91**, 126406.
18 Kim, Y.S. and Park, C.H. (2009) *Phys. Rev. Lett.*, **102**, 086403.
19 Wei, S.-H. (2004) *Comput. Mater. Sci.*, **30**, 337.
20 Yan, Y., Al-Jassim, M., and Wei, S.-H. (2006) *Appl. Phys. Lett.*, **89**, 181912.
21 Li, J., Wei, S.-H., Li, S.-S., and Xia, J.-B. (2006) *Phys. Rev. B*, **74**, 081201.
22 Yan, Y., Li, J., Wei, S.-H., and Al-Jassim, M. (2007) *Phys. Rev. Lett.*, **98**, 135506.
23 Wei, S.-H. and Zhang, S.B. (2002) *Phys. Rev. B*, **66**, 155211.
24 Wei, S.-H. and Krakauer, H. (1985) *Phys. Rev. Lett.*, **55**, 1200.
25 Kresse, G. and Hafner, J. (1993) *Phys. Rev. B*, **47**, 558.
26 Makov, G. and Payne, M.C. (1995) *Phys. Rev. B*, **51**, 4014.
27 Zhang, S.B., Wei, S.-H., and Zunger, A. (2001) *Phys. Rev. B*, **63**, 75205.
28 Rinke, P., Janotti, A., Scheffler, M., and Van de Walle, C.G. (2009) *Phys. Rev. Lett.*, **102**, 026402.
29 Oba, F., Togo, A., Tanaka, I., Paier, J., and Kresse, G. (2008) *Phys. Rev. B*, **77**, 245202.
30 Janotti, A., Varley, J.B., Rinke, P., Umezawa, N., Kresse, G., and Van de Walle, C.G. (2010) *Phys. Rev. B*, **81**, 085212.
31 Lee, W., Kang, J., and Chang, K.J. (2006) *Phys. Rev. B*, **73**, 024117.
32 Zhang, S.B. and Wei, S.-H. (2001) *Phys. Rev. Lett.*, **86**, 1789.
33 Park, C.H., Zhang, S.B., and Wei, S.-H. (2002) *Phys. Rev. B*, **66**, 073202.
34 Neugebauer, J. and van de Walle, C.G. (1995) *Phys. Rev. Lett.*, **75**, 4452.
35 Zhang, L., Yan, Y., and Wei, S.-H. (2009) *Phys. Rev. B*, **80**, 073305.
36 Limpijumnong, S., Zhang, S.B., Wei, S.-H., and Park, C.H. (2004) *Phys. Rev. Lett.*, **92**, 155504.
37 Kang, H.S., Ahn, B.D., Kim, J.H., Kim, G.H., Lim, S.H., Chang, H.W., and Lee, S.Y. (2006) *Appl. Phys. Lett.*, **88**, 202108.

38 Katayama-Yoshida, H. and Yamamoto, T. (1997) *Phys. Status Solidi B*, **202**, 763.
39 Heo, Y.W., Kwon, Y.W., Li, Y., Pearton, S.J., and Norton, D.P. (2004) *Appl. Phys. Lett.*, **84**, 3474.
40 Kalish, R., Saguy, C., Cytermann, C., Chevallier, J., Teukam, Z., Jomard, F., Kociniewski, T., Ballutaud, D., Butler, J.E., Baron, C., and Deneuville, A. (2004) *J. Appl. Phys.*, **96**, 7060.
41 Kajihara, S.A., Antonelli, A., Bernhole, J., and Car, R. (1991) *Phys. Rev. Lett.*, **66**, 2010.
42 Wang, L.G. and Zunger, A. (2002) *Phy. Rev. B*, **66**, 161202(R).
43 Goss, J.P., Jones, R., Heggie, M.I., Ewels, C.P., Briddon, P.R., and Oberg, S. (2002) *Phys. Rev. B*, **65**, 115207.
44 Goss, J.P., Briddon, P.R., Sque, S., and Jones, R. (2004) *Phys. Rev. B*, **69**, 165215.
45 Ohtomo, A., Kawasaki, M., Koida, T., Masubuchi, K., Koinuma, H., Sakurai, Y., Yoshida, Y., Yasuda, T., and Segawa, Y. (1998) *Appl. Phys. Lett.*, **72**, 2466.
46 Ryu, Y.R., Lee, T.S., Lubguban, J.A., Corman, A.B., White, H.W., Leem, J.H., Han, M.S., Youn, C.J., and Kim, W.J. (2006) *Appl. Phys. Lett.*, **88**, 052103.

14
Electrostatic Interactions between Charged Defects in Supercells
Christoph Freysoldt, Jörg Neugebauer, and Chris G. Van de Walle

14.1
Introduction

Theoretical calculations have revolutionized our understanding of the doping behavior in semiconductor materials for electronic and optoelectronic devices [1, 2]. The advent of density-functional theory (DFT), and, more recently, *ab initio* approaches beyond it such as many-body perturbation theory in the GW approximation [3] or quantum Monte-Carlo methods [4], has enabled us to study the microscopic details of point defects with almost no *a priori* assumptions. These calculations complement experiment in various ways: they not only help to interpret experimental findings and link them to an atomistic model of the relevant defects, but also provide additional data such as formation energies, geometric structures, or the character of the wavefunctions that cannot be obtained with present-day experimental methods.

The work-horse of these calculations has been DFT with local or semilocal functionals [1, 2]. The defect is usually modeled in a supercell, consisting of the defect surrounded by a few dozen to a few 100 atoms of the host material, which is then repeated periodically throughout space. This allows to employ the highly efficient and thoroughly tested computer codes developed for periodic solids [5]. Recent advances in the theoretical framework tend to maintain the supercell models for the same reason [3, 4]. However, it must be kept in mind that the use of supercells implies that the isolated defect is replaced by a periodic array of defects. Such a periodic array contains unrealistically large defect concentrations, resulting in artificial interactions between the defects that cannot be neglected. These interactions include overlap of the wavefunctions, elastic interactions, and – in the case of charged defects – electrostatic interaction. The focus of this contribution are the electrostatic interactions which typically dominate.

A large variety of approaches to control electrostatic artifacts exists in the literature, as has been reviewed recently by Nieminen [6]. Our aim is not to compare the different approaches, but to work out explicitly all assumptions that are made to estimate defect–defect interactions. This allows – at least in principle – to verify each of them for any defect or material under consideration, thereby greatly enhancing the reliability of the method.

Supercell calculations for formally charged systems must always include a compensating background charge, since the electrostatic energy of a system with a net charge in the unit cell diverges [7, 8]. It is most common to include a homogeneous background, which is equivalent to setting the average electrostatic potential to zero. By increasing the supercell lattice constant L, the isolated defect limit can be recovered in principle for $L \to \infty$. Nevertheless, the defect energy converges only slowly with respect to L. The origin of this effect lies in the unphysical electrostatic interaction of the defect with its periodic images and the background, decaying asymptotically as q^2/L, where q is the defect charge [7, 8]. Its magnitude can be estimated from the Madelung energy of an array of point-charges with neutralizing background [7]. Makov and Payne [8] proved for isolated ions that the quadrupole moment of the charge distribution gives rise to a further term scaling like L^{-3}. For realistic defects in condensed systems, however, such corrections, scaled by the macroscopic dielectric constant ε to account for screening, do not always improve the convergence [9–11]. Therefore, the prefactors have often been regarded as parameters to be obtained from fitting a series of supercell calculations [6, 9–12]. Unfortunately, such "scaling laws" require large supercells and may include higher order L^{-n} terms with not well defined physical significance. While it is relatively straightforward to determine the most slowly decaying terms of the relevant interactions (which give the leading terms in a $1/L$ expansion), higher-order terms can have a variety of functional forms. Focussing on electrostatic interactions only, contributions beyond the $1/L$ asymptotic limit arise from finite overlap of the defect charge densities (which have an asymptotically exponential decay), details of the microscopic screening (which decay faster than $1/L^2$, but exhibit oscillatory behavior), non-linear effects, and higher-order moments of the charge distributions. They can however not be separated from the remaining wavefunction overlap errors (asymptotic exponential decay) and the cell-shape variations usually present in "scaling law" approaches to increase the number of available supercells. Representing all these contributions by one L^{-n} term (or a few of them) is clearly a very strong reduction of the underlying complexity and and removes any physical meaning from the L^{-n} prefactor even if individual contributions exhibit such an asymptotic limit. The loss of significance is apparent already for the L^{-n} term in the standard scaling law approach: the fitted prefactor usually deviates from the predictions of macroscopic electrostatic theory, which – at least in the limit of sufficiently large supercells – is the physically correct limit for any localized defect.

Recently, a modified version of the Makov–Payne corrections has been proposed by Lany and Zunger [13, 14]. The approach has been employed very successful in practice. A potential drawback is that the approach proposed in Ref. 14 does not always recover the asymptotic $1/L$ limit of Makov–Payne theory. Significant efforts have been undertaken to assess the applicability of the existing correction schemes, but no clear picture regarding applicability and limitations has emerged so far [9–11, 15, 16].

As an alternative to the homogeneous-background approaches, several authors suggested to modify the computation of the electrostatic potential in the DFT calculation itself to remove the unwanted and unphysical interactions [17–20]. These approaches can also be regarded as the introduction of a neutralizing surface charge at the boundary of the supercell instead of the homogeneous background when the potential is calculated. Complications arise if the boundary cuts through the material, since the material will try to screen away the electric-field discontinuities.

Recently, we have proposed a scheme that accounts for the electrostatic screening of the defect right from the beginning. A major advantage is that all approximations are well defined and easily verified by the actual calculations. The aim of this paper is to provide a careful discussion of the underlying assumptions, provide examples how the approach can be used in practice, and analyze its performance with respect to supercell size convergence. The aim is not to give a detailed overview over all the alternative approaches, since this would go well beyond the scope of the present paper. A recent overview on alternatives can be found in Ref. [6].

The remainder of this chapter is organized as follows. In Section 14.2, electrostatics and screening in real materials is discussed. We will then summarize in Section 14.2.1 the key steps to derive an explicit and exact expression for the electrostatic artifacts in the supercell approach and discuss tractable approximations that yield a parameter-free correction scheme for these artifacts. Results for two model point defects that have been used already by other groups to study supercell size convergence, namely the Ga vacancy in GaAs and the vacancy in diamond will be discussed in Section 14.3.

14.2
Electrostatics in Real Materials

When a localized charge $q(r)$ (total charge q) is introduced into condensed matter, it attracts a screening charge of opposite sign that reduces its long-range potential to $\frac{q}{\varepsilon r}$ where r is the distance from the localized charge.[1] It immediately follows that the amount of the screening charge is $(1-\frac{1}{\varepsilon})q$. For a finite system, the total charge q is conserved, but the screening charge is expelled to the surface. In an infinite system, on the other hand, the screening charge is almost homogeneously taken from the host material, modulated only by the underlying atomic structure. It is worthwhile to note that in the idealized case of an isolated charge in an infinite covalently bound semiconductor, the total screening charge inside any *finite* distance is non-zero, or in other words the screening process does not conserve charge in any *finite* region. This surprising result is a direct consequence of the quantum-mechanical, truly non-local nature of screening in condensed matter [21, 22]. In an ideal ionic material with separated polarizable ions, the screening is effectuated by induced dipoles. Even though the charge on each ion is conserved, there is a net flow of charge along the radial axis. For large distances, the spherically averaged charge distribution, which is relevant for the distance-dependence of screening, approaches the one of an ideal

---

1) In the following, we will define the center of the charge such that the dipole moment of $(q(r)-q\delta(r))$ becomes zero.

homogeneous (jellium-like) semiconductor. The reason is that no matter how well the ions are separated in three-dimensional space, since the radial spacing between the ionic shells becomes arbitrarily small with increasing distance, they will necessarily overlap when projected onto the radial coordinate.

In a periodic array of defects the screening charge is taken from the supercell. In consequence, the average electron density far from the defect is shifted from its bulk value by $\frac{q(\varepsilon-1)}{\varepsilon\Omega}$ where Ω denotes the volume of the supercell. This effect is indeed visible in defect calculations in the framework of DFT. We illustrate the difference in the defect charge before and after screening in Figure 14.1 for the vacancy in diamond in the 2 – charge state. The unscreened charge density

$$q(\mathbf{r}) = 2|\psi_d(\mathbf{r})|^2 \tag{14.1}$$

results from filling the sixfold degenerate defect state ψ_d with two additional electrons. It is clearly localized. The screened charge density is obtained from the change in the self-consistent charge density with respect to the neutral state, and includes all screening effects. It is completely delocalized and indeed approaches the homogeneous limit $\frac{q(\varepsilon-1)}{\varepsilon\Omega}$ far from the defect.

In order to understand the quantum-mechanical nature of screening in a real material in more detail, it is illuminative to decompose the supercell error in the defect formation energy into the energy terms of the underlying electronic-structure calculation such as Hartree energy, kinetic energy, exchange–correlation energy, etc. We report such an analysis in Appendix A for a model system that avoids the electronic-structure complications of real defects. It reveals that the *electrostatic effects are not restricted to the energy contributions formally associated with electrostatics*, but are distributed to all parts of the total energy due to self-consistency. Recovering the electrostatic energy from the self-consistent electron density employing density-based expressions for electrostatic interactions might therefore be very difficult if not

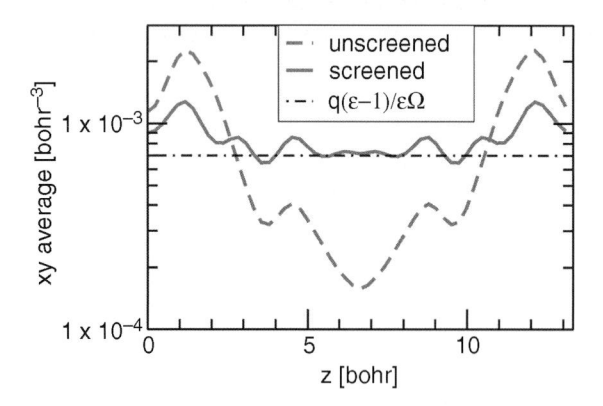

Figure 14.1 (online color at: www.pss-b.com) Comparison of defect charges before and after screening (see text) for the 2− vacancy in diamond in a 64-atom supercell. The defect is located at $z = 0$ with a periodic image at $z = 13.3$ bohr. The limit of a homogeneously distributed screening charge is indicated by the black dash-dotted line.

14.2.1
Potential-based Formulation of Electrostatics

In view of the aforementioned difficulties when working with the electron density, it turns out to be advantageous to express electrostatic interactions in terms of the unscreened charge density $q(r)$ and the electrostatic potential $V(r)$. To be more specific, let us consider the interaction of a single defect $q(r)$ with one of its periodic images $q_R(r) = q(r-R)$. Figure 14.2 depicts this situation graphically. The electrostatic interaction is given by

$$E = \int d^3r\, q(r) V[q_R](r). \tag{14.2}$$

Here, $V[q_R]$ is the potential due to the presence of the charge q_R including all the screening effects. At a sufficient distance from the defect center, the potential approaches its macroscopic value of $1/\varepsilon r$. At which length scale the macroscopic behavior is reached, depends on the localization of the charge and the characteristic screening length of the material under consideration. The potential is further modulated by the effects of microscopic screening (local field effects), in particular due to the underlying atomic structure. However, for the electrostatic energy according to Eq. (14.2) the microscopic details tend to average out. The same is true for the details of the charge distribution if the potential is sufficiently smooth. Note that the modulation amplitude decays faster than $1/r$. The potential therefore becomes smoother as the distance is increased. The electrostatic energy due to the unphysical interaction of the defect charge with its periodic image can then be estimated to good accuracy from a simplified model of the charge density and the long-range potential V^{lr}. The key advantage of this view on electrostatic interactions is that reasonable approximations to the charge distribution and the long-range

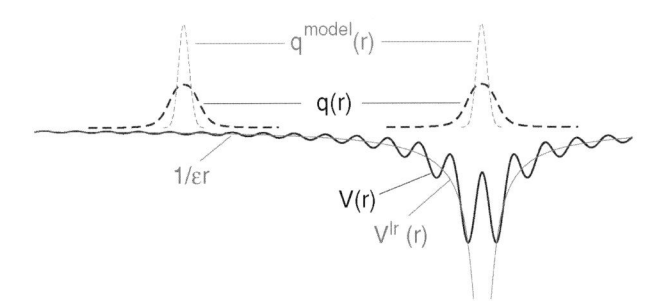

Figure 14.2 (online color at: www.pss-b.com) Potential-based formulation of defect–defect interactions (schematically). Dashed lines indicate the unscreened defect density of a defect (left) and its periodic image (right). The true potential (black solid line) of the image includes local-field effects (wiggles). The major part of the interaction can be captured by simplified models (thin red lines).

potential are easily found. Moreover, the electrostatic potential is a byproduct of a DFT calculation and therefore available at no additional computational cost.

Neglecting the influence of microscopic screening, which typically takes place at a length scale of a few bond lengths, the long-range potential is given by

$$V^{lr}(r) = \int d^3 r' \frac{q^{model}(r')}{\varepsilon |r-r'|}. \tag{14.3}$$

It remains to choose an appropriate model charge density q^{model}. For strongly localized defects, a point charge or a Gaussian with a small width ($\approx$1 bohr) is for most cases a reasonable choice. For very delocalized states, a more advanced model is needed and will be discussed in Section 14.3.2.

In addition to the interaction of the defect with its periodic images, also the interaction with the homogeneous background must be taken into account. For this, the potential-based expressions for the electrostatic energy are ideally suited. There is no need to introduce approximations here since the homogeneous background density and the defect-induced potential are known exactly. However, the defect images and the background cannot be considered separately as the individual energy contributions would diverge. Instead, we will separate all interactions into a long-range part, that is treated at the model level, and a remaining short-range part. The corresponding expressions will be derived in the following.

14.2.2
Derivation of the Correction Scheme

In this section, tractable expressions are derived for the electrostatic interactions introduced by the supercell approximation. The key idea is to exploit that long-range interactions can be captured with a simplified model as described in Section 14.2.1, and correct for the short-range interactions beyond this in a consistent way.

We start from the defect-induced potential

$$V = V^{els}(\text{charged defect}) - V^{els}(\text{reference}). \tag{14.4}$$

Here, V^{els} denotes the DFT electrostatic potential, i.e., the sum of the external (ionic) local potential and the Hartree potential. The scheme can of course be applied to any other electronic-structure method that provides the electrostatic potential. We will start with the neutral defect as reference and discuss the transition to the bulk reference below.

V can be formally split into a long-range and short-range part, i.e.,

$$V = V^{lr} + V^{sr}, \tag{14.5}$$

which implicitly defines V^{sr} from V and V^{lr}. The long-range potential V^{lr} is obtained from the model charge density q^{model} via Eq. (14.3). If q^{model} is well chosen and the supercell is large enough, V^{sr} decays to zero within the supercell. In this case, the short-range energy of the defect charge does not differ between the isolated defect and the periodic array of defects. For strongly localized defects, a Gaussian model charge is usually sufficient to ensure a fast decay of V^{sr}. However, the periodic array also includes a neutralizing background $n = -q/\Omega$, with a short-range interaction energy

$$\int d^3r\, n V^{sr}(r) = -q\left[\frac{1}{\Omega}\int d^3r\, V^{sr}(r)\right], \tag{14.6}$$

between the background and the defect.

The long-range potential for the periodic array (including the background) is obtained from the Fourier transform of Eq. (14.3) as

$$\tilde{V}^{lr}(\mathbf{G} \neq 0) = \frac{4\pi q^{model}(\mathbf{G})}{\varepsilon |\mathbf{G}|^2}, \qquad \tilde{V}^{lr}(0) = 0. \tag{14.7}$$

Note that the homogeneous background does not induce local-field effects; its only role is to cancel the divergence of the $\mathbf{G} = 0$ term [23]. The long-range interaction energy can then be estimated from the screened lattice energy (Madelung energy) of the model charge [23]. For spherical charge densities, it can be easily computed in reciprocal space via

$$E^{lat}[q^{model}] = \frac{2\pi}{\varepsilon\Omega} \sum_{\mathbf{G}\neq 0}^{|\mathbf{G}|\leq G_{cut}} \frac{\{q^{model}(|\mathbf{G}|)\}^2}{|\mathbf{G}|^2} - \frac{1}{\pi\varepsilon} \int_0^{G_{cut}} dg\, \{q^{model}(g)\}^2, \tag{14.8}$$

where $\mathbf{G}$ runs over the reciprocal lattice vectors. The first term in Eq. (14.8) is the energy of the periodic array in its own potential, with a prefactor of $\frac{1}{2}$ to account for double counting. The second term removes the electrostatic interaction energy of the model charge with itself, that is contained in the first term.

With these ingredients, the formation energy of a charged defect in a supercell can be expressed as [23]

$$E_f^q = E_f^0 + \Delta E^{iso}(q) + E^{lat}[q^{model}] - q\Delta, \tag{14.9}$$

where E_f^0 denotes the formation energy of the neutral defect in the same supercell and $\Delta E^{iso}(q)$ the difference in formation energy between an isolated charged defect and the neutral one. The alignment-like term

$$\Delta = \frac{1}{\Omega}\int d^3r\, V^{sr}(r) \tag{14.10}$$

is obtained from the short-range potential

$$V^{sr} = \tilde{V}^{els}(\text{charged}) - \tilde{V}^{els}(\text{neutral}) - \tilde{V}^{lr} + \Delta V, \tag{14.11}$$

where the alignment constant ΔV is chosen such that V^{sr} decays to zero in between the defects [23]. We demonstrate the alignment in Figure 14.3 for the Ga 3 – vacancy in GaAs referenced to the bulk potential (the change of reference from the neutral defect to the bulk is discussed below). For this, the potentials were averaged over the xy plane and plotted as a function of z. The defect-induced potential shows a parabolic shape in between the two defects [23]. This shape is well reproduced by the long-range

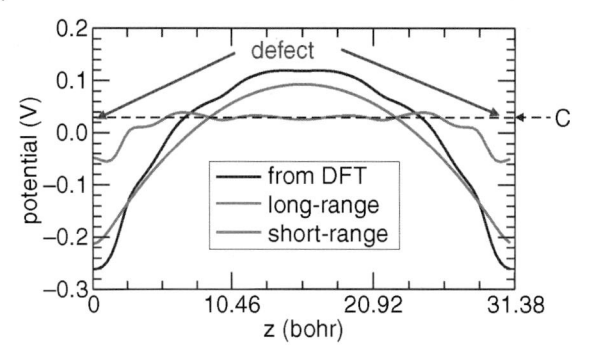

Figure 14.3 (online color at: www.pss-b.com) Potentials (see text, averaged along x, y) for a V_{Ga}^{3-} in a $3 \times 3 \times 3$ cubic GaAs supercell. The defect is located at $z = 0$ bohr with a periodic image at $z = 31.38$ bohr.

model, in this case a 1 bohr wide Gaussian. The difference between the two reaches a plateau at $C = -\Delta V = 0.03$ eV. The appearance of the plateau clearly demonstrates that the model has correctly reproduced the long-range tail of the defect potential. Any remaining curvature (apart from the unavoidable modulation due to the underlying atomic structure and/or the screening response) would indicate that the long-range modeling should be improved, e.g., by choosing a model charge that is derived from the actual defect wave function. It should be emphasized that the main approximation in the scheme lies in the neglect of the details of microscopic screening and their coupling to the details of the actual unscreened charge distribution beyond q^{model} for the long-range interactions. These neglected details are by definition cast into V^{sr}. The validity of the central approximation can therefore be easily controlled and checked by verifying that V^{sr} is well behaved.

We now proceed to replacing the neutral defect reference by the bulk, which constitutes the reference of interest in most practical applications. As shown in Appendix B, the potential alignment constant ΔV and the energy alignment constant Δ are related via

$$\Delta = \Delta V. \tag{14.12}$$

For the neutral defect reference discussed up to now, this can be seen as follows: the inclusion of the homogeneous background in the Hartree energy by setting $V(\mathbf{G} = 0)$ to zero implies that the average potential does not change between the neutral and the charged defect. Likewise, the average V^{lr} vanishes in this alignment convention. Eq. (14.12) then immediately follows from Eq. (14.11). Equation (14.12) remains valid if the neutral defect reference is replaced by another one, notably the bulk, even if the average alignment changes. The reason is that the alignment of the potentials reflects the dependence of the total energy (and derived quantities) when the formal charge is changed. In contrast to the original formulation [23], where it was incorrectly stated that Eq. (14.10) was to be used for any reference potential, using Eq. (14.12) for the alignment guarantees full internal consistency as outlined in Appendix B.

The formation energy of an isolated defect with charge q then becomes

$$E_f^{iso} = E^{DFT}(\text{defect}+\text{bulk}) - E^{DFT}(\text{bulk}) - E^{lat}[q^{model}]$$
$$+ q\Delta V - \sum n_s \mu_s + q(E^{Fermi} + \varepsilon^{vbm}), \qquad (14.13)$$

where we also included the reference chemical potentials for the chemical species s added ($n_s > 0$) or removed ($n_s < 0$) to form the defect. E^{Fermi} is the Fermi energy relative to the valence band maximum (vbm), and ε^{vbm} is the valence band maximum as obtained from the bulk reference calculation.

14.2.3
Dielectric Constants

The long-range modeling requires the dielectric constant ε of the material at hand. To be consistent with the theoretical framework, the dielectric constant must be computed. This can be done by perturbation theory [24, 25], or by a direct approach [22, 26]. The direct approach is straightforward to apply. For this purpose, a sawtooth potential V^{saw} is applied to an elongated cell [22, 26], typically a $1 \times 1 \times 6$ supercell of the simple-cubic bulk cell. The change in the effective potential ΔV^{SCF} then also shows a sawtooth-like shape, however reduced by the dielectric constant, see Figure 14.4. By comparing the slope of the applied and effective potentials between the turning points, the dielectric constant can be determined as

$$\varepsilon = \frac{\partial V^{saw}/\partial z}{\partial \Delta V^{SCF}/\partial z}. \qquad (14.14)$$

The direct approach has the advantage to provide immediate insight into the linearity and the screening length of the perturbation applied. It yields very accurate

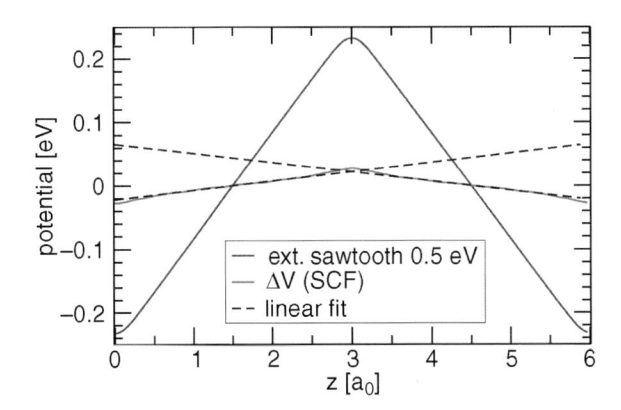

Figure 14.4 (online color at: www.pss-b.com) Determination of the dielectric constant of GaAs. Applying a (smoothened) sawtooth potential V^{saw} (blue) to the material induces a change ΔV^{SCF} in the self-consistent potential (red). The dielectric constant is given by $\frac{\partial V^{saw}}{\partial z} / \frac{\partial \Delta V^{SCF}}{\partial z} \approx 12.7$.

results (only 1–2% scatter between different cells and different amplitudes) if the following points are observed: (i) The total height of the sawtooth potential must not exceed the bandgap to avoid a dielectric breakdown, i.e., transfer of valence electrons from near the top of the potential to conduction band states near the bottom of the potential. (ii) The sawtooth potential must be adapted to the symmetry of the system. In particular, turning points should lie on symmetry planes to avoid the induction of additional dipoles at the turning points. If this is not observed, the rising and falling parts of the sawtooth potential may yield different apparent dielectric constants (even negative ones). (iii) The induced potential fluctuations are very soft modes. Therefore, the convergence criteria for the self-consistent field calculations must be very tight to yield accurate results. A helpful check is to compare the dielectric constants of the rising and falling parts of the potential. (iv) The induced potentials are modulated by the underlying atomic structure. In order to average out the modulations (also known as local-field effects), the slopes must be determined over full periods of these modulations. We also note from Figure 14.4 that there are deviations from the macroscopic screening behavior close to the turning points which extend over a range of $a_0/2$. The deviations reflect the finite screening length, and must of course be excluded from the slope fitting.

If ionic relaxations are taken into account for the defect calculation, the dielectric constant employed for the correction scheme must also reflect ionic screening. In the sawtooth approach, this implies that ionic relaxation must be included [27].

14.3
Practical Examples

In the following, we discuss the application and performance of the correction scheme for two representative examples. Specifically, we will consider the Ga vacancy in GaAs as a deep, well localized defect, and the carbon vacancy as a shallow, rather delocalized defect state to test and discuss the limits of the corrections. The calculations were performed in the framework of DFT in the local-density approximation, employing plane waves and normconserving pseudopotentials as implemented in the SPHInX code [28]. A plane-wave cutoff of 20 Ry for GaAs and 40 Ry for C was found to yield sufficient accuracy. In order to disentangle the electrostatic from strain effects, the ions were not relaxed. The supercell artifacts due to wavefunction overlap, on the other hand, cannot be avoided, but were minimized by a constant occupation scheme [2].

14.3.1
Ga Vacancy in GaAs

The first example is the Ga vacancy in GaAs. It shows a deep level in the band gap that supports – in its unrelaxed geometry – charge states between 0 and 3–. The corrections are obtained from a 1 bohr wide Gaussian model charge. The calculated formation energies for the four charge states are displayed in Figure 14.5 for a series of supercells. The uncorrected formation energies show a strong dependence on the supercell size. Since electrostatic interactions scale with q^2, it is most prominent for the $q=-3$ case shown in Figure 14.5a. This dependence is largely removed by including the correction energies according to Eq. (14.13). The remaining scatter in the data is ~0.1 eV, see Figure 14.5b. It does not scale with the charge state and is present even in the neutral state. This suggests that other than electrostatic effects are responsible. Since strain effects have been excluded by not considering ionic relaxation, it is probably caused by the overlap of the wavefunctions and the resulting Pauli repulsion. Clearly, corrections aiming at electrostatic interactions cannot and should not capture such effects.

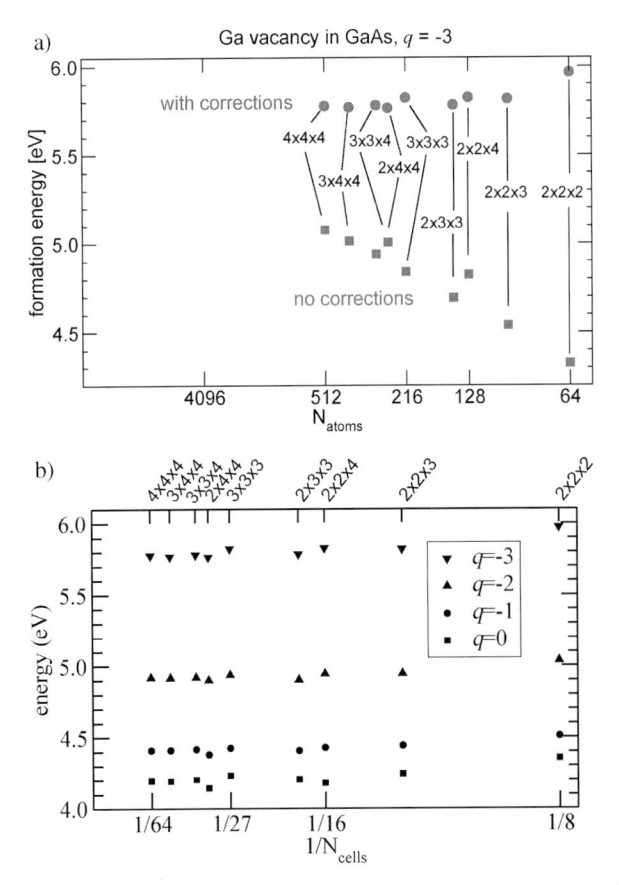

Figure 14.5 (online color at: www.pss-b.com) Formation energies of the Ga vacancy for a variety of supercells. The Fermi energy is set to the valence band maximum. (a) Comparison of uncorrected and corrected value for $q=-3$. (b) Corrected values for $q=0\ldots-3$.

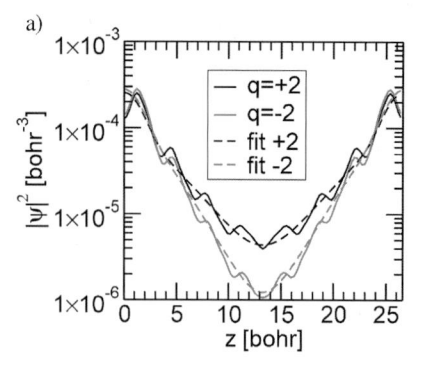

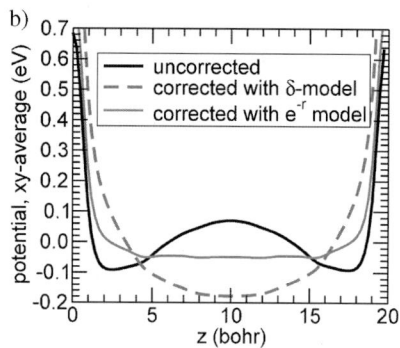

Figure 14.6 (online color at: www.pss-b.com) Localization analysis of the carbon vacancy. Shown is the xy-plane average of (a) the defect wavefunction, and (b) the defect-induced potentials. The defect is located at $z = 0$ with a periodic image at $z = 26.6$. (a) Shape of the defect state $|\psi(r)|^2$ for two different charge states averaged over the Brillouin zone and the three defect bands (solid lines). (b) Comparison of the defect-induced potential for the 2+ case without long-range corrections (black line), with long-range corrections from a 1 bohr wide Gaussian (red dashed), and from the fitted exponential-tail model (green).

14.3.2
Vacancy in Diamond

The second example is the vacancy in diamond. This defect has been used previously to study finite-size effects in supercells by Shim et al. [10]. The authors found that the 2− charge state is easily corrected by a Makov–Payne-like scheme, while the 2+ charge state shows a completely different behavior. This discrepancy was tentatively attributed to a qualitatively different screening for the two cases [10]. Such defect-induced changes in the electrostatic screening, however, are expected to converge proportional to the inverse volume of the supercell if the defect can be regarded as an impurity with a changed polarizability compared to the host material.

When the correction scheme of Section 14.2.2 is applied to the 2+ case, the alignment procedure immediately indicates problems to reproduce the long-range potential, see Figure 14.6b. The "short-range" potential shows a significant over-correction when a Gaussian charge model is employed. In order to derive an improved charge model, let us analyze the defect states. In the long-distance limit, every defect state shows an exponential decay behavior with increasing distance of the defect center. Indeed, as shown in Figure 14.6a, the shape of the 2+ and 2− defect states of the diamond vacancy reveals a significant contribution of this exponential tail to the overall wavefunction. While the states show a close match near the defect center, the exponential decay differs significantly for the two states. This can be attributed to the position of the levels in the band gap. The 2− Kohn–Sham level is located around midgap (cf. Figure 14.7, the Kohn–Sham level lies half-way between the −/2− and 2−/3− transition levels of the corresponding supercell). The state thus decays very quickly. The 2+ level is close to the valence band, and the associated state then starts to hybridize with the valence band states, leading to a delocalization that is sizeable and cannot be neglected even at the boundaries of the supercell.

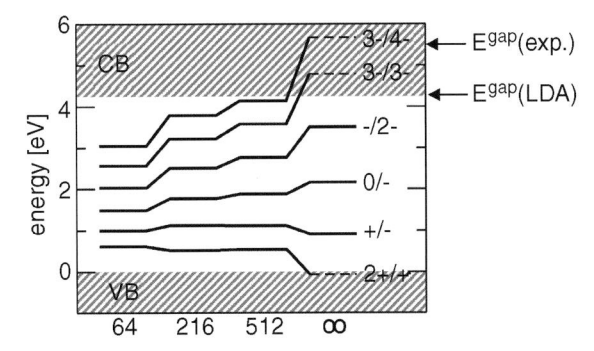

Figure 14.7 (online color at: www.pss-b.com) Charge transition levels for the unrelaxed vacancy in diamond as obtained from different supercells (size of 64, 216, 512 atoms) without correction compared to the corrected ones (∞).

To take the exponential tail into account, the defect charge can be modeled by the radial ansatz

$$q(r) = qxN_\gamma e^{-r/\gamma} + q(1-x)N_\beta e^{-x^2/\beta^2}. \tag{14.15}$$

N_β and N_γ denote the normalization constants for the exponential and the Gaussian, respectively. The decay constant γ and the tail weight x is obtained by fitting the wavefunctions of the $4 \times 4 \times 4$ (512 atom) cell. The exact value of β turns out to be relatively unimportant as long as the Gaussian stays localized; here, a value of 2 bohr was used. To ensure that this ansatz can be generally applied, we used it for all charge states, even though a Gaussian-only model works reasonably well for the more localized mid-gap states $1 \geq q \geq -2$. Interestingly, all states have essentially the same tail weight ($x = 54$–60%). The decay constants, on the other hand, sensitively depend on the energetic distance to the valence or conduction band edge. They are listed in Table 14.1 along with the uncorrected and corrected values for cubic supercells of 64, 216, and 512 atoms. It is apparent that the exponential-tail model can be equally applied to all charge states. The supercell corrections reduce the errors in the calculated formation energies from up to 7 eV (4−, 64 atoms) to $\sim$0.1 eV. The charge transition levels derived from the supercells without the corrections as well as from the extrapolated values for the isolated defect are visualized in Figure 14.7. It is noteworthy that the corrections push the transition levels to high charge states into the valence band (2+/+) and the conduction band (2−/3− and 3−/4−), respectively. The occurence of high charge states thus turns out to be an artifact of the small supercells. This tendency to push the charged levels away from the neutral one is a general feature of supercell corrections whenever the Madelung term dominates.

The examples shown here illustrate the importance of supercell corrections for predicting the correct position of defect levels, sometimes even qualitatively. The charge correction scheme presented here is able to remove the dominating electrostatic artifacts from finite-size supercell calculations. In combination with the constant-occupation scheme, the agreement between different supercells is typically on the order of 0.1 eV, and thus considerably smaller than the errors that arise from the DFT framework employed. We therefore expect that major improvements over

Table 14.1 Formation energies of a vacancy in diamond for $N \times N \times N$ supercells in different charge states q with and without supercell corrections. Column 2 lists the defect states' decay constants γ (see text). The Fermi energy is set to the valence band maximum.

q	γ	uncorrected			corrected		
		$N=2$	$N=3$	$N=4$	$N=2$	$N=3$	$N=4$
2+	2.41	6.65	6.49	6.63	7.69	7.45	7.44
+	2.06	7.26	7.11	7.17	7.43	7.33	7.37
0	1.85	8.26	8.23	8.28	8.26	8.23	8.28
−	1.77	9.75	10.00	10.16	10.40	10.38	10.44
2−	1.81	11.78	12.51	12.92	13.93	13.91	13.95
3−	2.03	14.35	15.74	16.50	18.77	18.73	18.74
4−	2.56	17.40	19.55	20.64	24.55	24.54	24.41

the present state-of-the-art in defect calculations will come from the application of advanced electronic-structure methods in the supercell approach.

14.4 Conclusions

In this paper, we have presented an analysis of supercell artifacts in charged point defect calculations arising from electrostatic interactions. For the electrostatics in real materials, an exact, potential-based formulation overcomes the limitation of previous correction schemes that rely on *a priori* simplifications such as the restriction to macroscopic screening and point-like charge densities. A new correction scheme emerges from this analysis. The scheme itself requires no empirical parameters or fitting procedures, and requires only a single supercell calculation. It employs certain simplifications to model the long-range interaction, but in contrast to other approaches, the validity of these approximations can be verified immediately by visually inspecting the short-range potential. If needed, the underlying charge model can be refined in a straightforward manner. It should be emphasized that the correction scheme does not rely on DFT, but can be applied to any electronic-structure theory that provides the electrostatic potential. We believe that this approach may help to reduce the errors induced by approximate correction schemes and thus extend the applicability of advanced methods to charged defects, even if the supercells that can be afforded with these methods introduce significant electrostatic interactions.

Acknowledgements

This work was supported in part by the German Bundesministerium für Bildung und Forschung, project 03X0512G, the NSF MRSEC Program under award no. DMR05-

20415, the IMI Program of the NSF under Award no. DMR04-09848, and the UCSB-MPG Program for International Exchange in Materials Science.

Appendix
(A) Energy Decomposition of Electrostatic Artifacts in DFT

In order to understand the screening behavior in more detail, let us consider a model system that avoids the electronic-structure complications of real defects. For this purpose, a Gaussian charge is placed in an interstitial site in GaAs. (As anti-bonding, tetrahedrally surrounded by As.) A charge of -1 ensures that this model charge repels the surrounding electrons, preventing the formation of any localized electronic bounding states. We then calculated the self-consistent total energy within DFT in 22 different supercells using norm-conserving, non-local pseudopotentials. The calculated defect formation energy

$$E_d = E_{tot}(\text{defect}) - E_{tot}(\text{bulk}) \tag{14.16}$$

depends on the supercell. Figure 14.8 shows a decomposition of how the error in the formation energy is related to the various contributions to the Kohn–Sham functional. For this, we plot the change in each energy contribution with respect to the corresponding bulk as a function of the error of the defect formation energy with respect to the isolated case, i.e., the supercell error. We note that positive supercell errors occur for elongated cells, negative for cubic supercells. The energy of the isolated case was obtained with the defect correction scheme described in Section 14.2.2. If a single energy contribution were responsible for the supercell error, it would appear as a straight line of slope $m=1$. It becomes apparent that the energy contributions vary non-linearly with the supercell error. The electrostatic part

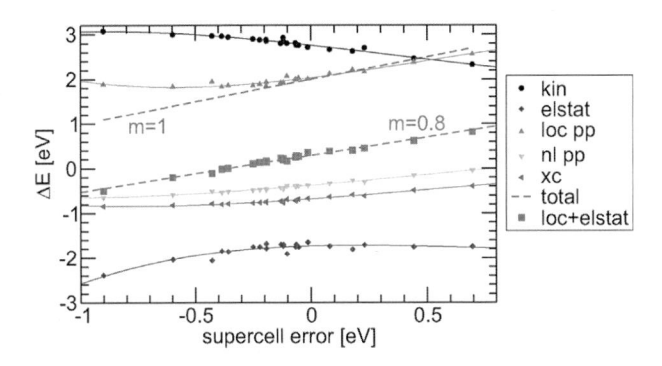

Figure 14.8 (online color at: www.pss-b.com) Decomposition of the defect formation energy into the various energy contributions to the Kohn–Sham functional for different supercells: kinetic (kin), Hartree + long-range pseudopotential (elstat), short-range local pseudopotential (loc pp), non-local pseudopotential (nl pp), exchange–correlation (xc). The supercell error corresponds to the sum of all contributions (red line), shifted by the isolated defect limit. The solid curves are polynomial fits to highlight the trends. Perfect correlation between an energy contribution and the supercell error corresponds to a line of slope $m=1$.

of the Kohn–Sham functional can be identified with the sum of Hartree and local pseudopotential contributions. This sum varies almost linearly with the supercell, but with a slope of only $m = 0.8$. In other words, the electrostatic part of the functional accounts for only 80% of the electrostatic supercell error. This highlights that the *electrostatic effects are not restricted to the energy contributions formally associated with electrostatics*, but are distributed to all parts of the total energy due to self-consistency.

(B) Alignment Issues in Supercell Calculations

Depending on the computer code implementation at hand, the treatment of the local pseudopotential may influence the potential alignment and thus the total energy for charged systems. It is common practice to split off the long-range part of the pseudopotential by subtracting the potential of a compensation charge (e.g., a Gaussian), adding the compensation charge to the charge density used for the Hartree energy, and correcting for the added self-energy of the compensation charges [29]. The remaining short-range pseudopotential may now be used for the electron density only [29] (which is also the case for the SPHInX pseudopotential code [28] employed for the examples in Section 14.3) or for the neutral "Hartree density" including the compensation charges, with a corresponding correction term for each atom [30]. In the latter case, one may alter the alignment of the periodic superposition of the short-range potentials, and in particular set its average to zero and thus include the neutralizing background. Only in this alignment convention, the energy of a charged system becomes independent of the choice of the compensation charge. Otherwise the potential shift acting on the electrons is only compensated in part by the self-energy of the compensation charges.

In this general case, Δ must be determined as follows. The alignment convention of setting the average of the periodic Hartree potential to zero is the consistent way to include a neutralizing background in the Hartree energy. For the neutral reference, Δ then becomes identical to ΔV: the Hartree alignment convention implies that the average potential does not change between the neutral and charged defect (the pseudopotentials and compensation charges are the same in both cases). Likewise, the average $\tilde{V}^{lr}$ vanishes in this alignment convention. The identity $\Delta = \Delta V$ (Eq. (14.12)) then immediately follows from Eq. (14.11). Using Eq. (14.12), we may now replace the neutral defect reference by any other, notably the bulk, even if the alignment changes. The crucial point is that the alignment is not changed arbitrarily, but consistent with the total-energy expressions and quantities derived from it. For instance, a change in the bulk alignment by an amount A will change ΔV by $+A$, which compensates the shift in the valence band energy $\varepsilon^{vbm} = \frac{\partial E}{\partial f_{vbm}}$ (f_{vbm} is the occupation number of the valence band maximum) appearing in Eq. (14.13) by $q^{electron} A = -A$. On the other hand, a shift of B in the defect potential (due to altered compensation charges) will change ΔV by $-B$, but also the total energy of the charged defect calculation by qB.

Moreover, Eq. (14.12) as the general definition of Δ allows for changing the $\tilde{V}^{lr}$ alignment convention defined by Eq. (14.7). In Eq. (14.7), the $\mathbf{G}=\mathbf{0}$ component was set to zero, consistent with excluding the $\mathbf{G}=\mathbf{0}$ component in Eq. (14.8). This choice corresponds to including the interaction of the model charge with the compensating background in the lattice energy. The lattice energies of non-overlapping charge densities therefore depend on the actual shape of the charge distribution. This is compensated for by a corresponding change in the depth of the short-range potential according to Eq. (14.11), and thus the $-q\Delta$ term in Eq. (14.9). An alternative is to modify the definition of the Madelung energy [Eq. (14.8)] and the associated potential [Eq. (14.7)] consistently. Using the analytic limit

$$V_0 = \lim_{G \to 0} V(G) = \frac{2\pi}{\varepsilon} \frac{\partial^2 q^{\text{model}}(g)}{\partial g^2}\bigg|_{g=0}, \quad (14.17)$$

for the alignment convention, the modified definitions are

$$\tilde{V}^{\text{lr}}(G=0) := V_0, \quad (14.18)$$

$$E^{\text{lat}} := E^{\text{lat}}[\text{Eq.}(15.8)] + qV_0. \quad (14.19)$$

The lattice energy of Eq. (14.19) is corrected for the interaction with the background density and coincides with the lattice energy of an array of point charges (provided that the model charges do not overlap in the periodic array). Using Eqs. (14.18) and (14.19) instead of Eqs. (14.7) and (14.8) removes the shape dependence of the individual contributions (lattice energy and alignment term) to the defect energy correction in the non-overlapping case. The realignment nicely highlights the need for a consistent treatment of energies and potentials in this context, but is not needed in the practical approach.

References

1 Seebauer, E.G. and Kratzer M.C. (2006) *Mater. Sci. Eng. R*, **55**, 57.
2 Van de Walle C.G. and Neugebauer J. (2004) *J. Appl. Phys.*, **95**, 3851.
3 Rinke, P., de Walle, C.G.V., and Scheffler, M. (2009) *Phys. Rev. Lett.*, **102**, 026402.
4 Needs, R.J. (2007) in: *Theory of Defects in Semiconductors, Topics in Applied Physics*, Vol. 104, (eds D.A. Drabold and S.K. Estreicher, (Springer, Berlin, Heidelberg, New York), p. 141.
5 http://www.dft.sandia.gov/Quest/DFT_codes.html.
6 Nieminen, R. (2009) *Modell. Simul. Mater. Sci. Eng.*, **17**, 084001.
7 Leslie M. and Gillan M.J. (1985) *J. Phys. C, Solid State*, **18**, 973.
8 Makov G. and Payne M.C. (1995) *Phys. Rev. B* **51**, 4014.
9 Wright A.F. and Modine N.A. (2006) *Phys. Rev. B* **74**, 235209.
10 Shim J., Lee E.K., Lee Y.J., and Nieminen R.M., (2005) *Phys. Rev. B*, **71**, 035206.
11 Castleton, C.W.M., Höglund, A., and Mirbt, S. (2006) *Phys. Rev. B*, **73**, 035215.
12 Castleton, C.W.M., Höglund, A., and Mirbt, S. (2009) *Modell. Simul. Mater. Sci. Eng.*, **17**, 084003.
13 Lany, S. and Zunger, A. (2008) *Phys. Rev. B*, **78**, 235104.
14 Lany, S. and Zunger, A. (2009) *Modell. Simul. Mater. Sci. Eng.*, **17**, 084002.
15 Lento, J., Mozos, J.L., and Nieminen, R.M. (2002) *J. Phys.: Condens. Matter* **14**, 2637.
16 Gerstmann, U., Deák, P., Rurali, R., Aradi, B., Frauenheim, T., and Overhof, H. (2003) *Physica B*, **340–342**, 190.
17 Carloni, P., Blöchl, P., and Parinello, M. (1995) *J. Phys. Chem.*, **99**, 1338.
18 Schultz, P.A. (2000) *Phys. Rev. Lett.*, **84**, 1942.

19 Rozzi C.A., Varsano D., Marini A., Gross E.K.U., and Rubio A. (2006) *Phys. Rev. B,* **73**, 205119.
20 Schultz, P.A. (2006) *Phys. Rev. Lett.,* **96**, 246401.
21 Resta, R. (1994) *Rev. Mod. Phys.,* **66**, 899.
22 Resta, R. and Kunc, K. (1986) *Phys. Rev. B,* **34**, 7146.
23 Freysoldt, C., Neugebauer, J., and Van de Walle, C.G. (2009) *Phys. Rev. Lett.,* **102**, 016402.
24 Baroni, S. and Resta, R. (1986) *Phys. Rev. B,* **33**, 7017.
25 Gonze, X. and Lee, C. (1997) *Phys. Rev. B,* **55**, 10355.
26 Kunc, K. and Resta, R. (1983) *Phys. Rev. Lett.,* **51**, 686.
27 Pham, T.A., Li T., Shankar, S., Gygi, F., and Galli, G. (2010) *Appl. Phys. Lett.,* **96**, 062902.
28 Boeck, S., Freysoldt, C., Ismer, L., Dick, A., and Neugebauer, J. (2011) *Comput. phys. commun.,* **182**, 543, Web page: http://www.sphinxlib.de.
29 Bockstedte, M., Kley, A., Neugebauer, J., and Scheffler, M. (1997) *Comput. Phys. Commun.,* **107**, 187.
30 Kresse, G. and Joubert, D. (1999) *Phys. Rev. B,* **59**, 1758.

15
Formation Energies of Point Defects at Finite Temperatures
Blazej Grabowski, Tilmann Hickel, and Jörg Neugebauer

15.1
Introduction

A crucial quantity for the *ab initio* study of point defects is the defect formation free energy $F^f(V,T)$ as a function of volume V and temperature T. The dominant contribution to F^f is due to the zero temperature formation energy $E^f(V) = F^f(V, T=0K)$, which can be calculated at a relatively low computational cost. The calculation of higher order contributions such as quasiharmonic excitations (= non-interacting harmonic vibrations + effect of thermal expansion; Section 15.2.2.3) and anharmonic excitations (= interacting vibrations; Section 15.2.2.4) significantly increases the required computational resources. Since these effects are also expected to yield only a comparatively small contribution to F^f they are typically neglected. Indeed, it is expected that their influence on defect properties in semiconducting materials is far smaller than the inaccuracies resulting from the band gap problem. Thus, the majority of defect studies in semiconductors (see Ref. [1] for a recent review) are based on E^f with only a few exceptions [2, 3].

The situation is likely to change in the near future. Recent progress in the development of new exchange-correlation functionals [4, 5] and methods going beyond density functional theory (DFT) [6, 7] allows for a highly accurate prediction of band structures. At present, such calculations are computationally too expensive to be routinely applicable to total energy calculations of defects. However, the steady progress in methodological development and hardware components will soon close this gap. Then, the determination of the above mentioned higher order contributions will become critical.

For metals which do not suffer from the band gap problem, the situation is different. The highly efficient screening in metallic materials removes a large part of the self-interaction error which is mainly responsible for the band gap problem in semiconductors. As a consequence, significantly more accurate defect formation energies are obtained even with common local or semi-local exchange-correlation functionals such as the local density approximation (LDA) or the generalized gradient approximation (GGA). Thus, in order to reach the next accuracy level in defect

Table 15.1 Representative *ab initio* studies of point defect calculations in unary metals for the specific case of vacancies. The abbreviations are: 3d/4d/5d: respective transition elements; xc: exchange-correlation functional; LDA: local density approximation; GGA: generalized gradient approximation; PWps: planewaves with pseudopotentials; FP-LMTO: full potential linearized muffin tin orbitals; PW-PAW: planewaves with projector augmented waves; V: rescaled volume approach; P: constant pressure approach; volOpt: volume optimized approach (Section 15.2.1.1); F^f: defect formation free energy; E^f: ($T = 0$ K) contribution to F^f; el/qh/ah: electronic/quasiharmonic/anharmonic contribution to F^f; 1s: first shell (around the defect) contribution to the dynamical matrix; emp: empirical potential approach.

year	ref.	elements	xc	potential	strain	E^f	el	qh	ah
1989	[8]	Al	LDA	PWps	V	x			
1991	[9]	Li	LDA	PWps	V	x			
1993	[10]	Al,Cu,Ag,Rh	LDA	FP-LMTO	V	x			
1995	[11]	3d,4d,5d	LDA	FP-LMTO	V	x			
1997	[12]	Al	LDA	PWps	P	x			
1998	[13]	W	LDA	PWps	P	x	x		
1999	[14]	Ta	LDA	PWps	P	x	x		
2000	[15]	Al	LDA/GGA	PWps	P	x		x^{1s}	x^{emp}
2003	[16]	Al	LDA/GGA	PWps	P	x		x^{1s}	x^{emp}
2009	[17]	Fe	GGA	PW-PAW	V	x		x	
2009	[18]	Al	LDA/GGA	PWps/PW-PAW	volOpt/P	x	x	x	x

calculations, the inclusion of finite temperature contributions to F^f is already now of importance. Indeed, a review of the related literature clearly reveals such efforts (Table 15.1): starting in the late 1980s with the seminal work by Gillan [8], DFT-based studies of point defects were limited to the $T = 0$ K contribution E^f. This situation persisted roughly until the beginning of the new century, when studies [13, 14] of the electronic contribution to F^f – of crucial importance for some metallic materials [19] – appeared. In 2000 and 2003, Carling *et al.* [15, 16] provided a first *ab initio* based assessment of the quasiharmonic contribution to the vacancy of aluminum. To make such a study feasible at that time, the authors had to restrict the dynamics of the system to the first shell around the vacancy, *i.e.*, to the atomic shell which experiences the largest effect as compared to the perfect bulk. An *ab initio* based evaluation of the anharmonic contribution was computationally prohibitive at that time, which made it necessary to resort to empirical potentials. Major methodological improvements and the boost in computer power provide now the opportunity to study all relevant free energy contributions of defect formation in a rigorous *ab initio* manner (*cf.* Table 15.1).

In the present paper, we review the methodology required to compute defect concentrations from *ab initio* including the electronic, quasiharmonic, and anharmonic contributions to the formation free energy (Section 15.2.2). For their correct evaluation and interpretation it is important to correctly treat the strain induced by the periodic array of defects in a supercell approach, since an improper treatment may lead to errors of the same order of magnitude. We therefore review first the possible strategies and available correction schemes (Section 15.2.1). We then present results demonstrating the quality and performance of the methods (Section 15.3). The focus will be on point defects in aluminum since this material system can be produced with high chemical purity and crystalline quality, thus providing accurate experimental data as needed for a critical comparison. On the theoretical side, a good performance of available exchange-correlation functionals can be expected due to the free electron character of Al. All these aspects render aluminum to be a particularly attractive system for evaluating the performance of *ab initio* simulations of point defects. Indeed, as shown in Table 15.1 most theoretical studies focused on this system.

15.2
Methodology

15.2.1
Analysis of Approaches to Correct for the Spurious Elastic Interaction in a Supercell Approach

In the literature, two major approaches have been proposed and employed to correct for the artificial strain fields in a supercell approach arising from a collective interplay of all periodic images: (i) the rescaled volume and (ii) the constant pressure approach. Within the rescaled volume approach [8], the volume of the supercell containing the defect is rescaled such as to account for the volume of the missing atom. In contrast, within the constant pressure approach [12], the volume of the defect supercell is adjusted such as to correspond to the same pressure as is acting on the perfect bulk supercell (commonly zero pressure is assumed). In the limit of asymptotically large supercell sizes both approaches will converge to the same result. For realistic finite sized *ab initio* supercells the two approaches give different results. It is commonly accepted that the constant pressure method is superior to (more accurate than) the rescaled volume one. This is due to the fact that the latter imposes additional constraints on the system while the constant pressure approach allows the system to relax along all degrees of freedom including the shape of the supercell. A disadvantage of the constant pressure approach is that additional relaxations are needed, which significantly increase computational effort. This fact is illustrated in Table 15.1: early calculations of defect properties, when computer power was severely limited, were solely based on the rescaled volume approach. Only at the end of the 1990s one was able to achieve the next level of accuracy and employ the constant pressure approach. For the accurate calculation of finite temperature contributions to the defect formation energy employing the more accurate constant pressure approach is mandatory. Recently a more general approach, the volume optimized scheme, was proposed [18]. It takes higher order terms in the concentration dependence of F^f into account thus going beyond the constant pressure treatment. A consequence is that the formation free energy becomes concentration dependent.

Approximating to first order in the defect concentration, the constant pressure approach is obtained. Performing an additional approximation in the volume of defect formation, the rescaled volume approach can be derived. The relation and hierarchy between these approaches is discussed in the following.

15.2.1.1 The Volume Optimized Aapproach to Point Defect Properties

The central quantity, which contains all thermodynamic information about the system such as e.g., the defect concentration, is the free energy surface $F(\Omega, T; N, n)$ of a macroscopic crystal. In general, it depends on the crystal volume Ω (we reserve the symbol V for the atomic volume introduced below), temperature T, the number of atoms N, and the number of defects n. For the following discussion, we consider a large fictitious supercell (Born–von Karman cell) representing this macroscopic crystal (Figure 15.1).

The term fictitious refers to the fact that an actual calculation of this supercell is not feasible (and as will be discussed not necessary). The basic assumption is that the presence of defects leads to two kinds of effects: (i) strong distortions of the atoms close to the defect away from their ideal perfect bulk positions and (ii) long ranged volumetric distortions affecting only the lattice constant. Around each defect, a cell/box is constructed which we call defect cell. The defect cell needs to be large enough to cover the first kind of effect but not necessarily the second kind since this will be accounted for by the volume optimization introduced below. The defect cell contains N^d atoms, has a volume Ω^d, and a free energy $F^d = F^d(\Omega^d, T; N^d)$. According to this construction, the crystal outside the defect cells can be described

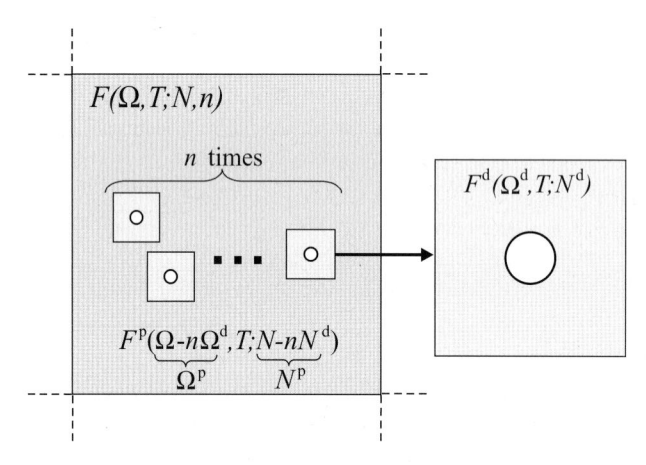

Figure 15.1 Schematic illustration of the concept to compute the free energy $F(\Omega, T; N, n)$ of a macroscopic crystal with defects. The larger box represents a supercell of volume Ω at temperature T containing N atoms and n defects. A light-gray shaded box with a white circle represents a cell of volume Ω^d, containing N^d atoms, exactly one defect, and having the free energy $F^d(\Omega^d, T; N^d)$. The dark-gray shaded region represents the perfect crystal without defects, with volume $\Omega^p = \Omega - n\Omega^d$, $N^p = N - nN^d$ atoms, and free energy $F^p(\Omega^p, T; N^p)$. The dashed lines indicate periodic boundary conditions.

by a perfect crystal with volume $\Omega^p = \Omega - n\Omega^d$, $N^p = N - nN^d$ atoms, and free energy $F^p = F^p(\Omega^p, T; N^p)$. The free energy of the fictitious supercell is then given by

$$F(\Omega, T; N, n, \Omega^d, N^d) = F^p(\Omega^p, T; N^p) + nF^d(\Omega^d, T; N^d) + F^{\text{conf}}(T, N, n), \quad (15.1)$$

where we have also explicitly indicated the dependence of F on the volume of the defect cell Ω^d and the number of atoms in the defect cell N^d. These dependencies and their treatment will be discussed in the following. Further in Eq. (15.1), F^{conf} is the configurational free energy of the defects in the dilute limit. It is approximated using Stirling's formula by [20] $F^{\text{conf}} \approx -k_B T[n - n\ln(n/N)]$, with the Boltzmann constant k_B. The volume optimized approach [18] is based on the key observation that in equilibrium the total free energy F is minimal with respect to changes in the volume of the defect cell,

$$\partial F / \partial \Omega^d \equiv 0, \quad (15.2)$$

so that the volume of the perfect crystal and of the defect cells will adjust self consistently, i.e., until the optimum volume for both is achieved when minimizing F. The actual result of the minimization procedure will depend on the free energy volume curves of the perfect crystal and of the defect cell.

Equation (15.1) can be further transformed. The free energy of the perfect crystal F^p scales with the number of atoms due to its extensivity property,

$$F^p(\Omega^p, T; N^p)/N^p = F^p(V^p, T; 1)$$
$$=: F^p(V^p, T) = F^p\left(\frac{V - c\Omega^d}{1 - cN^d}, T\right), \quad (15.3)$$

with

$$V^p = \frac{\Omega^p}{N^p} = \frac{V - c\Omega^d}{1 - cN^d}, \quad (15.4)$$

where we have defined the volume per atom $V = \Omega/N$ and the concentration of defects $c = n/N$. Using this property to rewrite Eq. (15.1) yields

$$F(\Omega, T; N, n, \Omega^d, N^d)/N = (1 - cN^d) F^p(V^p, T)$$
$$+ cF^d(\Omega^d, T; N^d) + F^{\text{conf}}(T, c) =: F(V, T; c, \Omega^d, N^d), \quad (15.5)$$

where the configurational free energy depends now on the concentration as $F^{\text{conf}}(T, c) = -ck_B T(1 - \ln c)$. Equation (15.5) is independent of N and defines the free energy per atom $F(V, T; c, \Omega^d, N^d)$ of the full supercell consisting of the perfect crystal and the defect cells. Now, a formation free energy F^f can be defined, which turns out to be concentration dependent

$$F^f(V, T; c, \Omega^d, N^d) = F^d(\Omega^d, T; N^d) - N^d F^p(V^p, T), \quad (15.6)$$

and thus:

$$F(V, T; c, \Omega^d, N^d) = F^p(V^p, T) + cF^f(V, T; c, \Omega^d, N^d) + F^{conf}(T, c). \quad (15.7)$$

Applying next the equilibrium condition with respect to the defect concentration, $\partial F/\partial c = 0$, to Eq. (15.7) yields an equation for the equilibrium defect concentration c^{eq}

$$-k_B T \ln c^{eq} = F^f(V, T; c^{eq}, \Omega^d, N^d) - \frac{v^f P^p}{c^{eq} N^d - 1}, \quad (15.8)$$

where the volume of defect formation $v^f = \Omega^d - N^d V$ and the pressure inside the perfect bulk $P^p = -\partial F^p/\partial V^p$ have been defined. It is straightforward to show that the latter equals the external pressure $P = -\partial F/\partial V$:

$$P = -\frac{\partial F}{\partial V} = -(1 - cN^d)\frac{\partial F^p}{\partial V^p}\frac{\partial V^p}{\partial V} = -\frac{\partial F^p}{\partial V^p} = P^p. \quad (15.9)$$

Further, it follows from Eq. (15.2) that

$$\frac{\partial F}{\partial \Omega^d} = (1 - cN^d)\frac{\partial F^p}{\partial V^p}\frac{\partial V^p}{\partial \Omega^d} + c\frac{\partial F^d}{\partial \Omega^d}$$

$$= c\left(\frac{\partial F^d}{\partial \Omega^d} - \frac{\partial F^p}{\partial V^p}\right) = c(-P^d + P^p) \equiv 0, \quad (15.10)$$

with $P^d = -\partial F^d/\partial \Omega^d$ the pressure inside the defect cell. Hence, the equilibrium defect cell volume $\Omega^{d,eq}$ is obtained when the pressure inside the defect cells equals the pressure inside the perfect cell which, according to Eq. (15.9), equals the external pressure, i.e., $P^d = P^p = P$.

Using $\Omega^{d,eq}$ and Eqs. (15.8, 15.9), Eq. (15.7) can be transformed to the final expression for the free energy of a crystal at (atomic) volume V and temperature T and with an equilibrium concentration of thermally excited defects:

$$F(V, T) = F(V, T; c^{eq}, \Omega^{d,eq}, N^d)$$
$$= F^p\left(\frac{V - c^{eq}\Omega^{d,eq}}{1 - c^{eq} N^d}, T\right) + \frac{c^{eq} v^f P}{c^{eq} N^d - 1} - c^{eq} k_B T. \quad (15.11)$$

The parameter N^d is determined by the specific supercell used for the defect calculation and has to be checked for convergence. Note that Eq. (15.8) [and thus Eq. (15.11)] cannot be solved for c^{eq} in closed form due to the dependence of F^f on c. The actual equilibrium defect concentration must be thus solved self consistently.

15.2.1.2 Derivation of the Constant Pressure and Rescaled Volume Approach

Employing well defined approximations, the constant pressure and rescaled volume approaches can be easily derived from the volume optimized approach. Let us first show the relation to the constant pressure approach. We therefore Taylor expand F in Eq. (15.5) as a function of c around $c = 0$:

$$F(V,T) = F^p(V,T) + c[F^d(\Omega^d, T; N^d) - N^d F^p(V,T) + Pv^f] + F^{conf}(T,c) + O(c^2), \tag{15.12}$$

where Eq. (15.9) has been used. Retaining only the terms linear in c, the expression for the free energy within the constant pressure approach is obtained [21]:

$$F(V,T) \approx F^p(V,T) + F^{conf}(T,c) + c \underbrace{[F^d(\Omega^d, T; N^d) - N^d F^p(V,T) + Pv^f]}_{=F^f_p,\ \text{formation free energy at const. pressure.}} . \tag{15.13}$$

The term in the square brackets defines the defect formation free energy at constant pressure, since Ω^d needs to be chosen such as to satisfy the pressure equality, Eq. (15.10). The term correctly includes the enthalpic Pv^f contribution. This contribution has been intensively discussed over many years in literature: in their textbook, Varotsos and Alexopoulos [21] stress that it has been frequently ignored in point defect studies. For a correct description at nonzero pressure it needs however to be included. Based on the above derivation, we can straightforwardly analyze the necessity of this contribution. It naturally arises from the Taylor expansion in Eq. (15.12) and needs to be taken into account since it is part of the first order term. Physically, the Pv^f term reflects the fact that the work needed to form a defect depends on the pressure and likewise on the volume changes it induces. Equation (15.13) can be simplified in a standard way [21] by using the defect equilibrium condition $\partial F/\partial c \equiv 0$ to yield $F(V,T) \approx F^p(V,T) - k_B T c^{eq}(V,T)$, with the equilibrium defect concentration $c^{eq}(V,T) = \exp[-F^f_p(V,T)/(k_B T)]$.

The rescaled volume approach can be easily derived from Eq. (15.12). We therefore note that $F^d(\Omega^d, T; N^d)$ is the zeroth order term of a Taylor expansion of $F^d(\Omega^d + v^f, T; N^d)$ in the volume of defect formation v^f around $v^f = 0$:

$$\begin{aligned}F^d(N^d V, T; N^d) &= F^d(\Omega^d + v^f, T; N^d) \\ &= F^d(\Omega^d, T; N^d) + Pv^f + O([v^f]^2).\end{aligned} \tag{15.14}$$

In the above equation the first equality follows from the definition of the volume of defect formation. Further, $-\partial F^d/\partial \Omega^d = P$ due to Eqs. (15.9) and (15.10) has been used. Approximating to first order in v^f, rearranging with respect to $F^d(\Omega^d, T; N^d)$, and plugging into Eq. (15.13) yields:

$$F(V,T) \approx F^p(V,T) + F^{conf}(T,c) + c\ \underbrace{[F^d(N^d V, T; N^d) - N^d F^p(V,T)]}_{=F^f_V,\ \text{formation free energy at rescaled volume.}} \tag{15.15}$$

The quantity in square brackets is the formation free energy of the rescaled volume approach [8]. To make this even more apparent, the extensivity property of F^p can be used to write F^f_V as

$$F^f_V(V,T; N^d) = F^d(N^d V, T; N^d) - \frac{N^d}{N^d \pm 1} F^p([N^d \pm 1]V, T; N^d \pm 1), \tag{15.16}$$

with the plus (minus) sign referring to vacancies (self interstitials). As for the constant pressure case, Eq. (15.15) can be simplified to $F(V, T) \approx F^p(V, T) - k_B T c^{eq}(V, T)$, with the equilibrium defect concentration given now by $c^{eq}(V, T) = \exp[-F_V^f(V, T)/(k_B T)]$.

The preceding derivations show that the two standard approaches arise as natural approximations of the volume optimized method. In particular, a hierarchy of approximations can be identified: first, the constant pressure approach arises by terminating the Taylor series in the defect concentration in Eq. (15.12) after the first order term. The rescaled volume approach needs a further approximation by terminating the Taylor series in the volume of defect formation in Eq. (15.14) likewise after the first order term. For practical purposes we note the following: The approximation in the defect concentration, Eq. (15.13), is well motivated, since the basic assumption of non interacting defects, *i.e.*, the dilute limit, is valid only for low defect concentrations. We therefore recommend to employ the constant pressure approach, also because of numerical instabilities in the volume optimized method when approaching $c^{eq} N^d \approx 1$ due to the denominator in Eq. (15.11). In contrast, the approximations needed to derive the rescaled volume approach are not appropriate for realistic supercell sizes and will result in sizeable errors even for low defect concentrations (see *e.g.*, Figure 15.7).

15.2.2
Electronic, Quasiharmonic, and Anharmonic Contributions to the Formation Free Energy

Let us now focus on the methodology needed to compute the electronic, quasiharmonic, and anharmonic contribution to the free energy surface. For a defect calculation, we need both the free energy of the defect cell $F^d(V, T)$ and the free energy of the perfect bulk $F^p(V, T)$ as a function of volume and temperature as outlined in the previous section. In the following we will derive the necessary steps with particular emphasis on numerical efficiency. The latter issue is crucial to allow a full *ab initio* determination of all contributions, specifically including the anharmonic one. Except for the remark at the end of Section 15.2.2.3 all considerations refer to both free energy surfaces (F^d and F^p) and we therefore use generically the symbol F.

15.2.2.1 Free Energy Born–Oppenheimer Approximation
Starting point is an expression for the free energy surface in which the ionic and electronic degrees of freedom are decoupled quantum mechanically (Born–Oppenheimer approximation), but which still contains the effect of thermodynamic electronic excitations on the ionic vibrations. For that purpose, the "standard" Born–Oppenheimer approximation [22] needs to be extended to the so called free energy Born–Oppenheimer approximation which was introduced by Cao and Berne [23] in 1993.

Within the standard Born–Oppenheimer approximation, the free energy F of a system consisting of electrons and ions is written as:

$$F = -k_B T \ln Z \quad \text{where} \quad Z = \sum_{\nu,\mu} e^{-\beta E_{\nu,\mu}^{\text{nuc}}}, \tag{15.17}$$

with

$$E_{\nu,\mu}^{\text{nuc}} = \langle \Lambda_{\nu,\mu} | (\hat{T}^{\text{nuc}} + \hat{1} E_\nu^{\text{el}}) | \Lambda_{\nu,\mu} \rangle, \tag{15.18}$$

and $\beta = (k_B T)^{-1}$. In Eq. (15.17), we have defined the partition function Z of the system in which the sums run over an electronic quantum number ν and an ionic quantum number μ. The energy levels $E_{\nu,\mu}^{\text{nuc}}$ and eigenfunctions $\Lambda_{\nu,\mu}$ are solutions to the nuclei Schrödinger equation with the Hamiltonian $(\hat{T}^{\text{nuc}} + \hat{1} E_\nu^{\text{el}})$, in which $\hat{T}^{\text{nuc}}$ is the ionic kinetic energy operator, $\hat{1}$ is the identity operator, and E_ν^{el} are potential energy surfaces generated by the electronic system, i.e., the solutions to the electronic Schrödinger equation (cf. Figure 15.2). (As commonly done, we include the nucleus–nucleus interaction into E_ν^{el}.) We can transform the partition function as

$$\begin{aligned} Z &= \sum_{\nu,\mu} e^{-\beta \langle \Lambda_{\nu,\mu} | (\hat{T}^{\text{nuc}} + \hat{1} E_\nu^{\text{el}}) | \Lambda_{\nu,\mu} \rangle} \\ &= \sum_{\nu,\mu} \langle \Lambda_{\nu,\mu} | e^{-\beta (\hat{T}^{\text{nuc}} + \hat{1} E_\nu^{\text{el}})} | \Lambda_{\nu,\mu} \rangle, \end{aligned} \tag{15.19}$$

since the $\Lambda_{\nu,\mu}$ are eigenfunctions of $(\hat{T}^{\text{nuc}} + \hat{1} E_\nu^{\text{el}})$. It would be now desirable to factorize the exponential to separate the E_ν^{el}. This factorization needs however to be performed

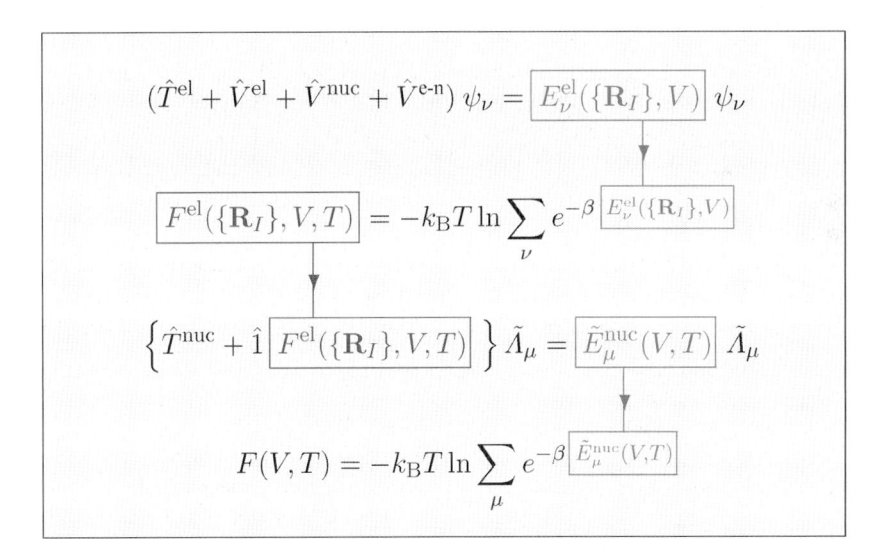

$$(\hat{T}^{\text{el}} + \hat{V}^{\text{el}} + \hat{V}^{\text{nuc}} + \hat{V}^{\text{e-n}}) \psi_\nu = \boxed{E_\nu^{\text{el}}(\{\mathbf{R}_I\}, V)} \psi_\nu$$

$$\boxed{F^{\text{el}}(\{\mathbf{R}_I\}, V, T)} = -k_B T \ln \sum_\nu e^{-\beta \boxed{E_\nu^{\text{el}}(\{\mathbf{R}_I\}, V)}}$$

$$\left\{ \hat{T}^{\text{nuc}} + \hat{1} \boxed{F^{\text{el}}(\{\mathbf{R}_I\}, V, T)} \right\} \tilde{\Lambda}_\mu = \boxed{\tilde{E}_\mu^{\text{nuc}}(V, T)} \tilde{\Lambda}_\mu$$

$$F(V, T) = -k_B T \ln \sum_\mu e^{-\beta \boxed{\tilde{E}_\mu^{\text{nuc}}(V, T)}}$$

Figure 15.2 (online colour at: www.pss-b.com) Key equations to compute the free energy Born–Oppenheimer surface. Here, ψ_ν are electronic wave functions and $(\hat{T}^{\text{el}} + \hat{V}^{\text{el}} + \hat{V}^{\text{nuc}} + \hat{V}^{\text{e-n}})$ is the electronic Hamiltonian with the electronic kinetic energy operator, electron–electron repulsion operator, nucleus–nucleus repulsion operator, and electron–nucleus attraction operator, respectively. The remaining quantities are defined in Section 15.2.2.1.

with caution since $\hat{T}^{\text{nuc}}$ and $\hat{1}\,E^{\text{el}}_\nu$ are non-commuting operators (both depend on the ionic coordinates). Therefore, we have to apply the so-called Zassenhaus formula [24]:

$$e^{-\beta(\hat{T}^{\text{nuc}}+\hat{1}\,E^{\text{el}}_\nu)} = e^{-\beta\hat{T}^{\text{nuc}}}\,e^{-\beta\hat{1}\,E^{\text{el}}_\nu}\,e^{\beta^2/2[\hat{T}^{\text{nuc}},\hat{1}\,E^{\text{el}}_\nu]}\,e^{-\beta^3/6(2[\hat{1}\,E^{\text{el}}_\nu,[\hat{T}^{\text{nuc}},\hat{1}\,E^{\text{el}}_\nu]]+[\hat{T}^{\text{nuc}},[\hat{T}^{\text{nuc}},\hat{1}\,E^{\text{el}}_\nu]])}\cdots \quad (15.20)$$

Here, the dots denote exponentials corresponding to higher orders in β and with increasingly nested commutators. An explicit formula for the higher order terms is given in Ref. [24]. In Ref. [23], it is shown that exponentials corresponding to orders β^2 and higher will be small if $m_e \ll M$ (m_e = electron mass, M = nucleus mass), i.e., under the same condition as assumed in the standard Born–Oppenheimer approximation. It is therefore justified to approximate

$$e^{-\beta(\hat{T}^{\text{nuc}}+\hat{1}\,E^{\text{el}}_\nu)} \approx e^{-\beta\hat{T}^{\text{nuc}}}\,e^{-\beta\hat{1}\,E^{\text{el}}_\nu}, \quad (15.21)$$

in any case in which the Born–Oppenheimer approximation is justified. Using Eq. (15.21), the fact that $(e^{-\beta\hat{T}^{\text{nuc}}}\,e^{-\beta\hat{1}\,E^{\text{el}}_\nu})$ corresponds to a block diagonal matrix and the invariance property of the trace (which allows to choose the same basis for different μ, e.g., $\nu = \nu'$; with a fixed ν';) yields:

$$Z = \sum_{\nu,\mu} \langle \Lambda_{\nu,\mu} | e^{-\beta\hat{T}^{\text{nuc}}}\,e^{-\beta\hat{1}\,E^{\text{el}}_\nu} | \Lambda_{\nu,\mu} \rangle$$

$$= \sum_{\nu,\mu} \langle \Lambda_{\nu',\mu} | e^{-\beta\hat{T}^{\text{nuc}}}\,e^{-\beta\hat{1}\,E^{\text{el}}_\nu} | \Lambda_{\nu',\mu} \rangle$$

$$= \sum_{\mu} \langle \Lambda_{\nu',\mu} | e^{-\beta\hat{T}^{\text{nuc}}} \left(\sum_\mu e^{-\beta\hat{1}\,E^{\text{el}}_\nu} \right) | \Lambda_{\nu',\mu} \rangle \quad (15.22)$$

$$= \sum_{\mu} \langle \Lambda_{\nu',\mu} | e^{-\beta\hat{T}^{\text{nuc}}}\,e^{-\beta\hat{1}\,F^{\text{el}}} | \Lambda_{\nu',\mu} \rangle,$$

with the electronic free energy defined by:

$$F^{\text{el}} := -k_B T \ln \sum_\nu \exp[-E^{\text{el}}_\nu/(k_B T)]. \quad (15.23)$$

In order to recombine the exponentials again, we need to apply the Baker–Campbell–Hausdorff formula which reads [24]:

$$e^{-\beta\hat{T}^{\text{nuc}}}\,e^{-\beta\hat{1}\,F^{\text{el}}} = e^{-\beta(\hat{T}^{\text{nuc}}+\hat{1}\,F^{\text{el}})-\frac{\beta^2}{2}[\hat{T}^{\text{nuc}},\hat{1}\,F^{\text{el}}]-\frac{\beta^3}{6}[[\hat{T}^{\text{nuc}},\hat{1}\,F^{\text{el}}],\hat{1}\,F^{\text{el}}-\hat{T}^{\text{nuc}}]\cdots}. \quad (15.24)$$

The terms in the exponential of order β^2 and higher correspond again to terms which are small if $m_e \ll M$, thus justifying the following approximation:

$$e^{-\beta\hat{T}^{\text{nuc}}}\,e^{-\beta\hat{1}\,F^{\text{el}}} \approx e^{-\beta(\hat{T}^{\text{nuc}}+\hat{1}\,F^{\text{el}})}. \quad (15.25)$$

In fact, as shown in Ref. [23] the terms neglected in Eq. (15.25) lead to contributions which are of the same order as the contributions from the neglected terms in

Eq. (15.21) having however the opposite sign. Therefore, the approximations performed in Eqs. (15.21) and (15.25) partially compensate each other. Inserting Eq. (15.25) into Eq. (15.22) yields for the partition function and thus the free energy:

$$F = -k_B T \ln Z, \tag{15.26}$$

with

$$Z = \sum_\mu \langle \Lambda_\mu | e^{-\beta(\hat{T}^{nuc} + \hat{1} F^{el})} | \Lambda_\mu \rangle. \tag{15.27}$$

We can rewrite this to a more convenient notation as

$$F = -k_B T \ln \sum_\mu e^{-\beta \tilde{E}_\mu^{nuc}}, \tag{15.28}$$

with

$$\left(\hat{T}^{nuc} + \hat{1} F^{el}(\{R_I\}, V, T)\right)\tilde{\Lambda}_\mu = \tilde{E}_\mu^{nuc} \tilde{\Lambda}_\mu, \tag{15.29}$$

where we have defined an effective nuclei Schrödinger equation with eigenfunctions $\tilde{\Lambda}_\mu$ and eigenvalues $\tilde{E}_\mu^{nuc}$. In Eq. (15.29), we have also explicitly written the dependence of F^{el} on the set of nuclei coordinates $\{R_I\}$ and the crystal volume V which is a consequence of the fact that each of the electronic potential energy surfaces E_ν^{el} depends on $\{R_I\}$ and V. The key equations summarizing the preceding derivation are collected in Figure 15.2.

The central step which allows for a separation into an electronic, quasiharmonic, and anharmonic part is a Taylor expansion of F^{el} in Eq. (15.28) in the $\{R_I\}$ around the $T = 0$ K equilibrium positions $\{R_I^0\}$:

$$F^{el} = F_0^{el} + \frac{1}{2}\sum_{k,l} u_k u_l \left[\frac{\partial^2 F^{el}}{\partial R_k \partial R_l}\right]_{\{R_I^0\}} + O(u^3). \tag{15.30}$$

Here, the zeroth order term is abbreviated as $F_0^{el}(V, T) := F^{el}(\{R_I^0\}, V, T)$, k and l run over all nuclei of the system and additionally over the three spatial dimensions for each nucleus, and $u_k = R_k - R_k^0$ is the displacement out of equilibrium. The expansion does not contain a first order term, since such a term relates to the atomic forces that are absent in an equilibrium structure. Each of the other terms in Eq. (15.30) corresponds to a different excitation mechanism as emphasized in Figure 15.3. The exact derivations are given in the following sections.

15.2.2.2 Electronic Excitations

It is convenient to decompose the electronic part $F_0^{el}(V, T)$ in Eq. (15.30) into a temperature independent part $E_{g,0}^{el}(V)$, which corresponds to the ground state of the potential energy surfaces E_ν^{el} at $\{R_I^0\}$, and a remainder $\tilde{F}_0^{el}(V, T)$ carrying the temperature dependence of the electronic system, *i.e.*, the electronic excitations: $F_0^{el}(V, T) = E_{g,0}^{el}(V) + \tilde{F}_0^{el}(V, T)$. The reason for this separation is that $\tilde{F}_0^{el}(V, T)$ can be accurately described with low order polynomials, while $E_{g,0}^{el}(V)$ can be parame-

$$F^{\text{el}}(\{\mathbf{R}_I\},V,T) = \underbrace{F_0^{\text{el}}(V,T)}_{\substack{\text{electronic contr.}\\ \to \text{Sec. 2.2.2}}} + \underbrace{\frac{1}{2}\sum_{k,l} u_k u_l \left[\frac{\partial^2 F^{\text{el}}(\{\mathbf{R}_I\},V,T)}{\partial R_k \partial R_l}\right]_{\{\mathbf{R}_I^0\}}}_{\substack{\text{quasiharmonic contribution} \to \text{Sec. 2.2.3}}} + \underbrace{\text{orders } u^3 \text{ and higher}}_{\substack{\text{anharmonic contribution}\\ \to \text{Secs. 2.2.4 and 2.2.5}}}$$

Figure 15.3 (online colour at: www.pss-b.com) Taylor expansion of the electronic free energy F^{el} as in Eq. (15.20) (see text for definitions). The color coding emphasizes the connection with the equations in Figure 15.2. The under braces indicate the type of contribution arising from each term and the section covering the issue.

trized using standard equations-of-state. To calculate the latter a standard DFT approach is sufficient. For the calculation of $\tilde{F}_0^{el}$, one needs instead to employ the finite temperature extension of DFT as originally developed by Mermin [25]. This approach is implemented in typical DFT codes and it amounts to using a Fermi–Dirac occupation distribution for the Kohn–Sham electronic energy levels,

$$\tilde{F}_0^{el}(V,T) \approx k_B T/2 \sum_i [f_i \ln f_i + (1-f_i)\ln(1-f_i)], \qquad (15.31)$$

with self consistently determined Kohn–Sham occupation numbers $f_i = f_i(V,T)$. Note that Eq. (15.31) is only approximately valid since there is a small contribution from the kinetic energy term, which is however fully accounted for in an actual finite temperature DFT calculation.

While the inclusion of F_0^{el} is important for metals at realistic temperatures, for semiconductors it is negligible except for narrow band gap semiconductors, where also partial occupations f_i may occur. In metals, $\tilde{F}_0^{el}$ can become significant particularly if the density of states shows a peak close to the Fermi energy as found for instance for d-states in Pt, Pd, Rh, or Ir [19].

15.2.2.3 Quasiharmonic Atomic Excitations

Neglecting for the moment the higher order terms in Eq. (15.29), the quasiharmonic approximation results. The necessary steps to compute this contribution are

$$D_{k,l}(V,T) := \frac{1}{\sqrt{M_k M_l}} \left[\frac{\partial^2 F^{el}(\{R_I\}, V, T)}{\partial R_k \partial R_l} \right]_{\{R_I^0\}}$$

$$\rightarrow D(V,T) w_i = \omega_i^2(V,T) w_i \qquad (15.32)$$

$$\rightarrow E_{\{n_i\}}^{qh}(V,T) = F_0^{el}(V,T) + \sum_i \hbar\omega_i(V,T)\left(n_i + \frac{1}{2}\right),$$

with n_i the number of phonons in state i. In Eq. (15.32), the dynamical matrix D (with elements $D_{k,l}$) has been defined, which corresponds to the second derivative of F^{el} scaled by the masses M_k of the nuclei. The eigenvalue equation of D with eigenvectors w_i and eigenvalues ω_i^2 defines the phonon frequencies ω_i. The energy $E_{\{n_i\}}^{qh}$ for a certain fixed phonon occupation configuration $\{n_i\}$ is given by a sum over the frequencies weighted by the corresponding occupation numbers. Note the important point that the phonon frequencies are not only volume dependent (quasiharmonic) but also explicitly temperature dependent through the temperature dependence of F^{el}. This temperature dependence does not correspond to an anharmonic atomic (i.e., phonon–phonon) interaction. To make this point explicit, we will use the notation T^{el} for this temperature, while the temperature determining the thermodynamics of the nuclei will be denoted by T^{nuc}. In fact, to speed up numeric convergence, T^{el} and T^{nuc} can be varied independently of each other. Of course, at the end of the calculation both have to be ensured to be equal to the actual external temperature, i.e., $T^{el} = T^{nuc} = T$.

The final step of the quasiharmonic approximation is to approximate the eigenvalues $\tilde{E}_\mu^{nuc}$ in Eq. (15.28) by the quasiharmonic energy $E_{\{n_i\}}^{qh}$ which yields (the tilde in

$\tilde{F}$ indicating the approximation at this stage)

$$\tilde{F}(V,T) = [\tilde{F}(V, T^{\text{el}}, T^{\text{nuc}})]_{T^{\text{el}}=T^{\text{nuc}}=T},\qquad(15.33)$$

with

$$\tilde{F}(V, T^{\text{el}}, T^{\text{nuc}}) \approx -k_B T \ln \sum_{\{n_i\}} e^{-\beta \bar{E}^{\text{qh}}_{\{n_i\}}} = \ldots = F_0^{\text{el}}(V, T^{\text{el}}) + F^{\text{qh}}(V, T^{\text{el}}, T^{\text{nuc}}),\qquad(15.34)$$

where the dots denote a series of straightforward transformations (see *e.g.*, Ref. [26]) and with

$$F^{\text{qh}}(V, T^{\text{el}}, T^{\text{nuc}}) = \frac{1}{N}\sum_i \left[\underbrace{\frac{\hbar}{2}\omega_i(V, T^{\text{el}})}_{T^{\text{nuc}}=0K\text{ zero point contr.}} + T^{\text{nuc}} \underbrace{k_B \ln[1-\exp\{-\beta\hbar\omega_i(V, T^{\text{el}})\}]}_{\text{(negative) entropic contribution.}} \right].$$

$$(15.35)$$

Here, N is an appropriately chosen scaling factor which is, *e.g.*, the number of sampled frequencies divided by three (number of spatial dimensions) if F^{qh} should refer to a "per atom" quantity. Note that F^{qh} fully contains quantum mechanical effects (*e.g.*, zero point vibrations) in contrast to a free energy obtained from a classical molecular dynamics run.

The quasiharmonic equations presented above are general and equally well applicable to defect and bulk cells. In practice, the calculation of the quasiharmonic free energy contribution of the defect cell requires special attention to ensure a consistent treatment with the corresponding perfect bulk cell. The reason for this is the break of translational symmetry introduced by the presence of the defect. As a consequence, the commonly applied [19] Fourier interpolation to generate dense wave vector meshes in the Brillouin zone cannot be employed to the defect cell. To nonetheless profit from the advantage of a Fourier interpolation for the perfect bulk (*e.g.*, a significantly improved description of the low temperature free energy) a correction scheme as *e.g.*, proposed in Ref. [18] should be used.

15.2.2.4 Anharmonic Atomic Excitations: Thermodynamic Integration

Let us now consider the higher order terms in Eq. (15.30). An intuitive and straightforward approach to include these terms would be an *ab initio* based classical molecular dynamics run. The fact that a direct computation of the free energy employing conventional molecular dynamics is not feasible (the free energy is an entropic quantity [27]) could be circumvented by calculating the inner energy and integrating it with respect to temperature. It turns out that such a "naive" phase space sampling leads to infeasibly long computational times [18]. Therefore, the development and application of highly efficient sampling strategies to perform the thermodynamic averages is crucial.

A fundamental concept is thermodynamic integration. The key idea is to start from a reference with an analytically known free energy or with a free energy that can be obtained numerically extremely fast. This reference is coupled adiabatically to the true *ab initio* potential energy surface and only the difference in free energies is sampled. Such an approach will be efficient, if one manages to construct a reference which closely approximates the true *ab initio* potential energy surface thus yielding a small free energy difference.

A good reference to perform thermodynamic integration is the quasiharmonic potential energy surface. Therefore, a system with free energy F_λ is introduced which couples the electronic + quasiharmonic system [having the free energy $\tilde{F}$ as given in Eq. (15.34)] to the full system (having free energy F) by the adiabatic switching parameter λ. The boundary conditions are defined to be

$$[F_\lambda]_{\lambda=1} = F \quad \text{and} \quad [F_\lambda]_{\lambda=0} = \tilde{F}, \tag{15.36}$$

so that the anharmonic free energy is given by

$$F^{\text{ah}} = F - \tilde{F} = [F_\lambda]_{\lambda=1} - [F_\lambda]_{\lambda=0}. \tag{15.37}$$

In principle, methods based on the quantum-classical isomorphism exist allowing to perform thermodynamic integration including quantum mechanical effects [28]. Unfortunately, their actual application requires very large computational efforts and at present only investigations employing empirical potential energy surfaces can be afforded [28]. In practice, a quantum mechanical treatment of F^{ah} is not expected to be necessary since the quasiharmonic free energy contains the major part of quantum mechanical effects (as derived in the previous section). The reason for this is that at low temperatures, where quantum effects are important, the quasiharmonic approximation is excellent thus fully accounting for such effects, while at high temperatures where the quasiharmonic approximation fails and anharmonicity becomes significant quantum effects become small/negligible. The anharmonic contribution can therefore be calculated in the classical limit (superscript "clas;" note that correspondingly a classical quasiharmonic reference is used):

$$F^{\text{clas,ah}} := F^{\text{clas}} - (F_0^{\text{el}} + F^{\text{clas,qh}}) = [F_\lambda^{\text{clas}}]_{\lambda=1} - [F_\lambda^{\text{clas}}]_{\lambda=0} = \int_0^1 d\lambda \frac{\partial F_\lambda^{\text{clas}}}{\partial \lambda}$$

$$\stackrel{\text{Eq. (15.39)}}{=} \int_0^1 d\lambda \left\langle \frac{\partial F_\lambda^{\text{el}}}{\partial \lambda} \right\rangle_{T,\lambda} \stackrel{\text{ergodicity hypothesis}}{=} \int_0^1 d\lambda \left\langle \frac{\partial F_\lambda^{\text{el}}}{\partial \lambda} \right\rangle_{t,\lambda}. \tag{15.38}$$

Here, F_λ^{el} is the λ dependent electronic free energy surface determining the classical motion of the nuclei in the coupled system, $\langle \cdot \rangle_{T,\lambda}$ denotes the thermodynamic, and $\langle \cdot \rangle_{t,\lambda}$ the time average at a given λ. To obtain the first equality in the third line, we have used:

$$\frac{\partial F^{\text{clas}}_\lambda}{\partial \lambda} = \frac{1}{Z^{\text{clas}}_\lambda} \int \frac{d\mathbf{R}_I}{\Omega^{\text{ph}}} \, e^{-\beta F^{\text{el}}_\lambda(\{\mathbf{R}_I\})} \frac{\partial F^{\text{el}}_\lambda}{\partial \lambda}$$

$$= \left\langle \frac{\partial F^{\text{el}}_\lambda}{\partial \lambda} \right\rangle_{T,\lambda}, \qquad (15.39)$$

with the standard definitions of a classical free energy F^{clas}, partition function Z^{clas}, and phase space volume Ω^{ph}. Besides the fixed boundary conditions, Eq. (15.36), any type of coupled system can be chosen. In practice, a simple linear coupling to the quasiharmonic reference (restoring the explicit notation for the volume and temperature dependencies),

$$F^{\text{el}}_\lambda(\{\mathbf{R}_I\}, V, T^{\text{el}}) = \lambda F^{\text{el}}(\{\mathbf{R}_I\}, V, T^{\text{el}})$$
$$+ (1 - \lambda) \left[F^{\text{el}}_0(V, T^{\text{el}}) + \sum_{k,l} \frac{\sqrt{M_k M_l}}{2} u_k u_l D_{k,l}(V, T^{\text{el}}) \right], \qquad (15.40)$$

yields computationally efficient results. Finally, the anharmonic free energy reads:

$$F^{\text{clas,ah}}(V, T^{\text{el}}, T^{\text{nuc}}) = \int_0^1 d\lambda \Bigg\langle F^{\text{el}}(V, T^{\text{el}}) - F^{\text{el}}_0(V, T^{\text{el}})$$
$$- \sum_{k,l} \frac{\sqrt{M_k M_l}}{2} u_k u_l D_{k,l}(V, T^{\text{el}}) \Bigg\rangle_{t,\lambda} . \qquad (15.41)$$

Note that the dependence of $F^{\text{clas,ah}}$ on T^{nuc} is hidden in the time average $\langle . \rangle_t$ which needs to be obtained e.g. from a molecular dynamics simulation. Combining the various contributions, the free energy of the system is given by:

$$F = E^{\text{el}}_{g,0} + \widetilde{F}^{\text{el}}_0 + F^{\text{qh}} + F^{\text{clas,ah}}$$
$$= [E^{\text{el}}_{g,0}(V) + \widetilde{F}^{\text{cl}}_0(V, T^{\text{el}}) + F^{\text{qh}}(V, T^{\text{el}}, T^{\text{nuc}})$$
$$+ F^{\text{clas,ah}}(V, T^{\text{el}}, T^{\text{nuc}})]_{T^{\text{el}} = T^{\text{nuc}} = T}. \qquad (15.42)$$

We stress that the dominant part of quantum mechanical effects contributing to F is accounted for by F^{qh}. For the case of (wide band gap) semiconductors the influence of T^{el} will be small (cf. Section 15.2.2.2) and the free energy can be approximated by:

$$F(V, T) \approx [E^{\text{el}}_{g,0}(V) + F^{\text{qh}}(V, 0\text{K}, T^{\text{nuc}}) + F^{\text{clas,ah}}(V, 0\text{K}, T^{\text{nuc}})]_{T^{\text{nuc}} = T}. \qquad (15.43)$$

15.2.2.5 Anharmonic Atomic Excitations: Beyond the Thermodynamic Integration

While the thermodynamic integration approach boosts the efficiency by a few orders of magnitude as compared to conventional molecular dynamics simulations, the calculation of a full free energy surface $F(V, T)$ from *ab initio* is still a formidable task even on today's high performance CPU architecture. Therefore, there are currently active efforts at exploring new methods to reduce computational time [18, 29, 30].

Wu and Wentzcovitch [29], e.g., developed a semi empirical ansatz based on a single adjustable parameter to describe anharmonicity. To explain their approach, it is useful to note first that already the quasiharmonic frequencies ω_i^{qh} contain an implicit temperature dependence due to thermal expansion: $\omega_i^{qh} = \omega_i^{qh}(V^{eq}(T))$. Here, $V^{eq}(T)$ denotes the thermal expansion, i.e., the equilibrium volume as a function of temperature T (for a fixed pressure). The thermal expansion is itself determined by the ω_i^{qh} so the problem must be solved self consistently.

The ansatz of Wu and Wentzcovitch [29] is to modify the implicit temperature dependence. Specifically, the following transformation to a renormalized frequency ω_i^{renorm} is proposed:

$$\omega_i^{qh}\left(V^{eq}(T)\right) \to \omega_i^{renorm}(T) := \omega_i^{qh}\left(V'(T)\right), \tag{15.44}$$

with a modified thermal expansion

$$V'(T) = V^{eq}(T)\left\{1 - C\left[\frac{V^{eq}(T) - V_0^{eq}}{V_0^{eq}}\right]\right\}, \tag{15.45}$$

where C is an adjustable parameter and $V_0^{eq} = V^{eq}(T = 0K)$. The renormalized frequencies ω_i^{renorm} replace then the quasiharmonic frequencies in the quasiharmonic free energy expression Eq. (15.35). The resulting free energy surface includes therefore anharmonicity in an approximative manner and can be used to derive anharmonic thermodynamic quantities such as thermal expansion or defect concentrations. In the original work [29], C is adjusted by fitting to experimental data. In a more recent work [30], it was shown how the parameter can be obtained using *ab initio* based thermodynamic integration. The main advantage of the approach is its effectiveness, since to determine C and thus to generate a complete anharmonic free energy surface only a single thermodynamic integration run at fixed V and T suffices. The disadvantage is however that the ω_i^{renorm} (and thus the anharmonic free energy) are assumed to have a certain temperature dependence, specifically that they scale linearly with the quasiharmonic thermal expansion. This assumption has been shown to work well for diamond and MgO [30] up to high temperatures, it cannot be however *a priori* assumed in general and further evaluations are necessary.

An alternative approach that is able to account for the full anharmonic free energy surface without assuming any specific temperature dependence was developed recently in Ref. [18] and termed upsampled thermodynamic integration using Langevin dynamics (UP-TILD) method. The key ideas are as follows. In a first step, a usual thermodynamic integration run for discrete values of λ is performed:

$$\left\langle \frac{\partial F_\lambda^{el}}{\partial \lambda} \right\rangle_{t,\lambda}^{low} = \left\langle F^{el}(V, T^{el}) - F_0^{el}(V, T^{el}) - \sum_{k,l} \frac{\sqrt{M_k M_l}}{2} u_k u_l D_{k,l}(V, T^{el}) \right\rangle_{t,\lambda}^{low}. \tag{15.46}$$

A critical insight gained in Ref. [18] is that instead of employing fully converged DFT parameters (*e.g.*, with respect to basis set or electronic k sampling) computationally much less demanding DFT parameters can be used (indicated by the superscript "low"). In particular, they need to be chosen such that the corresponding phase space distribution (termed $\{R_I\}_t^{\text{low}}$ in the following) closely resembles the phase space distribution $\{R_I\}_t^{\text{high}}$ as would be obtained from fully/highly converged parameters. As a consequence of using the low converged parameters, the thermodynamic integration is computationally very efficient and speed up factors of ≈ 30 are achievable. Applying this approach makes it therefore possible to sample various λ, V, and T values even on modest computer resources. The resulting free energy surface, *i.e.*, $\langle \partial F_\lambda^{\text{el}}/\partial\lambda\rangle_{t,\lambda}^{\text{low}}$, needs however to be corrected in a second step. The actual correction step is straightforward, provided that the above stated condition $\{R_I\}_t^{\text{low}}$ close to $\{R_I\}_t^{\text{high}}$ holds. In such a case, a typically small set of N^{UP} uncorrelated structures $\{R_I\}_t^{\text{low}}$ (indexed with t_u) is extracted from $\{R_I\}_t^{\text{low}}$ and the upsampling average $\langle\Delta F^{\text{el}}\rangle_\lambda^{\text{UP}}$ is calculated as:

$$\langle\Delta F_{\text{el}}\rangle_\lambda^{\text{UP}} = \frac{1}{N^{\text{UP}}}\sum_u^{N^{\text{UP}}} F^{\text{el,low}}\left(\{R_I\}_{t_u}^{\text{low}}\right) - F^{\text{el,low}}(\{R_I^0\}) \qquad (15.47)$$

$$- \left[F^{\text{el,high}}\left(\{R_I\}_{t_u}^{\text{low}}\right) - F^{\text{el,high}}(\{R_I^0\})\right].$$

Here, $F^{\text{el,low}}$ ($F^{\text{el,high}}$) refers to the electronic free energy calculated using the low (high) converged set of DFT parameters. The λ dependence of $\langle\Delta F^{\text{el}}\rangle_\lambda^{\text{UP}}$ is hidden in the trajectory $\{R_I\}_t^{\text{low}}$, which is additionally dependent on the volume and temperature. In the last step, the quantity of interest, the converged $\langle\partial F_\lambda^{\text{el}}/\partial\lambda\rangle_{t,\lambda}^{\text{high}}$ is obtained

$$\langle\partial F_\lambda^{\text{el}}/\partial\lambda\rangle_{t,\lambda}^{\text{high}} = \langle\partial F_\lambda^{\text{el}}/\partial\lambda\rangle_{t,\lambda}^{\text{high}} - \langle\Delta F^{\text{el}}\rangle_\lambda^{\text{UP}},$$

and thus the anharmonic free energy:

$$F^{\text{clas,ah}} = \int_0^1 d\lambda \langle\partial F_\lambda^{\text{el}}/\partial\lambda\rangle_{t,\lambda}^{\text{high}}. \qquad (15.48)$$

The efficiency of the UP-TILD method is a direct consequence of the fact that the upsamling average $\langle\Delta F^{\text{el}}\rangle_\lambda^{\text{UP}}$, which involves the computation of the CPU time consuming well converged DFT free energies $F^{\text{el,high}}$, converges extremely fast with respect to the number of uncorrelated configurations. In practice, less than 100 configurations are found to be sufficient to achieve an accuracy of better than 1 meV/atom in the free energy [18]. This small number of configurations has to be compared with the number of molecular dynamics steps needed to converge a thermodynamic integration run which is in the range of several thousands, *i.e.*, two orders of magnitude larger. Another advantage is that in many cases the λ dependence is invariant with respect to the upsampling procedure allowing to calculate the shift in Eq. (15.48) for a single λ value only. Using this invariance reduces computational costs further as illustrated in Figure 15.4.

The various methods needed to compute the electronic, quasiharmonic, and anharmonic contributions to the free energy surface are summarized in Figure 15.5.

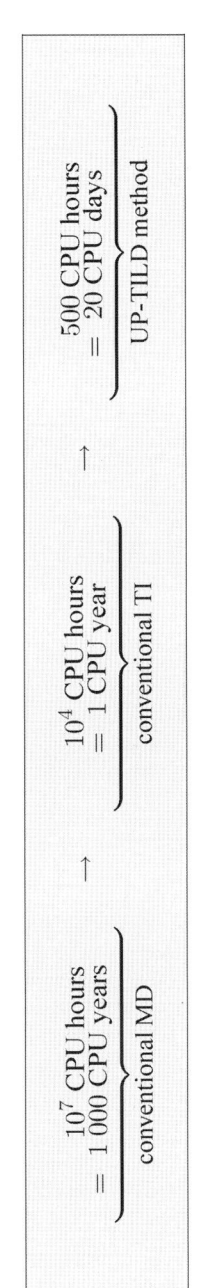

Figure 15.4 (online colour at: www.pss-b.com) CPU time reductions in calculating the anharmonic free energy as reported for the example of bulk aluminum including point defects (32 atomic fcc cell) [18]; MD: molecular dynamics; TI: thermodynamic integration; UP-TILD: upsampled TI using Langevin dynamics (see text for details).

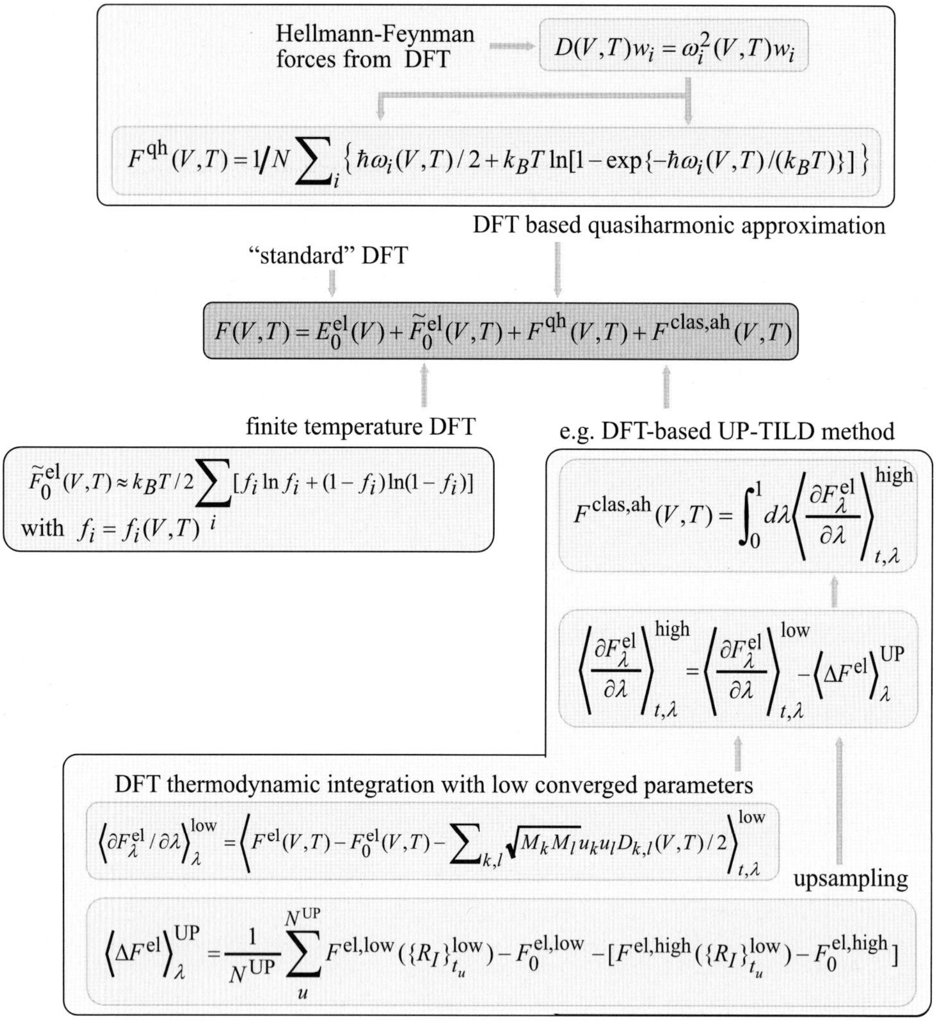

Figure 15.5 (online colour at: www.pss-b.com) Schematic illustration of the overall approach and its key equations to derive finite temperature contributions to the free energy presented in this article. The free energy surface $F(V,T)$ refers here either to the defect cell free energy surface $F^d(V,T)$ or the perfect bulk free energy surface $F^p(V,T)$ as used in Section 15.2.1. The corresponding methods for each component are indicated. Definitions are given in Sections 2.2.2–2.2.4.

15.3
Results: Electronic, Quasiharmonic, and Anharmonic Excitations in Vacancy Properties

To discuss the performance of the methodology described in Section 15.2, we consider point defects in aluminum. The focus will be on vacancies, since self interstitials are practically absent due to their high formation energy (3.4 eV) [31]. Computational details of the results presented in the following can be found in Refs. [18, 26].

Before applying the approach to defect properties, we will first consider bulk properties. This will allow a careful inspection and evaluation of the accuracy of the approach and the underlying exchange-correlation functionals. For the following discussion, we restrict on two properties that are highly sensitive to an accurate description of the free energy surface: The thermal expansion coefficient and the isobaric heat capacity. These quantities are first or higher order derivatives of the free energy surface being thus affected even by small changes in the free energy. As a consequence, to guarantee an unbiased comparison with experiment the error bar in the free energy has to be systematically kept < 1 meV/atom. This error bar is significantly lower than what is typically targeted at in defect calculations (≈ 0.1 eV) and particularly challenging to achieve at high temperatures. A major advantage of reaching this numerical accuracy is that the remaining discrepancy with experiment can be unambiguously related to deficiencies in the exchange-correlation functional. For the case considered here, fcc bulk aluminum, both of the most popular exchange-correlation functionals, the LDA and the GGA, show an excellent agreement with experiment for the expansion coefficient up to the melting point (Figure 15.6a) and for the heat capacity up to ≈ 500 K (Figure 15.6b). Above this temperature, the experimental scatter in the heat capacity becomes too large and prevents an unbiased comparison. Nonetheless, the *ab initio* results indicate a lower bound to experiment hinting at additional and apparently so far not controllable by experiment excitation mechanisms in the sample and/or experimental setup.

Having demonstrated the accuracy achievable by the method to describe bulk properties at finite temperatures, let us now turn to defect properties. Applying the same formalism as for the bulk and the constant pressure approach, the temperature and volume dependent isobaric defect formation free energy $F_P^f(V, T)$ can be calculated (see Section 15.2.1.2). Using it, the equilibrium defect concentration $c^{eq}(T, P) = \exp[-F_P^f(V^{eq}(T, P), T)/(k_B T)]$ is obtained. Here, $V^{eq}(T, P)$ is the equilibrium volume at temperature T and pressure P. The computed concentrations at finite temperatures and for zero pressure (the influence of atmospheric pressure is negligible) are shown together with available experimental data in Figure 15.7.

As can be seen, an excellent agreement is obtained. This agreement is particularly impressive when considering that errors in the defect formation free energy scale exponentially in the concentration. An important finding is that LDA and GGA provide an upper and lower bound to the experimental data and may thus be used as empirical (but *ab initio* computable) error bars. This interesting and highly useful behavior is not restricted to Al but has been systematically observed for a wide range of metals when considering temperature dependent material properties [19]. Figure 15.7 provides also a direct insight into the relevance of effects due to the finite size of practical supercells as discussed in Section 15.2.1. As can be seen, correcting these effects using the constant pressure approach increases the defect concentration by almost an order of magnitude. Therefore, the application of the constant pressure approach is critical to achieve the desired accuracy. A further extension of the

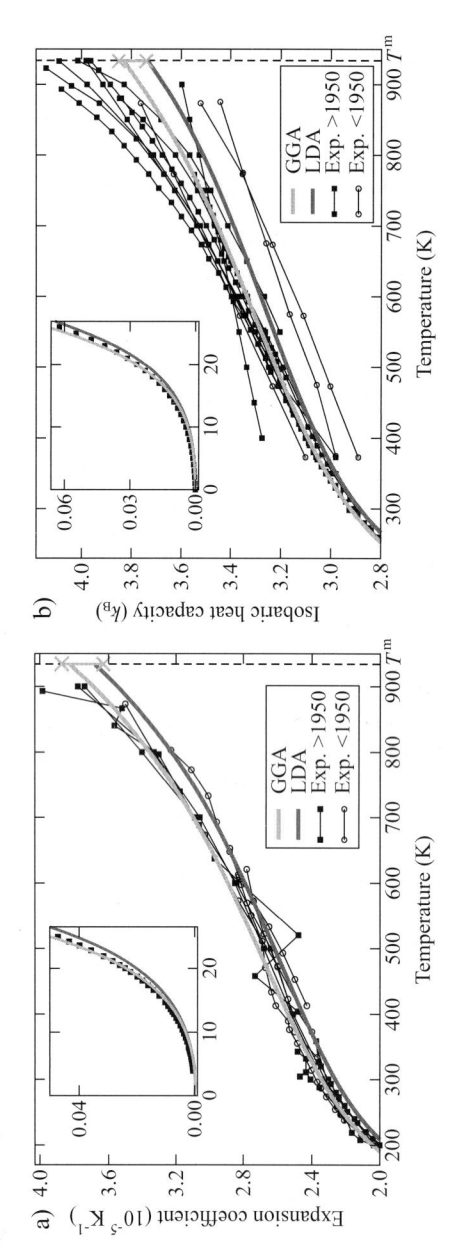

Figure 15.6 (online colour at: www.pss-b.com) (a) Thermal expansion coefficient and (b) isobaric heat capacity of aluminum including the electronic, quasiharmonic, anharmonic, and vacancy contribution compared to experiment. The melting temperature T^m of Al (933 K) is given by the vertical dashed line. At T^m, the crosses indicate the sum of all numerical errors (e.g., pseudopotential error or statistical inaccuracy; cf. Ref. [26]) in all contributions for GGA. The LDA error is of the same order of magnitude. Experimental data older than 1950 are indicated by open circles (Refs. [32–35] for the expansion coefficient and [36–38] for the heat capacity). The remaining experimental results are indicated by filled squares (Refs. [39–44] and [45–53]). The inset shows the low temperature region with experimental data from Refs.[54, 51].

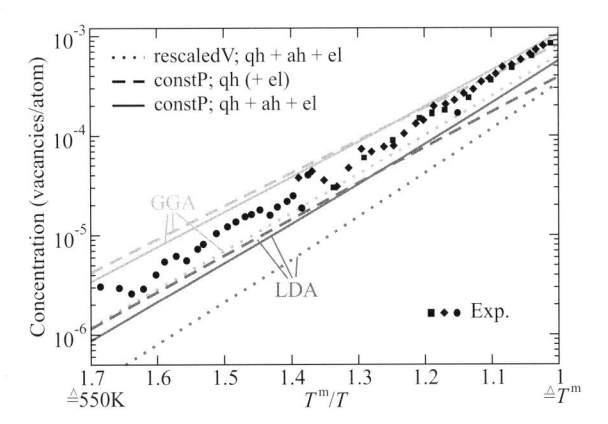

Figure 15.7 (online colour at: www.pss-b.com) Equilibrium vacancy concentration at zero pressure of aluminum as a function of the inverse temperature multiplied by the melting temperature T^m. Results for the rescaled volume and constant pressure approach are shown. The volume optimized approach yields concentrations which are identical to the constant pressure results on the shown scale. The electronic contribution yields a negligible contribution (indicated by the parenthesis). The squares indicate experimental values from Ref. [55] (differential dilatometry). The diamonds/circles indicate experimental values from Ref. [56] (differential dilatometry/positron annihilation).

methodology to the volume optimized approach has a negligible effect on the vacancy concentration shown in Figure 15.7.

To analyze the temperature dependence of the defect concentration in more detail, it is convenient to express the formation free energy in terms of the $T = 0$ K enthalpy and the entropy of formation. To provide a direct comparison with experimental data where only a rather small temperature window is available, we restrict on the experimental temperature interval and obtain both quantities from a linear regression in the log-$1/T$ plot (Figure 15.7). The results are summarized in Table 15.2. A surprising finding is that including anharmonicity has a major effect on the entropy and enthalpy: For LDA the entropy of formation increases from $0.2 k_B$ (quasiharmonic) to $2.2 k_B$ (anharmonic) and the enthalpy from 0.65 to 0.78 eV. These numbers are in excellent agreement with experimentally derived data of $2.4 k_B$ and 0.75 eV.

It is interesting to note that the substantial deviations due to anharmonic effects have little consequence on the absolute defect concentrations (see Figure 15.7). The changes in entropy and enthalpy largely compensate each other in this quantity. An important conclusion from this result is that the defect formation enthalpy derived from high temperature data (the only region where experimental data is available) can be substantially different from the true $T = 0$ K formation enthalpy. For the system considered here the difference is ≈ 0.1 eV. To guarantee an accurate comparison between theory and (high temperature) experimental data it is therefore critical to not restrict to the formation enthalpy (as usually done) but to use also the experimental entropy to compute and compare defect concentrations.

Table 15.2 The extrapolated formation energy E^f and averaged entropy of formation S^f for various approaches and combinations of the free energy contributions used for the calculation of vacancy properties of aluminum. ai/ep indicates values for the coupled *ab initio*-empirical potentials approach from Ref. [16]. The further values (also the experimental) are obtained by fitting the vacancy concentrations over the temperature range given in Figure 15.7 to the function $\exp[-(E^f - TS^f)/(k_B T)]$. The notation is as in Figure 15.7.

	E^f (eV)		S^f (k_B)	
	LDA	GGA	LDA	GGA
constP; qh	0.65	0.58	0.2	0.1
constP; qh + el	0.65	0.58	0.2	0.1
constP; qh + ah + el	0.78	0.68	2.2	1.5
volOpt; qh + ah + el	0.78	0.68	2.2	1.5
constV; qh + ah + el	0.85	0.75	2.5	1.9
ai/ep [16]	0.78	0.61	1.6	1.3
experiment		0.75		2.4

15.4
Conclusions

In this paper, we gave a brief overview about the challenges one encounters when including temperature effects in defect calculations beyond configurational entropy. The paper outlines strategies to compute all relevant free energy contributions arising due to electronic and vibronic (quasiharmonic and anharmonic) excitations. A major focus has been devoted to numerical performance since even on todays most powerful supercomputers such calculations quickly approach the limits of available resources when attempting a full *ab initio* description. As an example which numerical accuracy can be achieved, a simple yet instructive defect system has been considered: vacancies in fcc bulk Al. The results nicely illustrate that both bulk and defect properties in the absence of any band gap problem can be described with a surprising accuracy and predictive power. An interesting result of this study is that anharmonic contributions have a rather small effect on the vacancy concentration in the experimentally accessible temperature window, but have a drastic effect on the averaged/extrapolated entropy and enthalpy of formation. To achieve an accuracy of better than 0.1 eV the inclusion of anharmonic contributions is mandatory. With the advent of powerful approaches to overcome the band gap problem in semiconductor defect calculations, we believe that inclusion of finite temperature effects (possibly on the LDA/GGA level) becomes critical.

References

1 Estreicher, S.K., Backlund, D., and Gibbons, T. M. (2010) *Thin Solid Films*, 518, 2413.

2 Al-Mushadani, O.K. and Needs, R.J. (2003) *Phys. Rev. B*, 68, 235205.

3. Estreicher, S.K., Sanati, M., West, D., and Ruymgaart, F. (2004) *Phys. Rev. B*, **70**, 125209.
4. Heyd, J., Scuseria, G.E., and Ernzerhof, M. (2003) *J. Chem. Phys.*, **118**, 8207.
5. Heyd, J., Peralta, J.E., Scuseria, G.E., and Martin, R.L. (2005) *J. Chem. Phys.*, **123**, 174101.
6. van Schilfgaarde, M., Kotani, T., and Faleev, S. (2006) *Phys. Rev. Lett.* **96**, 226402.
7. Rinke, P., Qteish, A., Neugebauer, J., and Scheffler, M. (2008) *Phys. Status Solidi B*, **245**, 929.
8. Gillan, M. J. (1989) *J. Phys.: Condens. Matter*, **1**, 689.
9. Pawellek, R., Fähnle, M., Elsässer, C., Ho, K.M., and Chan, C.T. (1991) *J. Phys.: Condens. Matter*, **3**, 2451.
10. Polatoglou, H.M., Methfessel, M., and Scheffler, M. (1993) *Phys. Rev. B*, **48**, 1877.
11. Korhonen, T., Puska, M.J., and Nieminen, R.M. (1995) *Phys. Rev. B*, **51**, 9526.
12. Turner, D.E., Zhu, Z.Z., Chan, C.T., and Ho, K.M. (1997) *Phys. Rev. B*, **55**, 13842.
13. Satta, A., Willaime, F., and de Gironcoli, S. (1998) *Phys. Rev. B*, **57**, 11184.
14. Satta, A., Willaime, F., and de Gironcoli, S. (1999) *Phys. Rev. B*, **60**, 7001.
15. Carling, K. M., Wahnström, G., Mattsson, T. R., Mattsson, A. E., Sandberg, N., and Grimvall, G. (2000) *Phys. Rev. Lett.*, **85**, 3862.
16. Carling, K. M., Wahnström, G., Mattsson, T. R., Sandberg, N., and Grimvall, G. (2003) *Phys. Rev. B*, **67**, 054101.
17. Lucas, G. and Schäublin, R. (2009) *Nucl. Instrum. Methods B*, **267**, 3009.
18. Grabowski, B., Ismer, L., Hickel, T., and Neugebauer, J. (2009) *Phys. Rev. B*, **79**, 134106.
19. Grabowski, B., Hickel, T., and Neugebauer, J. (2007) *Phys. Rev. B*, **76**, 024309.
20. Keer, H. V. (1993) *Principles of the Solid State* (Wiley, New York).
21. Varotsos, P. A. and Alexopoulos, K. D. (1986) *Thermodynamics of Point Defects and Their Relation with Bulk Properties, Defects in Solids*, Vol. 14 (North-Holland, Amsterdam).
22. Born, M. and Oppenheimer, J. R. (1927) *Ann. Phys. (Leipzig)* **389**, 457.
23. Cao, J. and Berne, B. J. (1993) *J. Chem. Phys.*, **99**, 2902.
24. Suzuki, M. (1977) *Commun. Math. Phys.*, **57**, 193.
25. Mermin, N. D. (1965) *Phys. Rev.*, **137**, A1441.
26. Grabowski, B. (2009) Towards ab initio assisted materials design: DFT based thermodynamics up to the melting point, OT *PhD thesis*, Universität Paderborn.
27. Haile, J. M. (1997) *Molecular Dynamics Simulation: Elementary Methods* (Wiley, New York).
28. Ramirez, R., Herrero, C. P., Antonelli, A., and Hernández, E. R. (2008) *J. Chem. Phys.*, **129**, 064110.
29. Wu, Z. and Wentzcovitch, R. M. (2009) *Phys. Rev. B*, **79**, 104304.
30. Wu, Z. (2010) *Phys. Rev. B*, **81**, 172301.
31. Denteneer, P. J. H. and Soler, J. M. (1991) *J. Phys.: Condens. Matter*, **3**, 8777.
32. Honda, K. and Okubo, Y. (1924) *Sci. Rep. Tohoku Imp. Univ., Ser. 1* **13**, 101 (cited from Ref. [42]).
33. Uffelmann, F. L. (1930) *Philos. Mag.*, **10**, 633 (cited from Ref. [42]).
34. Nix, F. C. and MacNair, D. (1941) *Phys. Rev.*, **60**, 597.
35. Wilson, A. J. C. (1942) *Proc. Phys. Soc.*, **54**, 487.
36. Eastman, E. D., Williams, A. M., and Young, T. F. (1924) *J. Am. Chem. Soc.*, **46**, 1178.
37. Umino, S. (1926) *Sci. Rep. Tohoku Imp. Univ., Ser. 1* **15**, 597 (cited from Ref. [52]).
38. Avramescu, A. (1939) *Z. Tech. Phys. (Leipzig)* **20**, 213.
39. Nicklow, R. M. and Young, R. A. (1963) *Phys. Rev.*, **129**, 1936 (cited from Ref. [42]).
40. Strelkov, P. G. and Novikova, S. I. (1957) *Prib. Tekh. Eksp. (USSR)* **5**, 105 (cited from Ref. [42]).
41. Pathak, P. D. and Vasavada, N. G. (1970) *J. Phys. C, Solid State Phys.*, **3**, L44 (cited from Ref. [42]).
42. Touloukian, Y. S., Kirby, R. K., Taylor, R. E., and Desai, P. D. (1975) *Thermal Expansion: Metallic Elements and Alloys, Thermophysical Properties of Matter*, Vol. 12 (IFI/Plenum, New York).

43 von Guérard, B., Peisl, H., and Zitzmann, R. (1974) *Appl. Phys.*, **3**, 37, numerical differentiation of the given experimental expansion data; strongly scattering points were taken out.
44 Kroeger, F. R. and Swenson, C. A. (1977) *J. Appl. Phys.*, **48**, 853.
45 Pochapsky, T. E. (1953) *Acta Metall.* **1**, 747.
46 McDonald, R. A. (1967) *J. Chem. Eng. Data* **12**, 115.
47 Brooks, C. R. and Bingham, R. E. (1968) *J. Phys. Chem. Solids*, **29**, 1553.
48 Leadbetter, A. J. (1968) *J. Phys. C, Solid State Phys.*, **1**, 1481.
49 Schmidt, U., Vollmer, O., and Kohlhaas, R. (1970) *Z. Naturforsch. A* **A25**, 1258.
50 Marchidan, D. I. and Ciopec, M. (1970) *Rev. Roum. Chim.*, **15**, 1005 (cited from Ref. [52]).
51 Ditmars, D. A., Plint, C. A., and Shukla, R. C. (1985) *Int. J. Thermophys.*, **6**, 499.
52 Desai, P. D. (1987) *Int. J. Thermophys.* **8**, 621.
53 Takahashi, Y., Azumi, T., and Sekine, Y. (1989) *Thermochim. Acta*, **139**, 133.
54 McLean, K. O., (1969) Low Temperature Thermal Expansion of Copper, Silver, Gold, and Aluminum, OT *PhD thesis*, Iowa State University (cited from Ref. [42]).
55 Simmons, R. O. and Balluffi, R. W. (1960) *Phys. Rev.*, **117**, 52.
56 Hehenkamp, T. (1994) *J. Phys. Chem. Solids*, **55**, 907.

16
Accurate Kohn–Sham DFT With the Speed of Tight Binding: Current Techniques and Future Directions in Materials Modelling

Patrick R. Briddon and Mark J. Rayson

16.1
Introduction

The Kohn–Sham formalism [1] of density functional theory [2] (KSDFT) is one of the most widely used tools in the *ab initio* theoretical investigation of the properties of materials. Its success at providing quantitative comparison with experiment—given only atomic positions and species as input—combined with its favourable algorithmic prefactor and complexity accounts for this widespread usage and intensive efforts to improve the algorithms at the heart of KSDFT codes.

Here we describe how recent algorithmic advances in the computational kernel at the heart of one of these codes, *Ab Initio* Modelling PROgram (AIMPRO), will enable a new scale of calculation to be performed on inexpensive hardware. The modern algorithmic kernel, functionality, current advances and future perspectives will be discussed. In short, the aim of this work is to show how full KSDFT calculations can be performed in a time comparable to current tight binding implementations and further, to open a route to reaching the basis set limit in these calculations, essentially delivering plane wave accuracy in a time comparable to a tight binding calculation. In the following discussions we define the Kohn–Sham kernel as the calculation of energies and forces, leading to minimum energy structures (including lattice optimisation). Any other calculated quantity will be referred to as functionality. Readers interested in the details of algorithms used in the AIMPRO suite of codes are referred to Refs. [3–9].

As well as a description of the algorithms we also wish to address the wider audience of applications specialists. By discussing recent advances in terms of both their methodological context and relevance to the practitioner we hope this work will be of interest to a wide audience.

This chapter is organised as follows, Section 16.2 discusses the use of Gaussian orbitals and briefly describes the conventional AIMPRO kernel, Section 16.3 gives a brief overview of the functionality currently available, Section 16.4 describes recent improvements to the AIMPRO kernel and Section 16.5 discusses future research directions and perspectives based on recent advances. Readers familiar with conventional Gaussian algorithms, or those interested primarily in recent advances in the kernel, can skip straight to Section 16.4.

16.2
The AIMPRO Kohn–Sham Kernel: Methods and Implementation

In this section we briefly outline the major steps involved in a conventional electronic structure code, introducing notation that is needed when new innovations are introduced later.

The standard approach is to expand the Kohn–Sham levels in terms of basis function, $\phi_i(\mathbf{r})$:

$$\psi_\lambda(\mathbf{r}) = \sum_{\lambda=1}^{N} c_{i\lambda} \phi_i(\mathbf{r}), \tag{16.1}$$

which enables the Kohn–Sham equations to be recast in matrix form:

$$\sum_j H_{ij} c_{j\lambda} = \varepsilon_\lambda \sum_j S_{ij} c_{j\lambda} \quad \text{or} \quad \mathbf{Hc} = \mathbf{Sc}\Lambda, \tag{16.2}$$

where $\Lambda_{\lambda\lambda'} = \varepsilon_\lambda \delta_{\lambda\lambda'}$.

In this way the electronic structure problem is reduced into three components. The first is essentially one of quadrature, determining the Hamiltonian and overlap matrices:

$$\begin{aligned} H_{ij} &= \int \phi_i^*(\mathbf{r}) \hat{H} \phi_j(\mathbf{r}) d\mathbf{r}; \\ S_{ij} &= \int \phi_i^*(\mathbf{r}) \phi_j(\mathbf{r}) d\mathbf{r}. \end{aligned} \tag{16.3}$$

The second problem is the solution of the generalised eigenvalue problem (GEP) [Eq. (16.2)]. This will occupy discussion for the majority of this chapter, being the most computationally expensive part of the calculation. The final ingredient, which is not discussed in this chapter, is a method of iterating to self-consistency.

16.2.1
Gaussian-Type Orbitals

Uncontracted Cartesian Gaussian functions

$$\phi_i(\mathbf{r}) = (x - R_{ix})^{n_x} (y - R_{iy})^{n_y} (z - R_{iz})^{n_z} \exp[-a_i (\mathbf{r} - \mathbf{R}_i)^2]$$

are used to form our primitive set. For each atom, we typically use four different exponents a_i and we multiply each Gaussian function $\exp[-a_i(\mathbf{r} - \mathbf{R}_i)^2]$ by the Cartesian prefactors, including all combinations of n_x, n_y and n_z such that $n_x + n_y + n_z \leq \ell$. This produces four functions for $\ell = 1$ and ten functions for $\ell = 2$.

In our notation, exponents are arranged from lowest to highest (most diffuse Gaussian first) and the standard nomenclature is used to define the angular momentum. So, for example a *ddpp* basis has four exponents with $10 + 10 + 4 + 4 = 28$ functions. Such a basis set applied to carbon or silicon would be considered large for a routine quantum chemistry application.

We now consider the advantages and disadvantage of using Gaussian orbitals.

Advantages:

i) Memory. The size of the primitive basis is small, typically only 20–40 functions per atom. Furthermore, the storage of a primitive basis function requires only 3 integers and 2 double precision numbers.
ii) Adaptive. Basis functions can be placed where they are most needed.
iii) Gaussian functions are localised in both real and Fourier space and careful use of this fact enables matrix elements of the Hamiltonian to be found extremely efficiently (see Section 16.2.2).
iv) The integration error is independent from basis error leading to an internally consistent calculation. That is, even with a large basis error the matrix elements of the Hamiltonian are still evaluated to high precision. This is important when considering relative energies.

Adaptivity, coupled with rapid matrix element evaluation, allows chemical species with hard potentials to be treated almost as easily as species with soft potentials. Therefore, a single oxygen or hydrogen (typically having hard pseudopotentials) embedded in, say, a unit cell containing 1000 atoms of silicon (which has a fairly soft pseudopotential) would take a similar time to 1000 atoms of silicon.

Disadvantages:

i) Basis set superposition error (BSSE).
ii) Difficulty in running codes—a degree of experience is currently needed to choose a suitable basis set.
iii) Difficulty in proving convergence.

The main disadvantage is related to the relative complexity of the Gaussian basis set and the less obvious way in which basis sets can be augmented to move towards convergence in energy or some other property. Large basis sets generated to minimise the energy of a system can develop numerical linear dependencies as convergence is approached, with the result that a certain level of skill and experience is needed to work in this regime. The work in this chapter provides a first step to removing this as a legitimate concern.

It is important to note the effect of BSSE depends on the quantity of interest. When interested in relative energies the degree of difference between the systems must be considered. Therefore, two very different systems will suffer most from BSSE when considering their relative energy. Not quite as challenging is a calculation such as the formation energy of a vacancy—here the systems are fairly similar—though the vacancy calculation has fewer degrees of freedom, therefore this is a problem of intermediate difficulty. However, a slight perturbation of an atom (such as a numerical force calculation) will suffer far less from BSSE, this will be discussed further and demonstrated in Section 16.4.2. Since the large majority of computer time is spent in structural relaxations (and other quantities where derivatives of the energy are of paramount importance) a significant amount of computer time can be saved by correctly assessing the relevant impact of BSSE on a calculation. Although, at present, much of this must be done by the user, in principle, it can be automated. We will discuss work in this direction in Section 16.5.

16.2.2
The Matrix Build

The building of the Hamiltonian is achieved using standard techniques. The overlap matrix, and the matrix elements of kinetic energy and the non-local pseudopotential may be found analytically using recurrence relations reported in Ref. [10]. The matrix elements of the Kohn–Sham potential are found as described in Refs. [6] and [11].

An important difference between our approach and standard methods of quantum chemistry is our avoidance of four centre integrals. Our approach of quadrature using a set of equally spaced grids [6] has linear scaling with an acceptable prefactor. In doing this the charge density and the potential are expressed on an equally spaced grid in real space which, in plane wave parlance would have an exceptionally high cutoff—typical values would be 80 Rydbergs (silicon) and 300 Rydbergs (carbon). This is feasible as this expansion is done only for the charge density, and not for each individual Kohn–Sham level. A consequence is that the Hamiltonian matrix is determined essentially free of integration error, with arbitrarily high accuracy being achievable at modest cost. Timings for this are presented in passing in Section 16.4.1 when it is noted that this is a negligible contribution to run-times for large systems, being only 3 min when run in serial on a single core even for a system of 4096 atoms.

16.2.3
The Energy Kernel: Parallel Diagonalisation and Iterative Methods

Once the primitive Hamiltonian and overlap matrices—$\mathbf{H}$ and $\mathbf{S}$ respectively—have been evaluated these are then converted from sparse storage to a dense block-cyclic parallel distribution. The $N \times N$ GEP

$$\mathbf{Hc} = \mathbf{Sc\Lambda}, \tag{16.4}$$

is then solved (ScaLAPACK diagonalisation) to calculate the output density

$$n(\mathbf{r}) = \sum_{ij} b_{ij} \phi_i(\mathbf{r}) \phi_j(\mathbf{r}), \tag{16.5}$$

where

$$b_{ij} = \sum_{\lambda}^{m} f(\Lambda_{\lambda\lambda}) c_{i\lambda} c_{j\lambda}, \tag{16.6}$$

and m is the number of occupied states. A charge density mixing scheme [12] is used to iterate (the 'SCF cycle') towards the self-consistent density.

As well as diagonalisation, especially when the N/m ratio is high and good parallel scaling is important, an iterative algorithm—based on the direct inversion of the iterative subspace (DIIS)—is also used [5].

16.2.4
Forces and Structural Relaxation

It is occasionally argued that the determination of forces is more complex and time consuming with Gaussian orbitals as a consequence of the Pulay forces associated with atom centred basis functions. This, in fact, is not the case in reality. In fact, viewing Pulay forces as an approximation to incompleteness forces (see Section II C of Ref. [6] for a detailed discussion) it is more accurate to say that rather than being a burden, the ability to calculate Pulay forces (rather than their presence) is a distinct advantage as significant efficiencies can be obtained (see Section 16.4 for a more detailed discussion).

Forces are determined from the Hellmann–Feynman theorem as adapted for localised basis functions [3]:

$$\frac{\partial E}{\partial \mathbf{R}_\alpha} = \sum_{ij} \frac{\partial H_{ij}}{\partial \mathbf{R}_\alpha} b_{ij} - \sum_{ij} \frac{\partial S_{ij}}{\partial \mathbf{R}_\alpha} w_{ij},$$

where w_{ij} is the energy weighted density matrix

$$w_{ij} = \sum_\lambda^m f(\Lambda_{\lambda\lambda}) \Lambda_{\lambda\lambda} c_{i\lambda} c_{j\lambda}. \tag{16.7}$$

The first term, $\frac{\partial H_{ij}}{\partial \mathbf{R}_\alpha} b_{ij}$, is trivially evaluated in time scaling linearly with system size. Indeed the time for this is only marginally greater than the construction of the Hamiltonian itself, only ~45 s for a 1000 atom cell (see discussion of timings in Section 16.4.1). The construction of w_{ij} is likewise straightforward. Evaluating Eq. (16.7) directly using $\mathcal{O}(N^3)$ dense matrix operations imposes a negligible overhead—and, in principle, since only elements of w_{ij} that have corresponding non-zero elements in the Hamiltonian are required this step can be performed in $\mathcal{O}(N^2)$. In conclusion then, force determination is possible in a small fraction of the time for a single SCF step.

Movement of the atoms to attain equilibrium is then achieved using any standard scheme. We commonly use the conjugate gradient method [13], BFGS [13] and G-DIIS methods [14].

16.2.5
Parallelism

AIMPRO is parallelised using the message passing interface library (MPI). A library to handle the creation and destruction of multiple worlds—and levels of worlds—is

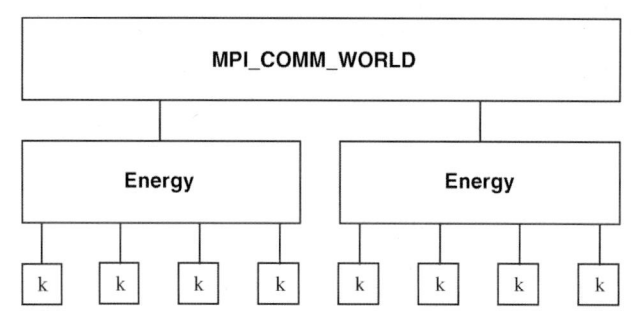

Figure 16.1 Schematic of parallel worlds (see text for details).

also implemented. A typical arrangement of these worlds for the calculation of the dynamical matrix is given in Figure 16.1. Each 'Energy world' could, for example, calculate a row of the dynamical matrix, furthermore within each energy calculation the calculation can further be split into separate 'k-point worlds'. Such flexible infrastructure such as this enables extra 'embarrassingly parallel' functionality to be included in the main algorithm itself.

16.3
Functionality

Although in this chapter, the emphasis is on the kernel of the calculation, and how this may be improved both in terms of speed and accuracy, the great utility of *ab initio* calculations over the last two decades has been their ability to link to an increasingly broad range of experiments, producing quantitatively accurate values for measurable quantities. In this section, we illustrate this by outlining some of the functionality incorporated into the AIMPRO code and the problems tackled.

16.3.1
Energetics: Equilibrium and Kinetics

The fundamental property given by these calculations is, of course, the total energy. In terms of defect physics, this is of outstanding importance with the formation energy controlling the equilibrium concentrations of defects. The energy barrier to motion of a defect through a material, gives information about kinetic motion and is as important as the formation energy in understanding the behaviour of defects in a material. For example, the result that the diffusion barrier of hydrogen in ZnO is under 0.5 eV has demonstrated that isolated H cannot be responsible for the residual *n*-type conductivity of this material [15] as had previously been thought. On the other hand, the fact we can show that H binds strongly to other impurities to produce thermally stable complexes can provide alternative explanations for this phenomena [16] and can also have important technological implications for other doping issues [17].

It is frequently the energetics of defects at the high temperatures at which material processing occurs that can determine the defects seen in materials. As such a free energy of formation should be calculated at the temperature at which the material is processed. Calculations of this requires treatment of vibrational modes for all atoms in a unit cell, once demanding but now becoming a more common calculation. It can often happen that a binding energy changes sign at high temperature leading to some defect complexes being absent in samples [18].

16.3.2
Hyperfine Couplings and Dynamic Reorientation

An accurate knowledge of the electron spin density enables the coupling with the magnetic moment of certain nuclei to be calculated, enabling a comparison with experimentally measured hyperfine coupling tensors. In the simplest case a comparison with experiment can be a powerful tool enabling the characterisation of defect centres [19]; in more complex cases low symmetry defects can re-orient dynamically at room temperature appearing experimentally as having a higher symmetry. In this case the ability to calculate both the energy barrier and the averaged hyperfine tensor is key [19]. The physics here is quite rich in variety with quantum tunnelling of hydrogen also being demonstrated [20].

16.3.3
D-Tensors

Defects with electron spin, $S > 1$ exhibit a zero field splitting, measured experimentally as the D-tensor. A method to calculate the first order contribution to the zero-field splitting tensor was presented in Ref. [9]. Again comparison of calculated tensors [21–25] with experiment aids in the characterisation of defect centres. The ability to perform a quantitative calculation has also shown that conclusions drawn from the phenomenological point dipole model frequently used to interpret the size of the D-tensor are not always reliable [21].

16.3.4
Vibrational Modes and Infrared Absorption

The vibrational modes associated with defects are readily measured experimentally, and may be calculated from the second derivatives of the energy surface. These modes have been some of the most fruitful methods of characterising defects [26, 27].

16.3.5
Piezospectroscopic and Uniaxial Stress Experiments

Calculation of piezospectroscopic (energy–stress) tensors of defects also provides a direct link with experiment [28]. The response of vibrational frequencies to uniaxial stress is also a valuable tool in the experimental determination of defect symmetry and these shifts can be calculated accurately from total energy calculations, providing a further aid in characterisation studies [29].

16.3.6
Electron Energy Loss Spectroscopy (EELS)

The simplest treatment of energy loss spectroscopies is based on the dipole matrix elements between Kohn–Sham states. This is an approximate model, but in many instances is sufficiently accurate for features in experimental spectra to be correlated with electronic states associated with regions of defects, particularly extended defects. Both low-loss [30, 31] and core–electron energy loss spectroscopy (EELS) experiments (or the theoretically similar XPS experiment [32]) can be modelled.

16.4
Filter Diagonalisation with Localisation Constraints

We now turn to the main topic of this chapter, namely, recent advances in the KSDFT kernel that enable such calculations to be performed in a time comparable to a tight binding calculation. The conventional AIMPRO kernel described in Section 16.2 has been dramatically improved upon recently [6]. The filter diagonalisation method with localisation constraints promises to allow calculations with larger primitive sets, thereby approaching the basis set limit, while the fundamental density matrix is only the size of a minimal (or tight binding like) basis density matrix. For a detailed account of the method and algorithmic details the interested reader is referred to Refs. [6, 7]. Here, we summarise the method to elucidate later discussions, however, it is the broader impact of the filtration algorithm we wish to concentrate upon.

Rather than a direct diagonalisation in the full primitive basis a subspace eigenproblem is constructed in a small basis of filtered functions, defined in terms of the primitive basis set $\{\phi_i\}$ as

$$\Phi_I(\mathbf{r}) = \sum_i k_{iI} \phi_i(\mathbf{r}). \tag{16.8}$$

For silicon, for example, using the pseudopotential approximation this reduces the size of the kernel eigenproblem from, say, ~28 (if using the *ddpp* basis described in Ref. [6]) functions per atom to only four functions per atom—a significant saving. The step that performs this reduction in basis size will be referred to as the filtration algorithm. A filtration radius (r_{cut}) is defined and the filtered basis on an atom at $\mathbf{R}_\alpha$ constructed using basis functions that lie on atoms at $\mathbf{R}_\beta$ where $|\mathbf{R}_\alpha - \mathbf{R}_\beta| < r_{cut}$. Henceforth, filtered basis sets will be referred to using the notation $\{\Phi^{(r_{cut})}\}$. A schematic of the filtration region for an atom if shown in Figure 16.2.

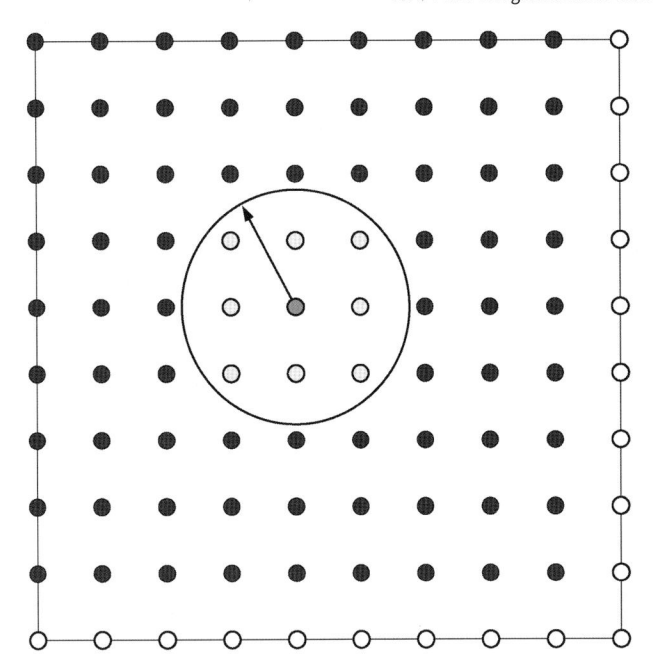

Figure 16.2 (online colour at: www.pss-b.com/) Schematic of the filtration region. Filled circles represent atoms and unfilled their periodic images. The green circle represents an atom for which filtered functions are to be calculated and the yellow circles represent atoms whose primitive functions will contribute to these filtered functions.

The filtration step, as outlined in Ref. [6], consists of the following operations on a trial function $|\tilde{t}\rangle$

$$|\tilde{k}\rangle = cf(\Lambda)c^T S|\tilde{t}\rangle, \qquad (16.9)$$

to obtain the vector of contraction coefficients $|\tilde{k}\rangle$. The remaining quantities are defined by the GEP

$$\mathbf{Hc} = \mathbf{Sc}\Lambda, \qquad (16.10)$$

where, here, H and S matrices only consisting of a subset of the rows and columns of the Hamiltonian and overlap matrix formed in the large primitive basis (see Ref. [6]). In other words they are the H and S matrices associated with the primitive functions within the filtration region (Figure 16.2). The filtration function f used in Ref. [6] and throughout this work is a high temperature ($kT \sim 3\,\mathrm{eV}$) Fermi-Dirac function, which has the desired effect of removing the unnecessary high eigenspace of the Hamiltonian. The GEP [Eq. (16.10)] can be transformed to an ordinary eigenproblem

$$\bar{\mathbf{H}}\mathbf{d} = \mathbf{L}^{-1}\mathbf{H}\mathbf{L}^{-T}\mathbf{d} = \mathbf{d}\Lambda, \qquad (16.11)$$

where L is a lower triangular matrix and

$$\mathbf{S} = \mathbf{LL}^T \text{ and } \mathbf{d} = \mathbf{L}^T\mathbf{c}. \tag{16.12}$$

From this we can express Eq. (16.9) as

$$|\tilde{k}\rangle = \mathbf{L}^{-T}\mathbf{d}f(\Lambda)\mathbf{d}^T\mathbf{L}^T|\tilde{t}\rangle = \mathbf{L}^{-T}f(\tilde{\mathbf{H}})\mathbf{L}^T|\tilde{t}\rangle. \tag{16.13}$$

The primitive space → subspace transformation is performed using the following sparse matrix multiplications;

$$\tilde{\mathbf{H}} = \mathbf{k}^T\mathbf{Hk} \text{ and } \tilde{\mathbf{S}} = \mathbf{k}^T\mathbf{Sk}. \tag{16.14}$$

From this one obtains the subspace GEP

$$\tilde{\mathbf{H}}\tilde{\mathbf{c}} = \tilde{\mathbf{S}}\tilde{\mathbf{c}}\tilde{\Lambda}. \tag{16.15}$$

Since the dimension of this eigenproblem is small—essentially the size of a tight binding Hamiltonian—it is, at present, solved with standard direct diagonalisation. However, it must be stressed that due to the filtered functions being localised this matrix will be sparse for large systems, and therefore alternatives to diagonalisation may be considered. After the solution of this eigenproblem is obtained the subspace density matrix is constructed

$$\tilde{b}_{IJ} = \sum_\lambda f(\Lambda_{\lambda\lambda})\tilde{c}_{I\lambda}\tilde{c}_{J\lambda}, \tag{16.16}$$

after which the subspace → primitive space transformation is performed

$$\mathbf{b} = \mathbf{k}\tilde{\mathbf{b}}\mathbf{k}^T, \tag{16.17}$$

and the calculation proceeds as normal. For calculations in silicon and similar materials we have used four functions produced using trial functions of s, and p type symmetry. The resulting functions $|\tilde{k}\rangle$ are plotted in Figure 16.3.

We should note that, although the number of functions we are using (four) has an obvious chemical significance for silicon, this is in no way a restriction of the algorithm. Indeed, we may choose to use more or less functions, with full convergence to the primitive basis being obvious as more functions are added. Using fewer than four functions can also give good results, but only if the filtration is performed at a low-enough temperature to permit this. Indeed, zero temperature filtration can produce an exact result with only two filtered functions per silicon atom, although this would not be practical as the functions would lose their localisation (in general), thereby removing any advantage of this method.

16.4.1
Performance

We now detail the performance of the current algorithm including latest developments [7] and also some very encouraging preliminary results from further

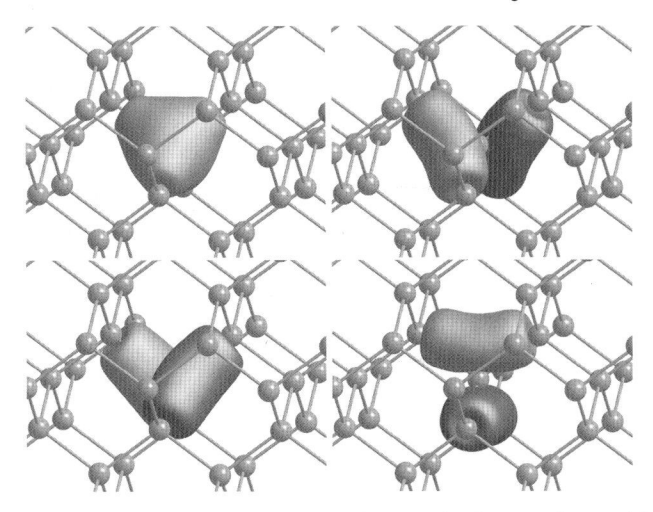

Figure 16.3 (online colour at: www.pss-b.com) Filter functions from trial functions of s, p_x, p_y and p_z (clockwise from top-left) symmetry.

optimisations of the algorithm. As a model system we look at unit cells of silicon created by forming $n \times n \times n$ arrays of the eight atom conventional unit cell, where $2 \leq n \leq 8$. It must be stressed, however, that the algorithm is not limited to wide-gap systems and is equally applicable to metals [6, 33]. The calculations use gamma point sampling and are performed on a single core of a 2.8 GHz Intel Xeon CPU.

Table 16.1 gives timings using an approximate filtration strategy—still good enough to give only a $\sim 10^{-4}$ Å error in relaxed final structures (see Section 16.4.2). Times are given for a single self-consistent iteration. For this system, SCF cycles

Table 16.1 Timings (s) of components of a self-consistent iteration for n^3 (simple cubic) cells of silicon ($n = 2, \ldots, 8$) on a single 2.8 GHz Intel Xeon core. The first row gives the number of atoms in each cell. See Sections 16.2 and 16.4 for a description of the algorithmic components. These calculations correspond to the approximate filtration scheme (see Table 16.2).

	64	216	512	1000	1728	2744	4096
matrix build	2.79	9.38	22.99	45.05	76.48	133.99	185.11
potential calculation	0.07	0.22	0.58	1.15	2.11	3.44	4.84
filtration kernel	2.00	6.74	16.00	31.19	53.94	85.97	128.12
primitive → subspace	0.48	3.82	12.41	27.61	48.76	77.71	116.10
subspace diagonalisation	0.03	0.47	7.73	52.58	270.63	1057.98	3740.67
density matrix build	0.00	0.07	0.86	6.11	31.88	124.13	414.33
subspace → primitive space	0.11	0.96	5.35	24.88	72.30	138.89	213.17
calculation of real space density	0.43	1.53	3.75	7.42	12.75	21.97	30.51
overhead	0.11	0.70	1.85	3.20	7.56	11.59	19.14
total	6.02	23.89	71.52	199.19	576.41	1655.67	4851.99

require fewer than ten iterations to converge the Hartree energy associated with the difference of input and output densities to less than 10^{-5} Ha. The most notable detail in the table is the bottom line. This shows that on a single core a self-consistency step for a 1000 atom system takes just 200 s and a 1728 atom system less than 10 min. Therefore initial total energies of these systems can be found in ∼30 min and ∼1.6 h respectively. Clearly, even modest parallelism over the 8 cores, which may typically be in a commodity dual processor PC, reduces these to remarkably small values, and enable even complex structural relaxations on inexpensive hardware.

Clearly for small systems (*e.g.* 216 atoms) the dominant time is that of the matrix build [Eq. (16.3)] together with the filtration kernel [Eq. (16.9)]. These have $O(N)$ complexity and are clearly unimportant for larger systems where the $O(N^3)$ subspace diagonalisation begins to dominate. One somewhat surprising feature of the timings is that the primitive to subspace transformation [Eq. (16.14)] and its inverse [Eq. (16.17)] are not significant at any system size, occupying at most 20% of the total time (in 216 atoms) and gradually reducing for larger systems. This is a consequence of the sparsity of k, H and S being well exploited together with reasonably efficient code (which achieves ∼25% of peak performance) to perform the block-multiplications.

As a final comment, it is seen that for the 1728 atom system, approximately half of the total time is spent solving the subspace matrix eigenvalue problem. As the size of this matrix is the same as in a tight binding calculation it may be supposed that an accurate full DFT calculation on this system size may be performed in twice the time of a tight binding calculation. The difference however diminishes to just 20% for the 4096 atom system and asymptotically will vanish entirely, if direct diagonalisation is used in both.

16.4.2
Accuracy

We now analyse the accuracy of the filtration method by comparing formation energies and relaxed structures to the parent primitive basis. The filtration algorithm has been previously shown to produce energies and forces which are in close agreement with those produced by the conventional algorithm [6]. We have subsequently looked at a variety of different systems including metals and wide band gap materials [33]. In this section some further results are given focusing particularly on the accuracy of equilibrium structures and the impact of filtration on the atomic co-ordinates.

We first present a comparison of the structures of single interstitial atoms in silicon. Three structures are presented: the 110 defect in which a pair of Si atoms straddle a lattice site, displaced from it in ⟨110⟩ directions; an atom placed at a tetrahedral interstitial site (T_d in the table below), and a hexagonal interstitial site, labelled H in the tables. The calculations were performed in unit cells containing 217 silicon atoms, using a *ddpp* primitive basis, the pseudopotentials of Hartwigsen *et al.* [34] and k-point sampling corresponding to $2 \times 2 \times 2$ Monkhorst–Pack grid [35].

Structures were optimised until all components of forces on atoms were less than 5×10^{-5} a.u.

Two filtration conditions were chosen. The first uses a cutoff radius of 12 a.u., our standard converged value for silicon as used in previous work [6]. For this, Table 16.2 illustrates the minimal impact filtration has on equilibrium structure, with maximum deviations from the unfiltered results of order 10^{-4} Å or less. To further illustrate the insensitivity of the structures produced by filtration, a second filtration strategy was adopted which used a smaller localisation radius (10 a.u.) together with an more approximate filtration kernel. This more approximate approach shifts the total energy of the system by 0.3 Ha, an immense change but the corresponding changes to equilibrium structure are still seen from Table 16.2 to be much less than 10^{-3} Å. The relative energies of the defects are also changed by only 20 meV by this. This is a clear demonstration of the arguments regarding BSSE given in Section 16.2.1. We find this result to be a general feature of our procedure.

Also shown in Table 16.2 are the relative energies of the different structures. It is seen that the error associated with filtration is <10 meV and even the approximate filtration invokes errors of only 20 meV. These results are typical of a number of systems we have looked at. Although in an application of this in materials science, we would not consider publishing the relative energy from the approximate filtration but include it here to illustrate an important behaviour of the filtration (indeed, atom centred localised basis set calculations in general)—even if an approximation is used which produces gross (of order 10 eV) shifts in the absolute energy of structures, the relative energies remain converged and the structure is almost unchanged, indeed if bond lengths are published to three decimal places, they would appear unchanged. This can be exploited if desired as a means of accelerating structural optimisation, which typically consumes the majority of computing time in a research project.

As a second example, we have considered the binding of an interstitial oxygen atom O_i to a vacancy oxygen VO centre in silicon:

$$VO + O_i \rightarrow VO_2,$$

Table 16.2 Relative energies (ΔE) and errors in relaxed structures (Δ^r)—both mean (avg) and maximum (max) errors—of various defects in silicon (see text for details).

	ΔE (eV)	Δ^r_{max} (mÅ)	Δ^r_{avg} (mÅ)
110 structure unfiltered	0.000	—	—
standard filtration	0.000	0.16	0.01
approx. filtration	0.000	0.68	0.06
T_d structure unfiltered	0.149	—	—
standard filtration	0.143	0.06	0.006
approx. filtration	0.162	0.41	0.05
H structure unfiltered	0.107	—	—
standard filtration	0.104	0.06	0.009
approx. filtration	0.125	0.28	0.046

where the VO$_2$ centre consists of two O atoms in a vacancy, each forming bonds to two of the silicon atoms and passivating the dangling bonds. In a calculation using Γ point sampling of the Brillouin zone, the binding energy is found to be 1.355 eV for the conventional algorithm and 1.354 eV for the filtered calculation, again showing a remarkable but nevertheless characteristic level of agreement between the two calculations.

Finally, we consider a defective metallic system, namely the ideal vacancy in aluminium. A 3^3 cell (108 atoms) with a relatively large primitive basis (*pddpp*) was used with a 10^3 uniform grid of points to perform integration over the Brillouin zone. To converge the formation energy to within ∼0.01 eV required only ∼50 min on a single core. We will shortly publish an extensive study of the accuracy of filtration algorithm for defects in a variety of systems.

16.5
Future Research Directions and Perspectives

In this section we wish to expand on previous discussions, assess their impact on the field and to address likely outcomes of on-going research in this exciting area. As well as significant improvements to speed and memory requirements, recent advances have presented a rich array of new questions and research directions that impact a very broad range of methods and applications. As this is a very rapidly evolving area we will spend sometime to discuss the impact of current research in the near-term.

There are some points that are immediately clear from the work presented or likely to be realised in the near-term:

i) Tight binding calculations, which use conventional $O(N^3)$ diagonalisation methods, are only marginally faster than the full-DFT algorithm presented here.
ii) Calculations involving structural relaxation or saddle-point location on systems containing ∼10 000 electrons are now comfortable on a single core.
iii) The fact that the size of the primitive basis is decoupled from the size of the filtered basis allows for larger, and therefore more accurate, primitive sets to be used. On-going improvements to this procedure are likely to enable calculations—essentially at the basis set limit—at a cost comparable to tight binding.

The possibility of using low-complexity [a subset of these approaches are more commonly referred to as 'linear-scaling' or '$\mathcal{O}(N)$'] subspace kernels (such as the recursive bisection density matrix method [36]) rather than diagonalisation leaves open the question as to the ultimate relative performance of tight binding approaches compared to a full KSDFT calculation in the limit of a large number of atoms. This is due to a complex range of factors, such as matrix sparsity and spectral width, and their relation to accuracy.

16.5.1
Types of Calculations

Now, we shall discuss some broad types of calculations not only to highlight the progress that has been made but also to emphasise challenges that remain.

16.5.1.1 Thousands of Atoms on a Desktop PC

The ideal computational limit for the use of Gaussian orbitals combined with localised filter diagonalisation is large systems on a small number of cores due to memory efficiency and low operation count. As quad-core processors are currently rather common we are already at the stage where $\sim$2000 silicon atoms can comfortably be handled on a single processor. Further algorithmic improvements coupled with the expected release of $8+$ core processor promises to facilitate $\sim$4000+ atom calculations on such machines in the near future. This computational limit also provides the perfect framework for calculations involving many independent subspace kernel calls, such as for separate **k**-points or calculation of the dynamical matrix. These can be effectively parallelised to allow the calculation of vibrational frequencies of $\sim$1000 atom systems.

16.5.1.2 One Atom Per Processor

The opposite limit to performing large calculations on an inexpensive desktop machine is the 'one atom per processor' limit, in which we are interested in obtaining energies/forces of systems extremely rapidly by way of outstanding parallel scaling. Here, the interest is in using large capability supercomputing facilities to perform many interdependent calculations—the classic example being long time-scale molecular dynamics. As supercomputing facilities and architectures are leading to a rapid growth in the number of cores available to a calculation this is an important area of research. The current filter diagonalisation approach still has a direct diagonalisation kernel and therefore is not suited to this computational limit. However, virtually all other aspects of the calculation scale very well. Alternative algorithms for the subspace kernel will have to be developed to tackle this important class of problems.

At larger system sizes, the use of low-complexity methods, which inherently scale well, will largely tackle the problem of scaling. However, the size of the system where this will become realistically advantageous is not clear, and there will likely be an intermediate size range (up to several thousand atoms) where other strategies will be needed.

16.5.2
Prevailing Application Trends

We now turn to the impact discussions in the Section 16.5.1 are likely to have on the prevailing types of calculations performed by users of KSDFT methods.

In a recent work [7] it was shown that for silicon—a difficult example due to the fairly high atom/electron number ratio—the filtration algorithm was competitive with the conventional algorithm (using an accurate filtration radius of 12 a.u.) at 216 atoms. Indeed, with more approximate filtration (typically acceptable for force calculations) the filtration algorithm is essentially faster at any system size (Table 16.1).

We would consider the modelling of a defect in a 216 atom unit cell to be a fairly small calculation, but such calculations can rapidly become large. For example, some defects can have surprisingly long ranged strain effects and the use of cells of this size has produced erroneous conclusions (see for example Ref. [37]). Even though atoms far from a defect typically do not relax far from their perfect lattice site the number of such atoms in a shell of a given radius is large and therefore they have an effect on formation and migration energies and the like [38].

However, even point defect calculations can become demanding when tests of cell size convergence are required—doubling the distance between point defects in a 216 atom system would require the use of the 1728 atom until cell discussed earlier. This would usually be regarded as a very 'large' (read 'time consuming') calculation by today's standards. However, a glance back at Table 16.1 and we see that such a calculation is tractable, with approximate filtration, on a single core. In the cases of certain types of problem (*e.g.* charged defects) correction schemes [39–41] have been popular to avoid performing large calculations (*i.e.* to avoid reaching the cubic bottleneck). Such schemes should be viewed as complementary to performing large-scale calculations not a replacement for them.

Maybe the ease at which ~1000–2000 atoms calculations can now be performed will lead to a re-evaluation of many point defect problems where questions remain over the size of supercell used. Moving away from point defects to extended defects—the need for a few thousand atoms becomes even more necessary. The performance improvements here will enable a far greater complexity of problem to be treated, for engineering problems associated with imperfect interfaces; the interaction of defects with complex environments; a more accurate treatment of dislocations; their motion and interaction with one another and other defects. So, one could see methods able to perform accurate calculations containing several thousand atoms as opening the afore mentioned systems to the same level of scrutiny presently reserved for point defects.

16.5.3
Methodological Developments

The filtration algorithm represents a significant shift, indeed a sea change, in the speed and convenience of accurate KSDFT calculations for a range of systems sizes—but especially for large systems. It is evident that, in many respects, calculations involving a few thousand electrons present a comparable challenge—in terms of computational burden—per SCF iteration to a tight binding calculation.

However, we wish to stress, these implementations are still rather new. A great deal of scope exists, provided by these recent developments, for further optimisations. Also, due to the recent dramatic advances [6] parts of the code that previously were insignificant, therefore did not warrant particular attention from the point of view of further optimisation (such as the matrix build) have once again become an issue and work is being done to address these topics.

From a methodological development standpoint—just as conventional algorithms have aided the development of $\mathcal{O}(N)$ and low-complexity algorithms—these new methods will, in turn, be of significant use in the development of new multi-length-scale approaches, such as hybrid QM/MM modelling and the like [42–44], as direct theory to theory comparisons will be possible. It remains to be seen whether current hybrid approaches will be successfully, in most cases, verified or whether a 'length scale drift' will occur [as has certainly been the case for many $\mathcal{O}(N)$ methods] and in a few years we may be using full KSDFT for calculations now seen as the purview of more approximate methods. Certainly where important questions remain with such approaches—such as (i) How does one link the QM and MM regions and decide their respective boundaries?, (ii) How does one treat highly complex structures with no obvious reference point for MM such as the prefect crystal?, (iii) How does one cope with a system where the QM region becomes very large?, (iv) How does one assess the propagation of errors as one drifts from the 'Kohn–Sham surface'? and (v) What is the chemical potential of the complete QM/MM system? In other words, how does one generate the correct QM state?—the use of a single, robust and tried and tested framework is certainly attractive, and worth striving for.

The fact that 1000 atom calculations are becoming comfortable in serial should, at once, put an end the oft expressed sentiment that Kohn–Sham density functional theory can only handle 'a few hundred atoms at most'.

Although the use of low-complexity methods certainly will become important, it is clear that for realistic calculations the much commented upon crossover point—where a low-complexity algorithm becomes quicker than a conventional approach—is significantly larger than suggested by the early optimism in the literature. In fact, it is, to a large extent, nonsense now to talk of a crossover point between low-complexity and conventional approaches—we have already seen (in Section 16.4) a cubically scaling algorithm benefit hugely from the inclusion of ideas from low-complexity methodologies.

We now return to the issue of BSSE and the impact of the filtration algorithm on this problem. Clearly, in the large system limit the subspace diagonalisation as the final remaining $\mathcal{O}(N^3)$ operation dominates the calculation. As this diagonalisation is independent of the size of the primitive set, in the large system limit, large primitive sets—previously not used because of their severe impact on performance—can be used. This will increase the effort required in the filtration step, but this is currently insignificant (for large enough systems), scaling linearly with system size and taking just 3% of the total time for the 4096 atom system considered earlier. On-going improvements in this part of the algorithm, already well in hand, will reduce the system size at which this becomes an issue and should enable these calculations with localised orbitals to be (almost) systematically converged, even in routine runs. The fact that, in many respects, plane-wave accuracy for the cost of tight binding seems to be achievable, and possibly even transparently to the user, makes this an exciting possibility.

16.6
Conclusions

A presentation of current and on-going developments in the AIMPRO code has been presented. The speed of KSDFT calculations using the latest state-of-the-art algorithms is comparable to a tight binding calculation. Even considering the extra memory and operations specimen examples such as 1000 silicon atoms on a single core are becoming comfortable and standard calculations.

Acknowledgement

The authors thank J. P. Goss and A. Lawson for useful discussions. M. J. R. gratefully acknowledges the support of the Alexander von Humboldt Foundation.

References

1. Kohn, W. and Sham, L.J. (1965) *Phys. Rev.*, **140**, A1133.
2. Hohenberg, P. and Kohn, W. (1964) *Phys. Rev.*, **136**, B864.
3. Jones, R. and Briddon, P.R., in: (1998), *Identification of Defects in Semiconductors*, Vol. 51A of *Semiconductors and Semimetals*, edited by M. Stavola, Academic Press, Boston, Chap. 6.
4. Briddon, P.R. and Jones, R. (2000) *Phys. Status Solidi B*, **217**, 131.
5. Rayson, M.J. and Briddon, P.R. (2008) *Comput. Phys. Commun.*, **178**, 128.
6. Rayson, M.J. and Briddon, P.R. (2009) *Phys. Rev. B*, **80**, 205104.
7. Rayson, M.J. (2010) *Comput. Phys. Commun.*, **181**, 1051.
8. Rayson, M.J. (2007) *Phys. Rev. E*, **76**, 026704.
9. Rayson, M.J. and Briddon, P.R. (2008) *Phys. Rev. B*, **77**, 035119.
10. Obara, S. and Saika, A. (1986) *J. Chem. Phys.*, **84**, 3963.
11. Lippert, G., Hutter, J. and Parrinello, M. (1999) *Theor. Chem. Acc.*, **103**, 124.
12. Kresse, G. and Furthmuller, J. (1996) *Phys. Rev. B*, **54**, 11169.
13. Vetterling, W.T. and Flannery, B.P. (2002) *Numerical Recipees*, second ed., edited by Press, and S.A. Teukolsky, (Cambridge University Press).
14. Farkas, O. and Schlegel, H.B. (2002) *Phys. Chem. Chem. Phys.*, **4**, 11.
15. Wardle, M.G., Goss, J.P., and Briddon, P.R. (2006) *Phys. Rev. Lett.*, **96**, 205504.
16. Wardle, M.G., Goss, J.P., and Briddon, P.R. (2005) *Phys. Rev. B*, **72**, 155108.
17. Wardle, M.G., Goss, J.P., and Briddon, P.R. (2005) *Phys. Rev. B*, **71**, 155205.
18. MacLeod, R.M., Murray, S.W., Goss, J.P., and Briddon, P.R. (2009) *Phys. Rev. B*, **80**, 054106.
19. Etmimi, K.M., Ahmed, M.E., Briddon, P.R., Goss, J.P., and Gsiea, A.M. (2009) *Phys. Rev. B*, **79**, 205207.
20. Shaw, M.J., Briddon, P.R., Goss, J.P., Rayson, M.J., Kerridge, A., Harker, A.H., and Stoneham, A. M. (2005) *Phys. Rev. Lett.*, **95**, 105502.
21. Goss, J.P., Coomer, B.J., Jones, R., Shaw, T.D., Briddon, P.R., Rayson, M., and Öberg, S. (2001) *Phys. Rev. B*, **63**, 195208.

22 Rayson, M.J., Goss, J.P., and Briddon, P.R. (2003) *Physica B*, **340**, 673.
23 Goss, J.P., Jones, R., Shaw, T.D., Rayson, M.J., and Briddon, P.R. (2001) *Phys. Status Solidi A*, **186**, 215.
24 Goss, J.P., Briddon, P.R., Rayson, M.J., Sque, S.J., and Jones, R. (2005) *Phys. Rev. B*, **72**, 035214.
25 Goss, J.P., Rayson, M.J., Briddon, P.R., and Baker, J.M. (2007) *Phys. Rev. B*, **76**, 045203.
26 Coutinho, J., Jones, R., Briddon, P.R., and Öberg, S. (2000) *Phys. Rev. B*, **62**, 10824.
27 Wardle, M.G., Goss, J.P., and Briddon, P.R. (2006) *Appl. Phys. Lett.*, **88**, 261906.
28 Goss, J.P., Jones, R., and Briddon, P.R. (2002) *Phys. Rev. B*, **65**, 035203.
29 Liggins, S., Newton, M.E., Goss, J.P., Briddon, P.R., and Fisher, D. (2010) *Phys. Rev. B*, **81**, 085214.
30 Fall, C.J., Jones, R., Briddon, P.R., Blumenau, A.T., Frauenheim, T., and Heggie, M.I. (2002) *Phys. Rev. B*, **65**, 245304.
31 Fall, C.J., Blumenau, A.T., Jones, R., Briddon, P.R., Frauenheim, T., Gutierrez-Sosa, A., Bangert, U., Mora, A.E., Steeds, J.W., and Butler, J.E. (2002) *Phys. Rev. B*, **65**, 205206.
32 Ewels, C.P., Van Lier, G., Charlier, J.C., Heggie, M.I., and Briddon, P.R. (2006) *Phys. Rev. Lett.*, **96**, 216103.
33 Briddon, P.R. and Rayson, M.J. (in preparation).
34 Hartwigsen, C., Goedecker, S., and Hutter, J. (1998) *Phys. Rev. B*, **58**, 3641.
35 Monkhorst, H.J. and Pack, J.D. (1976) *Phys. Rev. B*, **13**, 5188.
36 Rayson, M.J. (2007) *Phys. Rev. B*, **75**, 153203.
37 Goss, J.P., Briddon, P.R., and Eyre, R.J. (2006) *Phys. Rev. B*, **74**, 245217.
38 Sulimov, V.B., Sushko, P.V., Edwards, A.H., Shluger, A.L., and Stoneham, A.M. (2002) *Phys. Rev. B*, **66**, 024108.
39 Makov, G. and Payne, M.C. (1995) *Phys. Rev. B*, **51**, 4014.
40 Lany, S. and Zunger, A. (2008) *Phys. Rev. B*, **23**, 235104.
41 Freysoldt, C., Neugebauer, J., and Van de Walle, C.G. (2009) *Phys. Rev. Lett.*, **102**, 016402.
42 Csanyi, G., Albaret, T., Payne, M.C., and De Vita, A. (2004) *Phys. Rev. Lett.*, **93**, 175503.
43 Bulo, R.E., Ensing, B., Sikkema, J., and Visscher, L. (2009) *J. Chem. Theory Comput.*, **5**, 2212.
44 Gao, J.L. and Truhlar, D.G. (2002) *Annu. Rev., Phys. Chem.*, **53**, 467.

17
Ab Initio Green's Function Calculation of Hyperfine Interactions for Shallow Defects in Semiconductors
Uwe Gerstmann

17.1
Introduction

In semiconductor technology, besides hetero structures, shallow and deep defects are the key ingredients. The availability of mobile hole and electrons in dedicated functional regions of the material hinges on the ability to control the concentration of shallow defects. This requires a clear understanding of the physics of these important entities. In particular, the identification of defects present in a given material often is an important challenge. Here, magnetic resonance is the most sensitive experimental technique to address this essential problem. Structure identification, however, cannot be done by experiments alone. The measurements provide a set of parameters only, which have to be compared with theoretical predictions of these quantities. Only if theoretical modeling can exclude all but one structural model, then an unambiguous identification of defects has been achieved. For the modeling of the microscopic structure of the defects the electronic g-tensor and in particular their hyperfine (hf) tensors have been found to be crucial parameters as they contain detailed information on the wave function distribution on the central nucleus and the ligands, whereby the latter is often called *superhyperfine* (shf) interaction. With the development of the local spin density approximation (LSDA) of the density functional theory (DFT) [1] theoretical *ab initio* total energy methods have been introduced that describe defect structures in solids quantitatively. Nowadays many computational codes provide the possibility to calculate the hf structure of the experimental magnetic resonance spectra routinely, in most cases with the required degree of accuracy.

However, defects in semiconductors have to be divided in two groups: (1) Deep point defects seriously disturb the crystal in a small region centered at the defect site. Extremely long-ranged perturbations as *e.g.*, a Coulombic potential-tail for charged defects or strain fields are either ignored or treated as a correction. A self-consistent treatment of the deep defect is possible because it can be restricted to a small region in space, a cluster, a supercell, or to the perturbed region in a Green's function approach. (2) For shallow defects, on the other hand, the *wave function* of the defect-induced state typically extends over several hundred or even thousand unit cells and,

Advanced Calculations for Defects in Materials: Electronic Structure Methods, First Edition.
Edited by Audrius Alkauskas, Peter Deák, Jörg Neugebauer, Alfredo Pasquarello, and Chris G. Van de Walle.
© 2011 Wiley-VCH Verlag GmbH & Co. KGaA. Published 2011 by Wiley-VCH Verlag GmbH & Co. KGaA.

therefore, cannot be treated directly with *ab initio* supercell methods. Instead, the defect-induced change of the crystal potential is replaced by a model potential and the defect state is treated in an empirical one-electron approximation, the so-called effective mass approximation (EMA) [2–4]. In the simplest EMA for a substitutional donor, e.g., the defect wave function is expanded into the Bloch states close to the minimum of the lowest conduction band, for which the dispersion of the band can be approximated by a parabola defining the *effective mass*. However, an empirical correction is necessary to distinct between different atomic species: the so-called *central cell correction* is introduced to describe the local part of the potential. The best EMA are then able to reproduce the binding energies of shallow acceptors and donors. The empirical character of the EMA, however, prevents a prediction of the hf and shf parameters, decisive for an identification of the atomic structures. As a result, the wealth of information contained in the (s)hf interaction data for shallow dopants, the technologically most important class of defects, is completely obscured.

In this article, we show that within a Green's function approach the *central cell correction* as the empirical part of the EMA can be substituted by an *ab initio* calculation of this quantity. By this, a prediction of hf splittings becomes possible, within an accuracy comparable with that in case of deep defects. This article is organized as follows: first, we discuss the microscopic origin of the hf structure showing that DFT is perfectly suited to allow an accurate computation. Then a Green's function method is applied onto deep defects, whereby lattice relaxation is taken into account if calculating the hf splittings. Thereafter, based on a short review of the EMA and its empirical extensions, we present how the *ab initio* calculated local part of the potential is embedded via Green's functions into an EMA-like background. Finally, the approach is applied onto shallow donors in silicon and silicon carbide (SiC).

17.2
From DFT to Hyperfine Interactions

17.2.1
DFT and Local Spin Density Approximation

Our main target is the calculation of the hf splittings within a many-body problem given by the nuclei and the density of the surrounding electrons. In a first step we get rid of the nuclear degrees of freedom using the Born–Oppenheimer approximation. According the theorem by Hohenberg and Kohn [5] the (non-degenerate) ground state energy of the remaining many-electron system is a unique functional $E[n]$ of this electron density which, thus, provides the starting point of DFT. From a general many-electron wave function ψ it can be obtained via

$$n(\mathbf{r}) = \langle \Psi | \sum_{i=1}^{N_{el}} \delta(\mathbf{r}-\mathbf{r}_i) | \Psi \rangle. \tag{17.1}$$

The theory can easily be extended to include the spin polarization: Following von Barth and Hedin [6], we assume that an external magnetic field defines the direction of spin quantization with $\sigma = \uparrow, \downarrow$. The ground state energy is then a unique functional $E[n_\sigma]$ of the spin-polarized electron densities (the *spin densities*) or alternatively a unique functional $E[n, m]$ of the electron density

$$n(\mathbf{r}) = n_\uparrow(\mathbf{r}) + n_\downarrow(\mathbf{r}) \tag{17.2}$$

and the magnetization density

$$m(\mathbf{r}) = n_\uparrow(\mathbf{r}) - n_\downarrow(\mathbf{r}). \tag{17.3}$$

Kohn and Sham [7] have shown that the density $n(\mathbf{r})$ for interacting fermions can be mapped onto the density for a system of non-interacting particles that are subject to some extra energy, the exchange-correlation energy $E_{xc}[n, m]$. For non-interacting particles we know that the density $n_\sigma(\mathbf{r})$ can be expanded into the sum of squared single-particle orbitals $\varphi_{l,\sigma}(\mathbf{r})$

$$n_\sigma(\mathbf{r}) = \sum_{l=1}^{N_\sigma} |\varphi_{l,\sigma}(\mathbf{r})|^2. \tag{17.4}$$

For practical use these orbitals $\varphi_{l,\sigma}(r)$ are the solution of coupled single-particle equations (the *Kohn–Sham equations*), whereby the electrons move in a spin-polarized effective potential

$$V_{\text{eff},\sigma}(\mathbf{r}) = V_{\text{ext}}(\mathbf{r}) + e^2 \int \frac{n(\mathbf{r}')}{|\mathbf{r}-\mathbf{r}'|} d^3\mathbf{r}' + V_{xc,\sigma}^{\text{DFT}}[n, m], \tag{17.5}$$

given by the external potential V_{ext}, the Coulomb-potential of the electrons, completed by the exchange-correlation potential $V_{xc,\sigma}^{\text{DFT}}[n, m]$ that includes the many-particle contributions and that depends on the spin direction σ, on the electron density $n(\mathbf{r})$, and also on the magnetization density $m(\mathbf{r})$. The exact shape of this non-local exchange-correlation potential

$$V_{xc,\sigma}^{\text{DFT}}[n, m] = \frac{\delta E_{xc}^{\text{DFT}}[n, m]}{\delta n_\sigma(\mathbf{r})} \tag{17.6}$$

is unknown. For practical applications, however, there are useful parametrization schemes which approximate the exchange-correlation potential [8, 9] calculated for a homogeneous spin-polarized electron gas. The resulting $V_{xc,\sigma}^{\text{LSDA}}[n, m] = V_{xc,\sigma}^{\text{LSDA}}(n(\mathbf{r}), m(\mathbf{r}))$ depends only locally on the spin densities. This local density approximation (LDA) of DFT has proven to yield approximate results with an accuracy going far beyond the early expectations [1]. The perhaps most crucial shortcoming of the LSDA can be found in connection with the fundamental band gap of semiconductors as the gap of the single-particle energies turn out to be too small by about a factor of two. The reason for this is a discontinuity upon a change of the particle number that would be present in the exact exchange-correlation functional [10, 11], but is absent in the LSDA [12]. We will come back to this point later.

17.2.2
Scalar Relativistic Hyperfine Interactions

Within the LSDA it is also possible to calculate the hf interaction of the magnetic moments of the electrons with those of the nuclei. The influence of an external magnetic field B_0 (in the range of some 100 mT) typically leads to level splittings in the $10^{-12} \ldots 10^{-2}$ eV range. The smallness of the magnetic field-induced level splittings *simplifies* the computation considerably, and the influence of an external magnetic field can be described by perturbation theory.

Although there exist a non-relativistic derivation for the isotropic contact interaction (Fermi [13]), Breit has shown that the origin of the hf splitting can be only described correctly in a relativistic treatment [14]. The static magnetic field $B(\mathbf{r}) = \nabla \times \mathbf{A}(\mathbf{r})$ caused by the magnetic moment $\boldsymbol{\mu}_I = g_N \mu_N \mathbf{I}$ of a nucleus with gyromagnetic ratio g_N located at the origin is obtained using the vector potential

$$\mathbf{A}(\mathbf{r}) = \nabla \times \left(\frac{\boldsymbol{\mu}_I}{r}\right). \tag{17.7}$$

By replacing the momentum operator $\mathbf{p}$ in Dirac's equation

$$(c\boldsymbol{\alpha} \cdot \mathbf{p} + \beta mc^2 + V_{\text{eff}} - E_{\text{rel}})|\Psi\rangle = 0 \tag{17.8}$$

by the canonical momentum $\boldsymbol{\pi} = \mathbf{p} - e/c\mathbf{A}$ the expectation value of the hf interaction is, thus, given in first order perturbation theory by

$$E_{\text{HF}} = -e\langle\Psi|\boldsymbol{\alpha} \cdot \mathbf{A}(\mathbf{r})|\Psi\rangle. \tag{17.9}$$

Here, $\boldsymbol{\alpha}$ is a vector of 4×4 matrices constructed from the 2×2 Pauli spin matrices σ_x, σ_y, and σ_z, respectively, whereby $|\Psi\rangle = \begin{pmatrix}\Phi_L\\ \Phi_S\end{pmatrix}$ is given by the Dirac spinor, decomposing into the two-component Pauli spinors Φ_L and Φ_S. For light atoms, Φ_L is the dominant, large component whereas Φ_S turns out to be small. This leads to

$$E_{\text{HF}} = -e(\langle\Phi_L|\boldsymbol{\sigma} \cdot \mathbf{A}(\mathbf{r})|\Phi_S\rangle + \langle\Phi_S|\boldsymbol{\sigma} \cdot \mathbf{A}(\mathbf{r})|\Phi_L\rangle). \tag{17.10}$$

Thus, E_{HF} is a genuine relativistic term that couples large and small components of Dirac's equation. The small component Φ_S can be expressed in terms of the large component Φ_L as

$$\Phi_S = \frac{c\boldsymbol{\sigma} \cdot \mathbf{p}}{2mc^2 + E - V_{\text{eff}}(\mathbf{r})}\Phi_L = \frac{S(r)}{2mc^2}(\boldsymbol{\sigma} \cdot \mathbf{p})\Phi_L, \tag{17.11}$$

whereby $S(r)$ is the inverse relativistic mass correction. By this, the hf splitting

$$E_{\text{HF}} = E_{\text{contact}} + E_{\text{orb}} + E_{\text{dip}}, \tag{17.12}$$

is the sum over the following expectation values containing the large component only [15]:

$$E_{\text{contact}} = -\frac{8\pi}{3}\mu_B \langle \Phi_L | S(\mathbf{r})\boldsymbol{\mu}_I \cdot \boldsymbol{\sigma}\delta(\mathbf{r}) | \Phi_L \rangle$$
$$+ \langle \Phi_L | \frac{1}{r^4}\frac{\partial S}{\partial r}[\boldsymbol{\mu}_I \cdot \boldsymbol{\sigma} r^2 - (\boldsymbol{\mu}_I \cdot \mathbf{r})(\boldsymbol{\sigma} \cdot \mathbf{r})] | \Phi_L \rangle, \tag{17.13}$$

$$E_{\text{orb}} = -\frac{e}{mc}\boldsymbol{\mu}_I \cdot \langle \Phi_L | \frac{\mathbf{L}}{r^3} | \Phi_L \rangle, \tag{17.14}$$

$$E_{\text{dip}} = \mu_B \langle \Phi_L | \frac{1}{r^5}[\boldsymbol{\sigma} \cdot \boldsymbol{\mu}_I r^2 - 3(\boldsymbol{\sigma} \cdot \mathbf{r})(\boldsymbol{\mu}_I \cdot \mathbf{r})] | \Phi_L \rangle. \tag{17.15}$$

In the non-relativistic limit, since $S(\mathbf{r}) \to 1$, only the first term in Eq. (17.13) contributes to the isotropic contact term E_{contact}. By this, we obtain the results of the classical theory given by Fermi [13], that only the probability amplitude at the nucleus contributes. In the relativistic case, however, this first term does not contribute at all. It is the second term in Eq. (17.13) which is the relativistic analog to the contact interaction. For a pure Coulomb-potential around a given nucleus we obtain

$$V_{\text{eff}}(r) \approx \frac{-Ze^2}{r}, \tag{17.16}$$

and the derivative $\partial S(r)/\partial r$ is similar to a broadened δ-function

$$\delta_{\text{Th}}(\mathbf{r}) = \frac{1}{4\pi r^2}\frac{\partial S}{\partial r} = \frac{1}{4\pi r^2}\frac{r_{\text{Th}}/2}{\left[\left(1+\frac{E}{2mc^2}\right)r + \frac{r_{\text{Th}}}{2}\right]^2}. \tag{17.17}$$

In other words, the magnetization density of the electron in the relativistic theory is not simply evaluated at the origin, where it would be divergent for s-electrons, but is averaged over a sphere of radius

$$r_{\text{Th}} = \frac{Ze^2}{mc^2}, \tag{17.18}$$

which is the Thomson radius, about ten times the nuclear radius. As a result, the divergence of the s-electrons presents no problem. Also if we approximate the nuclear potential by that of a charged volume rather than that of a point charge [16], the divergence already disappears. However, it is important to note, that we would obtain divergent contact terms mixing the approximations, e.g., using (scalar)[1] relativistic orbitals in a non-relativistic formula.

If the ground state of the defect is a single determinantal orbital singlet state with total spin S, we have the simple case where the orbital angular momentum

---

[1] In the scalar relativistic treatment Φ_L is calculated solving Dirac's equation but thereby ignoring spin-orbit interactions. This leaves the electron spin as a "good" quantum number. Already in a scalar relativistic treatment, s-like wave functions diverge at the nuclear site (if the nucleus is taken to be a point charge).

is "quenched" [17]. The orbital state transforms like an angular momentum eigenfunction with quantum number $l = 0$. Hence, the expectation value of the angular momentum operator vanish, and there are no orbital contributions E_{orb} to the hf interactions. Writing the N-particle wave function $|\Phi_L\rangle = |S, M_S\rangle$ as a single Slater determinantal in real space representation, the hf interaction is then fully described by the matrix elements $\langle S, M_S | \mathcal{H}_{\text{HF}} | S, M'_S \rangle$ with respect to the spin Hamiltonian

$$\mathcal{H}_{\text{HF}} = \sum_k \{a_k \mathbf{S} \cdot \mathbf{I}_k + \mathbf{S} \cdot \mathbf{B}_k \cdot \mathbf{I}_k\}. \tag{17.19}$$

It is important to note that the matrix elements ($\mathbf{r}_k = \mathbf{r} - \mathbf{R}_k$)

$$a_k = \frac{1}{2S} \frac{8\pi}{3} \gamma_k \int m(\mathbf{r}) \delta_{\text{TH}}(\mathbf{r}_k) d^3\mathbf{r}, \tag{17.20}$$

$$\mathbf{B}_k = \frac{1}{2S} \gamma_k \int m(\mathbf{r}) \frac{3\mathbf{r}_k \otimes \mathbf{r}_k - r_k^2 \mathbf{1}}{r_k^5} d^3\mathbf{r}, \tag{17.21}$$

can be expressed as a sum over *single-particle* matrix elements with respect to the magnetization density without the need to *construct many*-body wave functions first. We have simply to insert the magnetization densities obtained from the self-consistent LSDA calculation and obtain the hf interactions with the central nucleus as well as with the ligands (ligand hf interactions or shf interactions). Since each of the $\mathbf{B}_k$ is traceless, its diagonal elements can be parametrized by two parameters b_k and b'_k, describing the axial and non-axial part of the anisotropy.

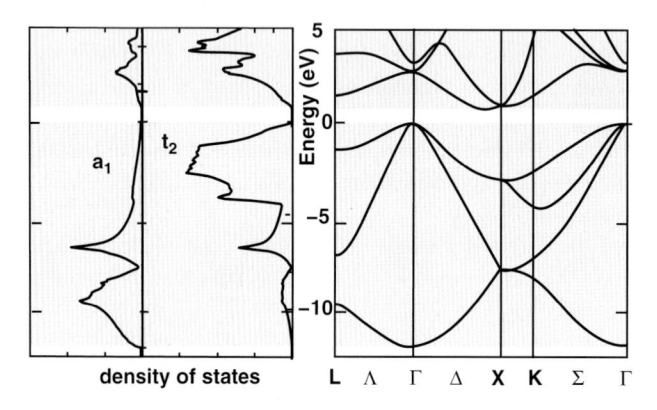

Figure 17.1 The density of states (DOS) distribution for the silicon crystal broken up into states transforming according to a_1 (left) and to t_2 (right), is compared with the energy bands plotted along the high-symmetry directions.

17.3
Modeling Defect Structures

17.3.1
The Green's Function Method and Dyson's Equation

By introducing a point defect into the otherwise perfect crystal we break the translational symmetry, and the periodicity of the ideal crystal is lost. Nevertheless, the method most frequently used for the computation of point defects in solids is the supercell method: instead of an isolated system with a single point defect one considers a three-dimensional lattice of clusters, each with a single point defect. This crystal can be treated theoretically by the known methods of energy band theory, however, with the cluster as the "supercell" unit cell.

The Green's function method [18–20] embeds a single defect into an otherwise perfect crystal (cf. Figure 17.2). For this system the effective potential V_{eff} is split into the effective potential V_{eff}^0 of the undisturbed crystal plus some short-ranged impurity-related potential ΔV

$$V_{eff} = V_{eff}^0 + \Delta V. \tag{17.22}$$

ΔV can be quite large at the defect site but will decay rapidly with the distance from the defect center. The Green's function, G_0, for a perfect crystal characterized by the Hamiltonian H_0 is defined as [21]

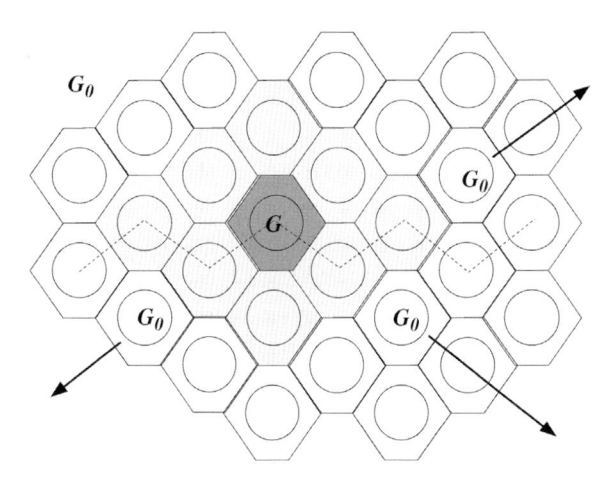

Figure 17.2 (online color at: www.pss-b.com) Dyson's equation: A perturbation ΔV is embedded into a reference system characterized by the "unperturbed" Green's function G_0. Within the perturbed region (colored) the Green's function G of the perturbed system is determined via Eq. (17.29): $G = (1 - G_0 \Delta V)^{-1} G_0$. Outside this region, G remains unchanged and matches G_0.

$$G_0(E) = \lim_{\varepsilon \to 0^+} \frac{1}{E - i\varepsilon - H_0}. \qquad (17.23)$$

A small imaginary part added to the energy prevents a singularity for $E = E_{n,k}$, i.e., within the valence and conduction bands (cf. Figure 17.1). G_0 can be expressed in terms of Bloch functions $|\varphi_{n,k,\sigma}\rangle$ for the spin state σ, which solve the Kohn–Sham equations for the perfect crystal:

$$G_0(E) = \lim_{\varepsilon \to 0^+} \sum_{n,k,\sigma} \frac{|\varphi_{n,k,\sigma}\rangle \langle \varphi_{n,k,\sigma}|}{E - i\varepsilon - E_{n,k,\sigma}}. \qquad (17.24)$$

The sum in Eq. (17.24) includes all states $\{n, k, \sigma\}$, not just the occupied states. Vice versa, the Green's function G for a given system includes all electronic ground state properties of this system. The full information of the density of states distribution (DOS)

$$D(E) = \frac{1}{\pi} \mathcal{I}m \, Tr\{G(E)\} \qquad (17.25)$$

and most important the (spin-polarized) electron densities are obtained by summing up the occupied bands only

$$n_\sigma(r) = \frac{1}{\pi} \mathcal{I}m \left\{ \int_{\text{occ.}} G^\sigma(E, r, r) dE \right\}. \qquad (17.26)$$

For a crystal containing a deep defect, the Green's function G corresponding to the full Hamiltonian H reads

$$G(E) = \lim_{\varepsilon \to 0^+} \frac{1}{E - i\varepsilon - H}, \qquad (17.27)$$

where the Hamiltonian now contains the full effective potential V_{eff}. The Green's function G is related to G_0 by a Dyson equation

$$G = G_0 + G_0 \Delta V G, \qquad (17.28)$$

which can be solved iteratively

$$G = (1 - G_0 \Delta V)^{-1} G_0. \qquad (17.29)$$

The solution of Eq. (17.29) is possible if we have to invert $(1 - G_0 \Delta V)$ only in the vicinity of the defect, the "perturbed region", where ΔV is non-negligible. In contrast, the Green's functions G_0 and G extend outside this perturbed region. Despite the localized shape of the "perturbation" ΔV, the approach via Dyson's equation is very flexible. Nowadays, it is a standard tool to describe transport properties in microscopic nanostructures [22], whereby ΔV is determined by the conductance electrons.

One result of a self-consistent calculation of Dyson's equation for a defect is the electron density $n(r)$. It can, of course, also be decomposed into contributions from different spin directions $n_\sigma(r)$. For the hf interactions we will need the magnetization density

$$m(\mathbf{r}) = \frac{1}{\pi} \mathcal{I}m \left\{ \int_{\text{occ.}} [G^\uparrow(E,\mathbf{r},\mathbf{r}) - G^\downarrow(E,\mathbf{r},\mathbf{r})] dE \right\}. \tag{17.30}$$

We have already noted in Section 17.2.1 that in fully converged LDA calculations the fundamental band gap turns out to be too small. For silicon, this difference is about 0.5 eV and for wide band gap semiconductors like GaAs, GaN, or SiC, the error is in the range of 1 eV or even larger. In some cases, defect-induced gap states, which should be located in the upper part of the gap, are calculated to be resonances in the conduction band. Also if this worst case scenario is not given, the hf interaction can be affected [23]. The Green's function method provides a way to circumvent this problem by a rigid shift of the crystalline conduction bands by some amount ΔE_{shift} with respect to the valence bands, before calculating G_0. This formalism, called scissor operator $\mathbf{Sc}\{E\} = E + \Delta E_{\text{shift}}$, was introduced by Baraff and Schlüter [24]. Hence, by substituting Eq. (17.24) by

$$G_0(E) = \lim_{\varepsilon \to 0^+} \sum_{n,\mathbf{k},\sigma} \frac{|\varphi_{n,\mathbf{k},\sigma}\rangle \langle \varphi_{n,\mathbf{k},\sigma}|}{E - i\varepsilon - \mathbf{Sc}\{E_{n,\mathbf{k},\sigma}\}}, \tag{17.31}$$

we are able to adjust the fundamental gap to a given experimental value. It is important to note here that this has to be done only once, namely if calculating the Green's function of the ideal crystal. Afterwards, since the Green's function approach is applied in real space without the need of periodic images, the band edges are retained, and each one-particle level is corrected automatically in a self-consistent way.

17.3.2
The Linear Muffin-Tin Orbital (LMTO) Method

For a periodic system like a crystal, one might consider a plane wave expansion to be the simplest computational method. And indeed, most supercell calculations use the pseudopotential method for which the inclusion of short-ranged lattice relaxations is relatively straightforward. Since by construction the pseudo-wave functions do not include the rapid oscillations in the core region, the resulting spin pseudo-wave densities are not directly applicable for the computation of hf interaction matrix elements, determined in the vicinity of the nuclei. Van de Walle and Blöchl [25] used a projector formalism to reconstruct the original wave function from the pseudo-functions. Alternatively, in the muffin-tin methods, V_{eff} is assumed to be spherically symmetrical within atomic spheres around R_j,

$$V(\mathbf{r}-\mathbf{R}_j) = \begin{cases} V(|\mathbf{r}-\mathbf{R}_j|) & \text{if } |\mathbf{r}-\mathbf{R}_j| \leq s_j \\ \text{const.} & \text{else,} \end{cases} \tag{17.32}$$

and the Kohn–Sham orbitals are expanded into spherical harmonics times a radial solution. Different muffin-tin methods are in use which differ in the technical procedure used to construct a regular Bloch wave from the partial wave solutions obtained within the atomic spheres: the Korringa–Kohn–Rostocker (KKR) method

[26, 27], the linear muffin-tin orbital (LMTO) method [28, 29], and the linearized augmented plane wave (LAPW) method [30]. Compared with the pseudopotential method the muffin-tin methods have the advantage that the correct electron and magnetization density in the nuclear region can be directly and very accurately obtained, which is decisive for the hf interaction. All muffin-tin methods, however, share the disadvantage that they are technically difficult and, due to the use of atomic spheres, less flexible with respect to larger lattice relaxations, although there are full-potential versions of all the muffin-tin methods, FP-KKR [31], FP-LMTO [32], and F-LAPW [33].

In this work, we use the LMTO-ASA method since it provides perhaps the easiest and straightforward way to realize the Green's function method via a muffin-tin approach [20]. In the atomic spheres approximation (ASA) the sphere radii for the integrals are chosen such that the unit cell volume equals to the sum of the ASA sphere volumes. In this approximation, the ASA spheres overlap slightly and the contribution of the neglected regions are assumed to be cancelled by the double-counted overlap. For open structures like semiconductors, additional "empty" spheres centered around the highly symmetrical interstitial sites of the lattice have to be inserted in order to optimize the volume-filling. We have extended the LMTO-ASA Green's function approach to include moderate lattice relaxations. For the resulting distorted structures the concept of space-filling ASA spheres is no more straightforward. Here, the concept of a Voronoi tessellation is very helpful (*cf.* Figure 17.3), whereby the decomposition of a metric space is determined by distances to a discrete set of points, the center of the Voronoi cells (either given by a nucleus or the center of an empty cell) [34]. The concept can be easily extended to hetero-atomic structures using a weighted decomposition. Here, we use the so-called Bragg–Slater radii [35], whereby the ASA condition of space-filling spheres can be easily fulfilled by constructing an ASA sphere $\mathcal{S}_j$ with radius s_j to each Voronoi cell $\mathcal{Z}_j$:

$$V(\mathcal{S}_j) = V(\mathcal{Z}_j). \tag{17.33}$$

Using the in this way extended ASA approximation, we solve Dyson's equation in LMTO matrix representation [20]

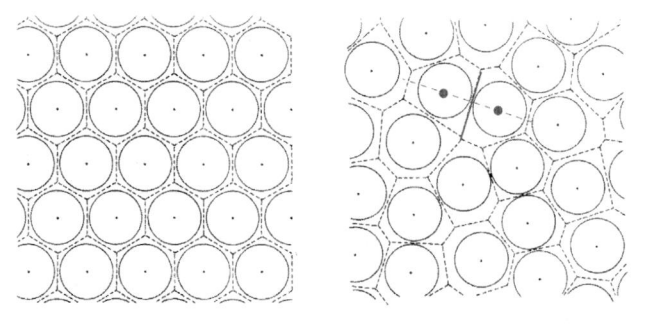

Figure 17.3 (online color at: www.pss-b.com) Wigner–Seitz cells of a two-dimensional regular structure (left) and in comparison a disordered structure divided into Voronoi cells (right). Possible non-overlapping muffin-tin radii are also given.

$$\{1 + g^0(E)[\Delta P(E) - \Delta S]\}g(E) = g^0(E), \tag{17.34}$$

within a "perturbed region" large enough to allow a proper description of ΔS and ΔP. Here, $\Delta P = P - P^0$ is the localized, diagonal perturbation of the so-called potential function [20] describing the electronic structure of the investigated atomic structure, and ensuring that the partial waves from the muffin-tin spheres fulfill the correct bounding conditions. $\Delta S = S_{R'L',RL} - S_{R'_0L',R_0L}$ is the relaxation-induced change of the LMTO-ASA structure constants

$$S_{R'L',RL} = (-1)^{l+1} 8\pi \sum_{l''} \frac{l! l'! (2l'')!}{(2l)!(2l')!l''!} \left(\frac{a}{d_{RR'}}\right)^{l+l'+1} \sum_{m''} C_{LL'L''} Y_{L''}(\hat{d}_{RR'}), \tag{17.35}$$

whereby $C_{LL'L''} := \int_\Omega Y_L(\hat{r}) Y^*_{L''}(\hat{r}) Y_{L'}(\hat{r}) d\Omega$ and $Y_L(\hat{r})$ as spherical harmonics with $L = (l, m)$. The calculation of the rather extended matrix ΔS is more elaborate: the long-ranged tails of the matrix elements are split up and treated by generalized Ewald sums [36], whereas the short-ranged contribution are calculated in a next-nearest neighbor approximation.

In the general case, the atomic position $\{\mathbf{R}\}$ of the relaxed structure can be taken from pseudopotential calculations. The electronic structure to the relaxed structures can then be calculated via Eq. (17.34), including the (s)hf parameters of paramagnetic states. We will see, however, that in some cases with rather moderate relaxation, the extended LMTO-ASA GF approach is also able to predict at least the nearest neighbor relaxation.

17.3.3
The Size of The Perturbed Region

Using a Green's function method, the spin densities arising from the defect states are not completely contained in the rather limited volume of the perturbed region. Surprisingly, this does not cause a problem as will be shown in the following, taking the isolated As_{Ga}^+ antisite in GaAs as a reference system (Table 17.1 where also well established experimental data is available):

The gap state, that mainly gives rise to the magnetization density (See Figure 17.4) and, thus, to the hf interaction of the paramagnetic As_{Ga}^+ charge state, transforms according to the a_1 irreducible representation of the group T_d. For the neutral and the two-fold positive, diamagnetic charge states, this state is unoccupied. The total energies, E_{tot}, and the single particle eigenvalues of this gap state, $E(a_1)$, also shown in Table 17.1 seem to depend on N_{atoms} in a rather unsystematic manner. However, the variation of E_{tot} for the E^{2+} and E^0 charge states is practically identical with that for E^+, and therefore, the charge transfer energies are essentially independent of N_{atoms}.

In Table 17.1 we also show the convergence of the corresponding hf interactions, calculated using perturbed regions of different sizes. The magnetization density is mainly concentrated on the central antisite atom, where it gives rise to the largest nearly isotropic hf splitting, and on its four nearest neighbors, where it is predom-

Table 17.1 The influence of the size of the perturbed region on the total energy, E_{tot}, the gap state energy $E(a_1)$, the magnetic moment of the gap state μ_{gap} and the total magnetic moment μ_{pert} within the perturbed region, as well as the hf interactions (MHz) (upper line for a, lower line for b) for the 3% outward relaxed isolated As_{Ga}^+ antisite in GaAs. Beside the hf values for the central As_{Ga}^+ nucleus, the shf splittings due to several neighbor shells of the crystal host are also given. N_{atoms} is the number of atomic ASA spheres in the perturbed region. Experimental data taken from Refs. [37, 38].

N_{atoms}	E_{tot}	$E(a_1)$	μ_{gap}	μ_{pert}	As_{Ga}^+ (0,0,0)	As (1,1,1)	Ga (2,2,0)	As (1,1,3)	As (3,3,1)
1	−107.83	1.478	0.097	0.113	2811.				
5	−107.47	0.988	0.469	0.499	2778.	187.7			
						46.1			
11	−108.09	1.080	0.572	0.598	2846.	175.3	8.8		
						45.8	1.5		
23	−108.04	1.063	0.614	0.638	2839.	175.8	7.4	0.3	
						45.7	1.5	−0.2	
47	−107.82	0.995	0.817	0.840	2879.	173.2	3.9	0.1	22.3
						46.6	1.4	−1.9	4.2
exp.					2650.	169.3	–	–	21.5
						53.2	–	–	2.2

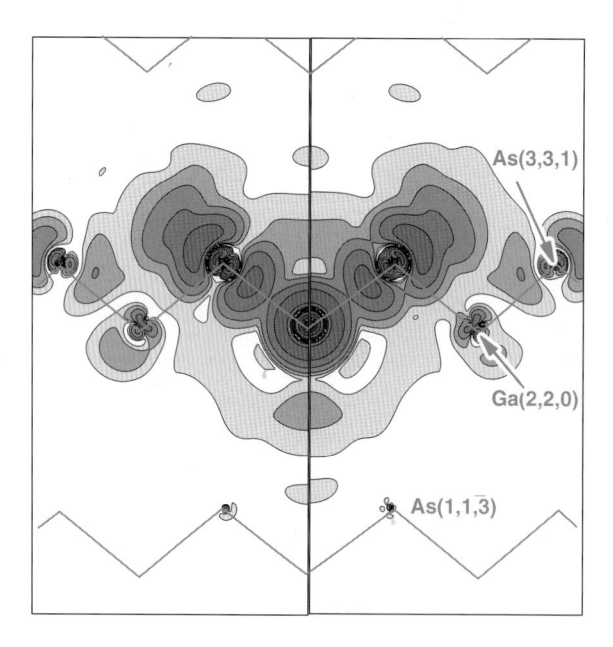

Figure 17.4 (online color at: www.pss-b.com) Contour plot of the magnetization density for the isolated As_{Ga}^+ antisite in a $(1\bar{1}0)$ plane in GaAs (taken from Ref. [44]). The left panel shows the contribution of the gap state, the right part gives the total magnetization density. The Ga (As) lattice sites are at the lower (upper) side of the zig-zag chain of nearest neighbor bonds, respectively.

inantly p-like. The smallest conceivable perturbed region consisting of the defect ASA sphere alone contains about 10% of the magnetic moment of the defect. Yet, we obtain a central contact hf interaction at the As_{Ga} antisite nucleus which is only 2% smaller than the value obtained for the largest perturbed region with 47 atoms. It is obvious and quite impressive that via the Green's function approach the defect is really embedded in an infinite background. Even if we do not present the data explicitly, the reader should believe that this procedure also works in the case of a vacancy: only an empty sphere is then necessary to obtain at least rough estimates. From Table 17.1 we see that for different sizes of the perturbed region the hf interactions with the nuclei at the "surface" of the perturbed region is slightly overestimated. With a further increase of N_{atoms} the corresponding values are reduced to better values. For the contact hf interactions of the more distant Ga (2,2,0) and As(1,1,$\bar{3}$) ligands the convergence apparently is quite poor, but here the magnetic moments for the ligand ASA shells are extremely small. The value of 4 MHz for the contact interaction with a [69] Ga nucleus corresponds to 2×10^{-4} of a single s-like spin only.

17.3.4
Lattice Relaxation: The As_{Ga}-Family

Experimentally well understood, the isolated As_{Ga} antisite is only one member of the technologically very important As_{Ga}-family: At least four different As_{Ga}-related defects with almost identical hf structure have been detected by magnetic resonance [37–39]. Their thermal stability is quite puzzling: the well established isolated As_{Ga} defect is obtained by low-temperature electron irradiation of semi-insulating GaAs and disappears at room temperature [40], when in electron-irradiated material the so-far unidentified As_{Ga}-X_1 defect is observed. At $T = 520$ K the As_{Ga}-X_1 defect disappears and the so-called EL2 becomes dominant. The latter defect is quite stable. It is the dominant defect in semi-insulating GaAs [41, 42] where it determines the position of the Fermi level. The EL2 can be eliminated by a rapid quench from 1100 °C [43] and is recovered by annealing the sample above 750 °C. Its paramagnetic properties strongly suggest the EL2 to be a nearly tetrahedral defect. If the EL2 is not the isolated antisite it should be, thus, at least some pair or complex with some other partners. However, the exact microscopic structure of the EL2 defect is still controversial (for a review, see Refs. [39, 44]).

Another interesting aspect of the members of the family of As_{Ga}-related defects is their metastability. Theoretical *ab initio* calculations [45–47] have shown that a lattice relaxation around the As_{Ga} antisite atom is responsible for the defect metastability. We have, thus, investigated the influence of the lattice relaxation onto the hf interaction (See Table 17.1 for the largest perturbed region). For the isolated tetrahedral point defects a symmetry-conserving relaxation of the nearest neighbors was included to determine the lattice relaxation from the minimum of the total energy. For the neutral isolated As_{Ga}^0 point defect we find a minimum of the total energy if the distance to the nearest neighbors is increased by 4.7% with respect to the bond-length in the unperturbed crystal. The energy gained by this relaxation is

0.32 eV. A very similar relaxation (4.0%, 0.33 eV energy gain) was reported by Dabrowski and Scheffler [45] for a 54-atom supercell calculation. For the defect in the singly positive, paramagnetic charge state the relaxation reduces to 3% (1.4% for the double positive charge state). For the relaxed defects the calculated charge transition energies are $E^{2+/+} = E_v + 0.98$ eV and $E^{+/0} = E_v + 1.18$ eV, if the band gap is adjusted to the experimental value by the Scissor operator, somewhat smaller than the results (1.25 and 1.5 eV, respectively) obtained by Baraff and Schlüter [48] and by Delerue [49]. Without such an adjustment of the gap, the charge transition energies would be $E_v + 0.37$ eV and $E_v + 0.55$ eV, respectively.

Figure 17.5 shows the calculated total energy for the isolated As_{Ga}^+ defect as a function of the nearest neighbor distance d for a relaxation that does not alter the tetrahedral defect symmetry. Also shown is the dependence of the hf interactions with the antisite nucleus and of the shf interactions with the nearest neighbors. As is the case for all deep donor states, the hf interaction with the donor nucleus is quite sensitively dependent on the nearest neighbor distance. In our case, the moderate 3% outward relaxation obtained for the minimum of the total energy leads to a 7% decrease of the hf interaction with the central nucleus (the same relaxation leads to a 9% decrease of the isotropic shf interaction and a 5% increase of the anisotropic shf interaction with the nearest neighbor nuclei).

Similar to the case of elemental semiconductors [50], the magnetization density is the subject of strong oscillations (See also Table 17.1): The shf splittings are rather small at the (2,2,0) and $(1, 1, \bar{3})$ neighbors, but again much larger at the more distant (3,3,1) neighbor, the fifth shell of neighbors. In particular, there are no larger interactions with the Ga nuclei, in agreement with the fact that these have not been

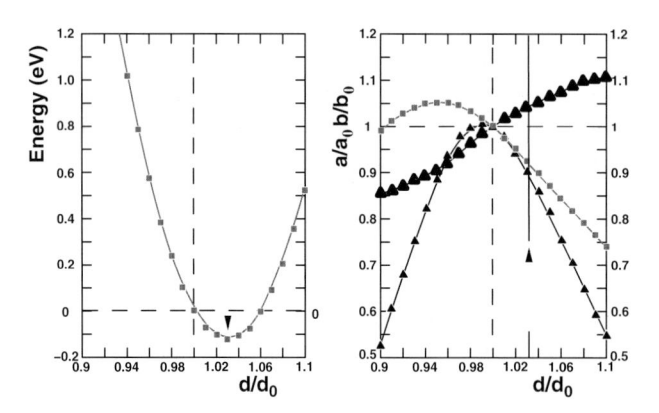

Figure 17.5 (online color at: www.pss-b.com) Calculated total energy for the As_{Ga}^+ antisite as a function of the nearest neighbor distance d (left). d_0 is the nearest neighbor distance for the unrelaxed GaAs crystal. The right panel shows the relative change of the hf interactions upon relaxation normalized to its respective unrelaxed values a_0 and b_0: Contact interaction with the antisite nucleus (square) and contact (full triangle) and dipolar (open triangle) interactions with the nearest neighbor nucleus.

detected by (optical detected) electron nuclear double resonance (OD)ENDOR. A detailed comparison with experimental data [37] shows a close agreement already for the interactions that had been calculated without taking into account a lattice relaxation. Nevertheless, the agreement is substantially improved if the lattice relaxation is included.

Since the hf interactions both with the central As nucleus and with the first shell of As ligands are strikingly similar for all members of the As_{Ga} family, it can be excluded that the experimentally observed paramagnetic states of any member of the As_{Ga} family are subject to a major lattice relaxation of the As_{Ga} nucleus. In Ref. [44] it has been furthermore shown that for the technologically important EL2 defect the As_{Ga}–As_i model can be excluded based on this argument and high-field OD(ENDOR) experiments [51]. Since a slight but definite deviation from tetrahedral symmetry is observed in these experiments, it appears that the *thermally most stable* defect in the As_{Ga} family is some defect aggregate. This at first view paradox observation has been proposed as the most likely solution, in which near room temperature the EL2 transforms to an isolated tetrahedral defect and that the deviations from tetrahedral symmetry observed experimentally are caused by the pairing with some other mobile defect, e.g., *shallow* acceptors or donors, which occur while cooling the sample.

17.4
Shallow Defects: Effective Mass Approximation (EMA) and Beyond

In the last section we have seen that shallow dopants, acceptors as well as donors can form complexes with intrinsic defects. In these complexes, the dopant levels appear in ionized form, so that the resulting complex form again a deep defect. Also the ionized charge state of an isolated donor is a deep defect, and the defect-induced change of the DOS is well localized. In the neutral charge state of the defect, however, the additional electron is rather extended. Only weekly bound, the donor electron provokes a hydrogen-like series of bound states with binding energies small compared to the fundamental band gap and with an effective Bohr radius that may exceed 100 Å. Whereas now a days supercell calculations of up to 1000 atoms nicely describe the ionization energies, the quantitative description of the spatial distribution of the wave function still remains a challenge. We will illustrate this problem taking conduction electrons in 4H-SiC as an example: Here, the delocalization of the electron wave function can be characterized by an effective Bohr radius of about 13 Å [52]. As a consequence, only 30% of the donor electron are found in a region containing 750 atoms around the donor atom (Figure 17.6). This is also demonstrated by recent 576-atom supercell calculations [53] where in comparison with experiment the localization of the donor wave function at a central P nucleus and the four neighboring ligands is overestimated by a factor of three (2.1% instead of 0.8%). In other words, the usual *ab initio* methods still cannot be used straightforward to treat the shf structure of these strongly delocalized defect states. Instead, we fall back on the *empirical* EMA as a standard tool to describe the *wave function* of shallow defects.

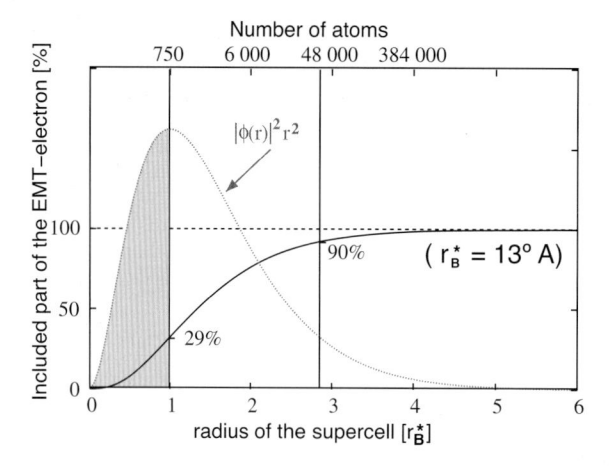

Figure 17.6 (online color at: www.pss-b.com) Delocalization of EMA-like donors: The radial probability distribution $|\Phi(r)|^2 r^2$ becomes maximum at the effective Bohr radius r_B^*. But only 29% of the electron are found in a sphere of radius r_B^* (solid curve). For 4H-SiC with $r_B^* = 13$ Å [52] e.g., huge supercells with more than 40 000 atoms would be necessary to reduce the artificial overlap of the periodic images of the wave functions to a more or less acceptable value below 10%.

The EMA "predicts" ionization energies and in addition the hydrogen-like series of bound states that agree sufficiently well with experimental data (for a review, see the classical article by Kohn [4] or the more recent article by Ramdas and Rodriguez [54]).

The comparison with experimental EPR data, however, shows the EMA results for the shf interactions to be at most qualitatively correct. In the following, this apparent failure of the EMA is discussed in comparison with results of an empirical pseudopotential calculation that provides accurate donor binding energies and corrected wave functions. Finally, we show that a Green's function based method allows an *ab initio* description of the magnetization density of shallow defects, including the resulting hf and shf splittings.

17.4.1
The EMA Formalism

The problem of a shallow defect can be divided into two parts: $D^0 = D^+ + e^-$. Providing a deep defect, the ionized donor D^+ can generally be treated using the standard methods for localized states. This deep defect gives rise to some potential ΔV_+ that apart from a local part ΔV_{local} asymptotically approaches the potential of a screened point charge:

$$\Delta V_+(\mathbf{r}) = -\frac{e^2}{\varepsilon_\infty r} + \Delta V_{local}(\mathbf{r}). \tag{17.36}$$

The extra electron e^- present in the neutral charge state D^0 is delocalized and moves within this potential, hardly disturbing the electron density of the deep state D^+.

Thus, the electron density for the extra electron is expected to coincide with the magnetization density of the donor state. Within the EMA [4] the extra electron is described by a single-particle wave function $\Psi(\mathbf{r})$ which obeys the Schrödinger equation

$$\left(-\frac{\hbar^2}{2m_e}\nabla^2 + V_{\text{host}}(\mathbf{r}) + \Delta V_+(\mathbf{r}) - E\right)\Psi(\mathbf{r}) = 0. \tag{17.37}$$

We expand $\Psi(\mathbf{r})$ into a complete orthonormal set of Bloch functions $\varphi_{n,\mathbf{k}}(\mathbf{r}) = u_{n,\mathbf{k}}(\mathbf{r})e^{i\mathbf{k}\cdot\mathbf{r}}$ leading to

$$\Psi(\mathbf{r}) = \sum_{n,\mathbf{k}} f_{n,\mathbf{k}}\varphi_{n,\mathbf{k}}(\mathbf{r}). \tag{17.38}$$

Moreover, only states near the minimum of the conduction bands are assumed to contribute to the expansion Eq. (17.38). Hence, the Bloch states obey $u_{n,\mathbf{k}}(\mathbf{r}) \approx u_{n,\mathbf{k}_0}(\mathbf{r})$ and their energies can be expanded around this extremum. For the simplest case of a non-degenerate conduction band edge at the Γ point of the Brillouin zone, we assume

$$E_{c,\mathbf{k}} = E_{c,\mathbf{k}_0} + \frac{\hbar^2}{2m^*}(\mathbf{k}-\mathbf{k}_0)^2, \tag{17.39}$$

with an isotropic conduction band mass m^*.

This brings us to an equivalent problem for the hydrogenic envelope function $\tilde{\Phi}(\mathbf{r})$: the effective mass equation (EME) reads

$$\left(-\frac{\hbar^2}{2m^*}\nabla^2 + \Delta V_+(\mathbf{r}) - (E - E_{c,\mathbf{k}_0})\right)\tilde{\Phi}(\mathbf{r}) = 0, \tag{17.40}$$

with m^* absorbing the periodic part of the potential. To proceed further we specify the potential ΔV_+. Far away from the impurity ΔV_+ is approximated by the potential of a point charge screened by the dielectric constant ε_∞. Anticipating that most of the particle density is delocalized, we approximate ΔV_+ by its asymptotic form $-\frac{e^2}{\varepsilon_\infty}\cdot\frac{1}{r}$ and neglect the specific local part of the potential completely.

With these approximations the eigenvalue problem (17.40) is identical to the elementary quantum mechanics textbook problem of the hydrogen atom. The solution for a particle of mass $m^* = \beta\, m_e$ moving in the screened Coulomb-potential can be written as

$$\tilde{\Phi}_{n,l,m}(\mathbf{r}) = \left(\frac{\beta}{\varepsilon_\infty}\right)^{3/2}\cdot R_{n,l}\left(\frac{\beta}{\varepsilon_\infty}|\mathbf{r}|\right)\cdot Y_{l,m}\left(\frac{\mathbf{r}}{|\mathbf{r}|}\right),$$

$$E_{n,l} = E_{c,\mathbf{k}_0} - \frac{\beta}{\varepsilon_\infty^2}\frac{\text{Ry}}{n^2}. \tag{17.41}$$

With $\varepsilon_\infty \sim 10$ and $m^* \sim 0.1 m_e$ we obtain an effective Rydberg energy of $\text{Ry}^* = 10^{-3}\,\text{Ry}$ and an effective Bohr radius $r_B^* = 10^2 r_B$. All approximations made appear to be valid under these conditions, except for the central cell, which contains a very small fraction of the extra electron density only. For the computation of hf interactions the hydrogenic envelope function $\tilde{\Phi}$ must be replaced by the true wave function Ψ.

In disagreement with the values of 53.73 meV, 45.53 meV, and 42.73 meV determined experimentally for As, P, and Sb in silicon [55], the best EMA predicts 31.27 meV *for all* the group V donors. This failure is of course a consequence of the neglect of the local part ΔV_{local} of the potential that would be necessary to distinct between different atomic species. A suitable central cell correction must be found to account for the so-called *chemical shift* within the binding energies.

17.4.2
Conduction Bands with Several Equivalent Minima

Besides the shollow electron centers in the silver halides AgCl and AgBr [56], we are not aware of experimental shf interaction data for shallow donors in a direct semiconductor with a conduction band minimum at the Γ point. EPR and ENDOR data are available for donors in the more conventional semiconductors Si and SiC. These have several equivalent conduction band minima far off the Γ point of the Brillouin zone. For silicon *e.g.*, the conduction band has six minima at $\mathbf{k}_0^{(1)} = 0.854 \frac{2\pi}{a}(1,0,0)$, along the so-called Δ axis near the boundary of the Brillouin zone. The solutions for the i different conduction band minima (the valleys) are degenerate. In order to construct realistic wave functions, symmetrical linear combinations of the single-valley wave functions from all equivalent valleys are required.

Whereas the single-valley solutions $\tilde{\Phi}_{1,0,0}^{(i)}(\mathbf{r})$ decay exponentially without nodes, the symmetrized wave function

$$\tilde{\Phi}_{1,0,0}^{(A_1)}(\mathbf{r}) = \sum_{i=1}^{6} \frac{1}{\sqrt{6}} \tilde{\Phi}_{1,0,0}^{(i)}(\mathbf{r}), \qquad (17.42)$$

describing the ground state and transforming according to the A_1 irreducible representation of the group T_d is oscillatory because of the $e^{i\mathbf{k}_0^{(i)} \cdot \mathbf{r}}$ factors in Eq. (17.38). Except at the donor site, the resulting magnetization density appears to be hardly related to the lattice structure (see right part of Figure 17.7). It shows an additional artificial mirror symmetry with respect to the horizontal (001) plane through the donor which is absent in the atomic positions. In addition, the first node of $\tilde{\Phi}_{1,0,0A_1}(\mathbf{r})$ nearly coincides with the nn ligand nucleus and, therefore, the isotropic shf interaction with the ^{29}Si nn nuclei (which one would naively assume to be largest) virtually vanishes.

17.4.3
Empirical Pseudopotential Extensions to the EMA

There were several attempts to find a central cell potential correction. Baldereschi [57] has pointed out that intervalley potential matrix elements are of particular importance. These are screened by the dielectric function $\varepsilon(\mathbf{q} = \mathbf{k}_0^{(i)} - \mathbf{k}_0^{(j)})$ rather than by the

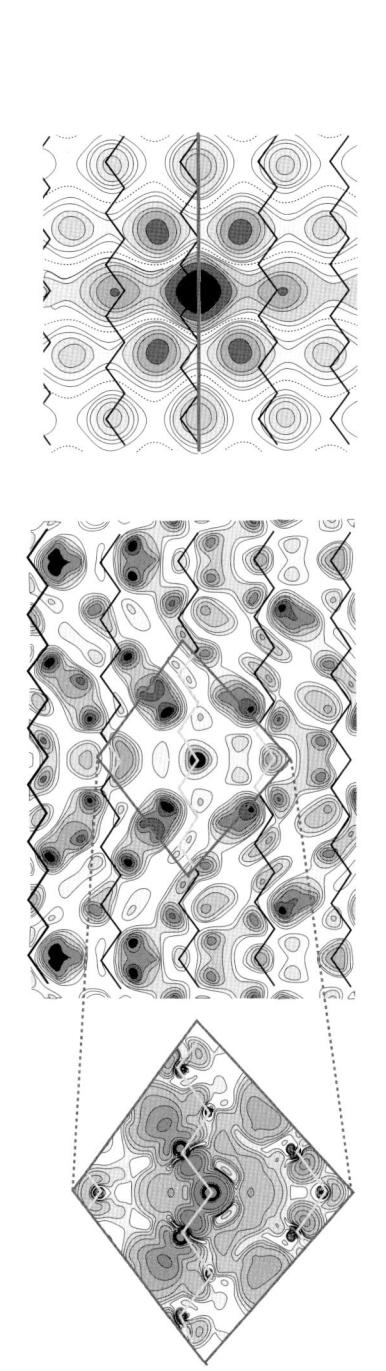

Figure 17.7 (online color at: www.pss-b.com) Contour plot of the electron density in the (1$\bar{1}$0) plane for As$_{Si}^0$ in Si calculated via the LMTO-GF scheme (left part). Right: Contour plot of the impurity electron density in the (1$\bar{1}$0) plane for a shallow donor in silicon according to the EMT of Hale and Mieher [64] (the donor atom is indicated by the black dot) and the same according the empirical pseudopotential theory Ivey and Mieher [60] (central).

full dielectric constant ε_∞. Several empirical pseudopotential schemes have been developed giving reliable ground state donor binding energies (for a review, see Pantelides [58]). Among these calculations, the calculation of Ivey and Mieher [59, 60] for group V donors in Si undertakes a calculation of the shf interactions. A model pseudopotential screened by the dielectric function is fitted to reproduce the experimental binding energy. In contrast to the EMA calculations, all **k** points throughout the Brillouin zone are sampled and $u_{n,\mathbf{k}}(\mathbf{r})$ is not approximated by some $u_{n,\mathbf{k}_0}(\mathbf{r})$. If compared to the EMA density, the resulting density of the donor electron has lost the mirror symmetry, retaining only the desired A_1 symmetry of the atomic structure (see also central part of Figure 17.7). The shf interaction of the nearest neighbors, however, are still by about two orders of magnitude too small. In consequence of this failure of the EMA and its extensions, the wealth of information contained in the shf interaction data for shallow donors is completely obscured. In the best case we need an *ab initio* calculation to unravel the experimental data. Without such a calculation we cannot identify a single ligand shell from its shf interaction data.

17.4.4
Ab Initio Green's Function Approach to Shallow Donors

In section 17.3.3 we have shown, that the Green's function method allows an accurate description of the hf interaction, already if the perturbed region contains only 10% of the magnetic moment of a deep defect. It is this observation that brings us to the idea that the same should also be possible in the case of shallow states where up to 90% of the delocalized electron are found outside the largest conceivable perturbed region. Hence, the basic idea is now to substitute the empirical part of the EMA, the central cell correction, by a first-principle description in which ΔV_{local} in Eq. (17.36) is calculated self-consistently and embedded via a Green's function approach into an otherwise periodic, EMT-like background [61].

Similar to the case of the As-antisite in GaAs, we solve Dyson's equation within a "perturbed region" that contains the donor and five shells of ligands (47 atoms in total) and six shells with 42 "empty" spheres to reduce the overlap of the ASA spheres. In the ENDOR experiments for group-V donors in Si no symmetry-lowering lattice distortions have been detected [62]. Minimizing the LMTO-ASA total energy by a symmetry-conserving relaxation of the nearest neighbor distances we find a minimum for a nearest neighbor distance that is decreased by 1% for P_{Si}^0, and increased by 3% As_{Si}^0 and by 6% for Sb_{Si}^0, respectively with respect to the distance in a perfect Si crystal. For P_{Si}^0 and As_{Si}^0, these values are reproduces by 216-atom supercell calculations [63]. For Sb_{Si}^0, however, a considerably larger outward relaxtion of 9% is predicted. We are of course more confident to the supercell geometry where all atoms are allowed to relax freely. Hence, in all what follows we use the 9% supercell-value for Sb_{Si}^0. By this, we obtain considerably improved values if compared with the values given in our original work [61].

Since in our approach we ignore the long-range tail of the Coulomb-*potential* for that part of the induced density that is not contained within the perturbed region, we do not find a shallow gap state but rather a resonance just above the onset of the conduction band. Thus, we cannot hope to obtain meaningful donor energies by this approach.

Figure 17.8 shows the change of the density of states (DOS) introduced by the defect (the "induced" DOS) for the three group-V donors in comparison with the DOS of the unperturbed crystal. In our approach we separate densities that arise from states transforming according to the different irreducible representation of the group T_d. In Figure 17.8 we display the a_1-like densities only, suppressing the t_2 and e-like resonances that are ascribed to excited states. The induced DOS for P^0_{Si} and As^0_{Si} show a relatively well-defined minimum near 1.6 eV above the valence band edge, for Sb^0_{Si} the resonance is much less pronounced. Starting from an effective one-particle picture, we shall consider the induced DOS below this minimum as a substitute for the shallow gap state. It contains about 15% of an electron within the perturbed region for P^0_{Si} and As^0_{Si}, while for Sb^0_{Si} we find as little as 8% of an electron. Identifying the resonance below the minimum with the extra electron, we calculate the spin polarization of all electrons within the LSDA. The resulting magnetization density plotted for As^0_{Si} in Figure 17.7 (left) is qualitatively different from the EMT result (right), in that it does not show the spurious inversion symmetry characteristic for the EMT. Instead it has some similarities with the envelope function obtained by Ivey and Mieher (central), with the clear distinction that there is no well-defined minimum at the nearest neighbors. A more detailed comparison of the different approaches is, however, possible via hf and shf data of ENDOR spectra, showing that our present approach is superior to the Ivey and Mieher (I-M) and the EMT methods (see Tables 17.2 and 17.3):

For the central donor nuclei the experimental values for the hf splitting are nicely reproduced within 5% for all donors – P, As, as well as Sb. Also for the nearest neighbor (1,1,1) shell, the isotropic and anisotropic shf interactions of our resonance

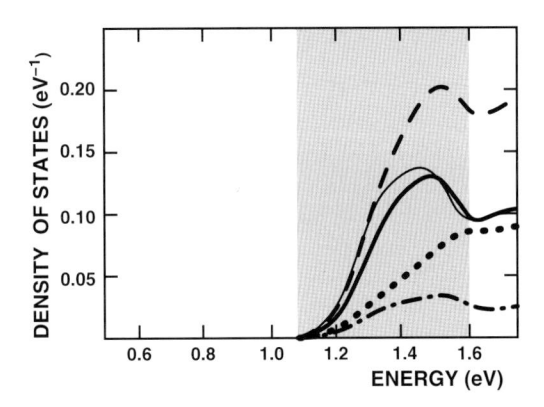

Figure 17.8 Density of states per impurity (DOS) that transform according the a_1 irreducible representation of the point group T_d for group-V donors in silicon. The (bold) full line denotes the induced density of the (un)relaxed As^0_{Si}, the dashed line is for P^0_{Si}, and the dash-dotted line is for the Sb^0_{Si} donors. The dotted line represents the a_1 density of the unperturbed Si crystal. The grey area denotes the energy interval of the a_1 resonance.

Table 17.2 Isotropic hf and shf interactions (in MHz) for group-V donors in Si. Experimental values from Ref. [62] are compared with theoretical results of the present LMTO-GF approach, the pseudopotential approach from Ivey and Mieher (I–M), and an EMT approach. All data shown have a negative sign. For Sb the values in parenthesis belong to a smaller ligand relaxation (6% instead of 9%).

shell	donor	exp.	this work	I-M	EMT
(0,0,0)	^{31}P	117.5	121.4	71.2	448.
	^{75}As	198.3	198.6	120.0	850.
	^{121}Sb	186.8	175.4	89.4	548.
			(66.8)		
(1,1,1)	P	0.540	0.518	0.036	1.524
	As	1.284	1.168	0.060	2.424
	Sb	0.586	0.405	0.090	1.232
			(2.053)		
(2,2,0)	P	–	0.115	0.608	0.861
		–	0.193	0.788	1.216
	Sb	–	0.312	0.532	0.734
			(0.587)		
(1,1,$\bar{3}$)	P	–	0.053	–	–
	As	–	0.179	–	–
	Sb	–	0.025	–	–
			(0.001)		
(0,0,4)	P	5.962	2.963	5.484	8.414
	As	7.720	3.160	7.606	11.400
	Sb	6.202	3.725	6.202	7.324
			(2.923)		
(3,3,1)	P	1.680	1.461	1.776	0.988
	As	2.242	2.351	2.590	1.290
	Sb	1.008	0.910	1.212	0.872
			(0.848)		

states compare favorably with the experimental data. The agreement is in fact much closer than for the I–M and EMT results for this shell. This becomes in particular clear if analyzing the ratio b/a which characterizes the hybridization at the (111) ligand shell (See also Table 17.3). For P and As, the values are rather insensitive to lattice relaxations. In case of Sb however, the 9% outward relaxation is necessary to predict a correct hybridization, $b/a = 1.09$ in comparison with the experimental ratio $(b/a)_{\text{exp}} = 0.89$. Note that for a reduced relaxation of 6% a much too small ansiotropy ratio of 0.12 is obtained [61]. The next two neighbor shells have not been identified experimentally, presumably because the isotropic shf constant is below about 600 kHz, the "continuum" of many overlapping ENDOR lines. This explanation is in line with our results. Note that for the (2,2,0) shell both I–M and EMT predict hf interactions that in contrast should be readily observed. For the (0,0,4) shell the isotropic shf data are predicted too small by a factor of 2. For the outermost (3,3,1) shell in the perturbed region our results compare again quite well with the experimental data.

Table 17.3 Anisotropic hf and shf interactions (the axial component b in MHz) for group-V donors in Si. In contrast to the pseudopotential approach from Ivey and Mieher (I-M) [60], our LMTO-GF values compare reasonably well with the experimental hybridization ratio b/a [62]. For Sb the values in parenthesis belong to a smaller ligand relaxation (6% instead of 9%).

shell	donor	exp.		this work		I-M	
		b	b/a	b	b/a	b	b/a
(1,1,1)	P	0.70	1.296	0.66	1.274	0.49	13.611
	As	1.26	0.981	1.14	0.976	0.93	15.500
	Sb	0.52	0.887	0.44	1.086	0.35	3.844
				(0.25)	(0.122)		
(2,2,0)	P	–	–	0.01	0.087	0.03	0.049
	As	–	–	0.02	0.104	0.03	0.038
	Sb	–	–	0.01	0.025	0.02	0.038
				(0.06)	(0.102)		
(0,0,4)	P	0.02	0.003	0.02	0.007	0.02	0.004
	As	0.03	0.004	0.02	0.006	0.02	0.003
	Sb	0.02	0.003	0.02	0.005	0.02	0.003
				(0.05)	(0.017)		
(3,3,1)	P	0.06	0.036	0.05	0.034	0.04	0.023
	As	0.08	0.036	0.09	0.038	0.08	0.031
	Sb	0.03	0.030	0.04	0.040	0.03	0.025
				(0.03)	(0.035)		

Altogether, the oscillating behavior of the shf splitting is qualitatively correct described. In contrast, the one-particle theories are by no means able to describe the correct order of the contribution of the different shells. Ivey and Mieher [60] have suggested that the discrepancy in their pseudopotential approach is due to the neglect of the lattice relaxations, an explanation that at least for P and As is not supported by our results. More important, according our Green's function calculation, the shf interactions are only to a small part due to the conduction band resonance: more than 75% of the isotropic shf is caused by the spin polarization of the valence band states. Such polarizations are not included in the one-electron approach of I–M, which may explain in part the striking discrepancy between the experimental data and the results of the one-electron theories.

The agreement between theoretical and experimental hf and shf data, confirms that the resonance is a valid representative of the ground state of the shallow defect state. One may wonder whether these interferences can be found in an approach where the Coulomb-potential that extends outside of the perturbed region has to be cut. However, we have to note again that there is a clear distinction between the long-ranged *wave function* and the long-range part of the *Coulomb-tail of the potential*, although the two quantities are of course not completely independent. Whereas the latter determines predominantly the ionization levels, it is the spatial distribution of the wave function that gives rise to the shf splittings. It is the specific benefit of the Green's function approach that it allows to describe the wave function of a defect correctly, although some parts of the long-ranged Coulomb-tail of the potential are ignored or approximated in a simple way. In order to prove that our shf results really do not suffer from termination errors we have calculated Green's functions for different perturbed regions. When decreasing the size of the perturbed region, the maximum of the resonance slightly shifts to higher energies, thereby decreasing the moduli of all hf and shf data monotonously. This decrease is not dramatic and amounts to less than 10% if we come down to a perturbed region that consists of the donor and 2 shells of ligands.

17.5
Phosphorus Donors in Highly Strained Silicon

Several approaches to built up solid-state based quantum computing hardware are actively pursued. The possible integration with existing microelectronics and the long decoherence times [65–67] are particular advantages if using the nuclear or electronic spins of phosphorus donors in group-IV semiconductors as qubits [68–71]. These concepts require gate-controlled exchange coupling between neighboring donors. However, to control the exchange coupling in semiconductors, the donor atoms have to be positioned with atomic precision [72] since the strength of the hf interaction, decisive for the rate at which two-qubit operations can be performed, varies strongly at the atomic scale due to Kohn–Luttinger oscillations of the donor wave function [69, 73, 74], already discussed in the last section. Under uniaxial compressive strain in [001]-direction, two conduction band minima are lowered in energy which is expected to suppress the oscillatory behavior in the (001) lattice plane [73].

In a recent work [75], the hf interaction of phosphorus donors in silicon was studied as a function of uniaxial compressive strain in thin layers of Si on virtual SiGe substrates, extending the regime investigated by Wilson and Feher [76] by a factor of 20 to higher strains. Fully strained 15 nm-thin P-doped ([P] $\simeq 1 \times 10^{17} \text{cm}^{-3}$) silicon epilayers were grown lattice matched on virtual relaxed $Si_{1-x}Ge_x$ substrates with Ge-contents $x = 0.07, 0.15, 0.20, 0.25$, and 0.30. The $Si_{1-x}Ge_x$ layer determines the strain of the Si epilayer: The higher lattice constant of SiGe alloys with respect to Si leads to biaxial tensile strain, accompanied by a compensating uniaxial compressive strain in growth direction (cf. inset in Figure 17.9), whereby the substrate with the highest Ge-content leads, of course, the largest strain. By high-resolution X-ray diffraction (XRD) it was shown that the compression in growth direction indeed follows linear elasticity theory.

To observe the P donors with high sensitivity, electrically detected magnetic resonance (EDMR) was used which monitors spin resonance via the influence of spin selection rules on charge transport processes [77–79]. The unstrained silicon layer provides the expected fingerprint of an isolated P-donor [80, 81] with an isotropic g-factor of $g = 1.9985$, whereby the characteristic hf-split satellite lines with a separation of $A_{hf} = 117.5$ MHz are clearly resolved. For the strained epilayers, in contrast, the hf splitting decreases monotonously with the applied strain. Simulta-

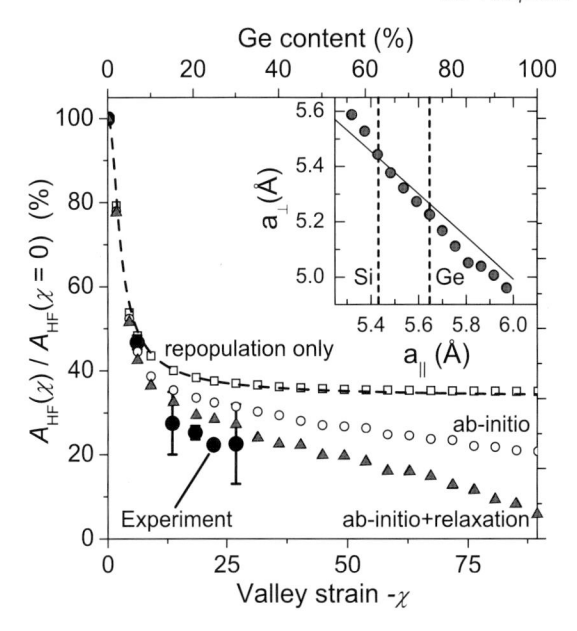

Figure 17.9 (online color at: www.pss-b.com) Relative hf splittings observed for P in fully strained epilayers on $Si_{1-x}Ge_x$ substrates as a function of the Ge content x and the resulting valley strain $\chi = \sim (a_{\parallel} - a_{\perp})$ (see also Ref. [75]). Black dots indicate the experimental data. The DFT-results for valley repopulation only are indicated by open squares and reproduce the prediction of Wilson and Feher (dashed line). The influence of strain-induced volume change is shown by open circles, whereby those additionally including nn relaxation are shown by triangles. The insert shows the out-of-plane lattice constant $a_{\perp}$ of thin Si layers (thickness 12 unit cells) as a function of the in-plane constant $a_{\parallel}$ predicted by supercell calculations compared to linear elasticity theory (strait line). Pure Si and Ge substrates are shown by dashed vertical lines.

neously, the resonance line becomes clearly anisotropic. For example, for the epilayer with $(a_{\perp}-a_{Si})/a_{Si} = -0.00729$ obtained by a substrate with 25% Ge content, the isotropic hf splitting shrinks to 26.9 MHz, while $\Delta g = (1.21 \pm 0.06) \times 10^{-3}$.

17.5.1
Predictions of EMA

For an isolated P donor atom in unstrained Si, the isotropic Fermi-contact hf interaction A_{hf} in the non-relativistic limit is proportional to the probability amplitude $|\psi(0)|^2$ of the unpaired electron wave function at the nucleus, giving rise to the hf satellite lines separated by $A_{hf} = 117.5$ MHz. To determine $|\psi(0)|^2$, we first note that the cubic crystal field leads to the formation of a singlet ground state and a doublet and a triplet of excited states instead of a six-fold degenerate ground state. Only the fully symmetric singlet ground state has a non-vanishing probability amplitude at the nucleus.

In Section 17.4.2 we have seen that the A_1 ground state wave function ψ is given by the symmetrical superposition $\psi(\mathbf{r}) = \sum_{i=1}^{6}(1/\sqrt{6})\Phi^{(i)}(\mathbf{r})$ of the six valleys contributing to the donor, whereby each $\Phi^{(i)}$ is a product of the corresponding conduction band Bloch wave function and a hydrogenic envelope function. The probability of the unpaired electron at the nucleus $|\psi(0)|^2$ becomes $1/6|\sum_{j=1}^{6}\Phi^{(i)}(0)|^2 = 6|\Phi(0)|^2$, since due to degeneracy $\Phi^{(i)}(0) = \Phi(0)$ for all i. Assuming that the only effect of strain is the change of relative population of the conduction band minima, we similarly find $\psi = \sum_{i=1}^{2}(1/\sqrt{2})\Phi_i$ and $|\psi(0)|^2 = 2|\Phi(0)|^2$ under high uniaxial strain, when only two conduction band minima contribute. Therefore, in the fully strained case, the hf interaction should be 1/3 of the unstrained case.

In contrast, in Figure 17.9 we already observe a reduction to 0.21 of the unstrained hf interaction A_{HF}, clearly below the 0.33 A_{HF} EMA-limit obtained above. Based on group and linear elasticity theory Wilson and Feher [76] evaluated the analytical dependence of $A_{hf}(\chi)$ from the so-called valley strain $\chi = -\frac{\Xi_u}{3\Delta_c}(a_\| - a_\perp)$, where $\Xi_u = 8.6$ eV is the uniaxial deformation potential [82], and $6\Delta_c = 2.16$ meV [54] is the energy splitting between the singlet and doublet state in the unstrained material. In Figure 17.9, a comparison of the prediction of Eq. (17.2) in Ref. [76] (dashed line) with the hf splittings determined experimentally (full circles) clearly shows, that pure valley repopulation is not able to describe the experimental data for $x > 0.07$. An empirical treatment of additional radial redistribution effects as discussed in Ref. [83] would lead to 0.29 for $\chi \to -\infty$, only a slight reduction of the repopulation-limit and, thus, still at strong variance with the experimental data.

17.5.2
Ab Initio Treatment via Green's Functions

Again, an *ab initio* prediction of hf interactions is necessary to clarify the situation. In the last section, our Green's function approach has been shown to describe the hf splittings in predictive accuracy for P$_{Si}$ in the unstrained case. In the case of a strained host material, however, the situation becomes more complicated since excited states are admixed to the former pure singlet ground state. An application of DFT is, thus, only possible in combination with linear elasticity theory. Due to the applied strain, the symmetry of the P donor is reduced, and the resonance at the bottom of the CB transfomring according the a_1 representation in the unstrained case, now shows admixtures of the b_1 and b_2 representations of D_{2d} symmetry (cf. Figure 17.10). The location of the P donor atom in their nodal planes implies a correlation of these b_1 and b_2-like orbitals with the admixed doublet state. Since, furthermore, only one component of the diamagnetic doublet state is contributing to the singlet ground state under strain [76], it is reasonable to construct the spin densities, which enter the self-consistent LSDA total energy calculations for a given valley strain, by $n^\sigma(r) = \left(1 - \alpha(\chi)\right) \cdot n_{a_1}^\sigma(r) + \alpha(\chi) \cdot n_{b_1}^\sigma(r)$, whereby $\alpha(\chi)$ is obtained from the strain-dependent admixture of the doublet states determined by linear elasticity theory (see Eq. (C6) in Ref. [76]). Figure 17.9 shows that the spin densities constructed this way allow a reasonable description of the pure valley repopulation effect: for an *unrelaxed*

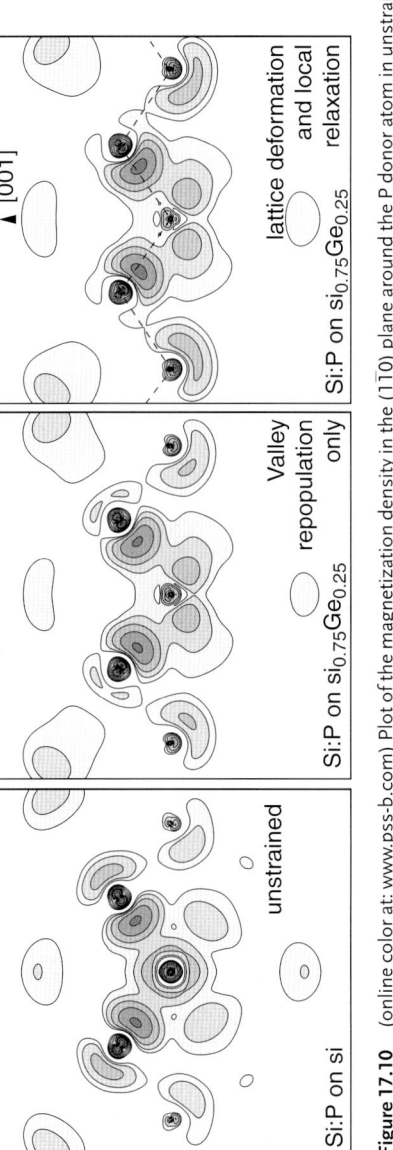

Figure 17.10 (online color at: www.pss-b.com) Plot of the magnetization density in the (1$\bar{1}$0) plane around the P donor atom in unstrained silicon (left) and in material strained by a Si$_{0.75}$Ge$_{0.25}$ substrate determined by valley repopulation (central) and including the change in the volume and local relaxation of the donor (right).

structure of an *ideal* silicon crystal, the results obtained by Wilson and Feher [76] are nicely reproduced after the self-consistent cycle (cf. open squares in Figure 17.9).

We are now able to take into account explicitly by *first principles* the strain of the silicon lattice as well as the relaxation around the P donors within. For the optimization of the strained Si cells, we used a supercell approach [84]. Optimization of long slabs (up to 12 unit cells along the (001)-direction) show an almost linear dependence of the compression along (001) as an answer to the tensile strain in the (001)-plane, effectively following linear elasticity theory as indicated in the inset of Figure 17.9. This result confirms that *linear* elasticity theory remains valid in the complete regime, even up to pure germanium as a substrate ($\chi \approx -89$). The hf parameters calculated with LMTO-GF under these assumptions (cf. open circles in Figure 17.9) already become smaller since the donor wave function becomes more delocalized as a result of the enhanced volume, lifting the high-stress limit of $0.33 A_{HF}$ obtained above. This tendency is strengthened, if local relaxation around the P donors is taken into account: According to total energy calculations on large explicitly strained supercells with 512 atoms, this relaxation is dominated by a slight reduction of the bond-length between the P donor and its nearest Si ligands by about 1%, nearly independent of the strength of the tensile strain in the plane of the Si epilayer. Re-calculating $A_{hf}(\chi)/A_{hf}(0)$ for this geometry with the LMTO-GF code, we find a further reduction (open triangles in Figure 17.9). In addition, being *e.g.*, in the unstrained case about 50 MHz too large before, also the absolute values A_{hf} are then in nice accordance with experiment. The hf interaction observed experimentally in the moderately strained P-doped silicon layers can, thus, be explained by the increased volume of the unit cell together with a slight inward relaxation of the nearest Si neighbors. Since already such a small relaxation has a huge influence on the predicted relative hf splittings for the strained material, the remaining discrepancy between experiment and theory can easily be explained by uncertainties due to the well-known flatness of the total energy surface in silicon [85]. Apparently, there exists no high-stress limit for the reduction of the P-related hf splitting. According to our DFT calculation the decrease of the central P-related hf interaction is accompanied by a remarkable increase of the shf interaction with neighboring ^{29}Si atoms.

17.6
n-Type Doping of SiC with Phosphorus

For high temperature and high frequency applications, silicon carbide (SiC) has been proven to provide in principle many advantages over silicon or gallium arsenide [86]. However, its technology, e.g. *n*-type doping, is considerably more difficult than that of silicon: although nitrogen is easily incorporated during growth, the electrical conductivity saturates at higher doping concentrations [87, 88]. The alternative shallow donor, phosphorus, cannot be easily introduced into the material by diffusion. One must fall back on other doping techniques like ion implantation, *in situ* doping during growth or neutron transmutation of ^{30}Si [89].

From the transmutation process $^{30}_{14}\text{Si} \xrightarrow{(n,\gamma)} {}^{31}_{14}\text{Si}^* \xrightarrow{\beta} {}^{31}_{15}\text{P}$ it was assumed for long time that the so created P-dopants enter the silicon sublattice. At first view, this assumption is also supported by total energy calculations: the formation energy of P_{Si} is essentially lower (by 1.5 eV) than for P incorporated at the carbon sublattice. However, the large number of different spectra observed in electron paramagnetic resonance (EPR) [90–94], at least six, can by no means be explained by one single defect at different lattice sites. Indeed, molecular dynamic (MD) simulations including the recoil process and the following annealing processes result in various P-related defects [95]. Due to kinetic effects during the recoil (770 eV after capture of thermal neutrons [96]) and subsequent annealing processes, an incorporation of P at the carbon sublattice becomes possible. The inconsistencies in early models for the P-related donors in SiC are a consequence of neglecting this possibility. Beside the already discussed donors (P_{Si}, P_C, $P_{Si} V_C$ [97]) alternative complexes with intrinsic defects (e.g. $P_C C_{Si}$) are predicted with high probability [95]. At usual annealing temperatures, the exchange mechanism toward isolated P_{Si} is hindered. Although providing much lower formation energy, due to a large activation barrier of about 5 eV, extremely high temperature annealing above 2000 K is expected to be necessary to achieve a reasonable incorporation of P at the Si-sublattice, and allows the observation of P_C in moderately annealed samples.

This scenario is supported by our *ab initio calculation* of the corresponding hf splittings. Like isolated P at the silicon site, P_C and $P_C C_{Si}$ are calculated to act as shallow donors. Thus, we use again the EMT embedded Green's function method to model the shallow defect states. As a first reference in the compound semi-conductor SiC, we applied our extended method to the well-known nitrogen donors N_C in 6H-SiC. As can be seen in Table 17.4, the experimental values are well reproduced: at the hexagonal site (h) minor additional relaxation occurs compared to the ionized donor N_C^+ retaining a tetrahedral arrangement of the ligands (slightly outward relaxed by about 5% of the bond-length). A rather small hf splitting below 5 MHz is the consequence. For the quasi-cubic sites (k_1, k_2) in contrast, a small distortion toward C_{3v}-symmetry (one ligand relaxes 7% away from the donor atom) yields an increased

Table 17.4 Calculated and observed hf parameters (MHz) for P-related EPR-spectra in 6H-SiC (see Ref. [94] for a review). Note that for the quasicubic sites (k_1, k_2) only minor differences in the calculated values (below 0.4 MHz) can be observed. Here, only the values for the k_2-site are given.

defect	site	a_{calc}	b_{calc}	center	a_{exp}	b_{exp}
N_C	h	4.49	0.00	N_h	2.47	0.13
N_C	k_1,k_2	28.48	1.05	N_{k_1,k_2}	33.40	0.01
P_{Si}	h	1.76	0.59	I_1	1.56	0.89
P_{Si}	k_1,k_2	8.75	2.78	I_2	8.70	4.20
P_C	h	156.1	0.00	P_1	145.0	0.00
P_C	k_1,k_2	147.2	5.04	P_2?	156.0	0.70
$P_C C_{Si}$		169.8	−0.10			
$P_{Si} V_C$		−20.6	3.54	P + V	22.0	1.87

hf splitting of 28.5 MHz (exp. 33.4 MHz). Note that the observation that more pronounced distortions occur at the quasi-cubic sites whereas the hexagonal sites mimic the lattice sites in the cubic material is also reported in case of the carbon vacancy in 4H-SiC [98, 99].

We obtain the same trend when applying the Green's function-based method to the isolated P-donors at the silicon sublattice (P_{Si}). However, compared to the P_1, P_2 spectra (originally assigned to P_{Si}), the values are by about a factor of 20 too small. The calculated values fit, instead, very well to the second set of spectra (I_1, I_2, see also Table 17.4 and the upper part of Figure 17.11). Large, rather isotropic hf splittings in the range of 150 MHz can only be obtained for P at a carbon site – an observation which is in line with the results of our MD calculations. The isotropic spectrum P_1 is most likely due to P_C at the hexagonal site (again retaining essentially a tetrahedral arrangement of the ligands but with a more pronounced outward relaxation of about 15%) whereas an explanation of the P_2 spectra with a small but non-vanishing anisotropic part requires C_{3v}-symmetry, either obtained by a distortion around the quasi-cubic sites or by a nearby carbon antisite (Figure 17.11). We no longer expect that the P_1, P_2 and P + V lines are caused by one center at three different lattice sites as argued in the more recent models [92, 93]. In this point, based on the calculated hf parameters, we come back to the original model of Veinger *et al.* [90], instead, and reassign the so-called P + V center to a $P_{Si}V_C$ pair (Table 17.4).

Further confirmations of our model have been later obtained from additional EPR measurements on the 3C-SiC and 6H-SiC polytypes [53] as well as from investigations on different new 4H-SiC samples: (1) the first was doped *in situ* with phosphorus grown by the PVT method using SiP_2O_7 as a source [89]; (2) the second, based on our theoretical results, have been ^{30}Si-enriched (50% instead of about 3% natural abundance) in order to reexamine neutron transmutation. In one point, both samples show a similar result: intense lines of the P + V center can be resolved, but even no trace of the P_1 and P_2 spectra. Hence, one has clearly to rule out any model in which P_1/P_2 differ from the P + V center by the lattice site only. The absence of the P_C-related spectra can be explained by high temperature annealing (sample 2) and by an incorporation of the P-donors close to the thermal equilibrium during growth (sample 1). However, only in the ^{30}Si-enriched, the I_1/I_2 spectra of P_{Si} can be easily resolved in comparable intensity, whereby in the sample *in situ* doped during growth the lines are expected being covered by the intense central resonance line of the N-donor.

17.7
Conclusions

We have shown that a Green's functions approach is able to calculate hf splittings of shallow donors in semiconductors in predictive accuracy, whereby the *ab initio* calculated local part of the potential, the central cell correction of effective mass theory (EMT), is embedded via Green's functions into a periodic, EMT-like background. The method was successfully applied onto shallow donors in Si and SiC: hf parameters of donors in silicon (P, As, Sb) are reproduced in quantitative agreement with exper-

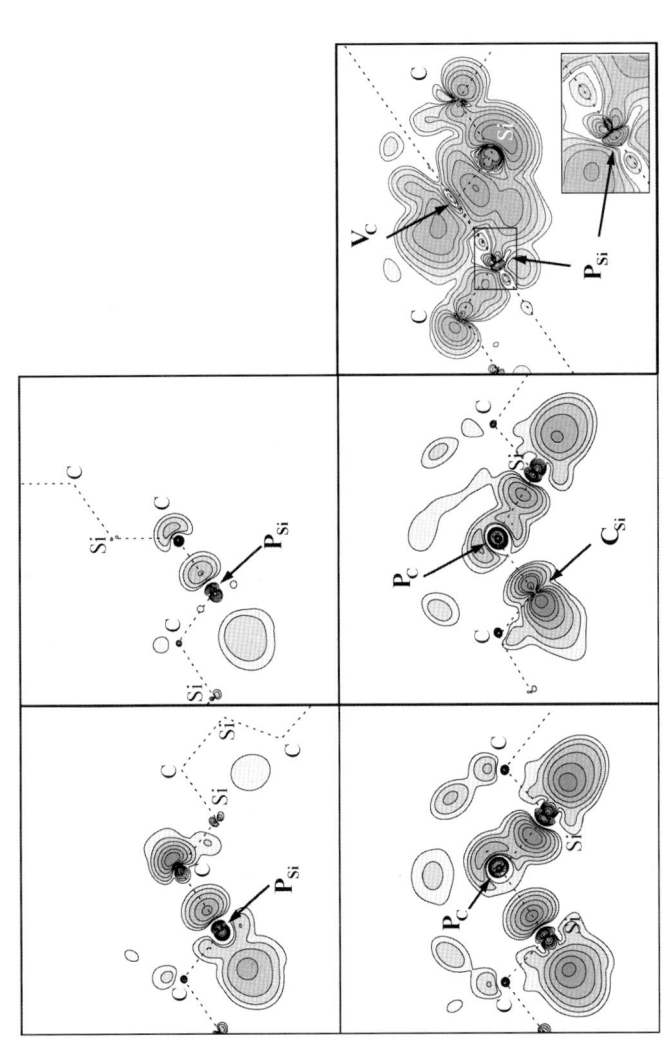

Figure 17.11 (online color at: www.pss-b.com) Magnetization density of some P-related donor states in 6H-SiC in a $(1\bar{1}0)$-plane: In case of P on the silicon sublattice (top: P_{Si} at a quasi-cubic site k_2 (left part) and at the hexagonal site h (right)), p-like spin densities at the donor nucleus occur, explaining the small, mainly anisotropic hf splittings of the I_1, I_2 spectra. In contrast, P_C and $P_C\,C_{Si}$ (bottom) provide s-like spin densities around the ^{31}P nuclei and, thus, rather isotropic hf parameters (Table 17.4). For comparison, the corresponding plot for the $P_C\,C_{Si}$ deep defect pair is also given (bottom right).

imental data including the so-called Kohn–Luttinger oscillations due to the neighbor shells. For P in strained silicon, a relaxation of the next nearest (nn) silicon neighbors is shown to be crucial to explain quantitatively the experimentally observed decrease of hf interaction, whereby our results indicate that in contrast to the prediction of the EMA, there exists no high-stress limit for the reduction of the P-related hf splitting. It is also confirmed that by strain the Kohn–Luttinger oscillations can be partially suppressed. Both observations are crucial for quantum computing applications since the spatial distribution of the hf interaction has a direct impact on the rate with which two-qubit operations can be performed. The application of the approach onto N and P donors in SiC confirms an incorporation of the P atoms onto both, the silicon as well as the carbon sublattice.

It is now possible to treat shallow donors without invoking a one-electron approximation and several fitting parameters. The presented *ab initio* calculations, although considerably more complex, are much more flexible and furthermore their application requires considerably less manpower than one of the usual, more involved one-electron EMA methods for shallow defects. Most probably our method cannot be extended directly to supercells containing a few hundred atoms, because here the periodic images of the defect superlattice are still superimposed into each cell. But it is not unlikely that for cluster calculations a similar extention will hold, whereby in principle the unperturbed Green's function can be obtained from a supercell calculation of the corresponding ionized donor. This would be very promising, since in a Green's function approach the *ab initio* calculation results in hf and shf interactions that for the center region of the defect are considerably more accurate than those obtained from the best empirical approaches.

Acknowledgements

This work is dedicated to Harald Overhof. Furthermore, I am grateful to all my collaborators, theoreticians as well as experimentalists, who made the scientific work and this article possible. The work was partially supported by the Deutsche Forschungsgemeinschaft (DFG).

References

1 Kohn, W. (1999) *Rev. Mod. Phys.*, **71**, 1253.
2 Kohn, W. and Luttinger, J.M. (1955) *Phys. Rev.* **97**, 1721. *Phys. Rev.*, **98**, 915 (1955).
3 Luttinger, J.M. and Kohn, W. (1955) *Phys. Rev.*, **97**, 969.
4 Kohn, W. (1960) Shallow impurity states in silicon and germanium, in: *Solid State Physics 5*, eds F. Seitz and D. Turnbull Academic Press, New York, p. 1.
5 Hohenberg, P. and Kohn, W. (1964) *Phys. Rev. B*, **136**, 864.
6 von Barth, U., and Hedin, L. (1972) *J. Phys. C* **5**, 1629.
7 Kohn, W. and Sham, L.J. (1965) *Phys. Rev.*, **140**, A1133.
8 Ceperley, D.M. and Alder, B.J. (1980) *Phys. Rev. Lett.*, **45**, 566.
9 Perdew, P. and Zunger, A. (1981) *Phys. Rev. B*, **23**, 5048.

10 Perdew, J.P. and Levy, M. (1983) *Phys. Rev. Lett.*, **51**, 1884.
11 Sham, L.J. and Schlüter, M. (1983) *Phys. Rev. Lett.*, **51**, 1888.
12 Sanna, S., Frauenheim, Th., Gerstmann, U., and (2008) *Phys. Rev. B*, **78**, 085201.
13 Fermi, E. (1930) *Z. Phys.*, **60**, 320.
14 Breit, G. (1930) *Phys. Rev.*, **35**, 1447.
15 Blügel, S., Akai, H., Zeller, R., and Dederichs, P.H. (1987) *Phys. Rev. B*, **35**, 3271.
16 Bethe, H.A. and Negele, J.W. (1968) *Nucl. Phys. A* **117**, 575.
17 Abragam, A. and Bleany, B. (1970) *Electron Paramagnetic Resonance of Transition Ions* (Clarendon Press, Oxford, (reprint: Dover Publications, New York, 1986).
18 Koster, G.F. and Slater, J.C. (1954) *Phys. Rev.*, **96**, 1208.
19 Baraff, G.A. and Schlüter, M. (1978) *Phys. Rev. Lett.*, **41**, 892; (1979) *Phys. Rev. B*, **19**, 4965.
20 Gunnarsson, O., Jepsen, O., and Andersen, O.K. (1983) *Phys. Rev. B*, **27**, 7144.
21 Economou, E.N. 1979 *Green's Functions in Quantum Physics* (Springer-Verlag, Heidelberg, New York).
22 Ferry, D.K. 2009 *Transport in Nonostructures* (Cambridge University Press, Cambridge, New York).
23 Spaeth, J.-M. and Overhof, H., 2003 *Point Defects in Semiconductors and Insulators*, Springer Series in Material Science, Vol. 53 (Springer Verlag, Heidelberg).
24 Baraff, G. and Schlüter, M. (1984) *Phys. Rev. B*, **30**, 3460.
25 Van de Walle, C.G., and Blöchl, P.E. (1993) *Phys. Rev. B*, **47**, 4244.
26 Korringa, J. (1947) *Physica* **13**, 392.
27 Kohn, W. and Rostoker, N. (1954) *Phys. Rev.*, **94**, 1111.
28 Andersen, O.K. (1975) *Phys. Rev. B*, **12**, 3060.
29 Skriver, H.L. 1984 *The LMTO method*, Springer Series in Solid-State Sciences, Vol. 41 (Springer Verlag Berlin, Heidelberg, New York).
30 Slater, J.C. (1937) *Phys. Rev.*, **51**, 846.
31 Drittler, B., Weinert, M., Zeller, R., and Dederichs, P.H. (1989) *Phys. Rev. B*, **39**, 930.
32 Methfessel, M., Rodriguez, C.O., and Andersen, O.K. (2009) *Phys. Rev. B*, **40**, 1989.
33 Blaha, P., Schwarz, K., Sorantin, P., and Trickey, S.B. (1990) *Comput. Phys. Commun.* **59**, 399.
34 Voronoi, G. (1908) *Crelle J.* **134**, 198.
35 Becke, A.D. (1988) *J. Chem. Phys.*, **88**, 2547.
36 Ohno, K., Esfarjani, K., and Kawazoe, Y. 1999 *Computational Materials Science* (Springer, Berlin, Heidelberg).
37 Krambrock, K. and Spaeth, J.-M. (1992) *Mater. Sci. Forum* **83/87**, 887.
38 Krambrock, K., Spaeth, J.-M., Delerue, C., Allan, G., and Lannoo, M. (1992) *Phys. Rev. B*, **45**, R1481.
39 Koschnick, F.K. and Spaeth, J.-M. (1999) *Phys. Status Solidi B*, **216**, 817.
40 Spaeth, J.-M. and Krambrock, K. 1994 in: *Festkörperprobleme/Advances in Solid State Physics*, Vol. 33, eds R. Helbig (Vieweg, Braunschweig), p. 111.
41 Holmes, D.E., Chen, R.T., Elliott, K.R., and Kirkpatrick, C.G. (1982) *Appl. Phys. Lett.*, **40**, 46.
42 Nissen, M.K., Villemaire, A., and Thewalt M. L. W. (1991) *Phys. Rev. Lett.*, **67**, 112.
43 Lagowski, J., Gatos, H.C., Kang, C.H., Skoronski, M., Ko, K.K., and Lin, D.G. (1986) *Appl. Phys. Lett.*, **49**, 892.
44 Overhof, H. and Spaeth, J.-M. (2005) *Phys. Rev. B*, **72**, 115202.
45 Dabrowski, J. and Scheffler, M. (1988) *Phys. Rev. Lett.*, **60**, 2183 ; *Phys. Rev. B*, **40**, 10391 (1989).
46 Chadi, D.J. and Chang, K.J. (1988) *Phys. Rev. Lett.*, **60**, 2187; (1988) *Phys. Rev. B*, **39**, 10 063.
47 Chadi, D.J. (2003) *Phys. Rev. B*, **68**, 193204.
48 Baraff, G.A. and Schlüter, M. (1985) *Phys. Rev. Lett.*, **55**, 2340.
49 Delerue, C. (1991) *Phys. Rev. B*, **44**, 10525.
50 Gerstmann, U. and Overhof, H. (2001) *Physica B*, **308–310**, 561.
51 Tkach, I., Krambrock, K., and Overhof, H., Spaeth, J.-M. (2003) *Physica B*, **340–342**, 353.
52 Duijn-Arnold, A.V., Zondervan, R., Schmidt, J., Baranov, P.G., and Mokhov, E.N. (2001) *Phys. Rev. B*, **64**, 085206.
53 Son, N.T., Henry, A., Isoya, J., Katagiri, M., Umeda, T., Gali, A., and Janzén, E. (2006) *Phys. Rev. B*, **73**, 075201.

54 Ramdas, A.K. and Rodriguez, S. (1981) *Rep. Prog. Phys.*, **44**, 1297.
55 Aggarwal, R.L. and Ramdas, A.K. (1965) *Phys. Rev.*, **140**, A1246.
56 Overhof, H. and Gerstmann, U. (2000) *Phys. Rev. B*, **62**, 12585.
57 Baldereschi, A. (1970) *Phys. Rev. B*, **1**, 4673.
58 Pantelides, S.T. (1978) *Rev. Mod. Phys.*, **50**, 797.
59 Ivey, J.L. and Mieher, R.L. (1972) *Phys. Rev. Lett.*, **29**, 176.
60 Ivey, J.L. and Mieher, R.L. (1975) *Phys. Rev. B*, **11**, 822; (1975) *Phys. Rev. B*, **11**, 849.
61 Overhof, H. and Gerstmann, U. (2004) *Phys. Rev. Lett.*, **92**, 087602.
62 Feher, G. (1959) *Phys. Rev.*, **114**, 1219.
63 Rockett, A., Johnson, D.D., Khare, S.V., and Tuttle, B.R. (2003) *Phys. Rev. B*, **68**, 233208.
64 Hale, E.B. and Mieher, R.L. (1971) *Phys. Rev. B*, **3**, 1955.
65 Castner, T.G., Jr (1962) *Phys. Rev. Lett.*, **8**, 13.
66 Tyryshkin, A.M., Lyon, S.A., Astashkin, A.V., and Raitsimring, A.M. (2003) *Phys. Rev. B*, **68**, 193207.
67 Gordon, J.P. and Bowers, K.D. (1958) *Phys. Rev. Lett.*, **1**, 268.
68 Kane, B.E. (1998) *Nature* **393**, 133.
69 Kane, B.E. (2000) *Fortschr. Phys.*, **48**, 1023.
70 Vrijen, R., Yablonovitch, E., Wang, K., Jiang H.W., Balandin, A., Roychowdhury, V., Mor, T., and DiVincenzo, D. (2000) *Phys. Rev. A* **62**, 012306.
71 Hollenberg, L. C. L., Dzurak, A.S., Wellard C., Hamilton, A.R., Reilly, D.J., Milburn, G.J., and Clark, R.G. (2004) *Phys. Rev. B*, **69**, 113301.
72 Schofield, S.R., Curson, N.J., Simmons, M.Y., Rueß, F.J., Hallam, T., Oberbeck, L., and Clark, R.G. (2003) *Phys. Rev. Lett.*, **91**, 136104.
73 Koiller, B., Hu, X., and Das Sarma, S. (2002) *Phys. Rev. B*, **66**, 115201.
74 Wellard, C.J., Hollenberg, L. C. L., Parisoli, F., Kettle, L.M., Goan, H.-S., McIntosh, J. A. L., and Jamieson, D.N. (2003) *Phys. Rev. B*, **68**, 195209.
75 Huebl, H., Stegner, A.R., Stutzmann, M., Brandt, M., Vogg, G., Bensch, F., Rauls, E., and Gerstmann, U. (2006) *Phys. Rev. Lett.*, **97**, 166402.
76 Wilson, D.K. and Feher, G. (1961) *Phys. Rev.*, **124**, 1068.
77 Schmidt, J. and Solomon, I. (1966) *Compt. Rend.* **263**, 169.
78 Stich, B., Greulich-Weber, S., and Spaeth, J.-M. (1996) *Appl. Phys. Lett.*, **68**, 1102.
79 Brandt, M.S., Goennenwein, S. T. B., Graf, T., Huebl, H., Lauterbach, S., and Stutzmann, M. (2004) *Phys. Status Solidi C* **1**, 2056.
80 Young, C.F., Poindexter, E.H., Gerardi, G.J., Warren, W.L., and Keeble, D.J. (1997) *Phys. Rev. B*, **55**, 16245.
81 Cullis, P.R. and Marko, J.R. (1975) *Phys. Rev. B*, **11**, 4184.
82 Tekippe, V.J., Chandrasekhar, H.R., Fisher, P., and Ramdas, A.K. (1972) *Phys. Rev. B*, **6**, 2348.
83 Fritzsche, H. (1962) *Phys. Rev.*, **125**, 1560.
84 Frauenheim, Th., Seifert, G., Elstner, M., Hajnal, Z., Jungnickel, G., Porezag, D., Suhai, S., and Scholz, R. (2000) *Phys. Status Solidi B*, **217**, 41.
85 Gerstmann, U., Rauls, E., Overhof, H., and Frauenheim, Th. (2002) *Phys. Rev. B*, **65**, 195201.
86 Choyke, W.J, Matsunami, H., and Pensl, G. eds, (2004) *Recent Major Advances in SiC* (Springer-Verlag, Berlin)
87 Pensl, G., Frank, T., Krieger, M., Laube, M., Reshanov, S., Schmid, F., and Weidner, M. (2003) *Physica B*, 340–342, 121.
88 Gerstmann, U., Rauls, E., Frauenheim, Th., and Overhof, H. (2003) *Phys. Rev. B*, **67**, 205202.
89 Semmelroth, K., Schmid, F., Karg, D., Pensl, G., Maier, M., Greulich-Weber, S., and Spaeth, J.-M. (2003) *Mater. Sci. Forum* **433–436**, 63.
90 Veinger, A.I., Zabrodskii, A.G., Lomakina, G.A., and Mokhov, E.N. (1986) *Sov. Phys. Solid State* **28**, 917.
91 Kalabukhova, E.N., Lukin, S.N., and Mokhov, E.N. (1993) *Sov. Phys. Solid State* **35**, 361.
92 Greulich-Weber, S. (1997) *Phys. Status Solidi A* **162**, 95.

93 Baranov, P.G., Il'in, I.V., Mokhov, E.N., von Bardeleben, H.J., and Cantin, J.L. (2002) *Phys. Rev. B*, **66**, 165206.

94 Pinheiro, M. V. B., Greulich-Weber, S., and Spaeth, J.-M. (2003) *Physica B*, **340–342**, 146.

95 Rauls, E., Pinheiro, M. V. B., Greulich-Weber, S., and Gerstmann, U. (2004) *Phys. Rev. B*, **70**, 085202.

96 Heißenstein, H. (2002) Dissertation, Universität Erlangen, Germany.

97 Gali, A., Deák, P., Briddon, P.R., Devaty, R.P., and Choyke, W.J. (2000) *Phys. Rev. B*, **61**, 12602.

98 Bockstedte, M., Heid, M., and Pankratov, O. (2003) *Phys. Rev. B*, **67**, 193102.

99 Umeda, T., Isoya, J., Morishita, N., Ohshima, T., and Kamiya, T. (2004) *Phys. Rev. B*, **69**, 121201(R).

18
Time-Dependent Density Functional Study on the Excitation Spectrum of Point Defects in Semiconductors
Adam Gali

18.1
Introduction

Density functional theory (DFT) has been proven to be extremely powerful method to study defects in solids. Nowadays, this is a standard tool to investigate their concentration in thermal equilibrium, their interaction with each other, their vibration modes or their hyperfine tensors [1–5]. All these properties are associated with the ground state of the defect. The success of the DFT calculations are based on the well-developed approximate (semi)local functionals [6–9] that made possible to study relatively large systems at moderate computational cost with a surprisingly good accuracy. We mention here that the commonly used (semi)local functionals suffer from the self-interaction error [7] which results in the underestimation of the band gap of semiconductors. Nevertheless, for many semiconductors the (semi) local DFT calculations could predict qualitatively well the adiabatic (thermal) ionization energies of defects [1]. However, in pathological cases (semi)local functionals could fail to describe the nature of the defect states correctly due to the self-interaction error. Recent studies have shown [10–14] that non-local hybrid functionals could improve the results at large extent, and could provide quantitatively good results for the thermal ionization energies of point defects [15]. The effect of hybrid functionals is discussed in detail in another chapter in this book. We claim that hybrid density functionals provide relatively accurate quasi-particle energies and states compared to (semi)local functionals but these quasi-particle energies are still within the mean field approximation. In order to calculate the excitation spectrum properly one must go beyond the mean field approximation. Time-dependent DFT (TD-DFT) goes beyond this approximation. TD-DFT method has been successfully applied recently to calculate the excitation spectrum of molecules and semiconductor nanocrystals [16]. In some earlier studies efforts have been made to address the excitation of defect states in solids by TD-DFT method where "solids" were modeled by extremely small finite clusters including 5–50 host atoms [17–21]. However, it is very questionable whether the electron

Advanced Calculations for Defects in Materials: Electronic Structure Methods, First Edition.
Edited by Audrius Alkauskas, Peter Deák, Jörg Neugebauer, Alfredo Pasquarello, and Chris G. Van de Walle.
© 2011 Wiley-VCH Verlag GmbH & Co. KGaA. Published 2011 by Wiley-VCH Verlag GmbH & Co. KGaA.

states associated with the defect in a solid are properly described in such small clusters, and no thorough study has been carried out so far even for a single defect to address this very important issue.

We note that TD-DFT excitation spectrum obtained by (semi)local functionals is reliable only for finite structures [16]. In a very recent study it has been shown that hybrid density functional in the TD-DFT kernel provides appropriate excitation spectrum for infinite semiconductors [22]. Here, we restrict ourselves into finite structures where both the local and non-local functionals can be consistently applied and validate the results for bulk systems. We study two representative defects in wide gap semiconductors: nitrogen-vacancy (NV) center in diamond and divacancy in silicon carbide (SiC). A common behavior of these defects is that they produce characteristic transition in the photoluminescence (PL) spectrum. We briefly summarize the known properties of these defects below.

18.1.1
Nitrogen-Vacancy Center in Diamond

Nitrogen-vacancy center in diamond has attracted a lot of attention in recent years, since it has been detected at a single defect level [23, 24] and provides a quantum bit for quantum computing applications [25–30]. Besides providing a single photon source for quantum cryptography [31, 32], the NV center is also a promising candidate as an optically coupled quantum register for scalable quantum information processing, such as quantum communication [33] and distributed quantum computation [34]. In addition, it has been recently demonstrated that proximal nuclear spins can be coherently controlled via hyperfine interaction [35] and used as quantum memory with an extremely long coherence time [36].

The electronic structure of the NV center in diamond has been discussed in detail in a recent paper [37]. The NV center was found many years ago in diamond [38]. The model of the NV center consists of a substitutional nitrogen atom near a vacancy in diamond [38–42] as Figure 18.1a shows. The NV center has a strong optical transition with a zero phonon line (ZPL) at 1.945 eV (637 nm) accompanied by a vibronic band at higher energy in absorption with the largest intensity at about 2.20 eV and lower energy in emission (see also Figure 18.4). Detailed analysis of the ZPL revealed that the center has trigonal, C_{3v} symmetry [39]. Previous *ab initio* calculation clearly supported the negatively charged NV defect for 1.945 eV ZPL center [43–45] as was originally proposed by Loubser and van Wyk [40]. In the NV defect three carbon atoms have sp^3 dangling bonds near the vacancy (with three back bonds in the lattice) and nitrogen atom has also three back bonds with one dangling bond pointing to the vacant site. Since nitrogen has five valence electrons the negatively charged NV defect has altogether six electrons around the vacancy. One can use the defect-molecule picture [46] together with group theory to find the canonical orbitals of this system. According to this analysis there are two fully symmetric one-electron states (a_1) and one doubly degenerate e state which should be occupied by six electrons [44]. It was found that the two a_1 states is deeper in energy than the e state. As a consequence, four electrons occupy the a_1 states and two

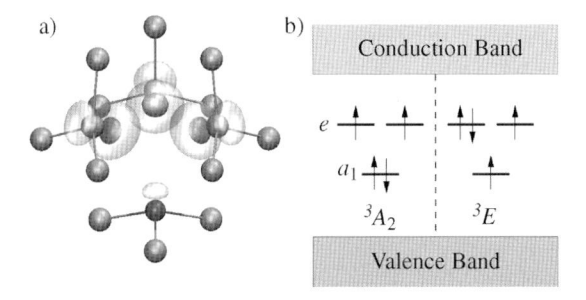

Figure 18.1 (online color at: www.pss-b.com) (a) The structure of N–V⁻ center in diamond; only first- and second-neighbor C (cyan spheres) and N (blue sphere) atoms to the vacant site are shown. The yellow and red lobes are contours of the calculated spin-density. (b) Schematic diagram of the defect states in the gap and their occupation in the $^3A_2$ (ground) and ^3E (excited) states.

electrons remain for the e state. Our calculated one-electron levels obtained by *ab initio* supercell (sc) calculations are shown in Figure 18.2.

As can be seen in Figure 18.2, the natural choice is to put the two remaining electrons parallel to the e level forming an $S = 1$ state (in analog to satisfy the Hund-rule for the p-orbitals of the isolated group IV elements in the periodic table). In the C_{3v} point group this total wave function has $^3A_2$ symmetry, where $3 = 2S + 1$ with $S = 1$. In our special case, we chose the $M_S = 1$, so both electrons are spin-up electrons on the e level. As can be seen, again in Figure 18.2, the lowest a_1 level is relatively deep in the valence band, so it seems to be a good approximation to assume that it does not contribute to the excitation process, and we do not consider it further. However, the next a_1 level in the gap is not very far from the e level. If one electron is excited from

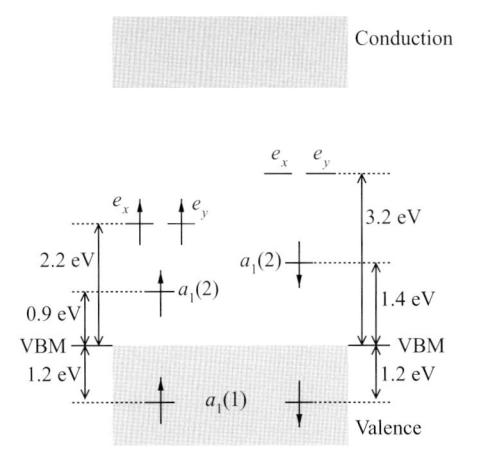

Figure 18.2 (online color at: www.pss-b.com) The calculated one-electron levels with respect to the valence band maximum in the ground state of the NV defect. The results obtained in a 512-atom supercell by applying the DFT within LDA functional in Ref. [44].

the this a_1 level into the e level then one will arrive at ^3E state (see Figure 18.1b). The ^3E is a doubly degenerate many-electron state, so it comprises two orthogonal many-electron states with the same eigenenergy. Both of them can be described by a single Slater-determinant: if the electron from a_1 level is promoted to e_x then the symmetry of the resulted many-electron state will be E_x, if it is promoted to e_y then the resulted many-electron state will be E_y. Thus, Figure 18.1b shows one of the true $M_S = 1$ eigenstates of the excited state under C_{3v} symmetry (see Ref. [44] and references therein).

The only allowed transition is $^3A_2 \rightarrow {}^3E$ in the first order. Thus, the excitation of this system may be explained by promoting an electron from the a_1 single particle state to the e state resulting in the ^3E excited state. This is certainly a simplified picture since the excited electron feels the presence of the hole left behind as a result of the Coulomb interaction between them, so they cannot be treated separately. Correspondingly, the wave function in the excited state, which should describe the motion of the correlated electron-hole pair, is *principally* not given by a simple product of electron and hole wave functions but requires a more general representation to account for energetic and spatial correlation between the two particles. Thus, we examine the excitation of NV center by TD-DFT theory which is able to address this complex phenomena. Before turning to the results we introduce another defect under consideration.

18.1.2
Divacancy in Silicon Carbide

Divacancies are common defects in semiconductors with consisting of neighbor isolated vacancies. The divacancy has been recently identified in hexagonal SiC polytypes [4]. The defect possesses C_{3v} symmetry in cubic SiC and also at on-axis configurations in hexagonal polytypes. The silicon vacancy part of the defect (C_{1-3} atoms) introduces three carbon dangling bonds while the carbon vacancy part of the defect (Si_{1-3} atoms) contributes with three silicon dangling bonds (see Figure 18.3). Again, group theory analysis revealed us [47] that the six dangling bonds will build two a_1 and two e defect levels in C_{3v} symmetry. Six electrons can occupy these states. According to our *ab initio* sc calculations [47] the two a_1 levels are lowest in energy and then the doubly degenerate e levels follow them in the hierarchy. Four electrons will occupy the a_1 states and the two remaining electron will occupy the degenerate e level. Again, by following the Hund-rule the natural choice is to place the electrons with parallel spins. Indeed, the ground state of the neutral divacancy is a high spin $S = 1$ state [4, 47]

A PL spectrum of about 1.0 eV is associated with this defect which can be also detected at low temperature infrared absorption [48]. The nature of the excitation is not well-understood. The position of the defect levels may reveal the possible excitation mechanism of the defect. The two doubly degenerate e defect levels occur in the fundamental band gap [47]. In the ground state, only the lowest e level is occupied by parallel spin-electrons. Two a_1 defect levels are in the valence band where the highest a_1 state is resonant with the valence band edge according to our *ab initio* sc

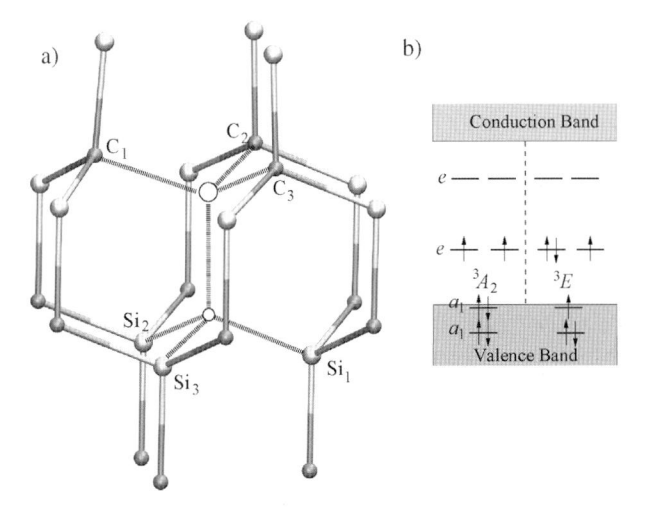

Figure 18.3 (online color at: www.pss-b.com) (a) The optimized geometry of the divacancy in the ground state. Cyan and yellow balls represent C and Si atoms, respectively. The open circles depict the vacant sites. (b) Schematic diagram about the defect states in crystalline environment in the $^3A_2$ ground and ^3E excited states.

calculations [47]. One possible model to explain the excitation is to promote an electron from the resonant a_1 state to the lowest e level in the fundamental gap [47] (Figure 18.3b). This will result in $^3A_2 \rightarrow {}^3E$ excitation. The sharp excitation between a resonant state and a defect state in the fundamental gap is a well-known process in semiconductors, e.g., similar process takes place for the isolated vacancy in diamond [49]. Other excitations may also occur for divacancy in SiC, for instance, between the e defect levels in the gap. We examine the lowest excitation energies of divacancy by TD-DFT calculations.

18.2 Method

18.2.1 Model, Geometry, and Electronic Structure

We embed the defects into a nanocrystal that contains 147 host crystal atoms and 100 hydrogen atoms for termination. The results are strictly valid for these nanocrystals but we discuss whether the obtained results could be valid in crystalline environment. We applied PBE [9] functional to optimize the geometry while the excitation spectrum is calculated both by the semi-local PBE and non-local hybrid PBE0 functionals. In PBE0 functional the Hartree–Fock exchange is mixed at 25% extent into PBE functional [50]. The optimization of the geometry has been done with numerical atomic basis set for diamond nanocrystal at double ζ polarized

(DZP) level which provided good results in scs compared to plane wave calculations [44]. We used the SIESTA code for this purpose [51]. Troullier–Martins pseudopotentials have been applied to model the effect of nuclei together with the core electrons in SIESTA calculations [52]. We used the relatively computationally expensive VASP code with plane wave basis set [53, 54] to study the divacancy in SiC. In the VASP calculations, we use a plane wave basis set of 420 eV (∼30 Ry) which is highly convergent with the applied projected augmentation wave (PAW) projectors for carbon, hydrogen, and silicon atoms [55, 56]. In the VASP calculations we applied the appropriate symmetry, and the energy of the ground state and the excited states are calculated by setting the appropriate occupation of the defect states in the gap as explained below. In the geometry optimization calculations, all the atoms were allowed to relax until the forces were below 0.01 eV/Å. In VASP code it is possible to set the occupation number of single particle states. This may be used to study the geometry change upon electronic excitation [12, 44]. In this scheme the excited state is described by promoting an electron from an occupied Kohn–Sham defect level to an unoccupied Kohn–Sham defect level of the ground state with allowing the nuclei to relax to find the optimum geometry with the charge density obtained from this fixed occupation of orbitals. This method is called *constrained* DFT briefly. The constrained DFT method is a computationally cheap method to find the ZPL transition energy within the Franck–Condon approximation as shown in Figure 18.4. In our experience this methodology *under special circumstances* provides reliable results regarding the *relaxation energy* due to electronic excitation [12] which is the Stokes-shift shown in Figure 18.4. The constrained DFT method may provide reliable results for the Stokes-shift upon the following conditions:

1) The excited state can be well-described by a single Slater-determinant and the symmetry of the ground and excited states are different with avoiding possible hybridization of states.
2) The nature of the excited and ground states is similar; for instance, they are originated from similar well-localized defect states.

As we show below we considered such defects here that can fulfill these criteria. Nevertheless, this statement cannot be proven by DFT based methods alone. We studied the nature of excitation by TD-DFT method that is capable of studying the above mentioned criteria. Next, we summarize the TD-DFT theory in nutshell with focusing on the approximations that make possible to use this method *in practice*.

18.2.2
Time-Dependent Density Functional Theory with Practical Approximations

In the TD-DFT within the Kohn–Sham formalism the system subject to a TD external potential:

$$\hat{v}_{\text{ext}}(\mathbf{r}, t) = \hat{v}_{\text{stat}}(\mathbf{r}) + \hat{v}_{\text{t}}(\mathbf{r}) f(t), \tag{18.1}$$

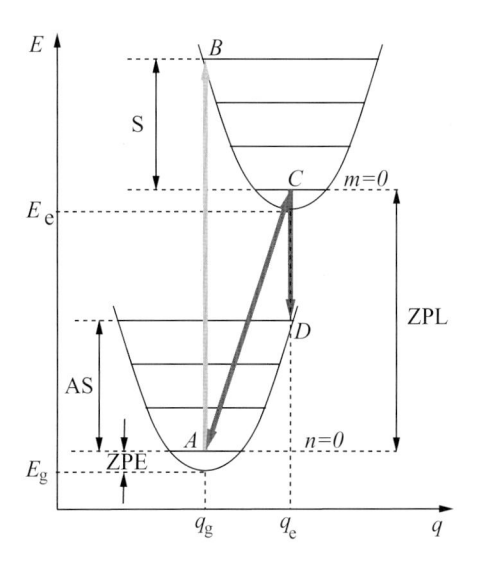

Figure 18.4 (online color at: www.pss-b.com) The energy (E) vs. configuration coordinate (q) diagram for the excitation process of a defect in the Franck–Condon approximation: E_g and E_e are the minima in the quasi-parabolic potential energy surfaces of the defect in the ground and excited states, respectively, and q_g and q_e are the corresponding coordinates. ZPE is the zero point energy (indicated only for the ground state). The energy ladders show the phonon energies with the phonon ground states at $n = 0$ (ground state of the defect) and $m = 0$ (excited state). At elevated temperatures the high-energy phonon states can be occupied by inducing transition $A \rightarrow B$ (vertical absorption, green arrow) and $C \rightarrow D$ (vertical emission, red arrow). Transition $A \leftrightarrow C$ corresponds to the zero-phonon line (ZPL, blue double arrow) both in absorption and emission. The energy of the Stokes-shift (S) and anti-Stokes-shift (AS) are also shown.

is mapped onto a effective one-particle (non-interacting) system (e.g., Refs. [57, 58]):

$$i\hbar \frac{\partial \varphi_i(\mathbf{r},t)}{\partial t} = \hat{H}_{\text{eff}}(t)\, \varphi_i(\mathbf{r},t), \tag{18.2}$$

where

$$\hat{H}_{\text{eff}}(t) = -\frac{1}{2}\nabla^2 + \hat{v}_{\text{ext}}(\mathbf{r},t) + \int d^3\mathbf{r}'\, \frac{\rho(\mathbf{r}',t)}{|\mathbf{r}-\mathbf{r}'|} + \hat{v}_{\text{xc}}[\rho](\mathbf{r},t). \tag{18.3}$$

Here $\hat{v}_{\text{ext}}(\mathbf{r},t)$ is the external potential, $\int d^3\mathbf{r}'(\rho(\mathbf{r}',t)/|\mathbf{r}-\mathbf{r}'|)$ the Hartree term, and $\hat{v}_{\text{xc}}[\rho](\mathbf{r},t)$ is the functional derivative of the TD exchange-correlation functional with respect to the TD density.

In the usual adiabatic approximation memory effects are neglected, i.e., $\hat{v}_{\text{xc}}[\rho](\mathbf{r},t)$ is approximated as: $\hat{v}_{\text{xc}}[\rho](\mathbf{r},t) = \hat{v}_{\text{xc}}(\rho(\mathbf{r},t))$. This approximation is proven to be very successful in many cases and easy to implement. By using Eq. (18.2), one can propagate the wave function in real time and calculate the full spectrum of the system driven by Eqs. (18.1–18.3).

In many cases the excitation can be taken as a perturbing potential where the density response ($\rho^{(1)}(\mathbf{r},t)$) is proportional to the perturbing potential (i.e., linear response theory):

$$\rho^{(1)}(\mathbf{r},t) = \int d^3\mathbf{r}' dt' \chi(t,t',\mathbf{r},\mathbf{r}') \hat{v}_t(\mathbf{r}') f(t'), \tag{18.4}$$

where $\chi(t,t',\mathbf{r},\mathbf{r}')$ is the full response function. It can be shown that the density response in the Kohn–Sham picture takes the following form (see Ref. [58]):

$$\rho^{(1)}(\mathbf{r},t) = \int d^3\mathbf{r}' dt' \chi_{KS}(t,t',\mathbf{r},\mathbf{r}')$$
$$\times \left[\hat{v}_t(\mathbf{r}') f(t') + \int d^3\mathbf{r}'' \frac{\rho^{(1)}(\mathbf{r}'',t')}{|\mathbf{r}'-\mathbf{r}''|} \right. \tag{18.5}$$
$$\left. + \int d^3\mathbf{r}'' \frac{\delta^2 E_{XC}}{\delta\rho(\mathbf{r}')\delta\rho(\mathbf{r}'')} \rho(\mathbf{r}'',t') \right],$$

where $\chi_{KS}(t,t',\mathbf{r},\mathbf{r}')$ is just the response function built from Kohn–Sham orbitals. Note that now the potential is not simply the TD part of Eq. (18.1), instead it also includes the Hartree-term and TD exchange-correlation potential, furthermore this equation has to be solved self-consistently since $\rho^{(1)}(\mathbf{r},t)$ appears on both sides.

Now taking the Fourier transform of both sides ($f(\omega) = \int e^{i\omega t} f(t) dt$), we arrive at the equation:

$$\rho^{(1)}(\mathbf{r},\omega) = \int d^3\mathbf{r}' \chi_{KS}(\omega,\mathbf{r},\mathbf{r})$$
$$\times \left[\hat{v}_t(\mathbf{r}') f(\omega) + \int d^3\mathbf{r}'' \frac{\rho^{(1)}(\mathbf{r}'',\omega)}{|\mathbf{r}'-\mathbf{r}''|} \right.$$
$$\left. + \int d^3\mathbf{r}'' \frac{\delta^2 E_{XC}}{\delta\rho(\mathbf{r}')\delta\rho(\mathbf{r}'')} \rho(\mathbf{r}'',\omega) \right] \tag{18.6}$$

In Fourier space the form of the response kernel is (where we also introduced spin dependence):

$$\chi_{KS,\sigma\sigma'}(\omega,\mathbf{r},\mathbf{r}') = \delta_{\sigma\sigma'} \sum_{i,a} \left[\frac{\varphi_{i\sigma}^*(\mathbf{r})\varphi_{a\sigma}(\mathbf{r})\varphi_{i\sigma}(\mathbf{r}')\varphi_{a\sigma}^*(\mathbf{r}')}{\omega-(\varepsilon_{a\sigma}-\varepsilon_{i\sigma})+i0^+} \right.$$
$$\left. - \frac{\varphi_{a\sigma}(\mathbf{r})\varphi_{a\sigma}^*(\mathbf{r})\varphi_{i\sigma}^*(\mathbf{r}')\varphi_{a\sigma}(\mathbf{r}')}{\omega+(\varepsilon_{a\sigma}-\varepsilon_{i\sigma})+i0^+} \right]. \tag{18.7}$$

In Eq. (18.7) $\varphi_{i\sigma}(\mathbf{r})$s are ground-state Kohn–Sham orbitals. From now on, i,j and a,b indices denote occupied and virtual (unoccupied) orbitals, respectively. k,l,m,n indices stand for general orbitals.

In order to solve Eq. (18.6), let us parameterize the density response as follows:

$$\rho_\sigma^{(1)}(\mathbf{r},\omega) = \sum_{i,a,\sigma}[P_{ia\sigma}(\omega)\varphi_{a\sigma}^*(\mathbf{r})\varphi_{i\sigma}(\mathbf{r})$$
$$+ P_{ai\sigma}(\omega)\varphi_{a\sigma}(\mathbf{r})\varphi_{i\sigma}^*(\mathbf{r})]. \qquad (18.8)$$

Using Eqs. (18.7) and (18.8) we arrive at the following coupled matrix equations for $P_{ia\sigma}(\omega)$ and $P_{ia\sigma}(\omega)$:

$$[\delta_{\sigma\tau}\delta_{ij}\delta_{ab}(\varepsilon_{a\sigma} - \varepsilon_{i\sigma} + \omega) + K_{ia\sigma,\,jb\tau}]P_{jb\tau}$$
$$+ K_{ia\sigma,\,bj\tau}P_{bj\tau} = -(\hat{v}t)ia\sigma, \qquad (18.9)$$

$$[\delta_{\sigma\tau}\delta_{ij}\delta_{ab}(\varepsilon_{a\sigma} - \varepsilon_{i\sigma} - \omega) + K_{ai\sigma,bj\tau}]P_{bj\tau}$$
$$+ K_{ai\sigma,\,jb\tau}P_{jb\tau} = -(\hat{v}t)ai\sigma, \qquad (18.10)$$

where we introduced the following short notation: $(\hat{v}t)_{ia\sigma} = \int d^3\mathbf{r}\,\varphi_{i\sigma}^*(\mathbf{r})\hat{v}_t(\mathbf{r})\varphi_{a\sigma}(\mathbf{r})$, and the K kernel:

$$K_{kl\sigma,mn\tau} = \int d^3\mathbf{r}\,d^3\mathbf{r}'\,\varphi_{k\sigma}^*(\mathbf{r})\varphi_{l\sigma}(\mathbf{r})$$
$$\times \left(\frac{1}{|\mathbf{r}-\mathbf{r}'|} + \frac{\delta^2 E_{xc}}{\delta\rho_\sigma(\mathbf{r}')\delta\rho_\tau(\mathbf{r}'')}\right)\varphi_{n\tau}^*(\mathbf{r}')\varphi_{m\tau}(\mathbf{r}'). \qquad (18.11)$$

Note that the kernel consists of two parts: the Hartree part is local in time (it causes the so-called *local field effects*), however the second part of the kernel is generally nonlocal both in space and time. In the adiabatic approximation derived above it is time independent but still space dependent. If the xc kernel is set to zero then this approximation is called the random phase approximation. Neglecting both terms in the kernel yield transitions between Kohn–Sham states.

Using $X_{ia\sigma} = P_{ia\sigma}(\omega)$, $Y_{ia\sigma} = P_{ai\sigma}(\omega)$, and $V_{ia\sigma} = (\hat{v}_t)_{ia\sigma}$ we arrive at the following matrix equation:

$$\left[\begin{pmatrix} L & M \\ M^* & L^* \end{pmatrix} - \omega\begin{pmatrix} -1 & 0 \\ 0 & 1 \end{pmatrix}\right]\begin{pmatrix} X \\ Y \end{pmatrix} = -f(\omega)\begin{pmatrix} V \\ V^* \end{pmatrix}, \qquad (18.12)$$

where

$$L_{ia\sigma,jb\tau} = \delta_{\sigma\tau}\delta_{ij}\delta_{ab}(\varepsilon_{a\sigma} - \varepsilon_{i\sigma}) + K_{ia\sigma,jb\tau}, \qquad (18.13)$$

$$M_{ia\sigma,jb\tau} = K_{ia\sigma,bj\tau}. \qquad (18.14)$$

In response theory, excitation energies are the poles of response function, thus by taking $f(\omega) = 0$ we find the following non-Hermitian eigenvalue problem:

$$\begin{pmatrix} L & M \\ M^* & L^* \end{pmatrix} \begin{pmatrix} X \\ Y \end{pmatrix} = \omega \begin{pmatrix} -1 & 0 \\ 0 & 1 \end{pmatrix} \begin{pmatrix} X \\ Y \end{pmatrix}, \tag{18.15}$$

This matrix equation is called Casida-equation [59]. We note that the term Y may be neglected (the so-called anti-resonance between the occupied and unoccupied states) that leads to the so-called Tamm-Dancoff approximation. We keep further the anti-resonance term in our derivation. If the ground-state density originates from a restricted Kohn–Sham calculation ($\varphi_{i\alpha}(\mathbf{r}) = \varphi_{i\beta}(\mathbf{r})$) then the size of the problem may be reduced by one half and a unitary transformation can help discriminate between singlet (s) and triplet (t) transitions:

$$u_{ia} = \frac{1}{\sqrt{2}}(P_{ia\alpha} + P_{ia\beta}), \quad v_{ia} = \frac{1}{\sqrt{2}}(P_{ia\alpha} - P_{ia\beta}). \tag{18.16}$$

It can be shown that if the orbitals can be chosen to be real (and that is true for finite structures) then the resulting matrices M^p, L^p will be also real where p labels either singlet (s) or triplet (t) excitations. The long expression for these matrices are given in the appendix of Ref. [60]. We note that these matrices contain the Hartree term and a modified expression for the xc term. In the case of hybrid functionals, the Hartree term is adjusted with respect to the case of (semi)local functionals. Now, we find the following matrix equation:

$$(M^p - L^p)(M^p + L^p)(X + Y) = \omega^2 (X + Y). \tag{18.17}$$

If $M^p - L^p$ is positive definite, Eq. (18.17) can be transformed to a hermitian eigenvalue problem:

$$(M^p - L^p)^{1/2}(M^p + L^p)(M^p - L^p)^{-1/2}(X + Y)' \\ = \omega^2 (X + Y)', \tag{18.18}$$

with

$$(X + Y)' = (M^p - L^p)^{-1/2}(X + Y). \tag{18.19}$$

Clearly, in order to obtain excitation energies, these matrices have to be build and the eigenvalue problem has to be solved. This is done using an iterative subspace method in Turbomole package [61]. We note that we obtain the transition energies in the frequency domain with this method which makes possible to restrict the calculations to the lowest excitation energies and to analyze the states contributing to the given excitation. This is a clear advantage over the time domain algorithms of TD-DFT where always the full spectrum is calculated and the analysis of the nature of transitions is not straightforward.

We applied the Turbomole package to carry out the TD-DFT calculations [61]. This is a cluster code with localized Gaussian basis sets that can utilize hybrid functional in the TD-DFT kernel. We applied an all-electron Gaussian DZP basis set for all the atoms in the system [62] at fixed coordinates supplied by Siesta or VASP DFT calculations. We calculated the adiabatic TD-DFT spectrum both with spin-polarized (semi)local DFT-PBE and non-local PBE0 functionals within the linear response theory and beyond the Tamm-Dancoff approximation. We particularly focused on the lowest excitation energies that may be manifested in the low temperature absorption and PL spectra.

18.3
Results and Discussion

18.3.1
Nitrogen-Vacancy Center in Diamond

In nanodiamond the gap opens up only slightly by about 0.1 eV even in our relatively small nanodiamond with a diameter of about 1 nm. Recently, we have found [63] that nanodiamond exhibits low-lying Rydberg states which results in only small change between the highest occupied molecular orbital (HOMO)–lowest unoccupied molecular orbital (LUMO) gap of nanodiamonds and the band gap of bulk diamond. This will have a serious consequence on the TD-DFT spectrum as we will show below.

We started with the geometry optimization of the negatively charged NV center by SIESTA DFT-PBE calculations using spin-polarization and without symmetry restriction. The defect automatically arrived at the $S=1$ state and retained the C_{3v} symmetry. It may be worthy to compare the relative positions of the defect states obtained in nanodiamond (labeled as "nd") and supercell (labeled as "sc") DFT-PBE calculations.

Two defect levels appear in the HOMO–LUMO gap: the a_1 and e levels like in the sc calculations. By comparing the PBE single particle Kohn–Sham levels obtained in the sc and nanodiamond (columns 2 and 3 in Table 18.1), we observe that the relative positions of the defect levels are very similar including the spin-polarization between a_1^α and a_1^β states or the energy difference between a_1 and e states within the same spin-channel. This implies that the defect levels are "fixed" relative to each other and do not change significantly by confining the crystalline states. In our sc calculations we found [44] that these defect states are strongly localized around the core of the defect, in other words, they "split" from the extended crystalline states that are not effected heavily due to quantum confinement. One may hope that the results obtained in the nanodiamond model may be relevant for crystalline environment as the excitation may occur between the localized defect states.

Next, we may check the quasi-particle shift due to non-local functionals or by G0W0 correction [64, 65]. In the case of sc calculations, we used the *screened* hybrid HSE06 functional with $\omega = 0.20\, 1/\text{Å}$ screening parameter [66, 67] which provided nice agreement with the experimental band gap of diamond and excitation energies of the NV center [12]. For instance, we obtained the experimental vertical energy of absorption (2.20 eV) very accurately by using constrained DFT calculation and ΔSCF procedure. In the case of nanodiamond we use unscreened hybrid PBE0 which gave us very accurate absorption spectra for tiny nanodiamonds [63]. We believe that the

Table 18.1 The relative positions of defect levels of NV center in the ground state in eV where α and β label the spin-up and -down channels, respectively. The PBE and HSE06 supercell (sc) calculations have been carried out in 512-atom sc yielding accurate excitation energies with HSE06 functional (Ref. [12]) while G0W0 sc calculations have been made in 256-atom sc with about 0.15 eV larger excitation energies together with BSE than the experimental values (Ref. [64]). "nd" labels the results obtained in our nanodiamond. a_1^α, a_1^β, and e^α states are occupied while e^β is unoccupied in the ground state. LUMO is a Rydberg state in nanodiamond. The values in the parentheses are the relative quasi-particle corrections defined as the following: in column HSE06 sc this is the difference between column HSE06 sc and PBE sc carried out with the same size of sc (but allowing relaxation with the two methods as explained in Ref. [12]); in column G0W0 sc the reference PBE values are taken from the same geometry and sc in Ref. [64]; in column PBE0 nd this is the energy difference between column PBE0 nd and PBE nd carried out in the same nanodiamond and with the same geometry. Lowest excitation energy (E_{exc}) is calculated by ΔSCF method for PBE sc and HSE06 sc methods, while with G0W0 + BSE method for G0W0 sc and TD-DFT for PBE nd and PBE0 nd. Exciton binding energy (E_X) is defined as an energy difference between the calculated excitation energy and the quasi-particle energy difference between the corresponding states in TD-DFT methods while BSE provides this value on the top of G0W0 sc calculation.

levels	PBE sc	PBE nd	HSE06 sc	G0W0 sc	PBE0 nd
$a_1^\beta - a_1^\alpha$	0.56	0.58	1.04 (0.48)	0.6 (0.2)	1.22 (0.64)
$e^\alpha - a_1^\alpha$	1.17	1.28	1.23 (0.06)	1.2 (0.1)	1.37 (0.09)
$e^\beta - a_1^\alpha$	2.45	2.44	4.00 (1.55)	3.3 (1.1)	4.67 (2.23)
$e^\beta - a_1^\beta$	1.89	1.84	2.96 (1.07)	2.7 (0.9)	3.45 (1.61)
LUMO-e^α		1.57			3.46
E_{exc}	1.91	1.57	2.21	2.32	2.20
E_X		0.0	0.75	0.30	1.25

unscreened hybrid functionals are more realistic for relatively small nanoclusters and molecules where the electron system cannot effectively screen the Coulomb potential like in an infinite crystal. By comparing the "HSE06 sc" and "PBE0 nd" columns we found that the quasi-particle shifts between occupied levels are very similar being but a bit (0.16 eV) larger for "PBE0 nd." However, there is a larger difference between the relative positions of the unoccupied e^β and the occupied states with the two methods. The quasi-particle shift for the unoccupied state is larger with the unscreened PBE0 functional which may be a natural consequence of the larger Fock-exchange term in the Hamiltonian. The G0W0 sc calculations have been carried out at the Γ-point in 256-atom sc [64] which may not be fully convergent for the single particle states but still it is worthwhile to compare the G0W0 quasi-particle levels with those obtained by hybrid functionals. The spin-polarization of a_1 state ($\sim$0.2 eV) is not enhanced such a large amount like with hybrid functionals (0.5–0.6 eV) after the quasi-particle correction. Interestingly, the quasi-particle shift difference relative to a_1^α is also larger for HSE06 sc than for G0W0 sc, thus the quasi-particle shift difference for ($e^\beta - a_1^\alpha$) will be similar.

Now, we turn to the investigation of the excitation energies. In the case of the TD-DFT calculation with PBE kernel we arrived at a very low value of 1.57 eV. According to our analysis the $e^\alpha \to$ LUMO transition is responsible for this absorption peak. As can be inferred in Table 18.1 the energy of LUMO is indeed close to the energy of the highest occupied spin-up defect level e^α. Apparently, this transition cannot occur in crystalline environment. The next lowest transition energy is about 1.88 eV which is close to the lowest excitation energy calculated in the sc by ΔSCF method (PBE sc column in Table 18.1). However, this excitation peak arises in only 39% from $a_1^\beta \to e^\beta$ transition but 57% from $e^\alpha \to$ LUMO + 1 transition. This indicates that the TD-DFT spectrum at PBE level cannot describe the nature of transition occurring in bulk diamond due to the low-lying empty Rydberg state of nanodiamond. We note that the calculated "exciton binding energy" of the lowest excitation is practically zero which is typical for TD-DFT spectrum with (semi)local functionals in the kernel. The situation for the TD-DFT calculation with PBE0 in the kernel is different. In PBE0 the LUMO level is about the same energy distance from the e^α state as the energy difference between a_1^β and e^β levels. The lowest excitation energy is 2.20 eV which is dominated by $a_1^\beta \to e^\beta$ transition as much as 92%. While the energy difference between these single particle states was large (3.45 eV) due to the large quasi-particle shift between these states relative to their PBE values (1.61 eV) the large excitonic effect (1.25 eV) could compensate this effect with providing an excitation energy (2.20 eV) which is very close to the experimental value. This number is also close to the values resulted from HSE06 sc and G0W0 + BSE (Bethe-Salpeter equation) sc calculations (Table 18.1). We believe that it is not fortuitous coincidence. HSE06 sc gave less quasi-particle shift than PBE nd and correspondingly the "exciton binding energy" in HSE06 sc is smaller than for PBE nd. In G0W0 + BSE sc calculation we found the smallest quasi-particle shift with again smaller "exciton binding energy." The unscreened PBE0 functional may overshoot the "quasi-particle shifts" relative to the PBE values but the large excitonic effect compensate this resulting in reasonable values for the excitation energies. This implies that the calculated "exciton binding energies" of defects in nanocrystals cannot reveal the "exciton binding energies" of defects in the bulk counterpart but the calculated excitation energies may be accurate after careful choice of the model and inspection the nature of the transition. From this investigation we can also conclude for NV center in diamond that the lowest excitation is indeed described by promoting an electron from the a_1^β level to the e^β level that can support the constrained DFT method to calculate the Stokes-shift for the PL process applied in Refs. [12, 64]. We note here that this transition describes the absorption of the light with perpendicular polarization to the symmetry axis which transforms as E irreducible representation in C_{3v} point group. The parallel polarized light transforms as A_1 under C_{3v} point group. According to the selection rules E polarization will be allowed for $^3A_2 \to {}^3E$ transition in the first order while A_1 polarization is forbidden. Our finding is in line with this group theory analysis.

18.3.2
Divacancy in Silicon Carbide

In our previous DFT-LDA sc calculation, we observed that neutral divacancy introduces two doubly degenerate defect levels in the gap (see Figure 18.3). Beside these states two a_1 level is associated with the dangling bonds according to the group theory [47]. The highest a_1 level is resonant with the valence band edge with ∼0.3 eV below the valence band top. In the ground state the lowest e level in the gap is occupied by two electrons. In spin-polarized calculation this e level splits by about 0.5 eV while the empty e level does not split practically.

In the DFT-PBE cluster calculation the "band gap" of SiC opens up with several eV compared to their bulk counterpart [68]. We note that we do not observe low-lying Rydberg states for SiC nanocluster like in diamond nanoclusters thus no such complications can occur in the analysis of the excitation spectrum of divacancy in SiC nanocrystals as for NV center in nanodiamond. The quantum confinement of crystalline states makes the resonant a_1 state visible that is localized strongly on the carbon dangling bonds. Because this a_1 state does not mix with the crystalline states, therefore the spin-polarization of this a_1 state is significant (∼0.7 eV) in contrast to the case of DFT-LDA sc calculation. The a_1^β state will pushes up the lowest empty e^β state by about 0.5 eV, thus the energy difference between the lowest and highest e^β states will be about 0.5 eV smaller than in sc calculation. It is important to notice that the relative position of the a_1^α and e^α states agree those in sc calculation within 0.1 eV. This may indicate that the localized resonant state also sticks with the other localized defect states and splits from the crystalline environment. In the C_{3v} symmetry we determined the excitation energies with different light polarizations. As expected, the excitation with E polarization is lowest in energy which allows to couple the $^3A_2$ ground state and the ^3E excited state. Indeed, in this case the excitation between a_1^β and lowest e^β states is dominant, however, there is a non-negligible contribution (13%) from the lowest and highest e^α states. The calculated absorption energy is 1.15 eV that is close to the measured ZPL transition of 1.0 eV. For the excitation with E polarization we obtained 1.43 eV which occurs between the lower and higher e^α defect states. We note that the corresponding "exciton binding energy" is zero as we found already for NV center in diamond.

Next, we studied this excitation by applying PBE0 hybrid functional in the TD-DFT kernel. The relative position between the occupied a_1^α and e^α defect levels remains about the same (∼0.7–0.8 eV). However, the energy differences between the lower occupied and higher empty e^α states and between the occupied a_1^β and empty e^β states are increased by about 1.7–1.8 eV. This quasi-particle shift is a natural consequence of the Fock-operator in the hybrid functional which opens the gap between the occupied and unoccupied states. Again, we calculated the lowest excitation energies with A_1 and E polarizations. The excitation energy with E polarization is 1.18 eV which almost coincides with the value obtained by the (semi) local PBE in the TD-DFT kernel. The basic difference between the two transitions that in TD-DFT with PBE0 in the kernel the transition is dominated (over 95%) by promoting the electron from the single particle a_1^β defect level to the lowest e^β level. The relative quasi-particle shift of about 1.8 eV was largely compensated by the excitonic effect of about 1.0 eV. Interestingly, the next excitation energy of 1.85 eV is due to the transition between the a_1^β state and the higher e^β state. The "exciton binding energy" of this process is again ∼1.0 eV. The following excitation energy of ∼2.35 eV is caused by transitions of E, A_1 polarizations between the lower occupied and higher empty e^α defect states where the resulting "exciton binding energy" is smaller (~0.8 eV).

We calculated the Stokes-shift due to electronic excitation by constrained DFT method. First, we studied the lowest excitation energy for which the transition between a_1^β and the lower e^β states is the dominant process resulting in the ^3E excited state as shown in Figure 18.3b. We set this occupation for the defect states and restricted the geometry optimization within C_{3v} symmetry. The calculated Stokes-shift is 0.075 eV. Because the e^β state is half occupied Jahn-Teller distortion may occur. By allowing C_{1h} geometry optimization we obtained a slightly deeper energy than for C_{3v} symmetry. The final value for this Stokes-shift is 0.12 eV. By combining the absorption energy (1.18 eV) and the Stokes-shift we arrived at 1.06 eV for the ZPL line which is very close to the experimental value found in bulk SiC. We found that the deviation from the C_{3v} symmetry is tiny, still it is not exactly C_{3v} symmetry. Thus, the strict selection rule of E polarization of light in the emission process may be relaxed and A_1 polarization may be slightly allowed. The situation is different for the second lowest excitation process. In that case the a_1^β electron is excited to the higher e^β state. That higher e^β state is localized on the Si dangling bonds that can overlap and interact with each other. The Jahn-Teller distortion for this excited state is much stronger (~0.2 eV) than for the previous process (~0.04 eV) with resulting in a large Stokes-shift of 0.46 eV. Combining this value with the calculated absorption energy (1.85 eV) one arrives at 1.39 eV ZPL energy. Here, the C_{3v} symmetry is considerably lowered to C_{1h} where the Si_2 and Si_3 atoms are bended to each other (see Figure 18.3a). Thus, both polarizations should be allowed for this PL process.

These results are strictly valid for the SiC nanocrystal. As both the resonant a_1 state and the lowest e state remains "fixed" going from crystalline environment to small nanocrystal one may hope that the calculated TD-DFT excitation energy is valid for the defect in bulk SiC. Indeed, the calculated lowest excitation energy in SiC nanocrystal is close to the experimental data recorded in bulk SiC. By assuming that the transition from the resonant defect state is the strongest to the lower e level we may claim that we could identify the physical process in the PL center associated with the divacancy [48]. Our calculations revealed that other type of excitations may also occur for divacancy. In crystalline environment the detection of the resonances occurring in SiC nanocrystals may be not feasible due to competing processes. For instance, from the sea of the valence electrons lying in the continuous valence band one can excite an electron to the lower e level instead of the resonant excitation between the a_1 level and the higher e level where the latter process is clearly observable in our calculation due to the discrete energy spectrum of the relatively small SiC nanocrystal.

By comparing the excitation and other properties of NV center in diamond and divacancy in SiC one can list the similarities of the two systems:

1) The defect exhibits $S=1$ high spin-state with C_{3v} symmetry.
2) The lowest excitation is due to $^3A_2 \rightarrow {}^3E$ transition.
3) The spin-density is localized close to the core of the defect.

Nevertheless, one can find also differences between the two systems:

1) Two highly localized well-separated defect states play the dominant role in the excitation of the NV center in diamond while one of the defect states is resonant with the valence band for the divacancy in SiC.
2) NV center has strict C_{3v} symmetry in the PL process while for divacancy in SiC it may be a bit lowered to C_{1h} symmetry.
3) NV center has a clear PL signal at room temperature while this has not yet been demonstrated by divacancy in SiC.
4) The spin-flip process in the PL process was already demonstrated for NV center in diamond while this has not been investigated in detail for divacancy in SiC, though electron paramagnetic resonance studies combined with photo-excitation indicate that similar process may occur for divacancy.

Further investigations are needed on divacancy in bulk SiC in order explore the fine details of its excited states. One may say from the present study that divacancy in SiC might be an alternative for realizing the concept of solid state quantum bit.

18.4
Summary

We investigated the excitation spectrum of NV center in nanodiamond and divacancy in SiC nanocrystal. We found that TD-DFT method with PBE0 in the kernel could reproduce the experimental data observed in their bulk counterparts. We discussed the validity of these results by careful inspection of the model and its electronic structure. Our results imply that TD-DFT method with non-local DFT functional together with nanocrystal models may be applied for studying the excitation of defects in bulk crystals when the excitation occurs between well-localized defect states.

Acknowledgements

AG acknowledges the support from Hungarian OTKA under grant no. K-67886 and the János Bolyai program of the Hungarian Academy of Sciences.

References

1 de Walle, C.G.V. and Neugebauer, J. (2004) *J. Appl. Phys.*, **95** (8), 3851–3879.
2 Estreicher, S.K., Backlund, D., Gibbons, T.M., and Doçaj, A., *Model. Simul. Mater. Sci. Eng.*, **17** (8), 084006 (2009).
3 Blöchl, P.E. (2000) *Phys. Rev. B*, **62**, 6158.
4 Son, N.T., Carlsson, P., Hassan ul, J., Janzén, E., Umeda, T., Isoya, J., Gali, A., Bockstedte, M., Morishita, N., Ohshima, T., and Itoh, H. (2006) *Phys. Rev. Lett.*, **96**, 055501.
5 Umeda, T., Son, N.T., Isoya, J., Janzén, E., Ohshima, T., Morishita, N., Itoh, H., Gali, A., and Bockstedte, M. (2006) *Phys. Rev. Lett.*, **96**, 145501.

6 Ceperley, D.M. and Alder, B.J. (1980) *Phys. Rev. Lett.*, **45**, 566.
7 Perdew, J.P., and Zunger, A. (1981) *Phys. Rev. B*, **23**, 5048.
8 Perdew, J.P. and Wang, Y. (1992) *Phys. Rev. B*, **45** (23), 13244–13249.
9 Perdew, J.P., Burke, K., and Ernzerhof, M. (1996) *Phys. Rev. Lett.*, **77** (18), 3865–3868.
10 Deák, P., Gali, A., Sólyom, A., Buruzs, A., and Frauenheim, T. (2005) *J. Phys.: Condens. Matter*, **17** (22), S2141.
11 Oba, F., Togo, A., Tanaka, I., Paier, J., and Kresse, G. (2008) *Phys. Rev. B*, **77** (24), 245202.
12 Gali, A., Janzén, E., Deák, P., Kresse, G., and Kaxiras, E. (2009) *Phys. Rev. Lett.*, **103** (18), 186404.
13 Lyons, J.L., Janotti, A., and de Walle, C. G. V. (2009) *Appl. Phys. Lett.*, **95** (25), 252105.
14 Clark, S.J., Robertson, J., Lany, S., and Zunger, A. (2010) *Phys. Rev. B*, **81** (11),115311.
15 Deák, P., Aradi, B., Frauenheim, T., Janzén, E., and Gali, A. (2010) *Phys. Rev. B*, **81** (15), 153203.
16 Onida, G., Reining, L., and Rubio, A. (2002) *Rev. Mod. Phys.*, **74** (2), 601–659.
17 Uchino, T., Takahashi, M., and Yoko, T. (2000) *Phys. Rev. Lett.*, **84** (7), 1475–1478.
18 Raghavachari, K., Ricci, D., and Pacchioni, G. (2002) *J. Chem. Phys.*, **116** (2), 825–831.
19 Zyubin, A.S., Mebel, A.M., and Lin, S.H. (2005) *J. Chem. Phys.*, **123** (4), 044701.
20 Zyubin, A.S., Mebel, A.M., Hayashi, M., Chang, H.C., and Lin, S.H. (2009) *J. Comput. Chem.*, **30** (1), 119–131.
21 Zyubin, A.S., Mebel, A.M., Hayashi, M., Chang, H.C., and Lin, S.H. (2009) *J. Phys. Chem. C*, **113** (24), 10432–10440.
22 Paier, J., Marsman, M., and Kresse, G. (2008) *Phys. Rev. B*, **78** (12), 121201.
23 Gruber, A., Drabenstedt, A., Tietz, C., Fleury, L., Wrachtrup, J., and Borczyskowski, C. (1997) *Science*, **276**, 2012.
24 Drabenstedt, A., Fleury, L., Tietz, C., Jelezko, F., Kilin, S., Nizovtsev, A., and Wrachtrup, J. (1999) *Phys. Rev. B*, **60**, 11503.
25 Wrachtrup, J., Kilin, S.Y., and Nizotsev, A.P. (2001) *Opt. Spectrosc.*, **91**, 429.
26 Jelezko, F., Popa, I., Gruber, A., Tietz, C., Wrachtrup, J., Nizovtsev, A., and Kilin, S. (2002) *Appl. Phys. Lett.*, **81**, 2160.
27 Jelezko, F., Gaebel, T., Popa, I., Gruber, A., and Wrachtrup, J. (2004) *Phys. Rev. Lett.*, **92**, 076401.
28 Jelezko, F., Gaebel, T., Popa, I., Dunham, M., Gruber, A., and Wrachtrup, J. (2004) *Phys. Rev. Lett.*, **93**, 130501.
29 Epstein, R.J., Mendoza, F., Kato, Y.K., and Awschalom, D.D. (2005) *Nature Phys.*, **1**, 94.
30 Hanson, R., Mendosa, F.M., Epstein, R.J., and Awschalom, D.D. (2006) *Phys. Rev. Lett.*, **97**, 087601.
31 Brouri, R., Beveratos, A., Poizat, J.P., and Gragier, P. (2000) *Opt. Lett.*, **25**, 1294.
32 Beveratos, A., Brouri, R., Gacoin, T., Poizat, J.P., and Grangier, P. (2002) *Phys. Rev. A*, **64**, 061802R.
33 Childress, L., Taylor, J.M., Sørensen, A.S., and Lukin, M.D. (2006) *Phys. Rev. Lett.*, **96**, 070504.
34 Jiang, L., Hodges, J.S., Maze, J.R., Maurer, P., Taylor, J.M., Cory, D.G., Hemmer, P.R., Walsworth, R.L., Yacoby, A., Zibrov, A.S., and Lukin, M.D. (2009) *Science*, **326**, 267–272.
35 Childress, L., Gurudev Dutt, M.V., Taylor, J.M., Zibrov, A.S., Jelezko, F., Wrachtrup, J., Hemmer, P.R., and Lukin, M.D. (2006) *Science*, **314**, 281.
36 Gurudev Dutt, M.V., Childress, L., Jiang, L., Togan, E., Maze, J., Jelezko, F., Zibrov, A.S., Hemmer, P.R., and Lukin, M.D. (2007) *Science*, **316**, 312.
37 Manson, N.B., Harrison, J.P., and Sellars, M.J. (2006) *Phys. Rev. B*, **74**, 104303.
38 du Preez, L. (1965) Ph.D. dissertation, University of Witwatersrand, Johannesburg .
39 Davies, G. and Hamer, M.F. (1976) *Proc. R. Soc. Lond., A*, **348**, 285.
40 Loubser, J. H. N. and van Wyk, J.P. 1977) *Diamond Research (London)* Industrial

Diamond Information Bureau, London, pp. 11–15.
41 Loubser, J. H. N. and van Wyk, J.A. (1978) *Rep. Prog. Phys.*, **41**, 1201.
42 Collins, A.T. (1983) *J. Phys. C*, **16**, 2177.
43 Goss, J.P., Jones, R., Breuer, S.J., Briddon, P.R., and Öberg, S. (1996) *Phys. Rev. Lett.*, **77**, 3041.
44 Gali, A., Fyta, M., and Kaxiras, E. (2008) *Phys. Rev. B*, **77** (15), 155206.
45 Larsson, J.A. and Delaney, P. (2008) *Phys. Rev. B*, **77** (16), 165201.
46 Watts, R.K. (1977) *Point Defects in Crystals* Wiley-Interscience Publication, New York/London/Sydney/Toronto.
47 Janzén, E., Gali, A., Henry, A., Ivanov, I.G., Magnusson, B., and Son, N.T. (2009), *Defects in SiC* Taylor & Francis Group, Boca Raton, FL, chap. 21, p. 615.
48 Magnusson, B. and Janzén, E. (2005) *Mater. Sci. Forum*, **483–485**, 341.
49 Lowther, J.E. (1993) *Phys. Rev. B*, **48** (16), 11592–11601.
50 Perdew, J.P., Ernzerhof, M., and Burke, K. (1996) *J. Chem. Phys.*, **105** (22), 9982–9985.
51 Sanchéz-Portal, D., Ordejón, P., Artacho, E., and Soler, J.M. (1997) *Int. J. Quantum Chem.*, **65**, 543.
52 Troullier, N. and Martins, J.L. (1991) *Phys. Rev. B*, **43**, 1993.
53 Kresse, G. and Hafner, J. (1994) *Phys. Rev. B*, **49**, 14251.
54 Kresse, G. and Furthmüller, J. (1996) *Phys. Rev. B*, **54**, 11169.
55 Blöchl, P.E. (1994) *Phys. Rev. B*, **50**, 17953.
56 Kresse, G. and Joubert, D. (1999) *Phys. Rev. B*, **59**, 1758.
57 Runge, E. and Gross, E. K. U. (1984) *Phys. Rev. Lett.*, **52** (12), 997.
58 Gross, E.K.U. and Kohn, W. (1990) *Adv. Quantum Chem.*, **21**, 255.
59 Casida, M.E. (1995) *Recent Advances in Density Functional Theory* World Scientific, Singapore, p. 155.
60 Bauernschmitt, R. and Ahlrichs, R. (1996) *Chem. Phys. Lett.*, **256** (4–5), 454–464.
61 Ahlrichs, R., Bär, M., Häser, M., Horn, H., and Klmel, C. (1989) *Chem. Phys. Lett.*, **162** (Oct), 165–169.
62 Schuchardt, K.L., Didier, B.T., Elsethagen, T., Sun, L., Gurumoorthi, V., Chase, J., Li, J., and Windus, T.L. (2007) *J. Chem. Inform. Model.*, **47**, 1045.
63 Vörös, M. and Gali, A. (2009) *Phys. Rev. B*, **80** (16), 161411.
64 Ma, Y., Rohlfing, M., and Gali, A. (2010) *Phys. Rev. B*, **81** (4), 041204.
65 Hedin, L. and Lundqvist, S. (1969) in: *Solid State Physics*, (eds H. Ehrenreich, F. Seitz and D. Turnbull), Academic, New York.
66 Heyd, J., Scuseria, G.E., and Ernzerhof, M. (2003) *J. Chem. Phys.*, **118** (18), 8207–8215.
67 Krukau, A.V., Vydrov, O.A., Izmaylov, A.F., and Scuseria, G.E. (2006) *J. Chem. Phys.*, **125** (22), 224106.
68 Vörös, M., Deák, P., Frauenheim, T., and Gali, A. (2010) *Appl. Phys. Lett.*, **96** (5), 051909.

19
Which Electronic Structure Method for The Study of Defects: A Commentary
Walter R. L. Lambrecht

19.1
Introduction: A Historic Perspective

The title of the CECAM workshop that gave rise to this compilation is "Which electronic structure method for the study of defects?" (see the Preface by Deák [1]). The first question, the reader might ask is why this question arises now. After all, the standard framework for dealing with defects in materials has been around for at least a few decades and has made significant contributions to our understanding of a wide variety of defects in semiconductors.

The "standard" toolkit includes density functional theory (DFT) combined with supercell band structure calculations. That is, the electronic structure problem is reduced to a standard band structure problem by using periodic boundary conditions. For surfaces, one uses a repeated slab geometry with vacuum regions to separate the surfaces; for interfaces, the artificial periodicity is introduced in only one dimension, while there still is a physical periodicity in the other two directions. For point defects, one needs to impose artificial periodicities in three dimensions, resulting in a faster increase of the size of the system that needs to be calculated. Dislocations, which are periodic in only one direction, are actually more complex because of their long-range strain fields and to avoid those, one either needs to use an opposing Burger's vector dipole pair of dislocations, or use a finite sample surrounded by vacuum regions for which one can then again restore periodicity in the two remaining dimensions. In any case, this compilation almost entirely focuses on point defects in the bulk of materials, not at surfaces or in nanostructures or not on extended defects, so we here restrict the discussion to point defects.

The main reason almost all calculations nowadays use periodic boundary conditions is a matter of convenience. One can use the general purpose computer programs that have been developed for band structure calculations, once these have been boosted to be able to deal with sufficiently large number of atoms. This is not the only choice, one could resort completely to finite models, or clusters, or one could embed the defect region into a perfect crystal by means of Green's function techniques.

Advanced Calculations for Defects in Materials: Electronic Structure Methods, First Edition.
Edited by Audrius Alkauskas, Peter Deák, Jörg Neugebauer, Alfredo Pasquarello, and Chris G. Van de Walle.
© 2011 Wiley-VCH Verlag GmbH & Co. KGaA. Published 2011 by Wiley-VCH Verlag GmbH & Co. KGaA.

Clusters are still popular with quantum chemists. Their drawback is that the surface of the clusters presents a more severe perturbation than the milder periodic boundary condition. One can avoid it to some extent by artificially satisfying the surfaces dangling bonds with pseudo hydrogens, but there is no obvious advantage to them compared to supercell techniques, except perhaps if one wants to use more advanced treatments of correlation than available in DFT, such as multiconfiguration interaction methods [2].

On the other hand, the most sophisticated approach to point defects, leading to an exact embedding of the local environment of the defect in the surrounding perfect crystal using correct boundary conditions is the Green's function method. Unlike in supercells, the band edges are defined precisely and one can calculate not only bound states in the gaps but also changes in the total and local densities of states within the bands, in the form of resonances and anti-resonances. One can make sure that the defect states, obey the exact symmetry of the defect site and are not influenced by the supercell geometry. One naturally treats an open system, into or out of which charge can flow as set by the chemical potential, without any uncertainties on how to define the chemical potential relative to the bands unlike the need in supercells to restore neutrality by artificial means such as a compensating homogeneous background charge density. In fact, there was a large amount of work done in the late 1970s–1980s to develop Green's function methods for point defects [3–13]. Strangely, these methods were abandoned in large part. The reason for this in my opinion is not because these methods were less intuitive or intrinsically less powerful, but rather their development could not keep up with the pace of the standard supercell approach. The original versions of these methods did not allow for relaxation of the structure of the defects and in many cases were even restricted to specific symmetries of the defect structure. On the other hand, practitioners using a standard all-purpose band structure method, were able to solve the problems that arose from experiment and required immediate attention. In particular, their ability to use the correct relaxed structure near the defects was a crucial advantage to get the essential physics correct. Meanwhile, practitioners of the Green's function methods, had to keep on generalizing their codes to handle such problems.

A second reason why the standard approach won the race is no doubt the almost universal use of plane wave basis sets as opposed to localized basis sets. For the latter it is notoriously more difficult to prove convergence and to calculate forces analytically. Their main advantages, a local "chemically intuitive" description and smaller basis sets were no match against the brute force plane wave approach as computer power increased and more importantly iterative minimization algorithms [14] for dealing efficiently with these intrinsically much larger basis sets were developed. In other words, it just became too easy to tackle the active research problems with the supercell technique. Why would one need to develop a special technique for point defects if the standard all-purpose method could do the job and provide answers the experimentalist needed?

One might thus have expected a lively discussion on which methods to use in the early 1980s, when the Green's function methods were still widely in use and cluster calculations were still competitive with the supercell approach. But nowadays, only a few practitioners of that approach remain. One at least is represented in this compilation (chapter 17, [15]). Development of methodologies for dealing with ever larger systems, has of course continued, in particular, the development of order-N methods. The large driving force for developing those methods has been the advent of nanostructures, but by and large they are not being applied very much to point defects in solids. It appears that the sizes we need to handle point defects satisfactorily are compatible with the computing power available without these new large scale system methods. Nonetheless, at least one article in this compilation discusses progress in developing methods to solve standard Kohn–Sham equations more efficiently for truly large systems (chapter 16, [16]).

At the same time, while it was long known that the local density approximation (LDA) (or its slight modification, the generalized gradient approximation, GGA) underestimates band gaps in semiconductors, the attitude of many studying point defects, has been to just put up with this problem by focusing on quantities which were supposedly not affected by this shortcoming. Thus the emphasis of point defect studies shifted from calculating one-particle energy levels or changes in densities of states in the bands, to total energies, energies of formation of defects and transition energies. As long as one avoided explicitly calculating excited state properties, the thinking went, we were safe. After all, transition energies are defined as the position of the chemical potential relative to the band edge, where one charge state becomes lower energy than another, so we keep focusing on the ground states, which is after all the legitimate quantity to calculate in DFT. Other quantities, which seemed safe are charge densities, and spin densities and those define such things as the hyperfine parameters, so useful to electron paramagnetic resonance (EPR) experimentalists, or local vibrational modes as measured by infrared spectroscopy.

Nonetheless, it appears that in the last few years, increasingly the "festering" underlying problems of the supercell plus LDA (or GGA) paradigm have become more apparent and have been increasingly discussed in the literature. In part, this is probably because a lot of the recent applications have been on more challenging systems. For example, wide band gap semiconductors and oxides, including transition metal oxides, present a new challenge that brings out these underlying problems. Transition metal impurities with strongly correlated d-states have received increasing interest in the context of dilute magnetic semiconductors. The underestimate of the band gap by LDA (and GGA) in these systems is often larger. Defect levels that should be and are experimentally in the gap, appear as resonances in the bands in the LDA calculations leading to qualitatively wrong descriptions of the defect behavior. The more ionic nature means that screening is reduced and brings out the effects of the Coulomb interactions more vividly. In particular, the self-interaction error of LDA and the orbital dependent correlation and exchange effects are being put in the spotlight in these systems. Ionic systems also exhibit stronger polaronic effects, which as will be seen below are strongly suppressed by LDA because of the incomplete cancellation of the self-interaction. Due to the reduced screening, spurious interactions involving charged defect states are also exacerbated.

In any case, whatever may have been the reason why these problems resurfaced, whether they were always there and they were just temporarily ignored while the community was absorbed in the successes of the standard approach, until our demands for accuracy and rigor overtook, or because the problems were becoming more apparent in the new systems to which the attention has shifted, this renewed debate on the methodologies is highly welcome.

At the same time, there has been a strong push in recent years to go beyond the limitations of DFT to ground state properties. The quasiparticle excitations can now be calculated in Hedin's GW approximation [17, 18] using pseudopotential plane wave [19–22], all-electron linearized augmented plane-wave (LAPW) [23, 24], linearized muffin-tin orbital (LMTO) [25–29], and projector augmented wave (PAW) implementations [30–32]. Even electron–hole interactions affecting optical properties can be treated by the Bethe–Salpeter equation approach [33–39]. Time dependent DFT provides an alternative way of dealing with excited states [36, 40]. Finally, the Quantum Monte Carlo method has continued to make strides and is represented in this compilation with chapter 2 by Hennig and coworkers [41]. While such methodologies are still computationally demanding and until recently only feasible for small systems like perfect crystal unit cells, parallelization of codes, and new algorithmic developments are now letting these methods make inroads in the defect world. At the same time, approximate methods to include some of the essential correlation or orbital dependent effects have continued to be developed, such as LDA + U (or GGA + U) and hybrid functionals.

Thus, the time is ripe not only to re-assess the accuracy of the standard approach now that we can push its limits by shear computational power but to incorporate these new methodologies beyond LDA into the world of point defects.

This compilation gives the reader a sampling of some of the problems under discussion. It will not provide definitive answers because there is as yet no consensus on many of these problems, but at least it will set the stage for further investigation and will allow a newcomer to the field to quickly get involved in the middle of the debate.

In the remainder of this article, I will comment on some of the highlights of the workshop and where the reader will find them in this issue.

19.2
Themes of the Workshop

Rather than commenting on the individual articles found in this book, the discussion is centered around a few themes that run throughout several articles and an attempt is made to place these in context and point out their connections. The themes are: (i) dealing with periodic boundary condition artifacts, (ii) dealing with the band gap underestimate by LDA, (iii) dealing with the self-interaction error of LDA, and (iv) developments of alternative methods to DFT and excited state methods.

19.2.1
Periodic Boundary Artifacts

Essentially the task is to extract the information on a single defect in the dilute limit from a calculation with periodically repeated defects at the smallest concentration one can handle computationally. Periodicity imposes several artifacts: the defect levels broaden into bands because of their interaction, the structural distortion around the defect may result in long-range elastic forces, and for charged defects, there is a spurious Coulomb interaction between the image charges and between them and the compensating background one introduces to enable a meaningful definition of total energies. The total energy of an infinite periodic system is only well-defined if it is overall neutral.

The direct band broadening effects are presumably easiest to avoid if the defects wave functions are exponentially localized, which is the case for deep defects, while for shallow defects, an accurate description of the binding energies relative to the bands is probably better attempted in the framework of effective mass approximation methods [42–48].

Nonetheless, some care is needed to deal with defect band dispersion. Aradi et al. [49] for example use a tight-binding fit to the defect band dispersion to derive the center of gravity of the actual isolated defect level. This approach was used earlier by Louie et al. [50] and is more important the smaller the cells are. Wei and Yan [51] in chapter 13 of this compilation discuss an approach in which the one-electron defect levels calculated at the Γ-point, which reflect the correct symmetry of the isolated defect levels, are combined with the transition state approach. The point is that one generally uses a special k-point set to calculate total energies and the differences in defect level position at the special k-point from that at Γ needs to be taken into account.

The elastic effects are expected to behave like $1/L^3$ while the image charge electrostatic effects are longer range and expected to behave like $1/L$ with L the characteristic length scale of the supercell, say $1/V^{1/3}$ with V the volume. The fact that the image charges in the neutral background leads to a spurious Madelung contribution $\alpha_M q^2/\varepsilon L$ to the total energy of the system is known since the work of Leslie and Gillan [52]. Makov and Payne [53] identified a correction describing the interaction of the quadrupole moment of the defect density with the background going as $1/L^3$. Nonetheless, these proposed corrections were not universally adopted by practitioners in the field. It was for example argued by Segev and Wei [54] that if the defect density becomes delocalized the Madelung model, invoking point charges in jellium, overestimates the corrections. This point of view is discussed in Wei's contribution in this compilation (chapter 13, [51]). Instead these authors proposed to use an extrapolation scheme proportional to $1/V$ or proportional to the number of atoms in the cell. Gerstmann et al. [55] also criticized the approach for the case of delocalized defect wave functions. Other more sophisticated approaches were introduced but because they are more difficult to implement or focus on periodic boundary calculations of molecules, were rarely used [56]. One confusing point here is that even if the defect wave function is delocalized, the total electrostatic density perturbation contains the δ-function part of the nuclear charges and thus certainly the total charge density contains a point-like monopole contribution. A major reason why it has not always been that clear in practical studies to see the pure q^2/L behavior, is that other finite size effects may be dominating. To clarify the situation, it is best to consider the effect in unrelaxed structures and one needs large enough cells of order several 100 atoms, before this behavior becomes apparent [57]. Attempts have been made by several authors to fit separately, $1/L$ and $1/L^3$ terms [58].

Recently, the problem was reformulated in a slightly different manner by Freysoldt *et al.* [59] and their method is described in additional detail in chapter 14 of this compilation [60]. Their analysis is based on plotting the defect minus perfect crystal potential and subtracting from it a long-range part, calculated with an assumed unscreened defect charge model, which in the simplest case is just a point charge. One important point arising from their analysis is that in doing so one must account for an arbitrary constant shift in the potential, a so-called alignment potential. The latter is chosen so that the remaining short-range part of the potential explicitly goes to zero far away from the defect. If necessary to make the potential flat in the region far away from the defect, a more sophisticated defect model charge density is introduced. This alignment term has the form $q\Delta_{q/b}$ and enters the defect formation energy together with the qE_F term which represents the chemical potential of the electron which one must add to describe correctly the change in Gibbs free energy of the charged defects, which is considered to be an open system in connection with an electron reservoir. Usually, this term is added separately but in Freysoldt *et al.*'s analysis, it emerges naturally from their consideration of the spurious interaction energies which one wishes to remove.

It is important to note that the alignment term goes as $1/V$ and was also emphasized by several other authors [57, 61–64]. It is being used by many other authors in some form or other, although the procedure used for determining it is not always explicitly mentioned in the literature. When a defect is created in charge state $+q$, the electron must be removed to the electron reservoir with energy its chemical potential $\mu_e = E_{vbm} + E_F$. Usually, one measures the Fermi level here relative to the valence band maximum (VBM). However, the question is how to calculate E_{vbm} of the perfect crystal in the defect containing cell. One might think this is just the highest occupied band (at the appropriate k-point), not counting the defect levels in the supercell itself, but the problem is that the defect may have perturbed the band edges in the supercell. It helps to plot the bands of the supercell, so one can recognize defect levels from host bands by their dispersion. The accepted alignment approach is to use a "local characteristic" of the potential, say the average of the electrostatic potential over an atomic sphere, or the potential at the muffin-tin radius in methods that use such spheres, or a core level. If one now knows the valence band energy in the perfect crystal relative to this "local potential marker," then all we need to determine is the same marker at atoms far away from the defect in the supercell containing the defect. Typically, one needs to average over a few atoms far away from the defect, and this is for instance illustrated in Lany and Zunger [65]. Freysoldt *et al.* [59, 60] essentially use a well defined separation approach of long and short range parts of the defect potential with a built-in check to make sure that after the long-range effects are subtracted the short-range potential indeed just becomes a constant. But essentially, it is just a different way of determining the alignment potential and the remaining correction is just the Leslie and Gillan [52] Madelung correction or the $1/L$ Makov–Payne term [53].

Now, there still remains the question of the Makov and Payne's quadrupole term. Lany and Zunger [57] recently pointed out that the net quadrupole of the defect charge density $Q \propto L^2$, i.e., it is not independent of the supercell size and the quadrupole correction $2\pi q Q/\varepsilon L^3$ then effectively behaves also as a $1/L$ term. The reason for this observed behavior is that Lany and Zunger's defect charge density from which the quadrupole moment is calculated includes the screening charge density which becomes almost constant at large distance from the defect in the supercell. The definition of Q involves an integral over $r^2 \varrho(r)$ and is thus dominated by large r. The same behavior in fact was pointed out earlier by Lento *et al.* [66]. Furthermore, Lany and Zunger [57] found that this reduces the monopole correction by a factor which is essentially independent of defect and amounts to about $-1/3$. Thus their prescription for the image charge correction becomes: take 2/3 of the point charge correction.

On the other hand, Makov and Payne [53] defined Q explicitly as "the second radial moment only of that part of the aperiodic density that does not arise from dielectric response or from the jellium, i.e., is asymptotically independent of L." This contradicts Lany and Zunger's statement that $Q \propto L^2$ but this is simply a question of whether or not one here includes the short-range dielectric screening. We here say short-range screening because the remaining long-range screening is included by dividing by the dielectric constant. It is not obvious how to determine Q according to Makov and Payne's strict definition and one might wonder why all (even short range) effects of the screening density should be excluded. Lany and Zunger define the defect charge density simply as the difference of the charge density in the cell with the defect minus the corresponding perfect crystal charge density calculated in the same cell.

In my opinion, this is closely related to the question, is it actually correct to include a compensating background density? In reality a charged defect is after all compensated by other defects far away and not by a homogeneous back ground. Near interfaces for example, it is well known that one has depletion layers which are actually charged. On the other hand, if we consider a finite small region around the defect, say one or two shells of neighbors, and consider this as an open system, then clearly charge can flow into this region. In principle it should be determined self-consistently in the presence of a given chemical potential of the electrons. The latter sets the energy up to which to integrate the local densities of states to determine the total charge inside the defect region. So, if a defect level is below it, it will be occupied and if above it, it will be empty. Even in a Green's function method, one would only treat a finite region of a few cells around the defect as the region where the potential is assumed to be different from bulk. But if the defect region retains a net charge, the potential outside still has a long-range Coulomb tail. Presumably that could be used as boundary condition for the potential in the defect region although I am not aware of calculations where this was explicitly done. Asymptotically a charged defect should behave as $q/(\varepsilon r)$ with ε the macroscopic dielectric constant, but this means that the net charge of the defect region is reduced to q/ε, meaning that $q[1 - (1/\varepsilon)]$ has indeed flown toward the defect. As long as ε is fairly large (perhaps a wrong assumption for very ionic oxides!) assuming local charge neutrality seems a reasonable approximation. In other words, the background density represents to some extent the physical screening charge density, more precisely, the short-range part of it.

Thus, the screening charge density if included in the definition of Q seems to arise in large part from the background density. The quadrupole correction of Makov and Payne is the interaction of this quadrupole with the background. So, it is somewhat puzzling whether this needs to be included since it appears to amount at least in part to an interaction of the background with itself. A detailed argument of Lany and Zunger's point of view can be found in Ref. [65].

A subtle point is to what extent the background density is really included in the codes that are commonly in use. In some codes, it appears the latter's presence is only assumed in order to give a well defined meaning to the reference electrostatic potential, which implicitly assumes charge neutrality but the interaction of the background density with the electrons and nuclei is not explicitly included in the total energy. This point is also discussed in [57, 65]. Other codes may explicitly add the uniform background density to the charge density in a systematic manner. This is for example the case in the FP-LMTO code [67].

Now, so far we have only considered corrections to the total energies of the system. In another recent paper, Lany and Zunger [68] showed that also one-electron energy levels are subject to finite size potential shifts.

On the other hand, it is worthwhile mentioning that some attempts have been made to go outside the standard practice of using a compensating background. Schultz [69–71] for example has advocated the use of a local moment counter charge approach, in which long-range effects of the net charge of the defect are treated with boundary conditions of a single isolated defect in calculating the electrostatic potential, while the remaining moment free (up to some moment order) density is treated with the usual periodic boundary conditions. The special role of an undetermined alignment potential also crops up in his theory. Its advantages and disadvantages versus the neutralizing back ground density have been studied by Wright and Modine [72] and by Lento *et al.* [66] One finds, in fact that the convergence with size of the system is slower than in the background charge approach and outside the defect region, still a classical continuum model polarization correction must be added. We cannot delve into the details of these other approaches here and conclude that likely, discussions of finite size effects will continue for some time.

As mentioned in Section 19.1, periodic boundary artifacts can in principle be entirely avoided by resorting to Green's function methods which provide an exact embedding of the defect region in the host crystal. A remarkable feat of such methods is that they are able to describe the correct defect wave functions well outside the region in which the potential (and possibly the structure) is perturbed. This is emphasized in the contribution by Gerstmann [15] in this monograph (chapter 17). It is evidenced by the degree of accuracy with which it can describe hyperfine and super-hyperfine interactions. While the Green's function methods were originally developed with deep defects in mind, they have now apparently found their most impressive performance on shallow defects. Even though in his approach, the long-range Coulomb tail of the charged defect is not included, and thus the shallow defect becomes a resonance in the conduction band, it is possible within this approach to accurately identify the resonance in the density of states and thus to reconstruct the charge density from the Green's function integrated over the energy range of the resonance.

Shallow defects have perhaps been overlooked for a while with the impression that the shallow defect problem was solved in terms of effective mass theory (EMT) long ago. Nonetheless Gerstmann's contribution here draws attention to the shortcomings of the EMT in terms of describing the so-called central cell correction. We also point out that some recent work is trying to revive the EMT with novel approaches to refine the central cell potential [46–48]. Perhaps a fruitful avenue will be to combine first-principles supercell or Green's function approaches to extract central cell potentials to be used in EMT.

19.2.2
Band Gap Corrections

As is well known, the Kohn–Sham eigenvalue gap is underestimated by the LDA. On the other hand, the prime question about point defects is where the defect levels, either one-electron levels or transition levels lie with respect to the band edges. Thus, a correction of the band gap is necessary before one can compare to experiment.

The most dramatic failure related to the band gap underestimate occurs when it leads to an erroneous occupation of host states rather than defect states because the defect level becomes a resonance in the bands. This situation for example occurs for the oxygen vacancy in ZnO [73], a defect that was discussed by several contributors at the workshop and has become a sort of benchmark [68, 74–78].

In the past, several approaches have been used: some are *a posteriori* corrections, some are addressing the problem at the level of the Hamiltonian by going beyond LDA in some way or other. *A posteriori* corrections come down to deciding what is the nature of the defect state. If the defect is essentially a shallow acceptor, the idea is that the defect level position relative to the VBM is correct, and one would then just shift up the conduction band minimum (CBM) without changing the defect level. If the defect is a shallow donor state, the defect level would be shifted up along with the CBM. If it is a deep defect, the intuitive idea is that the level would shift according to how much it is valence band or conduction band like. How to apply this intuitive idea in practice is another matter and different approaches give different results.

One approach goes by the name of a "scaled scissor correction." The approach consists in determining the projection of the defect state onto valence and conduction band edge states to determine its percentage valence and conduction band character [79]. However, its limitations were pointed out by Deák *et al.* [80] by comparing how much the one-electron levels shift in this approach compared to a GW quasiparticle calculation.

A related idea was used by Janotti and Van de Walle [77, 81] for the oxygen vacancy in ZnO. The LDA + U approach is designed originally to deal with localized orbitals such as semicore d states or open shell d and f systems [82, 83]. It produces a partial band gap correction because it reduces the p–d hybridization. Janotti and Van de Walle [81] reasoned that the extent to which this correction shifts the defect level, shows to what extent it is valence or conduction band like and thus used this as a basis for extrapolating to the full gap correction, even if the rest of the band gap correction is not arising from the p–d hybridization effect. However, it is not clear that one can decompose the defect wave function in CBM and VBM like host states. Certainly for deep defect levels, the idea of pinning the defect level to the host state at one dominant k-point seems incorrect.

The scissor shift is easily incorporated in Green's function methods [4, 8, 15]. In that case, one could even relatively easily include different shifts at different k-points instead of a uniform shift.

A better approach clearly is to use an energy functional or Hamiltonian for the host that reproduces the correct gap, or at least gives a better approximation to it that results in a qualitatively correct starting point. We already mentioned that the LDA + U approach at least partially opens the gap. An extension of this approach is applying it to the states that dominate the CBM, typically cation s-like states [73, 74]. Another is to add simply non-local external potentials that shift the appropriate states [57, 84], or to use modified pseudopotentials [77, 85]. The success of such approaches depends significantly on the details of the implementation [74]. While their main advantage is simplicity and computational efficiency, it is undesirable that one needs to adjust these potentials on a case by case basis. Much hope for a universally applicable approach is placed these days in hybrid functionals.

Hybrid functionals were discussed by several contributors to the conference [75, 77, 86–88]. Hybrid functionals essentially mix some Hartree–Fock with LDA or GGA exchange. Thereby they add an orbital dependence to the exchange correlation functional that is missing in the LDA. This is also what the LDA + U methods are essentially trying to mimic. While hybrid functionals were first explored by chemists for small molecules, e.g., the Becke B3LYP functional [89, 90], more recent versions seem to be rather successful to reproduce band gaps of standard tetrahedrally bonded semiconductors. The main new development however, is that they are now being implemented in the popular plane wave programs and can thus more readily be applied to the systems of interest. They still are typically much more time consuming than semilocal functionals.

Among hybrid functionals, we can distinguish those that mix a fraction α of unscreened Hartree–Fock with GGA, and those that use screened Hartree–Fock. The former approach is called PBE0 or PBEh [91] if the fraction is $\alpha = 0.25$ and added to the original PBE–GGA functional [92, 93]. This fraction was argued to be optimal based on many-body perturbation theory [91]. The drawback of including unscreened Hartree–Fock is the $1/r$ singularity of Hartree–Fock, which manifests itself as a $1/q^2$ singularity in reciprocal space. How to treat this carefully is discussed by Alkauskas et al. [86, 94] and by Kotani et al. [28] Among the screened HF approaches, the HSE [95, 96] introduced by Scuseria and coworkers is most popular. In that approach, the Hartree–Fock is divided in a long-range and short range (or rather medium range, as explained by Scuseria in his contribution to this volume [88]) by means of an error-function cut-off similar to the well-known Ewald procedure, and a fraction of the medium range part is included in the final functional. Roughly speaking, the idea is that truly long-range behavior will be canceled by corresponding correlation and should be avoided because it would lead to unphysical behavior in metals for example. At the same time, this makes the computational approach more easy to implement. A detailed discussion of why it is believed to improve band gaps and which band gaps is provided in chapter 6 [88]. This chapter also discusses to what extent one can expect that a universal materials independent range-separation of exchange and correlation can be expected to work.

An alternative implementation of mixing screened Hartree–Fock is the so-called screened exchange approach [97–99]. The latter is discussed here in the chapter by Clark and Robertson (chapter 5, [75]) and their recent papers [76, 100, 101] and is justifiable as a Generalized Kohn–Sham scheme [97]. In that case, the screening of the exchange is usually done by a Thomas–Fermi exponential screening and the screening length is not arbitrarily chosen but determined by the valence electron density (excluding d-electrons). The fraction of screened HF included is determined by the double counting correction in that case. A well defined LDA of the screened exchange exists and is subtracted from the usual LDA so that in practice again a mix of LDA and screened HF is effectively used.

The HSE approach has been implemented [102, 103] in Vienna *ab initio* simulation package (VASP) [104] and has begun to be tested for point defects by several contributors. Examples represented in this compilation are the chapters by Deák *et al.*, (chapter 8, [87, 105]) Van de Walle and Janotti (chapter 9, [77]), and Alkauskas *et al.* (chapter 7, [86]). Deák *et al.* (chapter 8, [87]) emphasize the importance of obtaining accurate defect energy levels as a prerequisite for obtaining accurate formation energies and other derived properties. They consider a variety of correction approaches of the one-electron levels and their impact on total energy properties. The SX approach was implemented by Clark and Robertson [75] in CASTEP [106] and has been much less used, so it is too early to compare with the HSE approach in terms of practical results. It has previously been implemented in FLAPW by Freeman and coworkers [99].

In spite of the successes of these hybrid functional approaches, there remain several issues under debate. One approach is to adjust the fraction α of mixing in (screened) HF so as to exactly adjust the band gap for a particular system. The other is to stick to the universal 0.25 mixing factor. The second freedom is what screening length or long-range cut-off parameter to use. Unfortunately, these choices still lead to significantly different results for defect levels relative to the band edges, notably again the oxygen vacancy in ZnO as can be seen by comparing results from the different groups [68, 74–77].

Another question is, if one adjusts the mixing parameter for each system, then what to do at an interface? This question is addressed by Alkauskas *et al.* [86, 107]. Another interesting point raised by these authors [108] is that at least for well localized defects, it appears that defect levels measured relative to the average electrostatic potentials are in much better agreement among different approaches than relative to the VBM. In other words, the problem appears to lie in determining the band edges rather than the defect levels! Thus, one needs to investigate not only how well these new functionals do on the band gaps but how they do on the individual band edges. This however is a tricky question because individual energy levels on an absolute scale cannot be defined in a periodic crystal [109]. Nonetheless, whether such a level has physical meaning or not, one can refer levels with respect to the average electrostatic potential as zero and ask how these differ in different approaches, such as GGA, HSE, and GW. Komsa *et al.* [110] recently discussed how the different mixing and screening parameters in hybrid functionals affect the band edges and defect levels relative to the electrostatic potential reference.

The most accurate approach for band gaps in semiconductors at present is the GW method. On the other hand, it is also the most computationally challenging to apply to defects. Secondly, GW does not readily fit into the DFT framework for total energies. It is a many-body perturbation theory for quasiparticle excitations. As such there is now a refocus of interest in one-electron energies [80]. One inventive approach by Rinke *et al.* [111] and discussed here in Giantomassi *et al.* [112] is to use a mixed approach of GGA and or hybrid functionals with GW. The idea is that GW quasiparticle energies give correctly the vertical (unrelaxed) total energy difference occurring in a change of charge state of a defect. After all, this occurs by transferring an electron from the defect level to a band state or vice versa. So, in principle, GW gives this excitation energy correctly for fixed geometry. The hybrid functional or GGA approach is then used subsequently to study the relaxation energy in a given charge state. This is a promising approach, which was recently also applied to the oxygen vacancy in ZnO by Lany and Zunger [68]. The GW method was well represented with several talks at the work shop and is further discussed in the next section.

While for some time calculations have focused on energies of formation and transition levels, i.e., the Fermi energies where the formation energies for different charge states cross, the use of GW brings back the one electron levels into focus. Relatedly Deák *et al.* [87] remind us that the one electron levels are a significant part of the total energy expression and thus changing these one-electron levels by for example gap corrections indirectly also changes the total energy derived quantities.

19.2.3
Self-Interaction Errors

A second important error of LDA that was widely discussed at the workshop is the self-interaction error. In Hartree–Fock, the exchange term exactly compensates the Coulomb interaction of an electron in a specific one-electron eigenstate with itself. LDA and GGA make approximations to the exchange functional and hence the self-interaction error (SIE) is not exactly canceled. This among other leads to a tendency of the semilocal functionals (LDA and GGA) to favor states in which defect electrons delocalize over several atoms. Hartree–Fock on the other hand misses correlation effects entirely and too strongly favors localization. This problem has been known for some time now to affect the structural relaxation primarily for deep acceptors [113–120]. It prevents the formation of a localized hole on a specific atom and the accompanying symmetry breaking relaxations, so-called Jahn–Teller distortions. Such localized relaxations trapping an electron or hole, are also called self-trapped polarons.

One of the earliest systems in which this problem was noticed is the Al acceptor in SiO_2. While LDA spreads the defect wave function equally over the four nearest neighbors, Hartree–Fock finds a localized state on one of the oxygens and it was recognized that this results from the spurious self-interaction error of LDA [113]. Subsequently, it was shown that a calculation including explicitly the self-interaction correction for the defect state solves the problem [114]. However, applying an *ad hoc* self-interaction correction for a specific defect state is cumbersome. Others found that LDA + U could also solve the problem [115]. Other defects were soon identified that show similar problems, for example the Zn vacancy in ZnO, the Li_{Zn} in ZnO [116, 120, 121].

The contribution in this compilation by Lany (chapter 11, [122]) discusses the problem and his cure for it. While LDA + U including Coulomb interactions U on the anion p states would help to localize the corresponding hole states, they have the disadvantage that applying a sufficiently strong U may perturb the host band structure. Lany and Zunger therefore construct a so-called hole state potential, which by construction is zero as long as the occupation of the anion p-orbitals is the same as that of the host. Only those orbitals on which a hole localizes, thus decreasing their occupation number feel a repulsive potential that further re-enforce the localization by pushing the level into the band gap. In order to answer the question, how strong should the localization potential be, they apply a so-called generalized Koopmans' theorem [123].

As shown by Perdew *et al.* [124] the total energy as function of occupation number in exact DFT for open systems with a continuously varying occupation number should be a piecewise linear function. The impacts of this on the delocalization problem were pointed out recently by Mori-Sánchez *et al.* [125]. Because of Janak's theorem [126] the linearity of the total energy implies constant Kohn–Sham eigenvalues as function of occupation numbers. LDA shows convex $d^2E/dn_i^2 > 0$ and Hartree–Fock concave $d^2E/dn_i^2 < 0$ behavior instead of linear behavior. A correct behavior leads to equality of the total energy difference (ΔSCF approach) with the Kohn–Sham eigenvalue for the defect levels, or satisfying a generalized Koopmans' theorem. Originally the theorem was derived within Hartree–Fock theory [123] but is only valid for delocalized states. Lany and Zunger thus adjust the strength of their hole-state potential so as to satisfy the generalized Koopmans' theorem and find that in many cases this leads to polaronic behavior where LDA fails to describe the correct relaxation of the system.

Others have used LDA + U and adjusted the U-value to satisfy the same criterion, for example in a study of polaronic trapping in TiO_2 [127]. Interestingly, the same considerations about the required linear behavior of the total energy as function of occupation number are at the heart of a recently proposed method for determining U in LDA + U methods [128]. The use of Janak's theorem to calculate transition energies in the context of LDA + U theory was also advocated by Sanna *et al.* [129]. A point that should not be forgotten in this context is that Janak's or Koopmans' theorem refer to a specific eigenstate, while in the LDA + U or hole-state potentials, one applies it to a basis set specific state or local atomic orbital. This is different from the explicit self-interaction correction by d'Avezac *et al.* [114].

While the correct physics, requiring an increased separation of empty and filled defect states is built into these methods and its strength is adjustable according to a well-described *ab initio* criterion, one might object against the *ad hoc* form of the hole state potential or worry whether applying such LDA + U corrections would not spoil other aspects of the electronic structure for the host system. It is therefore of interest to see how hybrid functionals do for such polaronic systems. For example, Clark and Robertson [75, 76] note that their screened exchange method correctly describes the polaron hole trapping at the Zn-vacancy in ZnO with localization on a single atom. Deák *et al.* [105] discuss the satisfaction of Koopmans theorem by the

HSE functional. The application of the B3LYP hybrid functional to bound polarons in ZnO was discussed at the workshop by Du and Zhang [120]. A comparison between the hole-state potential and the HSE hybrid functional for N_0 can be found in Lany and Zunger [117] and Lyons et al. [130].

19.2.4
Beyond DFT

Besides the direct applications to defects, the workshop contained a good deal of discussion of the underlying methodologies that go beyond LDA. The main issue is how to make the more advanced methods, such as GW, Bethe–Salpeter equations, time-dependent DFT efficient enough to become applicable to large systems as required for point defect studies.

Two examples represented in this compilation are the chapters by Umari et al. (chapter 4, [131]) and Giantomassi et al. (chapter 3, [112]). Umari et al. [131, 132] discuss the use of a separate small orthogonal basis set for expanding the polarizability operator. It should be recognized here that the bottleneck of GW calculations is the calculation of the wave vector and frequency dependent polarizability that goes into the calculation of the screened Coulomb interaction W. In most GW calculations this quantity is expanded in plane waves. Umari et al. discuss the construction of a separate small basis set that spans the space of products of Wannier functions. This approach reminds me of the product basis set approach introduced by Aryasetiawan and Gunnarsson [133]. While they construct product basis functions of muffin-tin orbitals rather than Wannier functions, the underlying idea is similar. We note that the product basis supplemented with plane waves forms the basis for expanding polarizability, screened and bare Coulomb interactions in the FP-LMTO implementation of GW by van Schilfgaarde et al. [27–29] Umari et al.'s approach shows great promise to speed up GW as they show by applying it to supercells containing a few 100 atoms.

One of the remaining problems is the need to sum over a large number of empty states. This problem was also tackled by Umari et al. [134] by reformulating the calculation in such a way that no explicit summation of empty bands is required. Instead one uses the completeness and rewrites the sum over empty states as one minus the projection operator over all filled states. This is essentially similar to the Sternheimer approach that has been so successful in linear response theory [135]. A very similar approach was recently introduced by Giustino et al. [136].

The chapter by Giantomassi et al. [112] on the other hand discusses approaches beyond the GW approximation. For example, they discuss the inclusion of so-called vertex corrections as well as the so-called quasiparticle self-consistent GW approach. After all, GW as conceived by Hedin is only the first approximation in a perturbation series. The key here is that usually GW theory is applied as a one-shot correction to some underlying one-electron theory. The latter is usually LDA or GGA but could also be HSE or LDA + U. The idea of QS–GW is to construct the best one-electron (generalized) Kohn–Sham starting point with a non-local exchange potential extracted from the GW self-energy itself. The method was introduced by van Schilfgaarde et al. [27–29]. and implemented with a FP-LMTO basis set and product basis sets.

The way forward beyond GW is still a matter of debate. Early results indicated that self-consistent GW without vertex corrections gave worse results that single-shot GW. Other complications arise from the effects of semicore orbitals and differences between all-electron and pseudopotential implementations. The QSGW appears to be the most successful among those approaches and leads to small remaining and highly systematic errors on the band gaps of a wide variety of systems, not only standard semiconductors. One of the remaining errors is the use of the random phase approximation (RPA) in the dielectric screening. Including electron–hole interaction effects in the latter is expected to further improve the method.

Giantomassi *et al.* [112] also discuss various technical aspects of the GW method, such as the summation over empty states, the plasmon–pole approximation, and the PAW implementation. They show applications to the problem of band-offsets at interfaces and to point defects along the lines discussed earlier.

While at present GW is used to calculate one-electron excitations, there is also progress to turn GW into a total energy theory. In principle, it is closely related to the RPA total energy [28, 137]. However, it is not yet clear how stable this approach is to calculating total energies and how good they are compared to LDA or GGA and experiment. A recent evaluation of RPA was made by Harl *et al.* [138, 139].

At the workshop some results were also presented by M. Rohlfing applying the Bethe–Salpeter approach including excitonic effects to defect problems. While no contribution of his is included here, we refer the reader to [140–142].

Relatedly, a time-dependent density functional approach to point defects is presented in this compilation by Gali (chapter 18, [143]). His approach is part of a new direction that explores the calculation of excited states of defects. This is an extremely important new direction because much of the experimental information on point defects relates to optical excitations, within the defect. This first of all requires one to use non-equilibrium occupations of the defect levels (constrained DFT) but secondly, electron–hole excitonic effects can be expected to be important as they also are in low-dimensional systems and molecules. The contribution here describes a combination of such approaches, from the constrained DFT implemented with hybrid functionals in VASP to a time-dependent DFT approach implemented in a cluster calculation. He discusses two specific systems: the N–V centers in diamond and divacancies in SiC.

Finally, as mentioned earlier Hennig and coworkers (chapter 2, [41]) discuss progress in Quantum Monte Carlo calculations. These calculations provide a benchmark for total energies for defects in relatively small supercells of order 16 atoms. The discussion is mostly on Diffusion Monte Carlo and the various types of controlled and uncontrolled error in this method and their application to self-interstitials in Si.

19.3
Conclusions

In this chapter, I have reviewed some of the main themes that were addressed during the CECAM workshop "Which electronic structure method for the study of defects?," held in Lausanne, June 8–10, 2009. It was not my intention to be complete in this review but rather to give the reader a vivid impression of the issues that are currently under discussion in this field and the progress that is being made. The commentary given reflects my own point of view as a practitioner in this field and not necessarily that of the papers mentioned. If my understanding is incomplete, I apologize to the authors. I have tried to put the articles that can be found in this compilation in the context of the current open literature. The articles in this compilation provide only a limited glimpse at what was presented at the conference. First, not all presenters at the workshop chose to contribute to this compilation and secondly, some chose to focus on a particular part of their presentation. In any case, no set of separate articles can ever capture the gist of the many lively discussions that followed the presentations at the workshop.

I have focused this commentary on the original question in the title, focusing on methodology rather than on specific applications. At the same time, it should be said that the reader of this compilation will encounter a wide variety of defect problems in materials, from defects in Si to wide band gap semiconductors and oxides and interfaces, showing that the field is very much alive and not at all in an impasse over unsolved methodological questions. I have also tried to draw some attention to some of the "forgotten" problems and approaches, which were not very much represented during the workshop. In particular, the problem of EMT of shallow defects may see some revival and the chapter included here on the Green's function approach [15] will hopefully remind some people of the promise held by this method.

The newcomer or outsider to this field might at first get the impression that the field is in turmoil with many conflicting opinions. However, the fact that these methodological questions are now being discussed in the open literature is very healthy for the field and will hopefully help newcomers to avoid common pitfalls in how to apply these methods. It is pretty clear now that image charge corrections need to be made for charged defects and attention has to be paid to the proper alignment. Also, there is a consensus now that band gap corrections are important, in particular in cases where the low LDA gaps would lead to incorrect band filling for certain charge states. The different approaches to achieve it have their pros and cons: hybrid functionals and GW are more expensive but unbiased, not empirical. Non-local external potentials or LDA + U or hole-state potentials are less computationally expensive but require careful selection of the associated parameters. The broader availability of these new approaches will assist in their rapid deployment and will help in ascertaining their success. In any case, it does not mean that all prior work with LDA only is invalidated by the new approaches. It all depends on the system under study and on what questions the calculation is trying to answer.

We should also be reminded of the continuing successes of the field. In the end the task of computational physics in this field is to assist experimentalists in extracting the maximum information and understanding from their experiments. The standard approach for defect calculations has provided a lot of guidance on which defects are likely to form under what circumstances and what the basic characteristics of specific defects are. It has also been able to provide significant guidance on how to overcome doping problems [51], and enables one to estimate defect concentrations accurately taking into account rather complex defect chemistry and reactions. With the new approaches, the accuracy with which defect level positions can be calculated with respect to the band edges is steadily improving. Theory still lags behind experiment: optics experiments can determine differences between sharp photoluminescence lines to better than a meV, and can purely spectroscopically distinguish different defects. However, determining the chemical indentity of defects is likely to remain a complex task requiring consistency between various experimental techniques, and computational results. The new focus in the theory on excited state properties which correlate more directly to optical studies is very promising. In fact, almost all experiments deal with excited states, in some way or another, whether optically activated EPR signals, or photoluminescence or optical absorption. The fact that we can now start addressing quasiparticle and optical (electron–hole pair) excitations including excitonic effects is an important advance in the theory. While the concepts of Franck–Condon diagrams have been around for a long time, we can now start actually calculating them just like for isolated molecules. Finally, defect wave functions, in particular the delocalized or localized character can now be addressed. In particular, great progress was made recently in realizing the importance of polaronic effects and such information can be directly tested by means of EPR fine structure. Even for delocalized shallow states, good agreement can be obtained between calculated wavefunctions and the hyperfine structure. Although not mentioned in this compilation, certain defects at or close to the surface can be visualized with scanning tunneling microscopy and this provides another testing ground for the theoretical capability to determine defect wave functions.

In short, the future of computational defect studies is bright. The theory can be increasingly applied to more complex solids, its accuracy is improving steadily and even defects in complex nanostructures are within reach.

Acknowledgements

I would like to thank P. Deák for the encouragement to write this summary and the authors for their willingness to let me comment on their work. I would like to thank Chris Van de Walle, Audrius Alkauskas, Alfredo Pasquarello, and Jörg Neugebauer to have the foresight to organize the workshop and inviting me to it. My own recent research on defect has been sponsored by AFOSR, NSF, and ARO.

References

1 Deák, P., *Phys. Status Solidi B*, in press.
2 Deák, P., Miró, J., Gali, A., Udvardi, L., and Overhof, H. (1999) *Appl. Phys. Lett.*, **75**(14), 2103–2105.
3 Baraff, G.A. and Schlüter, M. (1978) *Phys. Rev. Lett.*, **41**(13), 892–895.
4 Baraff, G.A. and Schlüter, M. (1979) *Phys. Rev. B*, **19**(10), 4965–4979.
5 Bernholc, J. and Pantelides, S.T. (1978) *Phys. Rev. B*, **18**(4), 1780–1789.
6 Bernholc, J., Lipari, N.O., and Pantelides, S.T. (1978) *Phys. Rev. Lett.*, **41**(13), 895–899.
7 Bernholc, J., Lipari, N.O., and Pantelides, S.T. (1980) *Phys. Rev. B*, **21**(8), 3545–3562.

8. Gunnarsson, O., Jepsen, O., and Andersen, O.K. (1983) *Phys. Rev. B*, **27**(12), 7144–7168.
9. Beeler, F., Andersen, O.K., and Scheffler, M. (1985) *Phys. Rev. Lett.*, **55**(14), 1498–1501.
10. Lindefelt, U. and Zunger, A. (1981) *Phys. Rev. B*, **24**(10), 5913–5931.
11. Lindefelt, U. and Zunger, A. (1982) *Phys. Rev. B*, **26**(2), 846–895.
12. Braspenning, P.J., Zeller, R., Lodder, A., and Dederichs, P.H. (1984) *Phys. Rev. B*, **29**(2), 703–718.
13. Zeller, R. and Dederichs, P.H. (1979) *Phys. Rev. Lett.*, **42**(25), 1713–1716.
14. Payne, M.C., Teter, M.P., Allan, D.C., Arias, T.A., and Joannopoulos, J.D. (1992) *Rev. Mod. Phys.*, **64**(4), 1045–1097.
15. Gerstmann, U. (2011) *Phys. Status Solidi B*, doi: 10.1002/pssb.201046237. (chapter 17 in this book).
16. Briddon, P.R. and Rayson, M.J. (2011) *Phys. Status Solidi B*, doi: 10.1002/pssb.201046147. (chapter 16 in this book).
17. Hedin, L. (1965) *Phys. Rev.* **139**(3A), A796–A823.
18. Hedin, L. and Lundqvist, S. (1969), Effects of electron–electron and electron–phonon interactions on the one-electron states of solids, in: *Solid State Physics, Advanced in Research and Applications*, edited by Seitz, D. Turnbull, and H. Ehrenreich (Academic Press, New York) pp. 1–181.
19. Hybertsen, M.S. and Louie, S.G. (1986) *Phys. Rev. B*, **34**(8), 5390–5413.
20. Godby, R.W., Schlüter, M., and Sham, L.J. (1988) *Phys. Rev. B*, **37**(17), 10159–10175.
21. Bruneval, F., Vast, N., and Reining, L. (2006) *Phys. Rev. B*, **74**(4), 045102.
22. Bruneval F. and Gonze X. (2008) *Phys. Rev. B*, **78**(8), 085125.
23. Aryasetiawan F. (1992) *Phys. Rev. B*, **46**(20), 13051–13064.
24. Friedrich C., Schindlmayr A., Blügel S., and Kotani T. (2006) *Phys. Rev. B*, **74**(4), 045104.
25. Aryasetiawan F. and Gunnarsson O. (1995) *Phys. Rev. Lett.*, **74**(16), 3221–3224.
26. Aryasetiawan F. and Gunnarsson O. (1998) *Rep. Prog. Phys.*, **61**(3), 237.
27. van Schilfgaarde, M., Kotani, T., and Faleev, S.V. (2006) *Phys. Rev. B*, **74**(24), 245125.
28. Kotani, T., van Schilfgaarde, M., and Faleev S.V. (2007) *Phys. Rev. B*, **76**(16), 165106.
29. Faleev, S.V., van Schilfgaarde, M., and Kotani, T. (2004) *Phys. Rev. Lett.*, **93**(12), 126406.
30. Lebègue, S., Arnaud, B., Alouani, M., and Bloechl, P.E. (2003) *Phys. Rev. B*, **67**(15), 155208.
31. Shishkin, M. and Kresse, G. (2006) *Phys. Rev. B*, **74**(3), 035101.
32. Shishkin, M. and Kresse, G. (2007) *Phys. Rev. B*, **75**, 235102.
33. Rohlfing, M. and Louie, S.G., (1998), *Phys. Rev. Lett.*, **80**(15), 3320–3323.
34. Rohlfing, M. and Louie, S.G. (1998) *Phys. Rev. Lett.*, **81**(11), 2312–2315.
35. Albrecht, S., Reining, L., Del Sole, R., and Onida, G. (1998) *Phys. Rev. Lett.*, **80**(20), 4510–4513.
36. Onida, G., Reining, L., and Rubio, A. (2002) *Rev. Mod. Phys.*, **74**(2), 601–659.
37. Schmidt, W.G., Glutsch, S., Hahn, P.H., and Bechstedt, F. (2003) *Phys. Rev. B*, **67**(8), 085307.
38. Fuchs, F., Rödl, C., Schleife, A., and Bechstedt, F. (2008) *Phys. Rev. B*, **78**(8), 085103.
39. Benedict, L.X., Shirley, E.L., and Bohn, R.B. (1998) *Phys. Rev. Lett.*, **80**(20), 4514–4517.
40. Reining, L., Olevano, V., Rubio, A., and Onida, G. (2002) *Phys. Rev. Lett.*, **88**(6), 066404.
41. Parker, D.W., Wilkins, J.W., and Hennig, R.G. (2010) *Phys. Status Solidi B*, DOI: 10.1002/pssb.201046149. (chapter 2 in this book).
42. Luttinger, J.M. and Kohn, W.(1955) *Phys. Rev.*, **97**(4), 869–883.
43. Pantelides, S.T. (1978) *Rev. Mod. Phys.*, **50**(4), 797–858.
44. Baldereschi, A. and Lipari, N.O. (1973) *Phys. Rev. B*, **8**(6), 2697–2709.
45. Baldereschi, A. and Lipari, N.O. (1974) *Phys. Rev. B*, **9**(4), 1525–1539.
46. Wang, H. and Chen, A.B. (2001) *Phys. Rev. B*, **63**(12), 125212.
47. Chen, A.B. and Srichaikul, P.(1997) *Phys. Status Solidi B*, **202**, 81.

48 Mireles, F. and Ulloa, S.E. (1998) *Phys. Rev. B*, **58**(7), 3879–3887.
49 Aradi, B., Gali, A., Deák, P., Lowther, J.E., Son, N.T., Janzén, E., and Choyke, W.J. (2001) *Phys. Rev. B*, **63**(24), 245202.
50 Louie, S.G., Schlüter, M., Chelikowsky, J.R., and Cohen, M.L. (1976) *Phys. Rev. B*, **13**(4), 1654–1663.
51 Wei, S.H. and Yan, Y. (2011) *Phys. Status Solidi B*, doi: 10.1002/pssb.201046292. (chapter 13 in this book).
52 Leslie, M. and Gillan, M.J. (1985) *J. Phys. C, Solid State Phys.*, **18**, 973.
53 Makov, G. and Payne, M.C. (1995) *Phys. Rev. B*, **51**(7), 4014–4022.
54 Segev, D. and Wei, S.H. (2003) *Phys. Rev. Lett.*, **91**(12), 126406.
55 Gerstmann, U., Deák, P., Rurali, R., Aradi, B., Frauenheim, T., and Overhof H. (2003) *Physica B*, **340**, 190.
56 Blöchl, P.E. (1995) *J. Chem. Phys.*, **103**, 7482.
57 Lany, S. and Zunger, A. (2008) *Phys. Rev. B*, **78**(23), 235104.
58 Castleton, C.W.M., Hoglund, A., and Mirbt, S. (2006) *Phys. Rev. B*, **73**(3), 035215.
59 Freysoldt, C., Neugebauer, J., and Van de Walle, C.G. (2009) *Phys. Rev. Lett.*, **102**(1), 016402.
60 Freysoldt, C., Neugebauer, J., and Van de Walle, C.G. (2011) *Phys. Status Solidi B*, doi: 10.1002/pssb.201046289. (chapter 14 in this book).
61 Laks, D.B., Van de Walle, C.G., Neumark, G.F., Blöchl, P.E., and Pantelides, S.T. (1992) *Phys. Rev. B*, **45**(19), 10965–10978.
62 de Walle, C. G. V. and Neugebauer, J. (2004) *J. Appl. Phys.*, **95**(8), 3851–3879.
63 Mattila, T. and Zunger, A. (1998) *Phys. Rev. B*, **58**(3), 1367–1373.
64 Persson, C., Zhao, Y.J., Lany, S., and Zunger, A.(2005) *Phys. Rev. B*, **72**(3), 035211.
65 Lany, S. and Zunger, A.(2009) *Model. Simul. Mater. Sci. Eng.* **17**,.
66 Lento, J., Mozos, J.L., and Nieminen, R.M. (2002) *J. Phys.: Condens. Matter* **14**,.
67 Kotani, T. and van Schilfgaarde, M. (2010) *Phys. Rev. B*, **81**(12), 125117.
68 Lany, S. and Zunger, A.(2010) *Phys. Rev. B*, **81**(11), 113201.
69 Schultz, P.A. (1999) *Phys. Rev. B*, **60**(3), 1551–1554.
70 Schultz, P.A. (2000) *Phys. Rev. Lett.*, **84**(9), 1942–1945.
71 Schultz, P.A. (2006) *Phys. Rev. Lett.*, **96**(24), 246401.
72 Wright, A.F. and Modine, N.A. (2006) *Phys. Rev. B*, **74**(23), 235209.
73 Paudel, T.R. and Lambrecht, W.R.L. (2008) *Phys. Rev. B*, **77**(20), 205202.
74 Boonchun, A. and Lambrecht, W.R.L. (2011) *Phys. Status Solidi B*, doi: 10.1002/pssb.201046328. (chapter 10 in this book).
75 Clark, S.J. and Robertson, J. (2010) *Phys. Status Solidi B*, doi: 10.1002/pssb.201046110. (chapter 5 in this book).
76 Clark, S.J., Robertson, J., Lany, S., and Zunger, A. (2010) *Phys. Rev. B*, **81**(11), 115311.
77 Van de Walle, C.G. and Janotti, A. (2011) *Phys. Status Solidi B*,, doi: 10.1002/pssb.201046384. (chapter 9 in this book).
78 Oba, F., Togo, A., Tanaka, I., Paier, J., and Kresse, G. (2008) *Phys. Rev. B*, **77**(24), 245202.
79 Castleton, C. W. M., Höglund, A., and Mirbt, S. (2006) *Phys. Rev. B*, **73**(3), 035215.
80 Deák, P., Frauenheim, T., and Gali, A. (2007) *Phys. Rev. B*, **75**(15), 153204.
81 Janotti, A. and Van de Walle, C.G. (2005) *Appl. Phys. Lett.*, **87**(12), 122102.
82 Anisimov, V.I., Zaanen, J., and Andersen, O.K. (1991) *Phys. Rev. B*, **44**(3), 943–954.
83 Anisimov, V.I., Solovyev, I.V., Korotin, M.A., Czyżyk, M.T., and Sawatzky, G.A. (1993) *Phys. Rev. B*, **48**(23), 16929–16934.
84 Lany, S., Raebiger, H., and Zunger, A. (2008) *Phys. Rev. B*, **77**(24), 241201.
85 Segev, D., Janotti, A., and Van de Walle, C.G. (2007) *Phys. Rev. B*, **75**(3), 035201.
86 Alkauskas, A., Broqvist, P., and Pasquarello, A. (2010) *Phys. Status Solidi B*, doi: 10.1002/pssb.201046195. (chapter 7 in this book).
87 Deák, P., Gali, A., Aradi, B., and Frauenheim, T. (2010) *Phys. Status Solidi B*, doi: 10.1002/pssb.201046210. (chapter 8 in this book).
88 Henderson, T.M., Paier, J., and Scuseria, G.E. (2011) *Phys. Status Solidi B*,

doi: 10.1002/pssb.201046303. (chapter 6 in this book).
89 Becke, A.D. (1993) *J. Chem. Phys.*, **98**(2), 1372–1377.
90 Becke, A.D. (1996) *J. Chem. Phys.*, **104**(3), 1040–1046.
91 Perdew, J.P., Ernzerhof, M., and Burke, K. (1996) *J. Chem. Phys.*, **105**(22), 9982–9985.
92 Perdew, J.P., Burke, K., and Ernzerhof, M. (1996) *Phys. Rev. Lett.*, **77**(18), 3865–3868.
93 Perdew, J.P., Burke, K., and Ernzerhof, M. (1997) *Phys. Rev. Lett.*, **78**(7), 1396.
94 Broqvist, P., Alkauskas, A., and Pasquarello, A. (2009) *Phys. Rev. B*, **80**(8), 085114.
95 Heyd, J., Scuseria, G.E., and Ernzerhof, M. (2003) *J. Chem. Phys.*, **118**(18), 8207–8215.
96 Heyd, J., Scuseria, G.E., and Ernzerhof, M. (2006) *J. Chem. Phys.*, **124**(21), 219906.
97 Seidl, A., Görling, A., Vogl, P., Majewski, J.A., and Levy, M. (1996) *Phys. Rev. B*, **53**(7), 3764–3774.
98 Bylander, D.M. and Kleinman, L. (1990) *Phys. Rev. B*, **41**(11), 7868–7871.
99 Asahi, R., Mannstadt, W., and Freeman, A.J. (1999) *Phys. Rev. B*, **59**(11), 7486–7492.
100 Robertson, J., Xiong, K., and Clark, S.J. (2006) *Phys. Status Solidi B*, **243**(9), 2071–2080.
101 Xiong, K., Robertson, J., Gibson, M.C., and Clark, S.J. (2005) *Appl. Phys. Lett.*, **87**(18), 183505.
102 Paier, J., Hirschl, R., Marsman, M., and Kresse, G. (2005)*J. Chem. Phys.*, **122**(23), 234102.
103 Paier, J., Marsman, M., Hummer, K., Kresse, G., Gerber, I.C., and Ángyán, J.G. (2006) *J. Chem. Phys.*, **124**(15), 154709.
104 http://cms.mpi.univie.ac.at/vasp/.
105 Deák, P., Aradi, B., Frauenheim, T., Janzén, E., and Gali, A. (2010)*Phys. Rev. B*, **81**(15), 153203.
106 http://www.castep.org/.
107 Alkauskas, A., Broqvist, P., Devynck, F., and Pasquarello, A. (2008) *Phys. Rev. Lett.*, **101**(10), 106802.
108 Alkauskas, A., Broqvist, P., and Pasquarello, A. (2008) *Phys. Rev. Lett.*, **101**(4), 046405.
109 Kleinman, L. (1981) *Phys. Rev. B*, **24**(12), 7412–7414.
110 Komsa, H.P., Broqvist, P., and Pasquarello, A. (2010) *Phys. Rev. B*, **81**(20), 205118.
111 Rinke, P., Janotti, A., Scheffler, M., and Van de Walle, C.G. (2009) *Phys. Rev. Lett.*, **102**(2), 026402.
112 Giantomassi, M., Stankovski, M., Shaltaf, R, M. Grüning, F., Bruneval, R.P., and Rignanese, G.M. (2010)*Phys. Status Solidi B*,, DOI: 10.1002/pssb.201046094. (chapter 3 in this book).
113 Lægsgaard, J. and Stokbro, K. (2001) *Phys. Rev. Lett.*, **86**(13), 2834–2837.
114 d'Avezac, M., Calandra, M., and Mauri, F. (2005)*Phys. Rev. B*, **71**(20), 205210.
115 Nolan, M. and Watson, G.W. (2006) *J. Chem. Phys.*, **125**(14), 144701.
116 Lany, S. and Zunger, A. (2009) *Phys. Rev. B*, **80**(8), 085202.
117 Lany, S. and Zunger, A. (2010) *Phys. Rev. B*, **81**(20), 205209.
118 Lany, S. and Zunger, A. (2010) *Appl. Phys. Lett.*, **96**(14), 142114.
119 Raebiger, H., Lany, S., and Zunger, A. (2009) *Phys. Rev. B*, **79**(16), 165202.
120 Du, M.H. and Zhang, S.B. (2009) *Phys. Rev. B*, **80**(11), 115217.
121 Carvalho, A., Alkauskas, A., Pasquarello, A., Tagantsev, A.K., and Setter, N. (2009) *Phys. Rev. B*, **80**(19), 195205.
122 Lany, S. (2011) *Phys. Status Solidi B*,, DOI: 10.1002/pssb.201046274. (chapter 11 in this book).
123 Koopmans, T.C. (1934) *Physica (Utrecht)*, **1**(1–6), 104–113.
124 Perdew, J.P., Parr, R.G., Levy, M., and Balduz, J.L. (1982)*Phys. Rev. Lett.*, **49**(23), 1691–1694.
125 Mori-Sánchez, P., Cohen, A.J., and Yang, W. (2008) *Phys. Rev. Lett.*, **100**(14), 146401.
126 Janak, J.F. (1978) *Phys. Rev. B*, **18**(12), 7165–7168.
127 Morgan, B.J. and Watson, G.W. (2009) *Phys. Rev. B*, **80**(23), 233102.
128 Cococcioni, M. and de Gironcoli, S. (2005) *Phys. Rev. B*, **71**(3), 035105.

129 Sanna, S., Frauenheim, T., and Gerstmann, U. (2008) *Phys. Rev. B*, **78**(8), 085201.
130 Lyons, J.L., Janotti, A., and de Walle, C. G. V. (2009) *Appl. Phys. Lett.*, **95**(25), 252105.
131 Umari, P., Qian, W., Marzari, N., Stenuit, G., Giacomazzi, L., and Baroni, S. (2011) *Phys. Status Solidi B*, doi: 10.1002/pssb.201046264. (chapter 4 in this book).
132 Umari, P., Stenuit, G., and Baroni, S. (2009) *Phys. Rev. B*, **79**(20), 201104.
133 Aryasetiawan, F. and Gunnarsson, O. (1994) *Phys. Rev. B*, **49**(23), 16214–16222.
134 Umari, P., Stenuit, G., and Baroni, S. (2010) *Phys. Rev. B*, **81**(11), 115104.
135 Baroni, S., de Gironcoli, S., Dal Corso, A., and Giannozzi, P. (2001) *Rev. Mod. Phys.*, **73**(2), 515–562.
136 Giustino, F., Cohen, M.L., and Louie, S.G. (2010) *Phys. Rev. B*, **81**(11), 115105.
137 Miyake, T., Aryasetiawan, F., Kotani, T., van Schilfgaarde, M., Usuda, M., and Terakura, K. (2002) *Phys. Rev. B*, **66**(24), 245103.
138 Harl, J. and Kresse, G. (2009) *Phys. Rev. Lett.*, **103**(5), 056401.
139 Harl, J., Schimka, L., and Kresse, G. (2010) *Phys. Rev. B*, **81**(11), 115126.
140 Ma, Y. and Rohlfing, M. (2008) *Phys. Rev. B*, **77**(11), 115118.
141 Ma, Y., Rohlfing, M., and Gali, A. (2010) *Phys. Rev. B*, **81**(4), 041204.
142 Kaczmarski, M.S., Ma, Y., and Rohlfing, M. (2010) *Phys. Rev. B*, **81**(11), 115433.
143 Gali, A. (2011) *Phys. Status Solidi B*, doi: 10.1002/pssb.201046254. (chapter 18 in this book).

Index

a

ab initio   259, 260, 261, 266, 272–275, 279, 282
Ab Initio Modelling PROgram.
   see AIMPRO
acceptor   183–185, 188–196
AIMPRO (Ab Initio Modelling PROgram)   285, 286, 289, 290, 292
alignment   118, 120, 122, 248, 256
alignment potential   172, 174, 364, 366
aluminium   260, 261, 279, 280, 282
amorphous silica   201, 210
amorphous states   205, 207
anharmonic excitations   259, 278–282

b

backflow   19, 22, 25, 26, 28, 29
band alignment   34, 219–220
band edges/band extrema, valence band maximum (VBM)/conduction band minimum (CBM)   119, 123, 124
band gap   79, 80, 82–85, 88–93, 98–108, 115, 307, 313, 318, 319
band gap underestimate   362, 367
band offsets   33, 48, 49, 118, 128
band structure   79, 80, 82–85, 89–91, 93, 168–171, 178, 359, 360, 371
band-gap problem   6–13, 112, 120, 121
barrier (Energy-)   194, 195, 197
benzene   69
Bethe-Salpeter equation (BSE)   353, 362, 372
$BiFeO_3$ 83, 92, 93
bipolar doping   213, 214
bloch (wave) function   306, 312, 313, 321, 330

c

calculation   79, 80, 84, 85, 88, 89, 91–93
carbon (diamond)   23, 24

Car-Parrinello method   72
central cell correction   306, 322,324, 333
charge localization   252
charge state   131, 315, 318–320
charge transition level   113, 114, 118–125, 129–132, 140, 144–146, 253
chemical potential   4, 9, 214, 220–223, 237
co-doping   229, 232–235
coherence time   342
compensation   83
concentration   1–4
configuration coordinate diagram   194, 195
constant pressure approach   260–262, 264–266, 279, 281
convergence   116
correlated materials   104, 108

d

d states   7, 8, 10–13
dangling bond   129
deep defect   305, 306, 312, 319, 320, 324, 335
deep level, -state   183, 189, 195, 196, 207–209
defect energy levels   33
defect formation energies   54, 86, 92, 214, 224
defect localized state   193
defect   79, 81, 82–89, 91, 92, 201, 207–209
density functional   183, 185
density functional theory (DFT)   17, 33, 34, 37–41, 48–52, 55–57, 79, 97–100, 155, 259, 281, 301, 305–307, 329, 330, 332, 341, 346
density of states (DOS)   310, 312, 319, 325
diamond   342, 343, 345, 351, 353–356
dielectric constant   249
diffusion   1, 4, 12, 131

Advanced Calculations for Defects in Materials: Electronic Structure Methods, First Edition.
Edited by Audrius Alkauskas, Peter Deák, Jörg Neugebauer, Alfredo Pasquarello, and Chris G. Van de Walle.
© 2011 Wiley-VCH Verlag GmbH & Co. KGaA. Published 2011 by Wiley-VCH Verlag GmbH & Co. KGaA.

Diffusion Monte Carlo (DMC) 17
Dirac's equation 308, 309
disordered tetrahedral networks 201
divacancy 342, 344, 345, 346, 353–356
doping asymmetry 213
doping limit rule 214, 218, 219
doping 83
Dyson's equation 311, 312, 314, 324, 326, 327, 332

e

effective mass approximation (EMA) 306, 319–324, 330, 336, 363
EL2 317, 319
electrically detected magnetic resonance (EDMR) 328
electron nuclear double resonance (ENDOR) 319, 322, 324–326
electron paramagnetic resonance (EPR) 178, 320, 322, 333, 334
electronic excitations 266, 269, 270
electronic properties 33
electronic structure 83, 107, 201, 202, 204–206, 209
electrostatic interactions 241, 247, 255
energy levels of neutral atom 227
entropy of defect formation 281, 282
exchange-correlation 37, 38, 52
exchange-correlation functional 259–261, 279
excitation energy 140
exciton binding energy 352–355
excitonic 373, 375
expansion coefficient 279, 280

f

filter diagonalisation 292, 299
finite size corrections 172, 173, 176, 178
finite size effects 29, 363, 366
fixed-node approximation 19, 22, 25, 26
formation energies (defect formation energies) 2–6, 8–12, 113, 122, 249, 259, 261, 279, 282
free energy 259–270, 272–279, 281, 282
free energy Born-Oppenheimer approximation 266–269
functional 359, 362, 368–374

g

GaAs 196, 313, 315
GaN 195–197
gap correction to levels 143
gap correction to total energy 143
gap error 139–141, 143–146, 151, 152

gap level 141, 143–148, 150
Generalized Koopman's Theorem 371
generalized-gradient approximation (GGA) 17, 18
graphitic systems 105
Green's Function 2, 305, 306, 311–317, 320, 324, 327, 328, 330, 333, 334, 359, 360, 365–368, 374
group IV semiconductors 149
group theory 342, 344, 353, 354
GW 12, 24, 128, 141, 143, 146, 152, 165, 168, 169–171, 178
GW appoximation 34, 36–38, 40, 61–64, 202, 204, 362
GW method 352

h

Hafnium oxide 121
Hartree-Fock method 184
H-assisted doping 223–224
heat capacity 279, 280
HfO_2 83, 91, 92
hole-state potential 184, 188–190
HSE (Heyd, Scuseria, and Ernzerhof) hybrid functional 97–108, 149, 156, 158, 159, 160, 162
hybrid 362, 367–374
hybrid density-functional 216
hybrid functional(s) 9–11, 13, 18, 29, 80, 112, 114, 125, 133, 134, 142, 144–146, 148, 165, 167, 178, 184, 185, 187, 189–193, 341, 350, 352, 354
hybrid scheme 215
hyperfine interaction 342

i

imaginary time step 20, 28
impurities 155, 161, 162
impurity-band 232, 234, 236
interface properties 33–57
inverse participation ratio 76
ionization potential 69–70

j

Jahn-Teller distortion 355
Janak's theorem 185, 186
Jastrow factor 19

k

Kohn-Luttinger oscillation 328, 336
Kohn-Sham formalism of density functional theory (KSDFT) 285, 286, 288, 292, 301
Koopman's theorem 187, 216

Koopman's Condition   185–188, 191, 193, 196
KSDFT (Kohn-Sham-formalism of density functional theory)   285, 292, 299, 301

l

large-scale calculations   299, 301
lattice relaxation   306, 313, 314, 317, 319, 326, 327
LDA+U   6–9, 165–180
linear elasticity theory   328–330, 332
linear muffin-tin orbital (LMTO)   313–315, 323, 324
local density approximation (LDA)   155, 165, 307, 313, 361
local field effects   204, 209
local moment counter charge   366
local spin density approximation (LSDA)   17, 18, 23, 25, 27, 305, 307, 308, 310, 325, 330
localization (electron- or hole-)   194, 197
localized states   121

m

magnesium oxide   23, 25
magnetic field   306, 308
magnetic ordering   107
magnetization density   307, 309, 310, 312, 314–316, 318, 320–322, 325, 331, 335
Makov and Payne correction   216, 242
Makov Payne quadrupole correction   364
many-body perturbation theory (MBPT)   34–38, 61
Mg-acceptor   195
modified pseudopotentials   12, 13
molecular doping   220, 222–223, 237
molecular dynamics   272, 274, 277
molecular dynamic (MD) silmulation   333, 334
Mott insulators   106

n

neutralizing background   246
neutron transmutation   332, 334
nitrogen-vacancy center (NV center)   342, 351

o

optical gaps   100–102
optimal polarizability basis   61–68
oxidation   131
oxygen vacancy   165, 173, 174, 179, 180

p

periodic boundary conditions   97
perturbed host state   194
phase stability   105, 106
photoluminescence   342
point defects   17, 18, 72, 155, 156, 158, 159, 161, 259–261, 277, 278
polaron   183, 188–190
potential alignment   248, 256
projector-augmented wave (PAW)   34
pseudopotential   19, 22–26, 28, 313–315, 320, 322–327
pseudopotential locality approximation   28

q

quantum bit   342, 356
quantum communication   342
Quantum Monte Carlo (QMC)   18, 19
quantum-espresso   68
quasiharmonic approximation   271, 273
quasiparticle   62, 362, 367, 370, 372, 375
quasiparticle corrections   34

r

Random Phase Approximation (RPA)   63
rare earths   106
relative energy of different configurations   144
relativistic   308, 309
resonances   147, 148, 152

s

sawtooth   249
Schottky defect   23, 25
scissor operator   142, 143
screened exchange   100, 103, 104, 134, 369, 371
screened hybrids   100, 103, 104
screening   243
self-interaction   184, 186
self-interaction error   361, 362, 370
self-interstitial   18, 23, 25
self-trapped polaron   370
semiconductor devices (or electronic devices)   131
shallow defect   363, 366, 367, 374
shallow donors   306, 322, 324, 333, 334, 336
shallow level, -state   183, 189, 195, 196
SiC   306, 313, 319, 320, 322, 332–336
silica   70, 201, 202, 210
silicon (Si)   121, 130, 306, 310, 313, 322, 325, 328, 331–336

silicon carbide   342, 344, 353
silicon nitride (vitreous)   72–77
silicon oxide/silicon dioxide (SiO$_2$)   120, 131
silicon(bulk)   18, 25, 26, 28, 70, 105, 106
solid state calculations   97
Space-Time method (STM)   63
spin-orbit interaction   309
Stokes shift   346, 347, 353, 355
supercell   183, 184, 189
supercell correction   246
supercell geometry   2, 5
supercell-size convergence   3
surfactant   220, 224–226, 237
symmetry and occupation of defect levels   217

t
thermodynamic integration   272–277
thermodynamics   262, 266, 271–277
time-dependent density functional theory (TD-DFT)   346–351
TiO$_2$ 10, 11

transition energy level   213–215, 229
transition levels   3–5, 7–12

u
universal approaches   232
UP-TILD method   275, 276

v
vacancy   4, 8–11, 23–25, 251
vacancy concentration   281, 282
vertex corrections   51–54
voronoi cell/tesselation   314, 316–318, 324

w
Wannier's functions   64–67
wave function   250, 252
wide band gap oxides   92, 93

z
zero-variance extrapolation   29
zinc oxide (ZnO)   4, 6–12, 79, 83–90, 124, 165, 167–171, 173, 176, 177, 183–185, 188–191, 193–196, 213–214, 220–237
ZnTe   196